S0-CAH-177

QUINONE METHIDES

Wiley Series of Reactive Intermediates in Chemistry and Biology

Steven E. Rokita, Series Editor

Quinone Methides
Edited by Steven E. Rokita

QUINONE METHIDES

Edited by

STEVEN E. ROKITA

A John Wiley & Sons, Inc., Publication

Published by John Wiley & Sons, Inc., Hoboken, New Jersey
Published simultaneously in Canada

For general information on our other products and services or for technical support, please contact our Customer Care Department within the United States at (800) 762-2974, outside the United States at (317) 572-3993 or fax (317) 572-4002.

Wiley also publishes its books in a variety of electronic formats. Some content that appears in print may not be available in electronic formats. For more information about Wiley products, visit our web site at www.wiley.com.

Library of Congress Cataloging-in-Publication Data:

Quinone Methides / [edited by] S.E. Rokita.
p. cm. – (Wiley series on reactive intermediates in chemistry and biology ; v. 1)
Includes index.
ISBN 978-0-470-19224-5 (cloth)
1. Intermediates (Chemistry) I. Rokita, Steven Edward.
QD476.R416 2009
547'.2–dc22

2008038605

Printed in the United States of America
10 9 8 7 6 5 4 3 2 1

CONTENTS

PREFACE TO SERIES

Most stable compounds and functional groups have benefited from numerous monographs and series devoted to their unique chemistry, and most biological materials and processes have received similar attention. Chemical and biological mechanisms have also been the subject of individual reviews and compilations. When reactive intermediates are given center stage, presentations often focus on the details and approaches of one discipline despite their common prominence in the primary literature of physical, theoretical, organic, inorganic, and biological disciplines. The *Wiley Series on Reactive Intermediates in Chemistry and Biology* is designed to supply a complementary perspective from current publications by focusing each volume on a specific reactive intermediate and endowing it with the broadest possible context and outlook. Individual volumes may serve to supplement an advanced course, sustain a special topics course, and provide a ready resource for the research community. Readers should feel equally reassured by reviews in their speciality, inspired by helpful updates in allied areas and intrigued by topics not yet familiar.

This series revels in the diversity of its perspectives and expertise. Where some books draw strength from their focused details, this series draws strength from the breadth of its presentations. The goal is to illustrate the widest possible range of literature that covers the subject of each volume. When appropriate, topics may span theoretical approaches for predicting reactivity, physical methods of analysis, strategies for generating intermediates, utility for chemical synthesis, applications in biochemistry and medicine, impact on the environmental, occurrence in biology, and more. Experimental systems used to explore these topics may be equally broad and range from simple models to complex arrays and mixtures such as those found in the final frontiers of cells, organisms, earth, and space.

Advances in chemistry and biology gain from a mutual synergy. As new methods are developed for one field, they are often rapidly adapted for application in the other. Biological transformations and pathways often inspire analogous development of new procedures in chemical synthesis, and likewise, chemical characterization and identification of transient intermediates often provide the foundation for understanding the biosynthesis and reactivity of many new biological materials. While individual chapters may draw from a single expertise, the range of contributions contained within each volume should collectively offer readers with a multidisciplinary analysis and exposure to the full range of activities in the field. As this series grows, individualized compilations may also be created through electronic access to highlight a particular approach or application across many volumes that together cover a variety of different reactive intermediates.

Interest in starting this series came easily, but the creation of each volume of this series required vision, hard work, enthusiasm, and persistence. I thank all of the contributors and editors who graciously accepted and will accept the challenge.

University of Maryland STEVEN E. ROKITA

INTRODUCTION

The term "quinone methide" first appeared in literature in 1942 to describe the quinone analogue in which one of the carbonyl oxygens is replaced by a methylene group. Reactivity associated with such a species is typically greater than that of the parent quinone but more moderate than that of the corresponding quinodimethane in which both carbonyl oxygens are replaced by methylene groups. The single methylene substitution is still quite sufficient to create a highly transient intermediate or at least the perception of one, and this perception likely discouraged its initial study. Investigations were at first limited to polymerization and photochemistry. These topics have continued to develop and gain greater sophistication as the subtleties of quinone methides have been revealed. Despite approximately 1400 literature contributions and many reviews on quinone methides as of 2008, the current book is the first devoted to this fascinating and useful intermediate.

Most laboratories did not begin to recognize the widespread occurrence and potential applications of quinone methides until 20 years after its first report. Now, with an ever-increasing appreciation of the structural dependence of quinone methide reactivity, its use has became more frequent and diverse as illustrated by the topics covered in this volume. Their role in lignin formation was recognized as early as 1960. Soon after, the first stable quinone methide was discovered in the natural products taxodione and taxodone and offered a stark contrast to the expectation of its fleeting existence. Although the quinone methide derived from the food preservative 2,6-di-*tert*-butyl-4-methylphenol was first characterized in 1963, its discovery as a product of oxidative metabolism was published 20 years later. Just prior to this, the concept of bioreductive alkylating agents was introduced to form quinone methide intermediates for treating hypoxic tumors. Both reductive and oxidative metabolisms form quinone methides have since become a very important topic for quinone methides in drug design as well as drug safety.

Quinone methides are associated with sclerotization, the natural tanning process that stabilizes insect cuticle, as well as reactions of vitamin K and tocopherols including vitamin E. Quinone methides have also been integral to the design of many mechanism-based inactivators of enzymes, which has been adapted most recently to screen for catalytic activity within antibody libraries. Perhaps the field of organic synthesis has become the most frequent benefactor of quinone methides now that reliable methods are available for their generation and control. Of the various approaches for manipulating quinone methide reactivity, its complexation with transition metals remains the most remarkable. Finally, the reversibility of quinone methide reaction has established an excellent basis for polymer and dendrimer disassembly to the likely benefit of numerous processes in material science, biology, and medicine. My own laboratory has also been intrigued by this reversibility and in particular by its ability to extend the potential lifetime of electrophiles in biological systems.

My involvement in quinone methides arose very much by chance and was neither planned nor anticipated as typical of the serendipity associated with the pursuit of basic research. Interest has since been sustained by the intellectual challenges of this topic and the community of investigators sharing its exploration. What had once been left to the realm of physical and polymer chemists soon became the province of organic, medicinal, and theoretical chemists, biochemists, toxicologists, entomologists, biologists, and those involved in forestry and food sciences. The scientific literature is so vast that we struggle to remain current even in just the literature of our immediate disciplines, and yet innovation is often found in complementary perspectives and methodology. By assembling this collection of topics, I hope to entice readers already familiar with quinone methides to look beyond their typical focus and discover new inspiration and opportunities in allied areas. Concurrently, I hope that the range of topics and perspectives provides a comfortable entry for readers from a broad range of backgrounds and interests.

The volume has been created as a snapshot of significant activity on quinone methides and it neither attempts to cover the entire range of topics nor present comprehensive reviews on a subset of topics. A variety of excellent reviews have already been published on many of the interesting and important details. The authors of this volume embody the breadth of research involving quinone methides, and I am very much indebted to their dedication to this volume and the field in general. These authors along with many others past and present are responsible for our current understanding of quinone methides. I hope this volume will incite an even greater interest in quinone methides that in turn will merit further reviews and monographs in the future.

University of Maryland STEVEN E. ROKITA

CONTRIBUTORS

Takuya Akiyama, Dairy Forage Research Center, USDA-Agricultural Research Service, Madison, WI, USA; and RIKEN Plant Science Center, Suehiro, Tsurumi, Yokohama, Kanagawa 230-0045, Japan

Stefan Böhmdorfer, Department of Chemistry, University of Natural Resources and Applied Life Sciences, Muthgasse 18, A-1190 Vienna, Austria

Judy L. Bolton, Department of Medicinal Chemistry and Pharmacognosy (M/C 781), College of Pharmacy, University of Illinois at Chicago, 833 S. Wood Street, Chicago, IL 60612-7231, USA

Filippo Doria, Department of Organic Chemistry, Pavia University, V. le Taramelli 10, 27100 Pavia, Italy

Rotem Erez, Department of Organic Chemistry, School of Chemistry, Raymond and Beverly Sackler Faculty of Exact Sciences, Tel Aviv University, Tel Aviv 69978, Israel

Mauro Freccero, Department of Organic Chemistry, Pavia University, V. le Taramelli 10, 27100 Pavia, Italy

Hoon Kim, Department of Biochemistry and Great Lakes Bioenergy Research Center, University of Wisconsin, Madison, WI, USA

Fachuang Lu, Department of Biochemistry and Great Lakes Bioenergy Research Center, University of Wisconsin, Madison, WI, USA

Matthew Lukeman, Department of Chemistry, Acadia University, Wolfville, Nova Scotia, Canada, B4P 2R6

David Milstein, Department of Organic Chemistry, The Weizmann Institute of Science, Rehovot 76100, Israel

Stephen F. Nelsen, Department of Chemistry, University of Wisconsin, Madison, WI, USA

Liping Pettus, Department of Chemical Research and Discovery, Amgen Inc; Thousand Oaks CA 91320, USA

Thomas Pettus, Department of Chemistry and Biochemistry, University of California at Santa Barbara, Santa Barbara, CA 93106, USA

Elena Poverenov, Department of Organic Chemistry, The Weizmann Institute of Science, Rehovot 76100, Israel

John Ralph, Department of Biochemistry and Great Lakes Bioenergy Research Center, University of Wisconsin, Madison, WI, USA; Department of Biological Systems Engineering, University of Wisconsin, Madison, WI, USA; and Dairy Forage Research Center, USDA-Agricultural Research Service, Madison, WI, USA

Michèle Reboud-Ravaux, Enzymologie Moléculaire et Fonctionnelle, FRE 2852, CNRS-Université Paris 6, T43, Institut Jacques Monod, 2 place Jussieu, 75251 Paris Cedex 05, France

Steven E. Rokita, Department of Chemistry and Biochemistry, University of Maryland, College Park, MD 20742, USA

Thomas Rosenau, Department of Chemistry, University of Natural Resources and Applied Life Sciences, Muthgasse 18, A-1190 Vienna, Austria

Paul F. Schatz, Dairy Forage Research Center, USDA-Agricultural Research Service, Madison, WI, USA

Doron Shabat, Department of Organic Chemistry, School of Chemistry, Raymond and Beverly Sackler Faculty of Exact Sciences, Tel Aviv University, Tel Aviv 69978, Israel

Edward B. Skibo, Department of Chemistry and Biochemistry, Arizona State University, Tempe, AZ 85287-1604, USA

John A. Thompson, Department of Pharmaceutical Chemistry, School of Pharmacy, University of Colorado, Denver, C238-L15, 12631 E. 17th Avenue, Aurora, CO 80045, USA

Michel Wakselman, Institut Lavoisier de Versailles, UMR 8180, CNRS-Université Versailles Saint-Quentin, 45 Avenue des Etats Unis, F-78035 Versailles, France

Qibing Zhou, Department of Chemistry, Virginia Commonwealth University, 1001 West Main Street, Richmond, VA 23284-2006, USA

1

PHOTOCHEMICAL GENERATION AND CHARACTERIZATION OF QUINONE METHIDES

MATTHEW LUKEMAN
Department of Chemistry, Acadia University, Wolfville, Nova Scotia, Canada B4P 2R6

1.1 INTRODUCTION

Quinone methides (QMs) are reactive intermediates that are commonly encountered in many areas of chemistry and biology. Quinone methides contain a cyclohexadiene core with a carbonyl and a methylene unit attached and are related to quinones, which have two carbonyl groups, and quinone dimethides, which have two methylene groups. Quinone methides are highly polar and are much more reactive than their relatives. *Ortho* (**1**) and *para* (**2**) quinone methides are the most commonly encountered isomers, although *meta*-quinone methides (**3**), which are non-Kekulé and must be drawn as zwitterionic (**3a**) or biradical (**3b**) structures, are known as well. Both *ortho*- and *para*-quinone methides can also have zwitterionic resonance structures that give these species both cationic and anionic centers, emphasizing their polarized nature and indicating that they may react with both nucleophiles (similar to carbocations) and electrophiles (similar to phenolates). Quinone methides are much more reactive than simple enones (such as α,β-unsaturated ketones) since nucleophilic attack on a quinone methide produces an aromatic alcohol, with aromatization of the ring being a significant driving force (Scheme 1.1). *Ortho*-quinone methides can also readily engage in [4 + 2] cycloadditions with electron-rich dieneophiles to give chroman derivatives, again leading to rearomatization of the ring (Scheme 1.1). The reaction rates exhibited by quinone methides are highly

Quinone Methides, Edited by Steven E. Rokita

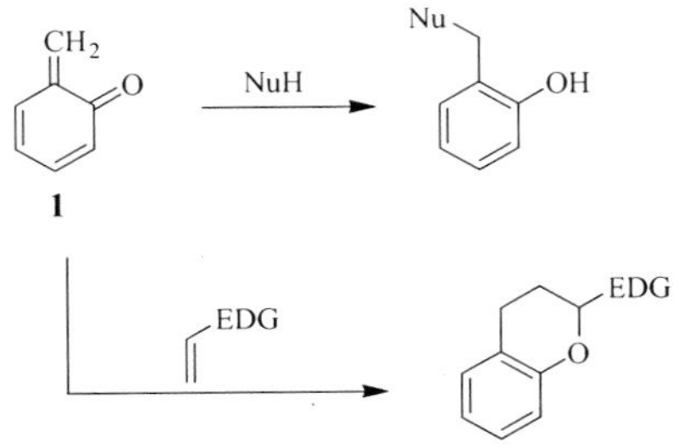

SCHEME 1.1

dependent on the substituents present; lifetimes can range from less than 1 ns to several seconds or minutes.

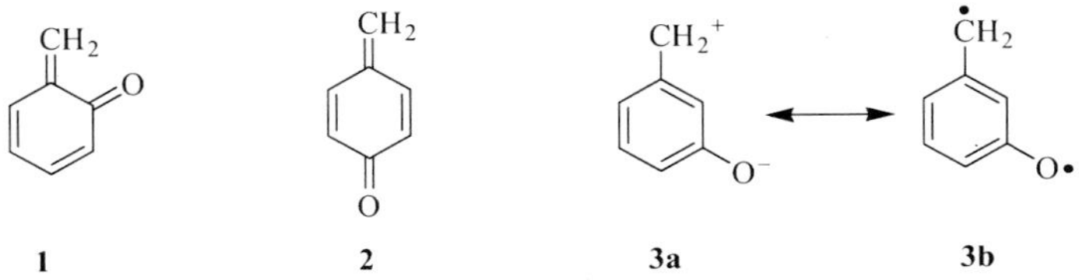

1 2 3a 3b

Quinone methides have been shown to be important intermediates in chemical synthesis,[1,2] in lignin biosynthesis,[3] and in the activity of antitumor and antibiotic agents.[4] They react with many biologically relevant nucleophiles including alcohols,[1] thiols,[5–7] nucleic acids,[8–10] proteins,[6,11] and phosphodiesters.[12] The reaction of nucleophiles with *ortho*- and *para*-quinone methides is pH dependent and can occur via either acid-catalyzed or uncatalyzed pathways.[13–17] The electron transfer chemistry that is typical of the related quinones does not appear to play a role in the nucleophilic reactivity of QMs.[18]

Much attention has been devoted to the development of methods to generate quinone methides photochemically,[1,19–20] since this provides temporal and spatial control over their formation (and subsequent reaction). In addition, the ability to photogenerate quinone methides enables their study using time-resolved absorption techniques (such as nanosecond laser flash photolysis (LFP)).[21] This chapter covers the most important methods for the photogeneration of *ortho*-, *meta*-, and *para*-quinone methides. In addition, spectral and reactivity data are discussed for quinone methides that are characterized by LFP.

1.2 QUINONE METHIDES FROM BENZYLIC PHOTOELIMINATION

1.2.1 Photoelimination of Fluoride

Seiler and Wirz carried out one of the earliest systematic studies of the photochemistry of phenols containing benzylic subsitutents.[22,23] They examined the photochemistry

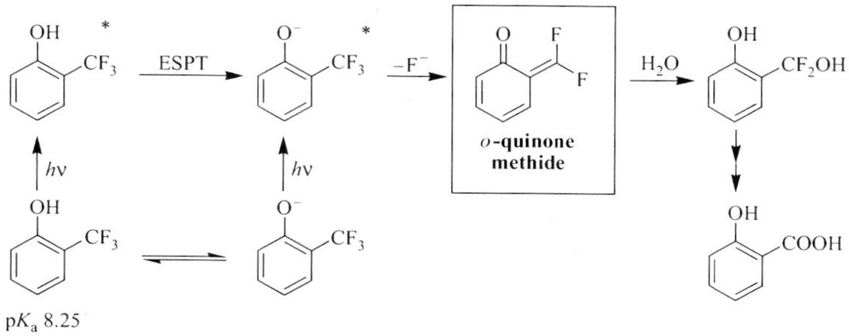

SCHEME 1.2

of 11 trifluoromethyl-substituted phenols and naphthols and found that all isomers underwent photohydrolysis in neutral or basic aqueous solution to give the corresponding hydroxybenzoic acid, some with efficiencies as high as $\Phi = 0.8$ (Scheme 1.2). A detailed mechanistic investigation led them to propose a common reaction mechanism for all 11 isomers: C–F bond heterolysis takes place from the singlet excited state of the phenolate, generated by either direct excitation of the ground-state phenolate (at $pH > pK_a$) or following excited-state proton transfer (ESPT) of the excited phenol (at $pH < pK_a$) (Scheme 1.2). Ejection of the fluoride gives an α,α-difluoroquinone methide as the first formed ground-state intermediate. Because the mechanism is common to all 11 reactive starting materials, this reaction can be considered a source of *ortho*-, *meta*-, and *para*-quinone methides, as well as a variety of naphthoquinone methides. The authors were able to detect one such naphthoquinone methide (**4**) using LFP and found that it absorbs light strongly in the 500–550 nm region and decays with a lifetime of 5 μs in phosphate buffer (pH 6.8). Although this work demonstrated for the first time that quinone methides can be efficiently photogenerated from phenols containing suitable benzylic substituents, the generality of this reaction would not be recognized until many years later.

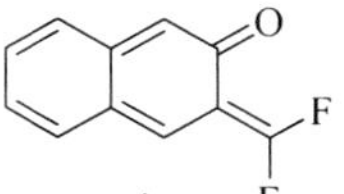

$\lambda_{max} = 500–550$ nm
τ(pH 6.8) = 5 μs

4

1.2.2 Photodehydration

Phenols containing benzyl alcohol side groups are much more accessible than the trifluoromethyl derivatives studied by Wirz, and their photochemistry has been studied extensively, beginning in the 1970s. Hamai and Kokubun[24,25] observed that solutions of **5** and **6** in hexane became highly colored when exposed to UV light due to their conversion to the corresponding α,α-diphenylquinone methides (*ortho*-fuchsones) **7** and **8** (Eq. 1.1). The mechanism of this reaction was not investigated, although the *ortho* arrangement of the phenol and the benzylic alcohol would permit excited-state intramolecular proton transfer (ESIPT) from the phenol OH to the benzyl alcohol via a six-membered transition state, facilitating the departure of the molecule of water. Methoxyfuchsone **8** was detected by UV–Vis spectroscopy and showed broad absorption bands at ~340 and 420 nm, which tailed off at 600 nm.

$h\nu$, *n*-hexane (1.1)

5 R = H
6 R = OCH_3
7 R = H
8 R = OCH_3

The photochemistry of the related *para*-hydroxytriphenylmethanols was first examined by Gomberg[26] in 1913. White powdered samples of **9** became yellow in color following exposure to sunlight, which was proposed to be due to dehydration to fuchsone **12** (Eq. 1.2). The photoreaction was revisited in the late 1970s by Lewis and coworkers,[27] who introduced related derivatives **10** and **11** to the study, which also undergo photodehydration to give the corresponding fuchsones **13** and **14** in the solid state. While direct ESIPT between the phenol and benzyl alcohol *within a given molecule* is not possible in the solid state due to the large distance between these groups, X-ray crystal structures revealed that each phenolic OH is aligned with the benzyl alcohol moiety of an adjacent molecule. One possible pathway for this reaction, then, is excited-state proton transfer (ESPT) from an excited phenol to the benzyl alcohol of its neighbor, thus facilitating the dehydration reaction.

$h\nu$, solid state (1.2)

9 R = H
10 R = Br
11 R = CH_3
12 R = H
13 R = Br
14 R = CH_3

Peter Wan was the first to investigate the photochemistry of hydroxybenzyl alcohols in a systematic way, and his work has established the foundation of our current understanding of their behavior. His work in this area began in 1986 when he investigated the photochemistry of *ortho*- and *meta*-hydroxybenzyl alcohol (**15** and **16**, respectively), along with several methoxybenzyl alcohols.[28] When irradiated in aqueous methanol, both **15** and **16** underwent conversion to the corresponding hydroxybenzyl ethers **17** and **18**, presumably via *ortho*- and *meta*-quinone methides **1** and **3**, although these were not detected in this work (Eq. 1.3). The reaction quantum yield for *ortho* derivative **15** was found to be much higher than that of the *meta* isomer. A mechanism was suggested that accounts for this difference: for **15**, excitation

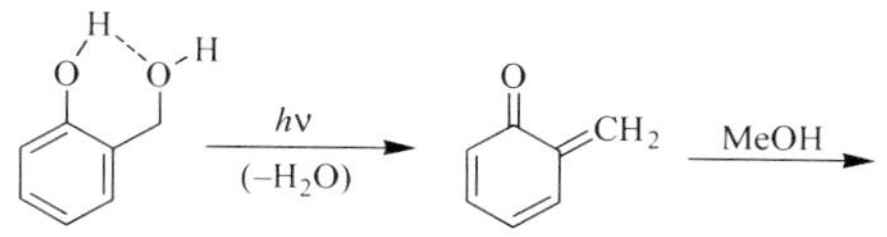

SCHEME 1.3

initiates an ESIPT from the phenol to the benzyl alcohol, which is facilitated by the presence of a hydrogen bond between these groups in the ground state, which loses water to give the quinone methide (Scheme 1.3). This intramolecular hydrogen bond is not present in the ground state for *meta* isomer **16**, and so this mechanism is not available to it leading to its lower efficiency. It was demonstrated in a subsequent paper that *ortho*-quinone methide **1** can also be photogenerated from **15** in alkaline solution, indicating that the phenolic OH is not necessarily a requirement for these reactions to take place.[29] Wan and coworkers[30] were later successful in detecting quinone methide **1** generated by photolysis of **15** using LFP. A broad absorption centered at 400 nm was obtained, although the spectrum was partially obscured by the simultaneous photogeneration of the phenoxyl radical of **15**, which absorbs in the same region.

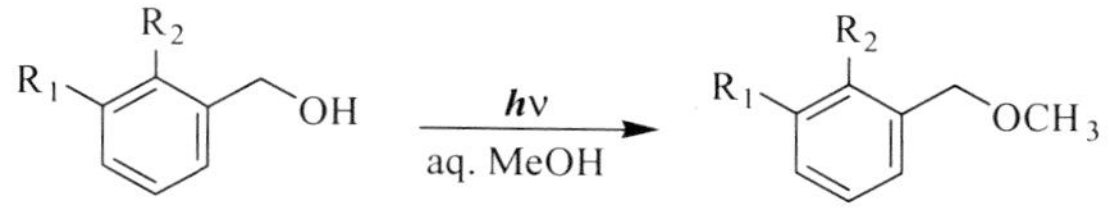

(1.3)

15 $R_1 = H, R_2 = OH$
16 $R_1 = OH, R_2 = H$
17 $R_1 = H, R_2 = OH$
18 $R_1 = OH, R_2 = H$

Almost a decade later, Wan greatly expanded his investigation into the photochemistry of hydroxybenzyl alcohols by examining **15** and **16** in more detail and adding derivatives **19–23**.[31] All hydroxybenzyl alcohols examined gave quinone methides on irradiation in aqueous solution, as evidenced by their trapping either as 4 + 2 cycloaddition chroman adducts with dienophiles (for *ortho*-quinone methides) or as their methyl ethers (via nucleophilic attack by solvent). The quantum yields observed for the *ortho* derivatives **15** and **19** were the highest, followed by the *meta* derivatives **16**, **20**, and **21**, and with *para* derivatives **22** and **23** reacting with the lowest efficiency (Table 1.1). Several of these derivatives were examined by LFP in an

TABLE 1.1 Quantum Yields of Photomethanolysis for Selected Hydroxybenzyl Alcohol Substrates

Substrate	Φ_R (in 1:1 $CH_3OH–H_2O$)
15	0.23
16	0.12
19	0.46
20	0.23
22	0.007
23	0.1

attempt to detect the suspected quinone methide intermediates. All of the α-phenyl derivatives (**19–21**, **23**) gave strongly absorbing transients in aqueous acetonitrile in the 300–600 nm region that showed quenching behavior characteristic of quinone methides. The spectra for the mono-α-phenyl quinone methides **24** (from **19**), **25** (from **20**), and **26** (from **23**) are shown in Fig. 1.1. *Ortho*- and *para*-quinone methides **24** and **26** were very long lived in aqueous acetonitrile, showing lifetimes longer than 5 s, while *meta*-quinone methide **25** decayed significantly faster with a lifetime of ~30 ns in the same solvent system, suggesting that its reactivity might be more closely related to that of benzylic cations.

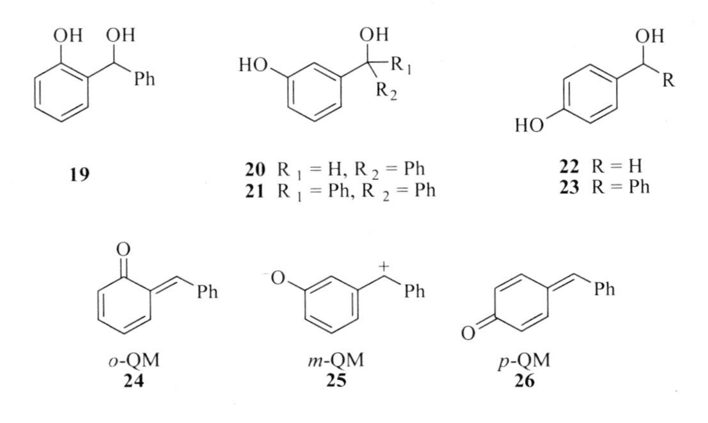

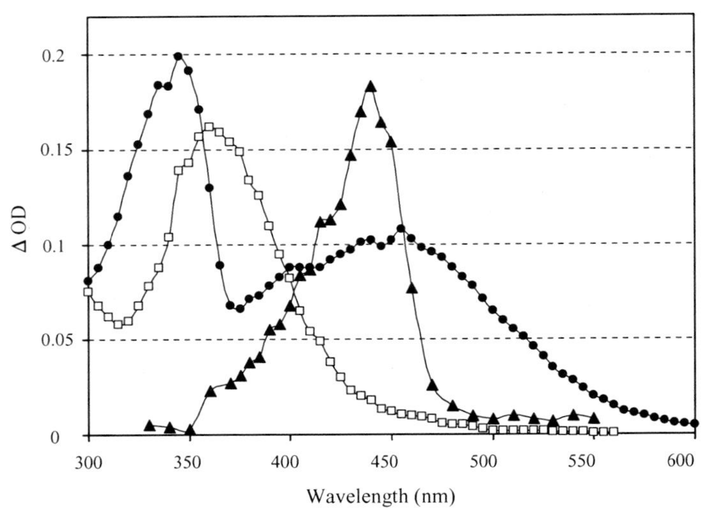

FIGURE 1.1 Absorption spectra of **24** (○), **25** (▲), and **26** (□) obtained by LFP of hydroxybenzyl alcohols **19**, **20**, and **23**, respectively, in aqueous acetonitrile. *Source*: Data taken with permission from Ref. [31].

Wan's group showed that the observed photodehydration of hydroxybenzyl alcohols can be extended to several other chromophores as well, giving rise to many new types of quinone methides. For example, he has shown that a variety of biphenyl quinone methides can be photogenerated from the appropriate biaryl hydroxybenzyl alcohols.[32,33] Isomeric biaryls **27–29** each have the benzylic moiety on the ring that does not contain the phenol, yet all were found to efficiently give rise to the corresponding quinone methides (**30–32**) (Eqs. [1.4–1.6]). Quinone methides **31** and **32** were detected via LFP and showed absorption maxima of 570 and 525 nm, respectively (in 100% water, Table 1.2). Quinone methide **30** was too short lived to be detected by LFP, but was implicated by formation of product **33** that would arise from electrocyclic ring closure of **30** (Eq. 1.4).

27 $\xrightarrow[(-H_2O)]{h\nu}$ [**30**] → **33** (1.4)

28 $\xrightarrow[(-H_2O)]{h\nu}$ **31** $\lambda_{max} = 570$ (1.5)

29 $\xrightarrow[(-H_2O)]{h\nu}$ **32** $\lambda_{max} = 525$ (1.6)

This concept was then extended to bulkier biaryl systems containing naphthalene rings.[34,35] Biaryls **34–36** were all observed to undergo photochemically

TABLE 1.2 Absorption and Lifetime (1/*k*) Data for Selected Quinone Methides

Quinone Methide	Solvent	λ_{max} (nm)	τ
24	Aq. CH_3CN	350, 460	>5 s
25	Aq. CH_3CN	445	30 ns
26	Aq. CH_3CN	360	>5 s
31	H_2O	570	400 ns
32	H_2O	525	67 μs
42	Cyclohexane	560	5 μs
43	Aq. CH_3CN	360, 580	230 μs
44	TFE	420	8.5 μs
45	Aq. CH_3CN	330, 425, 700	30 μs
48	Aq. CH_3CN	450	5–10 s
52	Aq. CH_3CN	410, 700	34 μs

induced dehydration and cyclization to go from the highly twisted biaryl starting materials to the (more) planar diarylpyrans **37–39** via biaryl quinone methides **40–42** (Eqs. [1.7–1.9]). These quinone methides were not detectable by LFP in aqueous acetonitrile, presumably because electrocyclic ring closure is very rapid since it regenerates two aromatic rings in all cases (and might relieve some steric congestion). However, Burnham and Schuster[36] observed a transient absorption ranging from 520 to 620 nm and centered at 560 nm, with a lifetime of 5.02 μs when examining **36** by LFP in cyclohexane. They assign the absorption to quinone methide **42**. Under the different experimental conditions of Burnham and Schuster, it is possible that some of the intermediate exists as the transoid form **42b**, which cannot readily cyclize to **39** and hence would be expected to persist for longer periods of time.

OH OH **34** $\xrightarrow[(-H_2O)]{h\nu}$ O **40** ⟶ O **37** (1.7)

OH OH **35** $\xrightarrow[(-H_2O)]{h\nu}$ O **41** ⟶ O **38** (1.8)

OH OH **36** $\xrightarrow[(-H_2O)]{h\nu}$ O **42** ⟶ O **39** (1.9)

O **42a** ⇌ O **42b**

Wan's group investigated a number of phenyl-substituted hydroxybiphenylbenzyl alcohols in the hope that the α-phenyl quinone methides photogenerated from them might show enhanced absorption and lifetimes, and thus be easier to characterize by LFP.[37,38] They were successfully able to photogenerate and characterize quinone methides **43–45** by LFP, and their absorption maxima and lifetimes (in 50% v/v aqueous acetonitrile) are presented in Table 1.2.

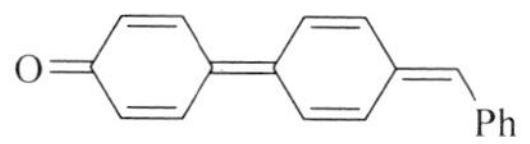

43 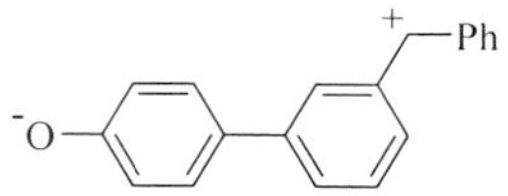 **44** 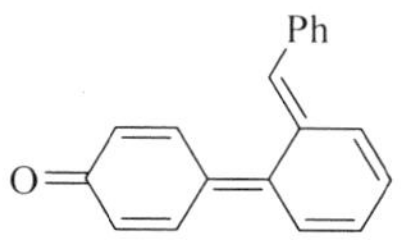 **45**

Hydroxy-9-fluorenols **46** and **47** have been similarly shown to undergo photodehydration in aqueous solution to give the corresponding fluorenylquinone methides **48** and **49** (Eqs. 1.10 and 1.11).[39] **49** was very reactive and not observable by LFP (presumably because its reaction regenerates two aromatic rings); however **48** was much more persistent, having a lifetime in the 5–10 s range, and was observable using a conventional UV–Vis spectrometer ($\lambda_{max} = 450$ nm). Formation of **48** was further confirmed by the isolation of its 4 + 2 cycloaddition products with ethyl vinyl ether (EVE).

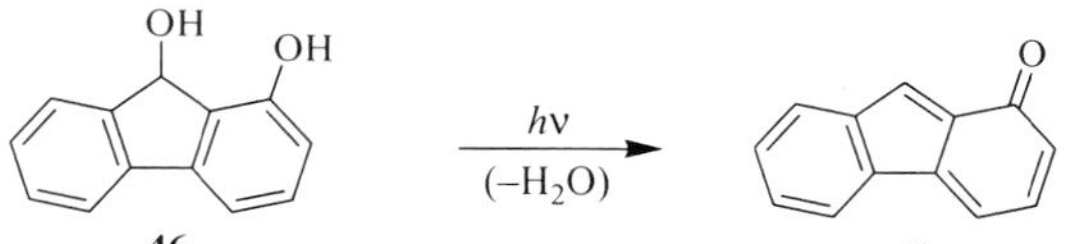

46 $\xrightarrow[(-H_2O)]{h\nu}$ **48** (1.10)

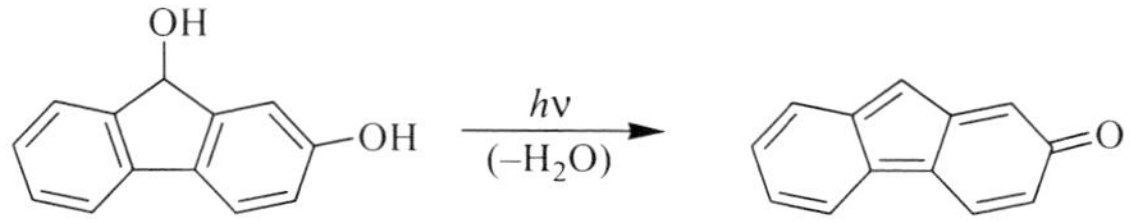

47 $\xrightarrow[(-H_2O)]{h\nu}$ **49** (1.11)

Wan also studied hydroxybenzyl alcohols based on the naphthalene chromophore.[40] Naphthols **50** and **51** were both examined for their ability to photogenerate naphthoquinone methides **52** and **53**, respectively (Eqs. 1.12 and 1.13). While **50** underwent very efficient photosolvolysis, presumably via naphthoquinone methide **52**, **51** was essentially unreactive when exposed to light. The inability of **51** to photogenerate **53** is a rare example where the generality of the photodehydration of benzyl alcohols fails. LFP of **50** yielded a very strong visible absorption ($\lambda_{max} = 410$ and 700 nm) that decayed in aqueous acetonitrile with a lifetime of 34 μs (Fig. 1.2). This transient was assigned to naphthoquinone methide **52** due in part to its efficient quenching when the nucleophilic ethanolamine was added.

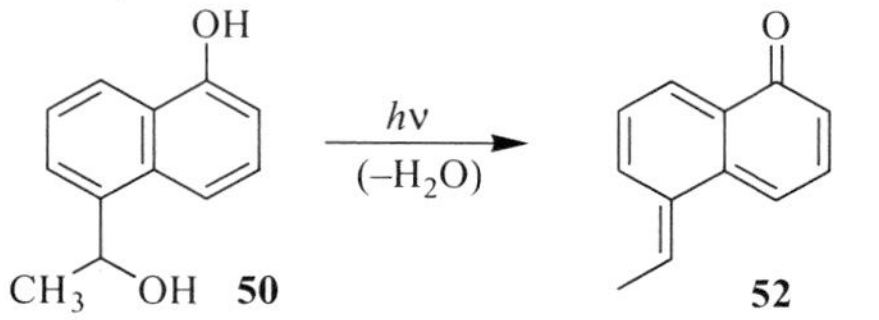

50 $\xrightarrow[(-H_2O)]{h\nu}$ **52** $\Phi = 0.84$ (1.12)

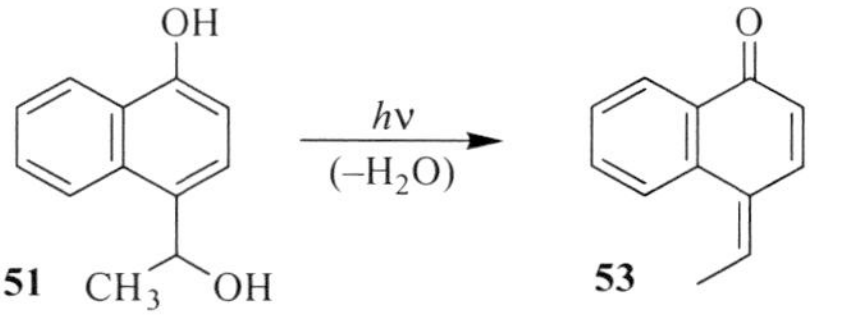

51 $\xrightarrow[(-H_2O)]{h\nu}$ **53** $\Phi < 0.01$ (1.13)

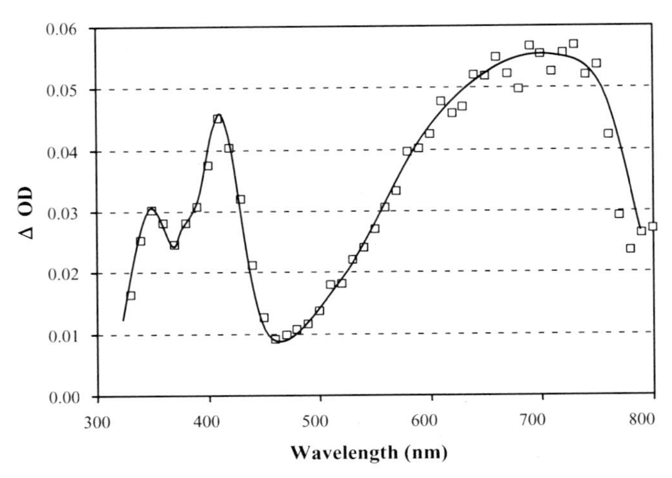

FIGURE 1.2 Absorption spectrum of **52** obtained by LFP of naphthol **50**, in aqueous acetonitrile. *Source*: Data taken with permission from Ref. [40].

Through these works, Wan has conclusively demonstrated that the photodehydration of hydroxybenzyl alcohols is a general reaction, and a wide variety of quinone methides can be photogenerated and detected using this method. Quinone methide photogeneration via this method has been shown to have importance in the photochemistry of Vitamin B_6[41,42] and in model lignins.[43]

1.2.3 Photoelimination of Quaternary Ammonium Salts

Saito and coworkers showed that a variety of *ortho*-quinone methides can be photogenerated from the Mannich bases of phenols and naphthols.[44] For example, irradiation of **54** gives *ortho*-quinone methide **55**, as evidenced by the isolation of the cycloaddition trapping product **56** (Eq. 1.14). Although quantum yields were not reported, the yields of quinone methides appeared to be much higher from the Mannich bases than from the analogous benzyl alcohols. The proposed mechanism involves initial adiabatic ESIPT from the phenol OH to the nitrogen to generate the singlet excited dialkylammonium zwitterion, which then undergoes loss of the neutral amine to generate the *ortho*-quinone methide (Scheme 1.4). Direct excitation of the ground-state phenolate did not lead to quinone methide formation, presumably because the nitrogen needs to be protonated by the phenolic OH prior to its departure.

54 (Ph, OH, NMe_2) $\xrightarrow{h\nu}$ **55** (Ph, O) $\xrightarrow{\text{EVE}}$ **56** (Ph, O, OEt) (1.14)

In one example described in the paper,[44] irradiation of biphenol **57** in the presence of EVE gives the bisquinone methide–EVE adduct **58** in 22% yield (Eq. 1.15). The

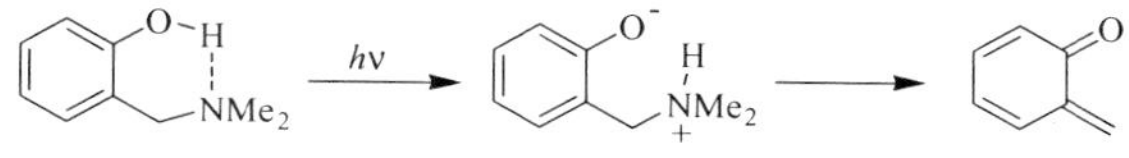

SCHEME 1.4

authors suggest bisquinone methide **59** as an intermediate in the reaction, although sequential photoreaction of the two halves of the molecule (to go through subsequent monoquinone methides) seems just as reasonable. Reasons for the modest yield are not stated, but a competing side reaction might be reaction of the intermediate quinone methide(s) with the nucleophilic nitrogen of unreacted **57**. Derivatives such as **57** that are capable of generating bisquinone methides are of particular interest since they can be used as DNA cross-linkers.

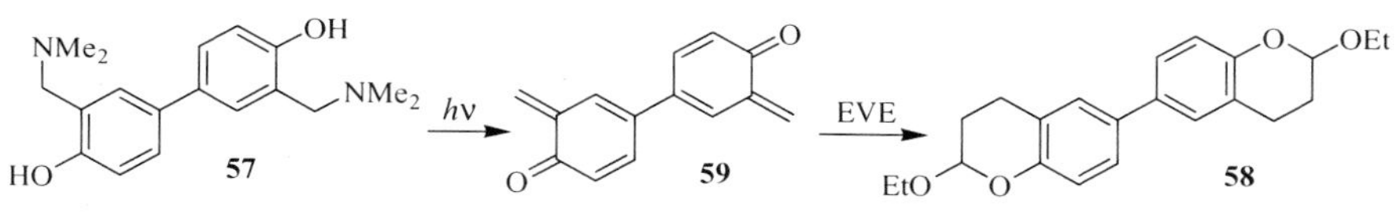

(1.15)

Freccero expanded on the work of Saito and showed that quaternary ammonium salts (in particular the trialkyliodide salts) are as reactive as the Mannich bases at neutral pH ($\Phi \sim 1$) and show several advantages including lower nucleophilicity, higher quantum yields at high pH, and superior water solubility.[45] The authors indicate that these features make such salts ideal photochemical precursors for the study of quinone methides by LFP. Indeed, the authors were able to photogenerate and detect *ortho*-quinone methide **1** on LFP of **60** (Eq. 1.16). The spectrum of **1** obtained in this way showed an absorption band centered at 400 nm (consistent with the results of Wan)[31] that persisted with a lifetime of ~3 ms in water (Fig. 1.3). The position of this peak is blue shifted by ~60 nm relative to its α-phenyl cousin **24**, due to reduced conjugation. The authors were able to quench the transient assigned to **1** with a variety of nucleophiles and determine the bimolecular reaction rate constants, which spanned 7 orders of magnitude.

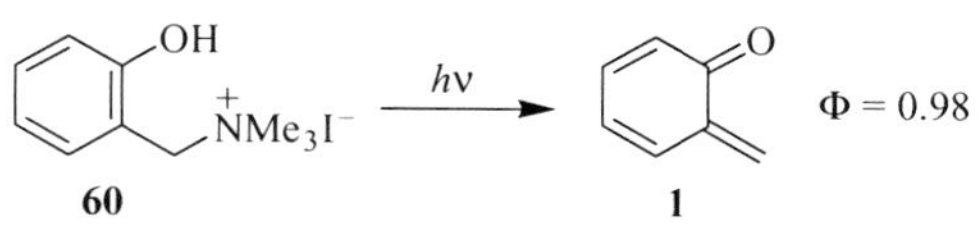

$\Phi = 0.98$ (1.16)

Using the appropriate benzylammonium precursors, Freccero and coworkers exploited this reactivity and successfully photogenerated *ortho*-quinone methides **61–65** and characterized them using LFP.[46] All of **61–65** gave rise to absorption bands centered between 400 and 440 nm, although the bimolecular rate constants for reaction with water span four orders of magnitude on changing the X group from the electron-donating methoxy group to the strongly electron-withdrawing nitro group. It is clear

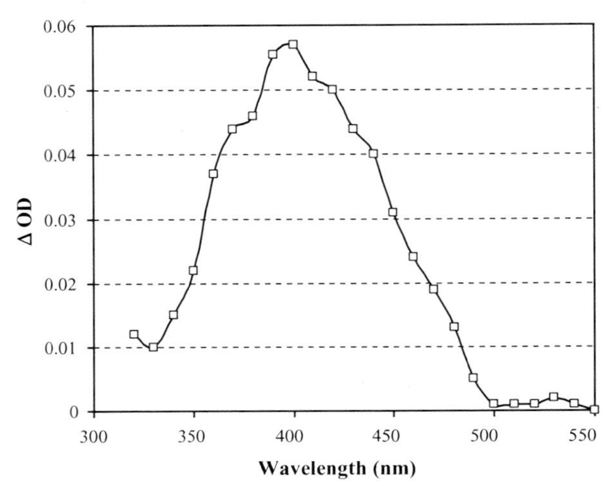

FIGURE 1.3 Absorption spectrum of **1** obtained by LFP of **60** in water. *Source*: Data taken from Ref. [45].

from this work that electron-donating groups reduce the reactivity toward nucleophilic attack and that electron-withdrawing groups increase their reactivity, showing that quinone methides are electron-poor species.

61 X = OCH_3
62 X = Cl
63 X = COOMe
64 X = CN
65 X = NO_2

Following this theme, DNA cross-linking agents based on the benzylammonium quinone methide precursors were developed by the groups of Zhou[47] and Freccero.[48,49] Phenols **66–69** were all shown to photochemically induce the cross-linking of DNA (albeit with variable efficiency), probably through sequential quinone methides and not through a bisquinone methide such as **59**.

66 **67** **68** **69**

1.2.4 Photoelimination of Alcohols and Esters

The demonstrated efficient photoelimination of fluoride,[22,23] hydroxide,[24–43] and amines[44–49] from the benzylic position of appropriate phenolic precursors suggests that the efficient photoelimination of other benzylic leaving groups might also be possible. The ability to photochemically release a wide array of benzylic leaving groups in this way could lead to the development of important classes of photolabile protecting groups, although the use of benzylic phenols in this context has to date been largely unexploited.

The Popik group has recently begun to explore the potential of phenols to photorelease ethers and alcohols attached to an *ortho*-benzylic group. For example, they have shown that esters and ethers of (3-hydroxymethyl)naphthalene-2-ol (i.e., **70**) efficiently release the corresponding alcohols and acids upon exposure to UV irradiation ($\Phi \sim 0.3$),[50] with formation of naphthoquinone methide **71** occurring in the process (Eq. 1.17).

X = OR, OC(O)R

70 —$h\nu$→ **71** + HX (1.17)

Kostikov and Popik[51] have also shown that 1,4-dihydroxybenzyl derivatives **72** and **73** undergo a similar reaction to efficiently expel aliphatic and aromatic alcohols and acids to give *ortho*-quinone methides **74** and **75**. Rather than being trapped by water, as was observed for **71**, these quinone methides instead underwent rapid tautomerization to give the benzoquinones **76** and **77** (Eqs. 1.18 and 1.19). The observed quantum yields were in the range of 0.2–0.5 indicating that these might be very promising for use as photolabile protecting groups.

72 —$h\nu$→ **74** → **76** (1.18)

73 —$h\nu$→ **75** → **77** (1.19)

Although Peter Wan's group was able to show that *para*-quinone methide **2** could be photogenerated from the corresponding hydroxybenzyl alcohol,[31] characterization of **2** by LFP was not possible due to its inefficient formation ($\Phi = 0.007$). Kresge and coworkers were interested in the thermal chemistry of **2** and sought an appropriate photochemical precursor so that the chemistry of **2** could be studied using LFP. They

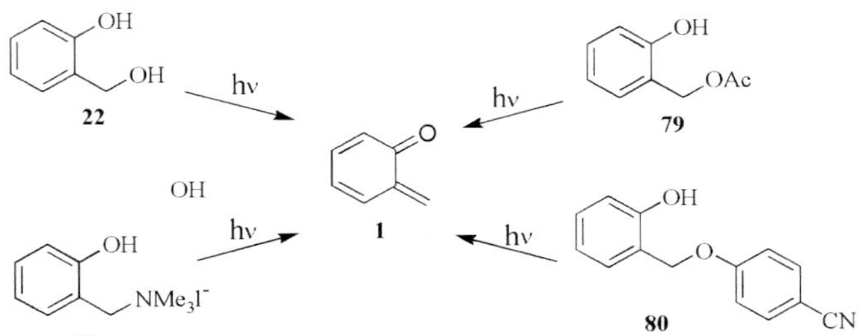

SCHEME 1.5

found that *p*-hydroxybenzyl acetate **78** was converted to **22** on irradiation via **2** (Eq. 1.20).[17] Moreover, LFP of **78** provided a long-lived (several milliseconds) transient absorption with $\lambda_{max} = 300$ nm that is assignable to **2**. Similarly, Kresge and coworkers were able to photogenerate *ortho*-quinone methide **1** from *o*-hydroxybenzyl acetate **79** and the analogous 4-cyanophenylether **80** (although photolysis of **80** gave unidentified side products that are possibly radical derived) (Scheme 1.5).[15]

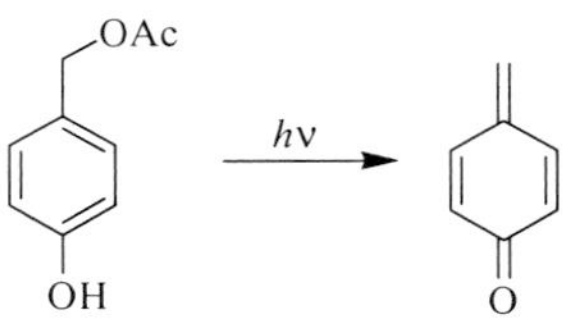

(1.20)

1.3 QUINONE METHIDES FROM ESIPT TO UNSATURATED SYSTEMS

1.3.1 Quinone Methides from ESIPT to Carbonyls

ESIPT from a phenol OH to a carbonyl oxygen is a well-known process that is very rapid and is usually wholly reversible to regenerate starting material.[52,53] The energy-wasting quality of this reaction has been exploited in the design of photostabilizers and sunscreens. For example, oxybenzone (**81**) is a common component of sunscreens and possesses a phenolic OH situated *ortho* to the ketone carbonyl. Upon absorption of light (UVA and UVB), ESIPT proceeds to give **82**, which is an *ortho*-quinone methide (Eq. 1.21). Most quinone methides photogenerated in this way undergo very rapid reverse proton transfer (sub-ps) and do not go on to give further products. The presence of quinone methides in a number of such systems produced by both intrinsic and solvent-mediated ESIPT has been suggested by laser-induced fluorescence and laser flash photolysis experiments.[54–60]

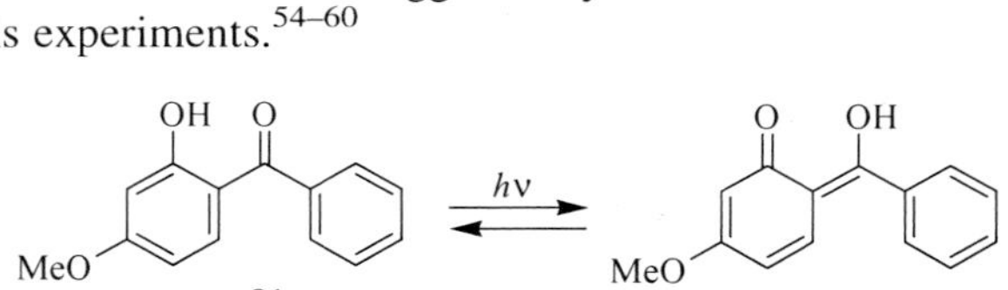

(1.21)

While the formation of new products is rare for quinone methides of this type, in 1975, Stermitz and coworkers reported the photogeneration of flavones **85** and **88** on irradiation of *o*-hydroxychalcone **83** and sorbophenone **86**, respectively (Eqs. 1.22 and 1.23).[61] Matsushima and coworkers[62,63] investigated this reaction using a variety of chalcone derivatives and observed that efficiencies were highest when carried out in a polar aprotic solvent and lowest in polar protic solvents. They proposed a mechanism in which ESIPT proceeds from the phenol OH to the carbonyl oxygen to give quinone methide intermediates (i.e., **84** from **83** and **87** from **86**), which can undergo ring closure to give the observed products. The groups of Matsushima[64–70] and others[71] have shown that related chemistry is responsible for the photocyclization of a variety of 2-hydroxychalcones such as **89** (to **90**) via quinone methide intermediates (Eq. 1.24).

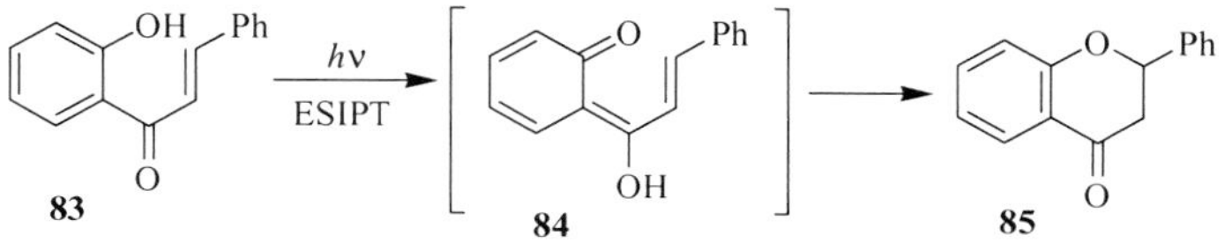

(1.22)

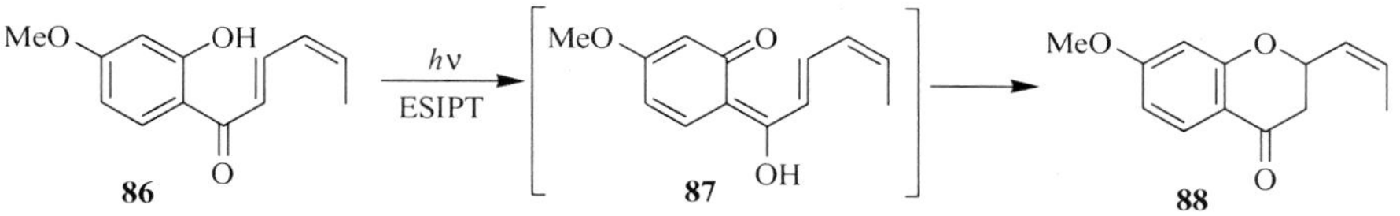

(1.23)

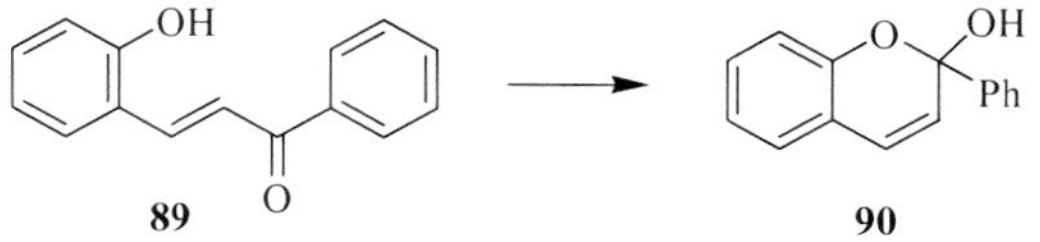

(1.24)

While ESIPT from a phenolic OH to a carbonyl group is a common and efficient process, the highly reversible nature of the reaction often precludes any further reactivity, except in rare cases. For this reason, this reaction is not commonly used to generate or study quinone methides.

1.3.2 Quinone Methides from ESIPT to Alkenes and Alkynes

Ferris and Antonucci[72] reported that irradiation of *o*-hydroxyphenylacetylene (**91**) gave *o*-hydroxyacetophenone (**93**), presumably resulting from Markovnikov photohydration of the alkyne via enol **92** (Eq. 1.25). Similarly, Yates and coworkers[73,74] reported that the alkene moiety of *o*-hydroxystyrene (**94**) undergoes Markovnikov photohydration to give **95** (Eq. 1.26). The reaction proved to be independent of solution pH between pH 0 and 7, prompting them to invoke a mechanism in which protonation of the alkene or alkyne takes place as a result of ESIPT from the phenol OH to the β-alkene or alkyne carbon, giving *ortho*-quinone methides **96** and **97** from **94** and **91**, respectively. The hydration products result from

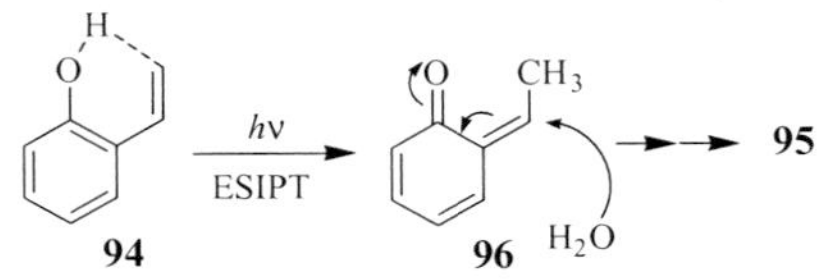

SCHEME 1.6

nucleophilic attack on **96** and **97** by water. The proposed mechanism for **96** is shown in Scheme 1.6.

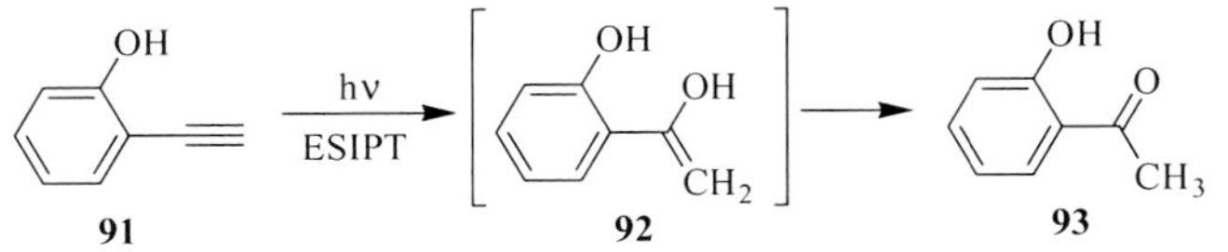

(1.25)

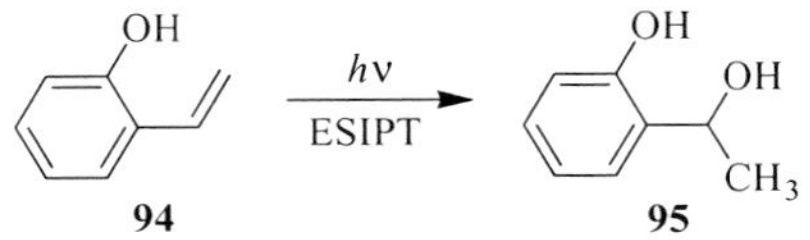

(1.26)

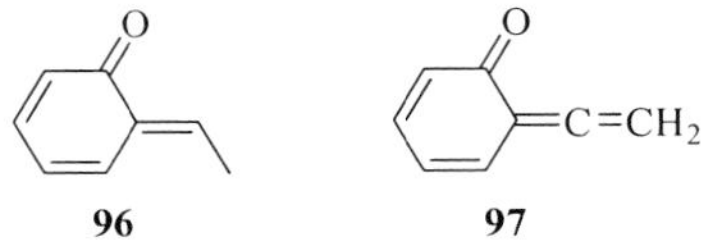

Wan and coworkers[75] examined the photochemistry of **98**, which is the α-phenyl derivative of **94**. Efficient photohydration ($\Phi = 0.13$) of **98** to give benzyl alcohol **100** via presumed *ortho*-quinone methide **99** was observed in aqueous solution (Eq. 1.27), as expected in analogy to the reported photochemistry of **94**. The additional conjugation provided by the α-phenyl group was expected to enhance the absorption of **99** relative to **96**, thus facilitating its study by LFP. Indeed, LFP of **98** in aqueous acetonitrile gave a strong absorption with peaks at 310 and 410 nm, and showed essentially no decay within the 100 μs window of the experiment, suggesting a lifetime that is at least several milliseconds. The authors attribute this signal to **99**. A very similar signal is observed on LFP of **98** in neat acetonitrile indicating that the proton transfer from the phenol to the alkene is intrinsic and does not require mediation by water. The blue shift observed for the peaks assigned to **99** relative to those of closely related *ortho*-quinone methide **24** might be due to reduced conjugation in the former resulting from twisting of the unsubstituted phenyl ring out of the plane because of the increased steric interactions with the methyl group.

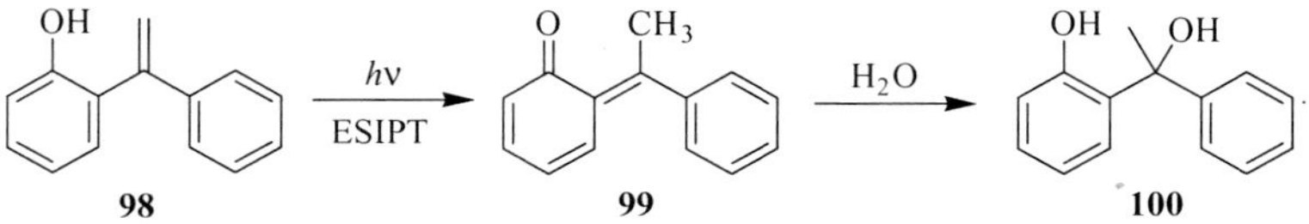

(1.27)

Further work by Wan's group[76,77] showed that ESIPT can also occur from phenols to alkenes substituted *meta* and *para* to one another. *m*-Hydroxystyrenes **101** and **102** and *p*-hydroxystyrene **103** all gave the Markovnikov photohydration products (**107–109**) with high efficiency ($\Phi = 0.22$, 0.24, and 0.1, respectively) via *meta*- and *para*-quinone methides **104–106** (Eqs. [1.28–1.30]). LFP of **101** and **102** gave rise to strong long wavelength absorptions centered at 425 and 415 nm, respectively. These bands resemble that observed for **20** both in terms of position (λ_{max}) and spectral breadth, and were thus assigned to **104** and **105**. For **101** and **102**, steady-state and time-resolved fluorescence quenching showed a cubic dependence with added water, prompting the authors to suggest that three water molecules are necessary to ferry the phenolic proton to the β-carbon of the *meta*-alkene during the ESIPT process.

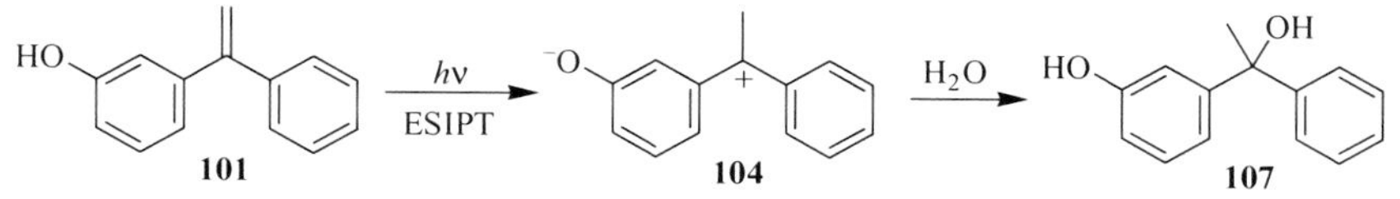

(1.28)

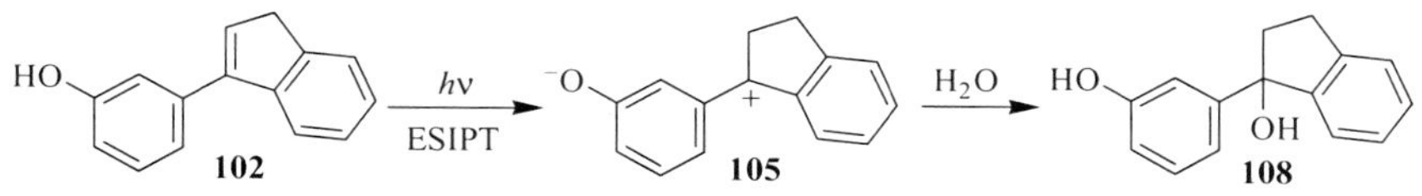

(1.29)

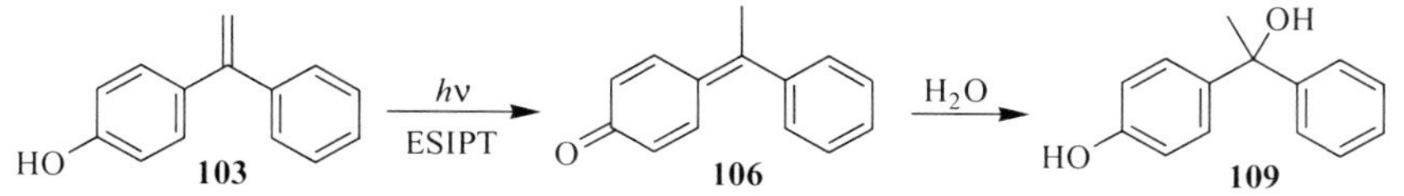

(1.30)

Using the same reaction, Cole and Wan later photogenerated *meta*-quinone methides **110–112** from the appropriate *m*-hydroxy-α-phenylstyrenes.[78] The three were characterized by LFP, with **110** and **111** showing strong and sharp absorptions at 430 and 450 nm, respectively, similar to those observed for **104** and **105**. The dimethoxy-susbstituted **112** showed a much broader absorption centered at ~420 nm. Addition of the electron-donating methoxy groups stabilized the quinone methides; the lifetimes of **110–112** in 1:1 aqueous acetonitrile were 5–200 times longer than that of **104**.

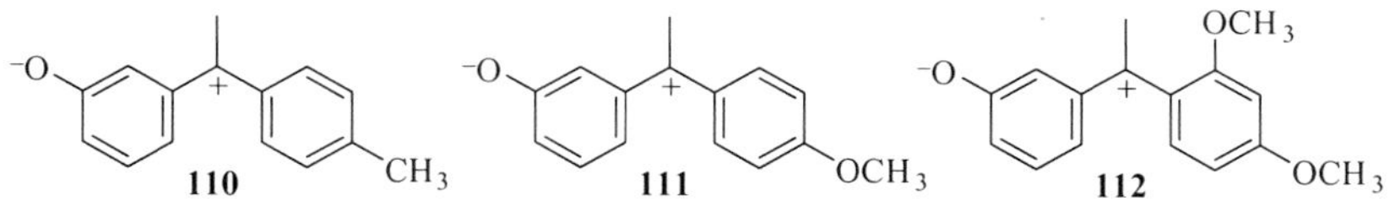

Photohydration was observed for biphenylalkenes **113** and **114** via biphenylquinone methides **115** and **116**, respectively (Eqs. 1.31 and 1.32).[37] *meta*-Biphenylquinone methide **116** was detected on LFP of **114** in TFE and showed a strong and sharp absorption band at 425 nm, consistent with spectra of previously observed *meta*-quinone methides. Detection of **115** on LFP of **113** proved to be more challenging, partly because of 10-fold lower reaction efficiency of **113** relative to **114**. Nevertheless, a transient was observed on LFP of **113** in acidic (pH 1) solution with bands at 360 and 580 nm and a lifetime of 23 μs that was assignable to **115**.

113 $\xrightarrow[\text{ESIPT}]{h\nu}$ **115** $\Phi = 0.013$ (1.31)

114 $\xrightarrow[\text{ESIPT}]{h\nu}$ **116** $\Phi = 0.1$ (1.32)

The photochemistry of the *ortho*, *meta*, and *para* derivatives of hydroxystilbene (**117–119**, respectively) has been investigated by the groups of Lewis and Arai.[79–81] The hydroxystilbenes represent an interesting series, since a wide variety of stilbenes are well known to undergo very rapid and efficient isomerization between the *cis* and *trans* forms. Only a very rapid ESIPT process can lead to photohydration of **117–119**, since this reaction must compete with the isomerization process. Irie was able to show that irradiation of the *trans* isomers of **117** and **119** led only to isomerization to their *cis* forms, while irradiation of *trans*-**118** gave the corresponding photohydration product ($\Phi = 0.15$) in addition to formation of the *cis* isomer. The authors did not indicate whether the regiochemistry of the photohydration was Markovnikov or anti-Markovnikov, although Markovnikov chemistry would be expected in analogy to the work of Peter Wan. While not explicitly addressed by the authors, Markovnikov photohydration of **118** would likely proceed via initial ESIPT to first give the *meta*-quinone methide **120** (Eq. 1.33).

117 **118** **119**

118 $\xrightarrow[\text{ESIPT}]{h\nu}$ **120** $\Phi = 0.15$ (1.33)

Uchida and Irie have reported a photochromic system based on ESIPT to an alkene carbon.[82] They observed that vinylnaphthol **121** isomerizes to the ring-closed **123** when irradiated with 334 nm light ($\Phi = 0.20$, Eq. 1.34). The reaction is photoreversible since irradiation of **123** (at 400 nm) regenerates the starting vinylnaphthol. The authors proposed a mechanism in which ESIPT from the naphthol OH to the β-alkenyl carbon gives intermediate *o*-quinone methide **122**, which undergoes subsequent electrocyclic ring closure. The reverse reaction is presumed to also go via **122**, indicating that quinone methides can be formed by electrocyclic ring-opening reactions.

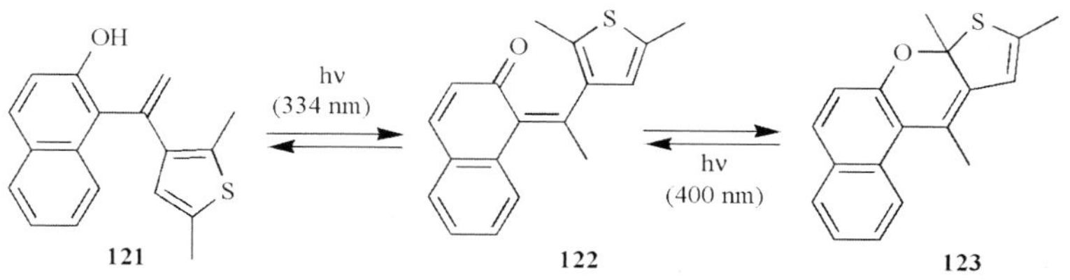

(1.34)

1.3.3 Quinone Methides from ESIPT to Aromatic Carbon

Tolbert et al. were the first to observe that irradiation of 1-naphthol (**124**) in acidic D_2O resulted in replacement of the aromatic C—H bonds at the 5- and 8-positions with C—D bonds (Eq. 1.35).[83] They proposed a mechanism in which the excited state of 1-naphthol (or 1-naphtholate) is directly protonated (or deuterated) by the acidic solvent. They identify naphthoquinone methide **125**-D as a necessary intermediate in the formation of **124**-5D, and by analogy, **126**-D would be the reactive intermediate responsible for the formation of **124**-8D. Lukeman et al.[40] examined the exchange reaction in greater detail and by looking at the solvent and pH (pD) dependence, were able to determine that acid catalysis of the exchange reaction only takes place below a pD of ~2. Between pD 2 and 10 (the pK_a of the naphthol), the exchange reaction was independent of solution pD, leading the authors to suggest a mechanism involving solvent-mediated ESIPT from the naphthol OH (OD) to the 5- and 8-positions to form **125** and **126**. Moreover, they were able to detect naphthoquinone methide **125** on LFP of 1-naphthol in aqueous acetonitrile, which showed $\lambda_{max} = 340$ and 530 nm (br), and persisted with a lifetime of 16 μs (Fig. 1.4). This transient was not visible in neat CH_3CN, which led the authors to the conclusion that the ESIPT was water mediated.

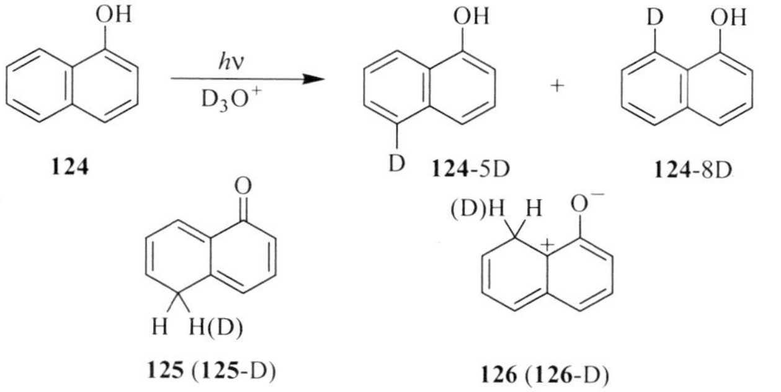

(1.35)

The ability to photogenerate quinone methides via ESIPT from a phenol OH to an aromatic carbon atom was explored for a wider number of substrates by Wan's

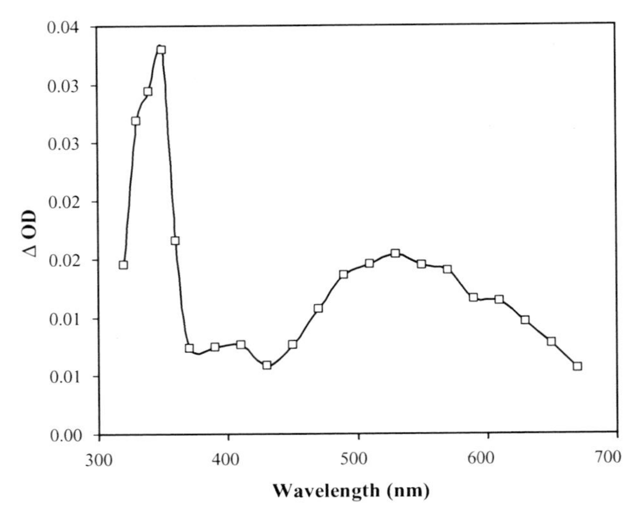

FIGURE 1.4 Absorption spectrum of **125** obtained by LFP of **124** in N_2 purged aqueous acetonitrile. *Source*: Data taken with permission from Ref. [40].

group.[84–88] 2-Phenylphenol (**127**) was shown to undergo moderately efficient ($\Phi = 0.08$) photoinduced deuterium exchange of the 2′- and 4′-positions of the ring not possessing the OH group (Scheme 1.7).[84,85] Biphenylquinone methides **128** and **129** were proposed as reaction intermediates, although attempts to detect either of them with LFP were unsuccessful, presumably because their decay takes place faster than the time resolution of the instrumentation (~20 ns). Solvent dependence studies indicated that different mechanisms were responsible for formation of the two quinone methides. A water-mediated ESIPT mechanism similar to the one that operates in 1-naphthol is responsible for the formation of **129**, while **128**

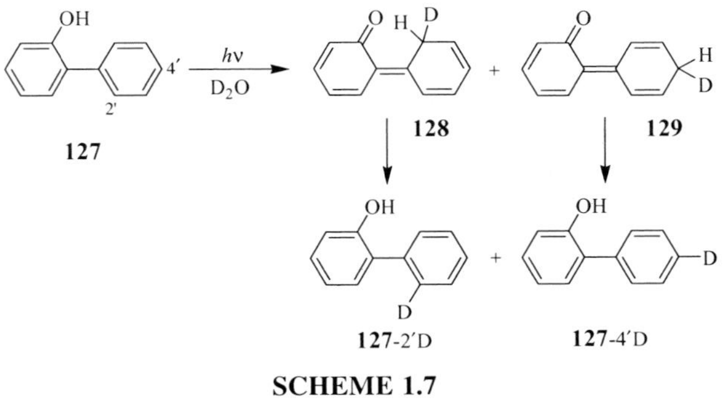

OH 4′ 2′ **127** $h\nu$ D_2O O H D **128** + O H D **129**

OH D **127**-2′D + OH D **127**-4′D

SCHEME 1.7

is generated by a novel mechanism in which a direct ESIPT from the phenol OH to the 2′-aromatic carbon takes place. The latter mechanism was shown to occur in various polar and nonpolar solvents, and even in the solid state, demonstrating that water is not required.

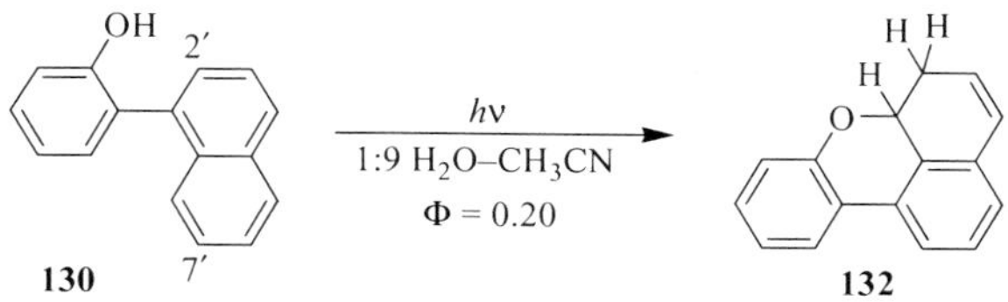

(1.36)

(1.37)

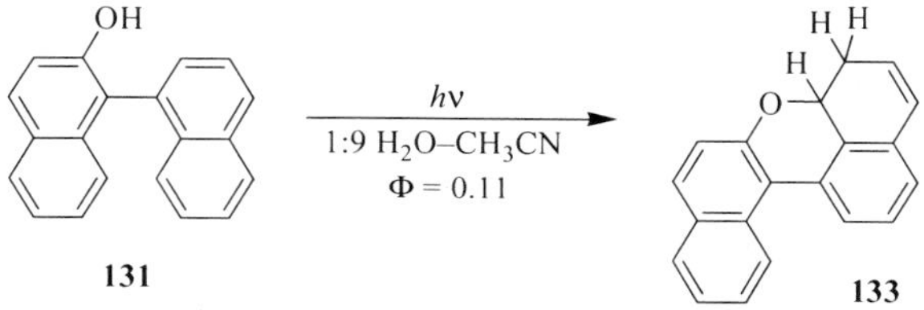

Naphthyl analogues **130** and **131** were shown to give dihydroxanthenes **132** and **133** when irradiated in 1:9 H_2O—CH_3CN (Eqs. 1.36 and 1.37).[86] When the aqueous portion of the solvent was replaced with D_2O, recovered starting material showed deuterium exchange at the 2′-position for both **130** and **131**. Biaryl naphthoquinone methides **134** and **135** were proposed to result from water-mediated ESIPT to the 7′-carbon of **130** and **131**, and undergo electrocyclic ring closing to give **132** and **133**, respectively. Quinone methides **136** and **137** are formed following water-mediated ESIPT from the phenol to the 2′-carbon of **130** and **131**, and these do not undergo ring closure. Instead, these intermediates undergo reverse proton transfer to regenerate starting material (deuterium labeled at the 2′-position if D_2O is present). This report represents the first examples where quinone methides generated by ESIPT to aromatic carbon give products (other than deuterium-labeled starting material).

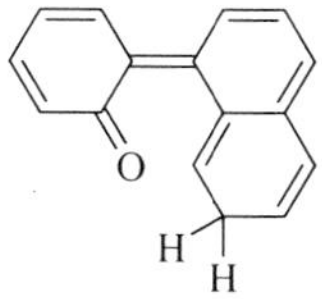

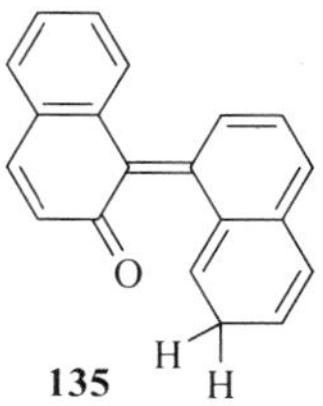

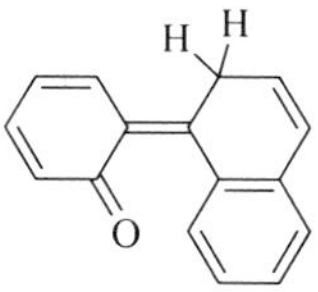

Similar reaction pathways were recently shown to be available to the widely used chiral ligand 1,1′-binaphthol (BINOL) (**138**).[87] Irradiation of BINOL in aqueous acetonitrile initiated ESIPT to the 4′-, 5′-, and 7′-ring carbons to give biaryl quinone

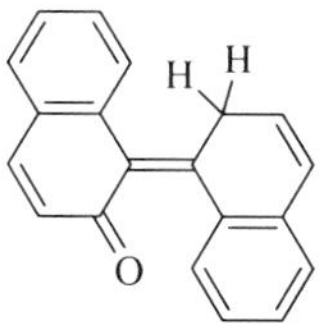

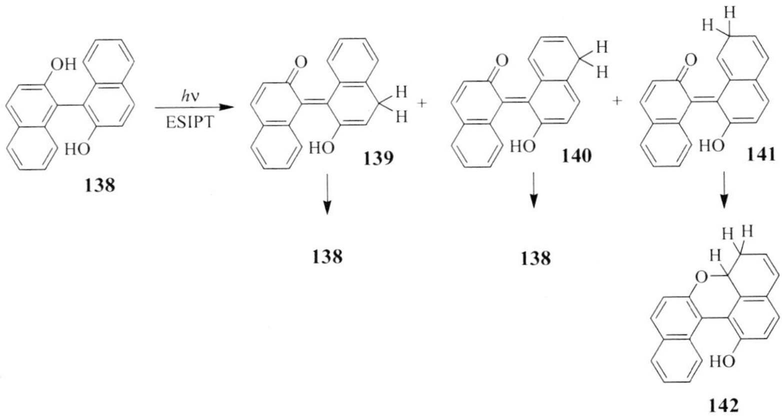

SCHEME 1.8

methides **139**, **140**, and **141**, respectively, with a total efficiency of 0.15 (Scheme 1.8). Quinone methides **139** and **140** returned to the starting material (leaving deuterium labels when the solvent contained D_2O), while **141** (formed in much smaller amounts) underwent rapid electrocyclic ring closure to give the isolable dihydrobenzoxanthene **142**. Starting material recovered from photoreactions performed in D_2O did not indicate deuterium incorporation at the 7′-position, suggesting that reverse proton transfer cannot compete with ring closing for **141**. It was observed that irradiation of a single enantiomer of **138** led to its racemization with efficiency similar to the quinone methide formation. It was proposed that formation of planar quinone methides **139** and **140** and their subsequent return to starting material is the process that gives rise to the photoracemization.

A third reaction pathway for quinone methides generated following ESIPT to aromatic carbon (in addition to H–D exchange and cyclization) was observed following the examination of the photochemistry of 9-(2′-hydroxyphenyl)anthracene **143**. Irradiation of **143** in acetonitrile containing water or alcohols lead to the formation of the photoaddition products **145** (R = H, Me, *i*Pr), which were sufficiently stable to be isolated and characterized (Eq. 1.38).[88] A reaction mechanism was proposed in which ESIPT from the phenolic OH to the 10-carbon gives rise to quinone methide **144**, which gives **145** following nucleophilic attack. LFP of **143** gave rise to a long-lived transient that absorbed between 280 and 550 nm (although the region between 330 and 390 nm was obscured by starting material bleaching in this region) and had a lifetime estimated to be between 10 ms and 1 s. This transient was assigned to QM **144**. Derivatives of **143** possessing substituents at the 10-position were also shown to give quinone methides analogous to **144**, as well as other ESIPT products that are not formed via quinone methide intermediates.[89] Attempts to detect quinone methides containing methyl and phenyl substituents at the 10-position were unsuccessful, primarily due to spectral interference from T–T absorptions, negative signals from starting material bleaching, and the presence of radical cations (formed via two-photon processes).

143 $\xrightarrow[\text{ESIPT}]{h\nu}$ 144 $\xrightarrow{\text{ROH}}$ 145 (1.38)

1.4 OTHER PHOTOCHEMICAL ROUTES TO QUINONE METHIDES

The methods of quinone methide photogeneration detailed in the previous sections all employ phenolic precursors, and most show wide generality and good efficiency. Several other reactions are available that have been shown to photogenerate quinone methides, although these reactions have not enjoyed widespread application for several reasons such as inaccessibility of precursors, lack of reaction generality, formation of byproducts, or low reaction efficiency. This section aims to highlight a number of selected examples, although an exhaustive review is beyond the scope of this chapter.

A number of carbonates and lactones have been shown to give rise to quinone methide intermediates on irradiation.[90–92] For example, Padwa and coworkers[92] demonstrated that 3-phenylisocoumaranone (**146**) will extrude a molecule of carbon monoxide when irradiated in methanol to generate *ortho*-quinone methide (**24**) with moderate efficiency ($\Phi = 0.058$, Eq. 1.39). This intermediate is subsequently trapped by the methanol solvent to give **147**.

146 $\xrightarrow[\text{- CO}]{h\nu}$ 24 $\xrightarrow{CH_3OH}$ 147 (1.39)

Kresge and coworkers[93] have recently reported that suitably substituted diazo compounds can give rise to quinone methides via photochemical extrusion of nitrogen (N_2). They observed that irradiation of *o*-hydroxyphenyldiazo acetate (**148**) gives *o*-quinone α-carbomethoxymethide **149** in aqueous solution, which gave the hydration product **150** (Eq. 1.40). LFP permitted detection of **149**, which showed a strong absorption band at 420 nm, similar to that previously observed for **1**. A related reaction had been previously reported in an argon matrix at 10 K in which diazo compound **151** (itself photogenerated *in situ*) gives *o*-quinone methide **1** on irradiation.[94]

OH N2 OCH3 O **148** —$h\nu$, $-N_2$→ O OCH3 O **149** —H_2O→ OH OH OCH3 O **150** (1.40)

OH N2 **151**

Pyrans and napthopyrans (chromenes) are photochromic compounds that undergo photochemically induced electrocyclic ring opening to give colored *ortho*-quinone methides.[95–98] For example, chromene **153** opens on irradiation to give **154** (Eq. 1.41).

O CH3 CH3 **153** —$h\nu$→ O CH3 CH3 **154** (1.41)

p-Quinone methides **156** and **157** were generated on irradiation of the *cis* and *trans* forms of **155**, respectively (Eqs. 1.42 and 1.43).[99,100] The mechanism is believed to involve a radical ring opening of the cyclopropane ring, followed by hydrogen or methyl migration.

O *t*-Bu *t*-Bu H_3C CH_3 *cis*-**155** —$h\nu$→ O *t*-Bu *t*-Bu H_3C CH_2CH_3 **156** (1.42)

O *t*-Bu *t*-Bu H_3C CH_3 *trans*-**155** —$h\nu$→ O *t*-Bu *t*-Bu H CH_3 CH_3 **157** (1.43)

1.5 CONCLUSIONS AND OUTLOOK

Photochemical methods enable the efficient generation of a wide variety of quinone methides under very mild conditions. The use of light as a "reagent" provides the experimentalist with great control over when and where the quinone methides are formed. The use of flash photolytic techniques (such as LFP) allows for relatively large concentrations of quinone methides to be generated in solution within a short period of time, greatly facilitating their spectroscopic detection.

A number of methods have been developed for the photogeneration of quinone methides, and the best developed of these is the photodehydration reaction of hydroxybenzyl alcohols (Section 1.2.2). This reaction employs readily accessible substrates and can be used for the photogeneration of *ortho*-, *meta*-, and *para*-quinone methides, as well as polyaromatic quinone methides. For a much narrower selection of systems, trialkyl ammonium quinone methide precursors have shown several advantages including quantum yields close to unity and superior water solubility (Section 1.2.3). Efforts to further expand the generality of this reaction including extension to *para* and *meta* systems may lead to this becoming the preferred method of quinone methide photogeneration. ESIPT from a phenol OH to an appended alkenyl group is another general method for the formation of *ortho*-, *meta*-, and *para*-quinone methides (Section 1.3.2). Recent extension of this reaction to aromatic systems has enabled photogeneration of some unusual quinone methides with interesting reactivity (Section 1.3.3).

REFERENCES

1. Wan, P.; Barker, B.; Diao, L.; Fischer, M.; Shi, Y.; Yang, C. Quinone methides: relevant intermediates in organic synthesis. *Can. J. Chem.* 1996, 74, 465–475.
2. Van De Water, R. W.; Pettus, T. R. R. α-Quinone methides: intermediates underdeveloped and underutilized in organic synthesis. *Tetrahedron* 2002, 58, 5367–5405.
3. Shevchenko, S. M.; Apushkinskii, A. G. Quinone methides in the chemistry of wood. *Russ. Chem. Rev.* 1992, 61, 105–131.
4. Thompson, D. C.; Thompson, J. A.; Sugumaran, M.; Moldeus, P. Biological and toxicological consequences of quinone methide formation. *Chem.-Biol. Interact.* 1993, 86, 129–162.
5. Ramakrishnan, K.; Fisher, J. 7-Deoxydaunomycine quinone methide reactivity with thiol nucleophiles. *J. Med. Chem.* 1986, 29, 1215–1221.
6. Bolton, J. L.; Turnipseed, S. B.; Thompson, J. A. Influence of quinone methide reactivity on the alkylation of thiol and amino groups in proteins: studies utilizing amino acid and peptide models. *Chem.-Biol. Interact.* 1997, 107, 185–200.
7. Awad, H. M.; Boersma, M. G.; Boeren, S.; van Bladeren, P. J.; Vervoort, J.; Rietjens, I. M. C. M. Quenching of quercetin quinone/quinone methides by different thiolate scavengers: stability and reversibility of conjugate formation. *Chem. Res. Toxicol.* 2003, 16, 822–831.
8. Lewis, M. A.; Graff Yoerg, D.; Bolton, J. L.; Thompson, J. A. Alkylation of 2′-deoxynucleosides and DNA by quinone methides derived from 2,6-di-*tert*-butyl-4-methylphenol. *Chem. Res. Toxicol.* 1996, 9, 1368–1374.

9. Pande, P.; Shearer, J.; Yang, J.; Greenbern, W.; Rokita, S. Alkylation of nucleic acids by a model quinone methide. *J. Am. Chem. Soc.* 1999, 121, 6773–6779.
10. Weinert, E. E.; Rokita, S. E. Kinetic and trapping studies of 2′-deoxynucleoside alkylation by a quinone methide. *Chem. Res. Toxicol.* 2005, 18, 1970–1970.
11. Thompson, D. C.; Perera, K.; London, R. Spontaneous hydrolysis of 4-trifluoromethylphenol to a quinone methide and subsequent protein alkylation. *Chem.-Biol. Interact.* 2000, 126, 1–14.
12. Zhou, Q.; Turnbull, K. D. Phosphodiester alkylation with a quinone methide. *J. Org. Chem.* 1999, 64, 2847–2851.
13. Chiang, Y.; Kresge, A. J.; Zhu, Y. Reactive intermediates. Some chemistry of quinone methides. *Pure Appl. Chem.* 2000, 72, 2299–2308.
14. Chiang, Y.; Kresge, J.; Zhu, Y. Kinetics and mechanism of hydration of *o*-quinone methides in aqueous solution. *J. Am. Chem. Soc.* 2000, 122, 9854–9855.
15. Chiang, Y.; Kresge, J.; Zhu, Y. Flash photolytic generation of *ortho*-quinone methides in aqueous solution and study of its chemistry in that medium. *J. Am. Chem. Soc.* 2001, 123, 8089–8094.
16. Chiang, Y.; Kresge, J.; Zhu, Y. Flash photolytic generation of *o*-quinone α-phenylmethide and *o*-quinone α-(*p*-anisyl)methide in aqueous solution and investigation of their reactions in that medium. Saturation of acid-catalyzed hydration. *J. Am. Chem. Soc.* 2002, 124, 717–722.
17. Chiang, Y.; Kresge, J.; Zhu, Y. Flash photolytic generation and study of *p*-quinone methide in aqueous solution. An estimate of rate and equilibrium constants for heterolysis of the carbon–bromine bond in *p*-hydroxybenzyl bromide. *J. Am. Chem. Soc.* 2002, 124, 6349–6356.
18. Ofial, A. R.; Ohkubo, K.; Fukuzumi, S.; Lucius, R.; Mayr, H. Role of electron-transfer processes in reactions of diarylcarbenium ions and related quinone methides with nucleophiles. *J. Am. Chem. Soc.* 2003, 125, 10960–10912.
19. Wan, P.; Brousmiche, D. W.; Chen, C. Z.; Cole, J.; Lukeman, M.; Xu, M. Quinone methide intermediates in organic photochemistry. *Pure Appl. Chem.* 2001, 73, 529–534.
20. Lukeman, M.; Wan, P. In *CRC Handbook of Organic Photochemistry and Photobiology*, 2nd edition; Horspool, W.; Lenci, F., Eds.; CRC Press: Boca Raton, FL, 2004; Chapter 39.
21. For a detailed discussion of nanosecond laser flash photolysis, see Scaiano, J. C. Nanosecond laser flash photolysis: a tool for physical organic chemistry. In *Reactive Intermediate Chemistry;* Moss, R. A.; Platz, M. S.; Jones, M., Jr., Eds.; Wiley: New York, 2005.
22. Seiler, P.; Wirz, J. The photohydrolysis of eight isomeric trifluoromethylnaphthols. *Tetrahedron Lett.* 1971, 20, 1685–1686.
23. Seiler, P.; Wirz, J. Struktur und photochemische Reaktivität: photohydrolyse von trifluormethylsubstituierten phenolen und naphtholen. *Helv. Chim. Acta* 1971, 55, 2693– 2712.
24. Hamai, S.; Kokubun, H. Photochromism of *o*-hydroxytriphenylcarbinol. *Z. Phys. Chem.* 1974, 88, 211.
25. Hamai, S.; Kokubun, H. Thermal decay of the colored form in photochromism of 2-hydroxy-4-methoxytriphenylmethanol in *n*-hexane. *Bull. Chem. Soc. Jpn.* 1974, 47, 2085.

26. Gomberg, M. On triphenylmethyl. XXIII. Tautomerism of the hydroxyl-triphenyl carbinols. *J. Am. Chem. Soc.* 1913, 35, 1035–1042.
27. Lewis, T. W.; Curtin, D. Y.; Paul, I. C. Thermal, photochemical and photonucleated thermal dehydration of *p*-hydroxytriarylmethanols in the solid state. (3,5-Dimethyl-4-hydroxyphenyl)diphenyl methanol and (3,5-dibromo-4-hyrdoxyphenyl)diphenyl methanol: X-ray crystallography of (4-hydroxyphenyl)diphenyl methanol and its 3,5-dimethyl derivative. *J. Am. Chem. Soc.* 1979, 101, 5717–5725.
28. Wan, P.; Chak, B. Structure–reactivity studies and catalytic effects in the photosolvolysis of methoxy-substituted benzyl alcohols. *J. Chem. Soc., Perkin Trans. 2* 1986, 1751–1756.
29. Wan, P.; Hennig, D. Photocondensation of *o*-hydroxybenzyl alcohol in an alkaline medium: synthesis of phenol–formaldehyde resins. *J. Chem. Soc., Chem. Commun.* 1987, 939–941.
30. Barker, B.; Diao, L.; Wan, P. Intramolecular [4 + 2] cycloaddition of a photogenerated *o*-quinone methide in aqueous solution. *J. Photochem. Photobiol. A: Chem.* 1997, 104, 91–96.
31. Diao, L.; Yang, C.; Wan, P. Quinone methide intermediates from the photolysis of hydroxybenzyl alcohols in aqueous solution. *J. Am. Chem. Soc.* 1995, 117, 5369–5370.
32. Huang, C.-G.; Beveridge, K. A.; Wan, P. Photocyclization of 2-(2′-hydroxyphenyl)benzyl alcohol and derivatives via *o*-quinonemethide type intermediates. *J. Am. Chem. Soc.* 1991, 113, 7676–7684.
33. Shi, Y.; Wan, P. Charge polarization in photoexcited alkoxy-substituted biphenyls: formation of biphenyl quinone methides. *J. Chem. Soc., Chem. Commun.* 1995, 1217–1218.
34. Shi, Y.; Wan, P. Photocyclization of a 1,1′-bisnaphthalene: planarization of a highly twisted biaryl system after excited state ArOH dissociation. *J. Chem. Soc., Chem. Commun.* 1997, 273–274.
35. Shi, Y.; Wan, P. Solvolysis and ring closure of quinone methides photogenerated from biaryl systems. *Can. J. Chem.* 2005, 83, 1306–1323.
36. Burnham, K. S.; Schuster, G. B. A search for chiral photochromic optical triggers for liquid crystals: photoracemization of 1,1′-binaphthylpyran through a transient biaryl quinone methide intermediate. *J. Am. Chem. Soc.* 1998, 120, 12619–12625.
37. Brousmiche, D. W.; Xu, M.; Lukeman, M.; Wan, P. Photohydration and photosolvolysis of biphenyl alkenes and alcohols via biphenyl quinone methide-type intermediates and diarylmethyl carbocations. *J. Am. Chem. Soc.* 2003, 125, 12961–12970.
38. Xu, M.; Lukeman, M.; Wan, P. Photogeneration and chemistry of biphenyl quinone methides from hydroxybiphenyl methanols. *Photochem. Photobiol.* 2006, 82, 50–56.
39. Fischer, M.; Shi, Y.; Zhao, B.-P.; Snieckus, V.; Wan, P. Contrasting behaviour in the photosolvolysis of 1- and 2-hydroxy-9-fluorenols in aqueous solution. *Can. J. Chem.* 1999, 77, 868–874.
40. Lukeman, M.; Veale, D.; Wan, P.; Munasinghe, V. R. N.; Corrie, J. E. T. Photogeneration of 1,5-naphthoquinone methides via excited-state (formal) intramolecular proton transfer (ESIPT) and photodehydration of 1-naphthol derivatives in aqueous solution. *Can. J. Chem.* 2004, 82, 240–253.
41. Brousmiche, D. W.; Wan, P. Photogeneration of an *o*-quinone methide from pyridoxine (vitamin B_6) in aqueous solution. *J. Chem. Soc., Chem. Commun.* 1998, 491–492.

42. Brousmiche, D. W.; Wan, P. Photogeneration of quinone methide-type intermediates from pyridoxine and derivatives. *J. Photochem. Photobiol. A: Chem.* 2002, 149, 71–81.
43. Forest, K.; Wan, P.; Preston, C. Catechin and hydroxybenzhydrols as models for the environmental photochemistry of tannins and lignins. *Photochem. Photobiol. Sci.* 2004, 3, 463–472.
44. Nakatani, K.; Higashida, N.; Saito, I. Highly efficient photochemical generation of *o*-quinone methide from Mannich bases of phenol derivatives. *Tetrahedron Lett.* 1997, 38, 5005–5008.
45. Modica, E.; Zanaletti, R.; Freccero, M. Alkylation of amino acids and glutathione in water by *o*-quinone methide. Reactivity and selectivity. *J. Org. Chem.* 2001, 66, 41–52.
46. Weinert, E. E.; Dondi, R.; Colloredo-Melz, S.; Frankenfield, K. N.; Mitchell, C. H.; Freccero, M.; Rokita, S. E. Substituents on quinone methides strongly modulate formation and stability of their nucleophilic adducts. *J. Am. Chem. Soc.* 2006, 128, 11940–11947.
47. Wang, P.; Liu, R.; Wu, X.; Ma, H.; Cao, X.; Zhou, P.; Zhang, J.; Weng, X.; Zhang, X.-L.; Qi, J.; Zhou, X.; Weng, L. A potent, water-soluble and photoinducible DNA cross-linking agent. *J. Am. Chem. Soc.* 2003, 125, 1116–1117.
48. Richter, S. N.; Maggi, S.; Colloredo-Mels, S.; Palumbo, M.; Freccero, M. Binol quinone methides as bisalkylating and DNA cross-linking agents. *J. Am. Chem. Soc.* 2004, 126, 13973–13979.
49. Verga, D.; Richter, S. N.; Palumbo, M.; Gandolfi, R.; Freccero, M. Bipyridyl ligands as photoactivatable mono- and bis-alkylating agents capable of DNA cross-linking. *Org. Biomol. Chem.* 2007, 5, 233–235.
50. Kulikov, A.; Arumugam, S.; Popik, V. V. Photolabile protection of alcohols, phenols, and carboxylic acids with 3-hydroxy-2-naphthalenemethanol. *J. Org. Chem.* 2008. 73, 7611–7615.
51. Kostikov, A. P.; Popik, V. V. 2,5-Dihydroxybenzyl and (1,4-dihydroxy-2-naphthyl) methyl, novel reductively armed photocages for the hydroxyl moiety. *J. Org. Chem.* 2007, 72, 9190–9194.
52. Le Gourrierec, D.; Ormson, S. M.; Brown, R. G. Excited-state intramolecular proton transfer. Part 2: ESIPT to oxygen. *Prog. React. Kinet.* 1994, 19, 211–275.
53. Formosinho, S. J.; Arnaut, L. G. Excited-state proton transfer reactions. II. Intramolecular reactions. *J. Photochem. Photobiol. A: Chem.* 1993, 75, 21–48.
54. Wolfbeis, O. S.; Schipfer, R.; Knierzinger, A. pH-dependent fluorescence spectroscopy. Part 12. Flavone, 7-hydroxyflavone, and 7-methoxyflavone. *J. Chem. Soc., Perkin Trans. 2* 1981, 1443–1448.
55. Wolfbeis, O. S.; Füerlinger, E. pH-dependent fluorescence spectroscopy. 15. Detection of an unusual excited-state species of 3-hydroxyxanthone. *J. Am. Chem. Soc.* 1982, 104, 4069–4072.
56. Itoh, M.; Adachi, T. Transient absorption and two-step laser excitation fluorescence studies of the excited-state proton transfer and relaxation in the methanol solution of 7-hydroxyflavone. *J. Am. Chem. Soc.* 1984, 106, 4320–4324.
57. Itoh, M.; Yoshida, N.; Takashima, M. Transient absorption and two step laser-excitation fluorescence studies on the proton transfer in the ground and excited states of 3-hydroxyxanthone in alcohols. *J. Am. Chem. Soc.* 1985, 107, 4819–4824.

58. Itoh, M.; Hasegawa, K.; Fujiwara, Y. Two-step laser excitation fluorescence study of the ground- and excited-state proton transfer in alcohol solutions of 7-hydroxyisoflavone. *J. Am. Chem. Soc.* 1986, 108, 5853–5857.
59. Itoh, M.; Mukaihata, H.; Nakagawa, T.; Kohtani, S. Picosecond and two-step LIF studies of the excited-state proton transfer in 3-hydroxyxanthone and 7-hydroxyflavone methanol solutions: reinvestigation of tautomer and anion formations. *J. Am. Chem. Soc.* 1984, 116, 10612–10618.
60. Brousmiche, D. W.; Wan, P. Excited state (formal) intramolecular proton transfer (ESIPT) in *p*-hydroxyphenyl ketones mediated by water. *J. Photochem. Photobiol. A: Chem.* 2000, 130, 113–118.
61. Stermitz, F. R.; Adamovics, J. A.; Geigert, J. Synthesis and photoreactions of sorbophenones: a photochemical synthesis of flavones. *Tetrahedron* 1975, 31, 1593–1595.
62. Matsushima, R.; Hirao, I. Photochemical cyclization of 2′-hydroxychalcones to 4-flavones. *Bull. Chem. Soc. Jpn.* 1980, 53, 518–522.
63. Matsushima, R.; Kageyama, H. Photochemical cyclization of 2′-hydroxychalcones. *J. Chem. Soc., Perkin Trans. 2* 1985, 743–748.
64. Matsushima, R., Miyakawa, K., Nishihata, M. Photochromic properties of 2-hydroxychalcones. *Chem. Lett.* 1988, 1915–1916.
65. Matsushima, R.; Suzuki, M. Photochromic properties of 2-hydroxychalcones in solution and polymers. *Bull. Chem. Soc. Jpn.* 1992, 65, 39–45.
66. Matsushima, R.; Mizuno, H.; Kajiura, A. Convenient chemical actinometer with 2-hydroxy-4′-methoxychalcone. *Bull. Chem. Soc. Jpn.* 1994, 67, 1762.
67. Matsushima, R.; Mizuno, H.; Itoh, H. Photochromic properties of 4-amino-substituted 2-hydroxychalcones. *J. Photochem. Photobiol. A: Chem.* 1995, 89, 251–256.
68. ?A3B2 twb=.25we?>Matsushima, R.; Suzuki, N.; Murakami, T.; Morioka, M. Chemical actinometer with 2-hydroxy-4′-dimethylaminochalcone. *J. Photochem. Photobiol. A: Chem.* 1997, 109, 91–94.
69. Matsushima, R.; Fujimoto, S.; Tokumura, K. Dual photochromic properties of 4-dialkylamino-2-hydroxychalcones. *Bull. Chem. Soc. Jpn.* 2001, 74, 827–832.
70. Matsushima, R.; Kato, K.; Ishigai, S. Photochromism of 2-hydroxychalcones in solid hydrogel matrices. *Bull. Chem. Soc. Jpn.* 2002, 75, 2079–2080.
71. Dewar, D.; Sutherland, R. G. The photolysis of 2-hydroxychalcone and its possible implication in flavonoid biosynthesis. *J. Chem. Soc., Chem. Commun.* 1970, 272–273.
72. Ferris, J. P.; Antonucci, F. R. Photochemistry of *ortho*-substituted benzene derivatives and related heterocycles. *J. Am. Chem. Soc.* 1974, 96, 2010–2014.
73. Isaks, M.; Yates, K.; Kalanderopoulos, P. Photohydration via intramolecular proton transfer to carbon in electronically excited states. *J. Am. Chem. Soc.* 1984, 106, 2728–2730.
74. Kalanderopoulos, P.; Yates, K. Intramolecular proton transfer in photohydration reactions. *J. Am. Chem. Soc.* 1986, 108, 6290–6295.
75. Foster, K. L.; Baker, S.; Brousmiche, D. W.; Wan, P. *o*-Quinone methide formation from excited state intramolecular proton transfer (ESIPT) in an *o*-hydroxystyrene. *J. Photochem. Photobiol. A: Chem.* 1999, 129, 157–163.
76. Fischer, M.; Wan, P. *m*-Quinone methides from *m*-hydroxy-1,1-diaryl alkenes via excited-state (formal) intramolecular proton transfer mediated by a water trimer. *J. Am. Chem. Soc.* 1998, 120, 2680–2681.

77. Fischer, M.; Wan, P. Nonlinear solvent water effects in the excited-state (formal) intramolecular proton transfer (ESIPT) in *m*-hydroxy-1,1-diaryl alkenes: efficient formation of *m*-quinone methides. *J. Am. Chem. Soc.* 1999, 121, 4555–4562.

78. Cole, J. G.; Wan, P. Mechanistic studies of photohydration of *m*-hydroxy-1,1-diaryl alkenes. *Can. J. Chem.* 2002, 80, 46–64.

79. Murohoshi, T.; Kaneda, K.; Ikegami, M.; Arai, T. Photoisomerization and isomer-specific addition of water in hydroxystilbenes. *Photochem. Photobiol. Sci.* 2003, 2, 1247–1249.

80. Lewis, F. D.; Crompton, E. M. Hydroxystilbene isomer-specific photoisomerization versus proton transfer. *J. Am. Chem. Soc.* 2003, 125, 4044–4045.

81. Lewis, F. D.; Sinks, L. E.; Weigel, W.; Sajimon, M. C.; Crompton, E. M. Ultrafast proton transfer dynamics of hydroxystilbene photoacids. *J. Phys. Chem. A* 2005, 109, 2443–2451.

82. Uchida, M.; Irie, M. A new photochromic vinylnaphthol derivative. *Chem. Lett.* 1991, 2159–2162.

83. Webb, S. P.; Philips, L. A.; Yeh, S. W.; Tolbert, L. M.; Clark, J. H. Picosecond kinetics of the excited-state, proton-transfer reaction of 1-naphthol in water. *J. Phys. Chem.* 1986, 90, 5154–5164.

84. Lukeman, M.; Wan, P. Excited state intramolecular proton transfer (ESIPT) in 2-phenylphenol: an example of proton transfer to a carbon of an aromatic ring. *J. Chem. Soc., Chem. Commun.* 2001, 1004–1005.

85. Lukeman, M.; Wan, P. A new type of excited-state intramolecular proton transfer: proton transfer from phenol OH to a carbon atom of an aromatic ring observed for 2-phenylphenol. *J. Am. Chem. Soc.* 2002, 124, 9458–9464.

86. Lukeman, M.; Wan, P. Excited-state intramolecular proton transfer in *o*-hydroxybiaryls: a new route to dihydroaromatic compounds. *J. Am. Chem. Soc.* 2003, 125, 1164–1165.

87. Flegel, M.; Lukeman, M.; Wan, P. Photochemistry of 1,1′-bi-2-naphthol (BINOL)—ESIPT is responsible for photoracemization and cyclization. *Can. J. Chem.* 2008, 86, 161–169.

88. Flegel, M.; Lukeman, M.; Huck, L.; Wan, P. Photoaddition of water and alcohols to the anthracene moiety of 9-(2′-hydroxyphenyl)anthracene via formal excited state intramolecular proton transfer. *J. Am. Chem. Soc.* 2004, 126, 7890–7897.

89. Basaric, N.; Wan, P. Competing excited state intramolecular proton transfer pathways from phenol to anthracene moieties. *J. Org. Chem.* 2006, 71, 2677–2686.

90. Chapman, O. L.; McIntosh, C. L. Photochemical decarbonylation of unsaturated lactones and carbonates. *J. Chem. Soc., Chem. Commun.* 1971, 383–384.

91. Padwa, A.; Lee, G. A. Photochemical transformations of small ring heterocyclic compounds. L. Tautomeric control of the photochemistry of 3-phenylbenzofuran-2-one. *J. Am. Chem. Soc.* 1973, 95, 6147–6149.

92. Padwa, A.; Dehm, D.; Oine, T.; Lee, G. A. Competitive keto-enolate photochemistry in the 3-phenylisocoumaranone system. *J. Am. Chem. Soc.* 1975, 97, 1837–1845.

93. Chiang, Y.; Kresge, A. J.; Zhu, Y. Generation of *o*-quinone α-carbomethoxymethide by photolysis of methyl 2-hydroxyphenyldiazoacetate in aqueous solution. *Phys. Chem. Chem. Phys.* 2003, 5, 1039–1042.

94. Tomioka, H.; Matsushita, T. Benzoxetene. Direct observation and theoretical studies. *Chem. Lett.* 1997, 399–400.

95. Laatsch, H.; Schmidt, A. J.; Haucke, G. Photochromism of furofurans. *J. Inf. Rec. Mater.* 1994, 21, 599–600.

96. Padwa, A.; Lee, G. A. Photochemical transformations of 2,2-disubstituted chromenes. *J. Chem. Soc., Chem. Commun.* 1972, 795–796.

97. Lenoble, C.; Becker, R. S. Photophysics, photochemistry and kinetics of photochromic 2*H*-pyrans and chromenes. *J. Photochem.* 1986, 33, 187–197.

98. Lenoble, C.; Becker, R. S. Photophysics, photochemistry and kinetics of indolinospiropyran derivatives and an indolinothiospiropyran. *J. Photochem.* 1986, 34, 83–88.

99. Pirkle, W. H.; Smith, S. G.; Koser, G. F. Stereospecificity and wavelength dependence in the photochemical rearrangement of spiro[2.5]octa-4,7-dien-6-ones to quinone methides. *J. Am. Chem. Soc.* 1969, 91, 1580–1582.

100. Schuster, D. I.; Krull, I. S. Photochemistry of unsaturated ketones in solution. XIX. Photochemistry of spiro[2.5]octa-4,7-dien-6-one. 2. Mechanistic aspects and the relationship to the photochemistry of quinone methides. *Mol. Photochem.* 1969, 1, 107–133.

2

MODELING PROPERTIES AND REACTIVITY OF QUINONE METHIDES BY DFT CALCULATIONS

MAURO FRECCERO AND FILIPPO DORIA
Department of Organic Chemistry, Pavia University, V. le Taramelli 10, 27100 Pavia, Italy

2.1 INTRODUCTION

The prototype *o*-quinone methide (*o*-QM) and *p*-quinone methide (*p*-QM) are reactive intermediates. In fact, they have only been detected spectroscopically at low temperatures (10 K) in an argon matrix,[1] or as a transient species by laser flash photolysis.[2] Such a reactivity is mainly due to their electrophilic nature, which is remarkable in comparison to that of other neutral electrophiles. In fact, QMs are excellent Michael acceptors, and nucleophiles add very fast under mild conditions at the QM exocyclic methylene group to form benzylic adducts, according to Scheme 2.1.[2a,3]

Experimental data suggest that reactivity and selectivity in the reaction of QMs with biological nucleophiles can be highly sensitive to (i) structure modification (substituents X, Y, and R on the QM ring, Scheme 2.1)[4] and to (ii) the protonation of the carbonyl moiety.[5] Reactivity and selectivity of QM benzylation reactions also seem to be affected by hydrogen bonding involving the QM carbonyl oxygen[6] by either a protic solvent such as water[7] or Brønsted acid[8] and acid hydrogens in peptides and in DNA bases.[9] Shielding of the carbonyl oxygen from such a solvent interaction has been suggested as the cause of the low reactivity of crowded *p*-QMs with hydrophobic substituents R at the 2- and 6-positions.[10]

Quinone Methides, Edited by Steven E. Rokita

o-QM +HNu → ... *p*-QM +HNu → ...

X, Y = H, CH_3, OCH_3, $COOCH_3$, CN, NO_2 **R** = H, CH_3, $C(CH_3)_3$

SCHEME 2.1 *o*-QM and *p*-QM reactivity as Michael acceptors.

Research on QMs as biological benzylating agents has been very active on the experimental side for the past 15 years,[1–10] but, with the exception of the pioneering investigation by Soucek et al. in the eighties, based on the Hückel molecular orbital (HMO) method[11] and semiempirical CNDO,[12] a systematic computational investigation on properties and reactivity of *p*- and *o*-QMs only started at the beginning of the new millennium. Such a delay is surprising, because a computational approach could provide a solid model to describe and predict QM reactivity and the stability of the resulting adducts with nucleophiles. The delay of reliable computational models describing QM reactivity in solution can be explained by the fact that only in the last decade computationally affordable methods in the frame of density functional theory (DFT) started to be coupled with simple but effective polarizable continuum solvation models (PCMs). Activation Gibbs energies calculated by DFT methods in solution have allowed a reliable comparison to the available kinetic experimental data, offering a valuable benchmark for the computational tools. These computationally generated kinetic data suggested and in some cases anticipated to the experimentalists new and unexpected chemical pathways, providing evidences of the unique features of QMs as useful electrophiles for biological and synthetic applications. From 2001, several aspects have been tackled by DFT computational tool to rationalize and predict the chemical behavior of both QMs and their resulting adducts. In this chapter, we will review in detail the results from DFT computational calculations related to the properties and reactivity of QMs as:

1. *Alkylating agents*, clarifying the role of water and more generally the role of the general acid catalysis on QM reactivity.[13–15]
2. *Heterodienes* in [4 + 2] cycloaddition reactions.[16]
3. *Nucleophiles*, at the exocyclic methylene when bonded to Ir, Rh, and Co cyclopentadienyl complexes.[17]

In addition, we reviewed the computational data regarding both the generation and the reactivity of QMs involved as intermediates in the polymerization processes leading to lignin[18] and eumelanine[19–21] and in the oxidation of toluene at high temperature (partial combustion processes).[22, 23]

2.2 QM REACTIVITY AS ALKYLATING AGENTS

2.2.1 Computational Models

The properties and reactivity of the prototype *o*-QM as alkylating agent have been studied by quantum chemical methods in the frame of DFT in reactions modeling its reactivity in water with simple nucleophiles. Ammonia and hydrogen sulfide have been considered as prototype nucleophiles of amines, and thiols in competition with water as solvent.[13] The computational analysis exploring the conjugate addition reactions of NH_3, H_2O, and H_2S to *o*-QM (both free and H-bonded at the oxygen QM to an ancillary water molecule) defined a reliable computational model capable of taking into account both the specific and bulk effect of the solvent water on *o*-QM properties and reactivity as benzylating agent under mild conditions.

The effects of specific and bulk solvation on the stationary points (reactants, transition structures TSs, and products) geometries and on the activation energy of the reactions under study was investigated in two consecutive steps. First, a specific water molecule was explicitly included in the gas phase computation. Then, the solvent was considered as a macroscopic and continuum medium. Each stationary point was optimized within DFT using the Becke3-LYP hybrid functional (B3LYP)[24] and several basis sets, mostly the 6-311 + G(d,p) one. The B3LYP hybrid functional consists of the nonlocal exchange functional of Becke's three-parameter set[24a] and the nonlocal correlation function of Lee, Yang, and Parr.[24b] Solvent effects were modeled by the conductor PCM version of the polarizable continuum model (C-PCM)[25] implemented in the Gaussian 98 package.[26a]

The contributions of bulk solvent effects to the activation free energy* of the reactions under study were calculated via the self-consistent reaction field (SCRF) method using the same solvation model. The version of PCM used was the United Atom for Hartree–Fock (UAHF)[27] model to build the cavity for the solute. In this solvation model, the solute cavity is defined through interlocking van der Waals (vdW) spheres centered on heavy (i.e., nonhydrogen) elements only (united atom approach). The vdW radius of each atom is a function of atom type, connectivity, overall charge of the molecule, and the number of attached hydrogen atoms. The reaction field is represented through point charges on small regions (tesserae) located on the surface of the molecular cavity. Such a model includes the nonelectrostatic terms [cavitation (G_{cav}), dispersion and repulsion energy (G_{dr})] in addition to the classical electrostatic contribution (G_{es}), calculating the molecular free energy in solution (G_{sol}) as the sum over three terms:

$$G_{sol} = G_{es} + G_{dr} + G_{cav}. \tag{2.1}$$

For all PCM–UAHF calculations, the number of initial tesserae/atomic sphere has been set to 60 by default. For comparative purposes, C-PCM calculations of the

*The Gibbs free energy (computed in the harmonic approximation) were converted from the 1 atm standard state into the standard state of molar concentration (ideal mixture at 1 mol L^{-1} and 1 atm).

Reactants **I** → TS **S** → Product **P**

HNu	Reactants I	TS S	Product P
NH_3	**I-1**	**S-1**	**P-1**
H_2O	**I-2**	**S-2**	**P-2**
H_2S	**I-3**	**S-3**	**P-3**

Complex reactants $-H_2O$ → TS → Product

HNu	Complex reactants $-H_2O$	TS	Product
NH_3	**I-4**	**S-4**	**P-4**
H_2O	**I-5**	**S-5**	**P-5**
H_2S	**I-6**	**S-6**	**P-6**

SCHEME 2.2 Stationary points along the reaction coordinate for the conjugate additions of NH_3, H_2O, and H_2S to *o*-QM, with and without and ancillary water molecule H-bonded to the QM.

solvation energies with 60 initial tesserae were also performed for the TSs on the oxygen alkylation pathway, using Bondi's and Pauling's set of atomic radii (options radii = Pauling and radii = Bondi) in the PCM version implemented in Gaussian 98, and no significant differences on activation energies in solution were detected (Scheme 2.2).

Optimizations of all the stationary points (reactants, **I1–I3**; reactant complexed to water, **I4–I6**; transition structures, **S1–S6**; and products **P1–P6**) were performed using the hybrid functional B3LYP, with several basis sets [6-31G(d), 6-311 + G(d, p), adding d and f functions to the S atom, 6-311 + G(d,p),S(2df), and aug-cc-pVTZ]. The suitability of DFT for a reliable modeling of hydrogen-bonded systems has been the subject of many investigations,[28] and such methods have proved quite useful in studying hydrogen-bonded complexes.[29] The B3LYP functional in particular has proven quite effective, at least as long as an appropriate basis set is used.[30] Basis set extensions with polarization function also for hydrogen as well as introduction of diffuse functions on heavy atoms [i.e., B3LYP/6-311 + G(d,p)] were used to properly describe lone pairs and hydrogen bonding interactions, which are very important in controlling the polarizability and reactivity of the *o*-QM in the alkylation reactions of nitrogen, oxygen, and sulfur nucleophiles. The extension of the S atom basis set with further d and f functions [i.e., 6-311 + G(d,p),S(2df)] had provided a better computational treatment of the stationary points containing S atoms.[31]

2.2.1.1. Basis Set Choice The most relevant geometrical parameters of several stationary points, with and without an explicit water molecule, located using several

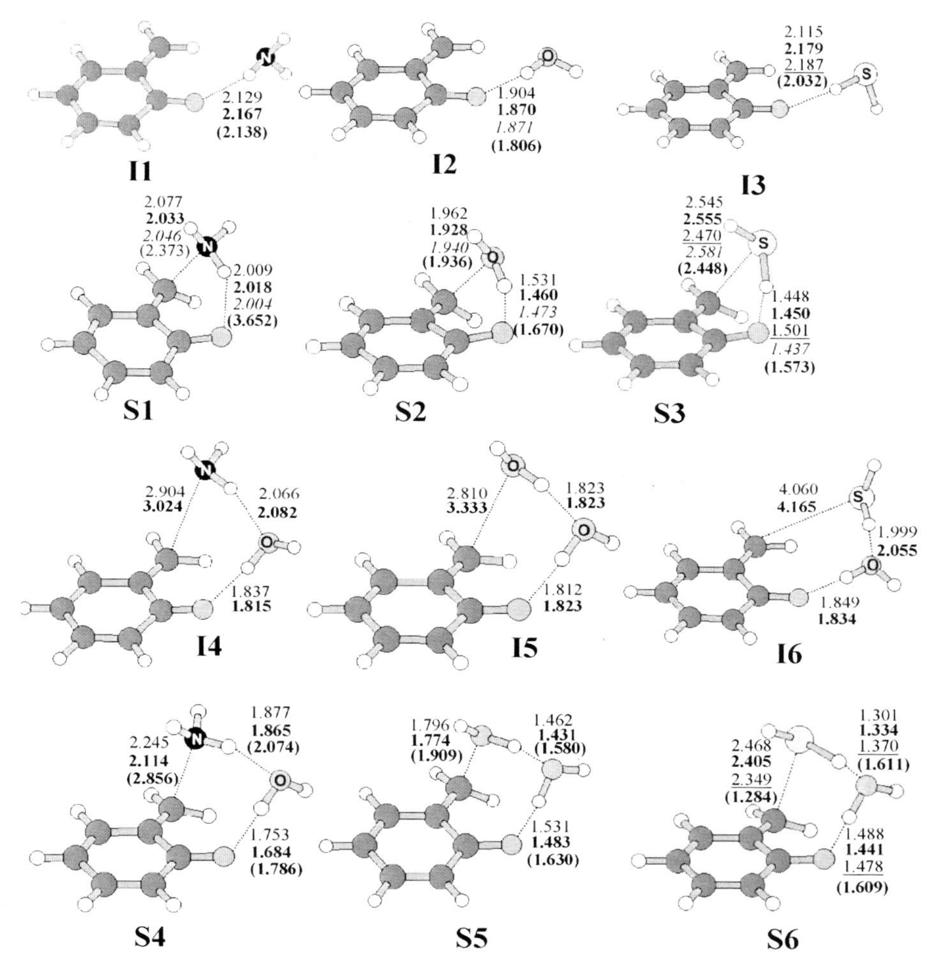

FIGURE 2.1 Geometries of the reactants (**I1–I6**) and transition structures (**S1–S6**) at B3LYP level of theory in gas phase, using 6-31G(d), 6-311 + G(d,p) (bold characters), 6-311 + G(d,p), S(2df) (underlined), aug-cc-pVTZ (italic) basis sets, and in aqueous solution at B3LYP-C-PCM/6-311 + G(d,p) level of theory (bold characters in parenthesis). Bond length data have been taken from Ref. [13].

basis sets [6-31G(d), 6-311 + G(d,p), 6-311 + G(d,p),S(2df), and aug-cc-pVTZ] on the potential energy surfaces (PESs) describing the conjugate addition of ammonia, water, and hydrogen sulfide to the *o*-QM are summarized in Fig. 2.1.

From a geometrical point of view, the enlargement of the basis set from 6-31G(d) to 6-311 + G(d,p) does not decrease TS forming bond lengths by more than 0.13 Å, while prereaction cluster geometries (**I1–I6**) are more affected by basis set choice, exhibiting larger differences (up to 0.52 Å). Moreover, on passing from 6-31G(d) to

6-311 + G(d,p), both intermediate and TS geometries become more reliable, at least as judged from a comparison with the corresponding geometries achieved using the augmented correlation-consistent polarized valence triple-zeta (aug-cc-pVTZ) basis sets.[32]

In fact, it has been demonstrated that geometrical features of stationary points (including the hydrogen bonded complexes) optimized at 6-311 + G(d,p) are very similar to those obtained with the very large, but too time-consuming, aug-cc-pVTZ basis.[†]

Energies, as expected, are much more basis set dependent than geometries. In fact, activation energies of TSs and formation energies of the intermediates change considerably on passing from 6-31G(d) to 6-311 + G(d,p) basis sets. However, a further enlargement of the basis set, namely, on passing from 6-311 + G-(d,p) to aug-cc-pVTZ basis, is accompanied by a much smaller variation in activation energies (less than 2.5 kcal/mol) and, even more important, the difference between activation energies of TSs (i.e., **S1**, **S2**, and **S3**) does not change appreciably. Moreover, it is remarkable that B3LYP/aug-cc-pVTZ//B3LYP/6-311 + G(d,p) single-point activation energies for these TSs differ from fully optimized B3LYP/aug-cc-pVTZ activation energies by less than 0.03 kcal/mol.

This observation suggested that refining energies by single-point calculations with the aug-cc-pVTZ basis set on 6-311 + G(d,p)-optimized geometries is a reliable practice.

2.2.1.2. Energetics of the Benzylation by o-QM in the Gas Phase and in Aqueous Solution The Gibbs energy profiles for the NH_3, H_2O, and H_2S addition reactions to *o*-QM, in the gas phase and in aqueous solutions, both in the presence (water-catalyzed mechanism) and in the absence of an ancillary water molecule (uncatalyzed mechanism) have been explored, and are displayed in Scheme 2.3.[13]

These profiles clearly show that in the gas phase the alkylations of both ammonia and water by *o*-QM are assisted by an additional water molecule H-bonded to *o*-QM (water-catalyzed mechanism), since **S4** and **S5** TSs are favored over their uncatalyzed counterparts (**S1** and **S2**) by 5.6 and 4.0 kcal/mol [at the B3LYP/6-311 + G(d,p) level], respectively. In contrast, the reaction with hydrogen sulfide in the gas phase shows a slight preference for a direct alkylation without water assistance (by 0.8 kcal/mol).

These data suggest that *o*-QM reactivity is heavily affected by general acid catalysis in the gas phase or in low polar medium. The general acid catalysis can be provided by a water molecule for nucleophiles bearing weak acid hydrogens (such as those in ammonia).

The bulk solvent effect on the reaction energy, described by the lower portion of Scheme 2.3, significantly modifies the relative importance of the uncatalyzed and water-assisted alkylation mechanism by *o*-QM in comparison to the gas phase.

[†]Aug-cc-pVTZ basis sets have been suggested (in the epoxidation of allylic alcohols by peroxy acids) to be a good basis sets choice for hydrogen bonding description (Adam, W.; Bach, R. D.; Dmitrenko, O.; Saha-Moller, C. R. *J. Org. Chem.* **2000**, *65*, 6715). However, they are highly time consuming in the optimization of stationary points.

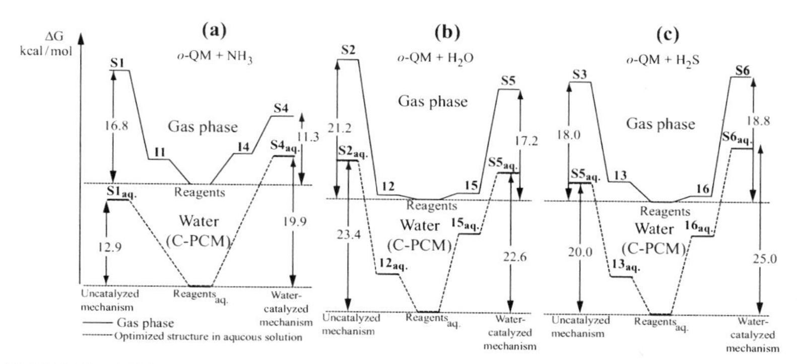

SCHEME 2.3 Gibbs energy profiles for the benzylation of NH_3 (a), H_2O (b), and H_2S (c) by *o*-QM in the gas phase (continuous line), water-catalyzed (**S4**–**S6**) and uncatalyzed (**S1**–**S3**), and in aqueous solution (dotted line, $\mathbf{S1_{aq}}$–$\mathbf{S6_{aq}}$) optimizing both reagents and TSs in aqueous solution [B3LYP-C-PCM/6-311 + G(d,p)]. Data are taken from Ref. [13].

The uncatalyzed mechanism becomes highly favored over the catalyzed one in the alkylation reaction of ammonia (by 7.0 kcal/mol) and hydrogen sulfide (by 4.0 kcal/mol). This was due in part to the higher dipole moment of the TSs **S1** (7.4 D) and **S3** (4.5 D) in comparison to the water-catalyzed counterparts **S4** (6.6 D) and **S6** (2.8 D), respectively, and to the cavitation term that destabilized less the smaller (uncatalyzed) TSs. In contrast, activation induced by water catalysis still plays an important role in the *o*-QM hydration reaction in water as solvent.[13] These results suggest that although the QM reactivity is strongly affected by general acid catalysis in gas phase and likely in low polar medium, the solvation effects of an aqueous solution on *o*-QM reactivity toward amines and thiols can be reliably described with a simple polarizable continuum models (PCMs), neglecting the specific interactions with the solvent. Such a conclusion has been important to explore the *o*-QM reactivity toward purine and pyrimidine bases in aqueous solvent using a very simple solvation model.[14,15] In contrast, the hydration reaction requires a solvent model where both specific and bulk solvation effects have to be taken into account.[13]

2.2.2 H-Bonding and Solvent Effects in the Benzylation of Purine and Pyrimidine Bases

2.2.2.1. Cytosine Benzylation under Kinetic Control The nucleophilicity of cytosine and 1-methyl cytosine (as model of deoxycytidine) at the NH_2, N3, and O^2 centers (see Scheme 2.4 for numbering) toward a model *o*-QM as alkylating agent has been studied using DFT computational analysis [at the B3LYP/6-311 + G(d,p) level] in aqueous solution.[14] The aim was to compare from a kinetic and thermochemical point of view the three competing benzylating pathways at the (i) NH_2, (ii) N3, and (iii) O^2, as depicted in Scheme 2.4.

Bulk effects of the aqueous solvent have been evaluated by C-PCM solvation model. The specific effect of the solvation on the alkylation pathways has also been

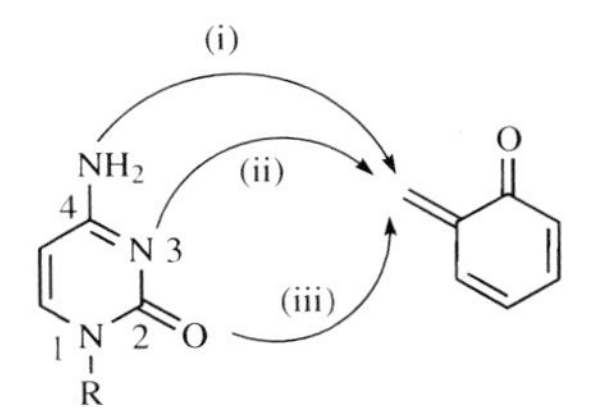

SCHEME 2.4 Competing reaction pathways in the benzylation of cytosine (R = H) and 1-methyl cytosine (R = Me) explored by computational DFT means.

investigated by analyzing the competing reaction mechanism with an ancillary water molecule H-bonded to the QM carbonyl oxygen (Fig. 2.2). Such a water molecule may participate bifunctionally in cyclic hydrogen-bonded transition structures (see **S–O–H_2O, S–NH_2–H_2O**, and **S–N3–H_2O** transition structures in Fig. 2.2) as a proton shuttle from the HNu to the QM carbonyl oxygen. A comparison of two reaction mechanisms, the unassisted one and the water-assisted reaction mechanisms, has been made on the basis of activation Gibbs free energies ($\Delta G^{\ddagger}$), which have been computed localizing the TSs depicted in Fig. 2.2.

The unassisted alkylation becomes the preferred mechanism for the reaction at both the exocyclic (NH_2) and the heterocyclic (N3) nitrogen atoms, in aqueous solution, through the transition structures **S–NH_2′** and **S–N3**. By contrast, alkylation at the cytosine oxygen atom is a water-catalyzed process, since the water-assisted mechanism is still favored, exhibiting a lower activation energy (**S–O–H_2O** versus **S–O** and **S–O′** TSs). Such an evidence is consistent with the QM-hydration mechanism, which is a water-catalyzed process, with one water molecule reacting as a nucleophile at the exocyclic methylene group and the second one acting as proton shuttle (**S5** in Fig. 2.1).[13] As far as competition between O, NH_2, and N3 centers is concerned, among all the possible mechanisms, these calculations unambiguously suggest that the most nucleophilic site of cytosine both in gas phase and in water solution is the heterocyclic N3 nitrogen atom in agreement with experimental product distribution analysis.[33] These computational data rationalize the high reactivity and nucleophilicity of the cytosine N3 moiety toward *o*-QM-like structures due to both the intrinsic N3 nucleophilicity and the strong H-bonding (HB) interactions between the NH_2, hydrogen, and the carbonyl oxygen of the quinone methides in the TS **S–N3** (Fig. 2.2). The DFT investigation also provides a solid computational model to predict the nucleophilicity of nucleic acid toward QMs in aqueous solution, describing the solvation effects of an aqueous solution only by bulk effects through PCMs.

2.2.2.2. Stability/Reactivity of the QM-Cytosine Conjugates The very same theoretical approach has been exploited to evaluate the stability of the resulting benzylated adducts, computing their free energy in gas phase (ΔG_{gas}) and aqueous solution (ΔG_{aq}) relative to the reactants. These thermochemical data combined with the activation free energy data for the benzylation pathways in Fig. 2.2 allowed a kinetic analysis of the reversal of the benzylation process through the calculation of the activation free energies ($\Delta G^{\ddagger}_{Rev\text{-}aq}$) for each adduct.

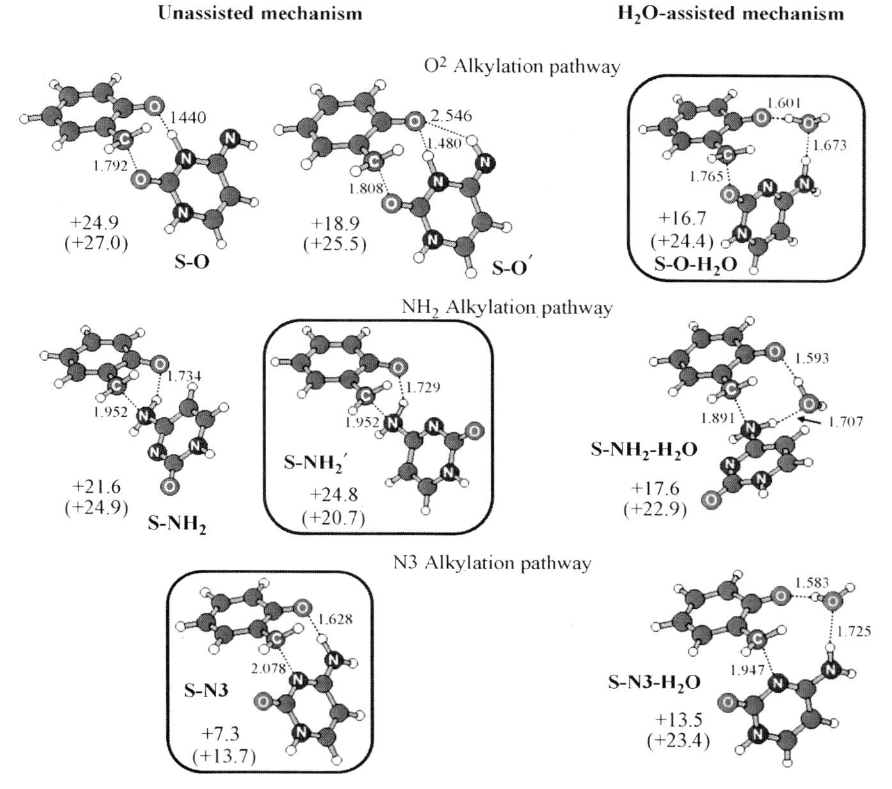

FIGURE 2.2 Optimized TS geometries of the cytosine alkylation reaction by *o*-QM, without and with water assistance. Bond lengths (in Å) and activation Gibbs free energies (in kcal/mol) in the gas phase and in aqueous solution (in parentheses) at the B3LYP/6-311 + G(d,p)// B3LYP/6-31G(d) level of theory have been taken from Ref. [14]. (See the color version of this figure in Color Plates section.)

The data in Table 2.1 suggest that the *O*-benzylated adduct cannot be isolated since it is less stable than reactants. The N3-benzylated adduct should be generated faster, but it should also decompose under mild conditions into free reactants, because the activation free energy in aqueous solution for the decomposition into free QM and methylcytosine is only 21.4 kcal/mol.[14] In other words, these data suggested that the QM-N3-cytosine conjugate could act as QM-carrier, few years before the experimental data related to the stability of QM-conjugates became available.[4]

2.2.2.3. Purine Bases Benzylation: Kinetic and Thermodynamic Aspects The reactivity of *o*-QM as benzylating agent toward 9-methyladenine (**MeA**, Scheme 2.5) and 9-methylguanine (**MeG**, Scheme 2.6), as prototype substrates of deoxyadenosine and -guanosine, was investigated in the gas phase and in aqueous solution, using a similar computational approach used for cytosine [B3LYP/6-31 + G(d,p) in the gas phase and in aqueous solution by using the C-PCM solvation model].[15]

TABLE 2.1 Stability of the QM-Cytosine Conjugates Relatively to Free Reactants in the Gas Phase (ΔG_{gas}) and in Aqueous Solution (ΔG_{aq}). Activation Gibbs Energy for Their Decomposition into QM and Cytosine ($\Delta G^{\ddagger}_{Rev\text{-}aq}$)

Adduct	R	ΔG_{gas} [a,b]	ΔG_{aq} [b,c]	$\Delta G^{\ddagger}_{Rev\text{-}aq}$ [b,c]
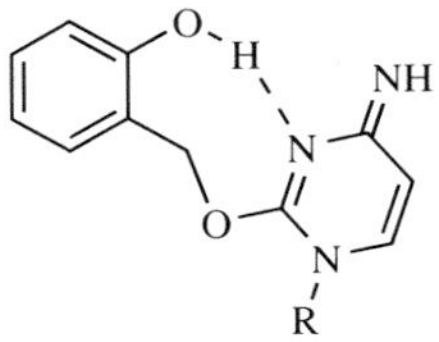	H	+8.4	+13.3	12.2
	CH_3	+9.9	+14.2	11.2
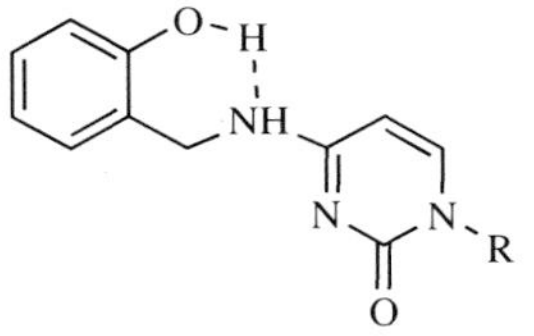	H	−11.2	−11.3	32.0
	CH_3	−11.4	−11.5	32.2
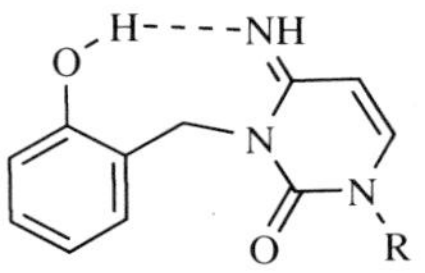	H	−14.9	−8.1	21.8
	CH_3	−14.1	−7.2	21.4

Source: Data taken from Ref. [14].

[a]B3LYP/6-31+G(d,p).

[b]Gibbs energy in kcal/mol.

[c]C-PCM B3LYP/6-31+G(d,p).

The effect of the aqueous medium on the reactivity and on the stability of the resulting adducts has been investigated to assess which adduct arises from the kinetically favorable path or from an equilibrating process. The calculations indicate that the most nucleophilic site of the methyl-substituted nucleobases in the gas phase is the guanine oxygen atom, followed by the adenine N1, while other centers exhibit a substantially lower nucleophilicity (see activation Gibbs energies in Table 2.2).

The bulk effect of water as a solvent is rather dramatic since it causes a drastic reduction of the nucleophilicity of 9-methyladenine N1 and even more of 9-methylguanine O6. As a result, there is a reversal of the nucleophilicity order of the purine bases passing from gas phase to aqueous solution. In fact, in solution, methyladenine is more nucleophilic than methylguanine. Moreover, oxygen and N7 nucleophilic centers of 9-methylguanine compete almost on the same footing in solution (Table 2.2) and also the reactivity gap between N1 and N7 of 9-methyladenine is highly reduced in comparison to the gas phase.

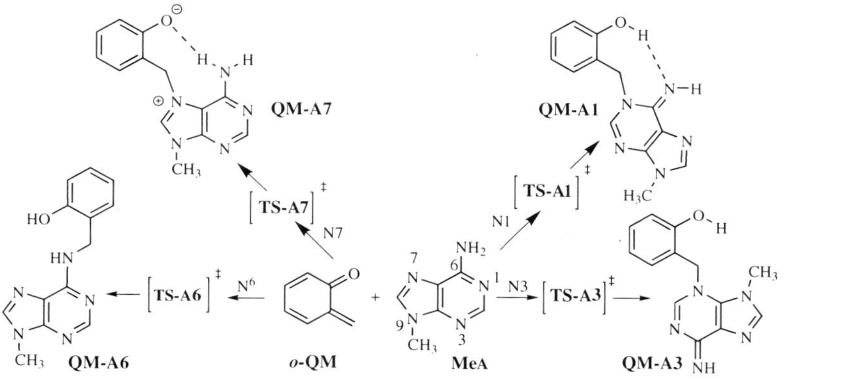

SCHEME 2.5 Reaction pathways for 9-methyladenine alkylation by *o*-QM, investigated by DFT.

Regarding product stability, DFT calculations predict that only two of the adducts of *o*-QM with 9-methyladenine, **QM-A1** and **QM-A6** are lower in energy than reactants, both in the gas phase and in water (Table 2.3). However, the adduct at N1 (**QM-A1**) can easily dissociate in aqueous solution, exhibiting an activation energy for the reversal of the benzylation process of 19.7 kcal/mol.

The adducts arising from the covalent modification of 9-methylguanine are largely more stable than reactants in the gas phase, and the stability of all of them but **QM-G2** is markedly reduced in water. In particular, the oxygen alkylation adduct (**QM-G6**) becomes slightly unstable in water, and the N7 alkylation product (**QM-G7**) remains only moderately more stable than free reactants (Table 2.3). These data show that site alkylations at the N1 of **MeA** and at the N7 of **MeG** in water are the result of kinetically controlled processes and that the selective modifications of the *exo*-amino groups of **MeG** (**QM-G2**) and adenine NH_2 (**QM-A6**) are generated by thermodynamic equilibrations. These computational results rationalize the product distribution of dA and dG obtained experimentally and independently in aqueous DMF by Rokita and coworkers.[34]

The ability of *o*-QM to form several metastable adducts with pyrimidine (at cytosine N3) and purine bases (at guanine N7 and adenine N1) in water suggested that the above adducts may be exploited as *o*-QM carriers under mild conditions, anticipating that *o*-QM could actually migrate along the structure of an oligonucleotide.[35]

2.3 REACTIVITY AS HETERODIENE

The reactivity of the prototype *o*-QM as heterodiene in Diels–Alder cycloaddition reactions with several substituted alkenes such as methyl vinyl ether (**MVE**), styrene,

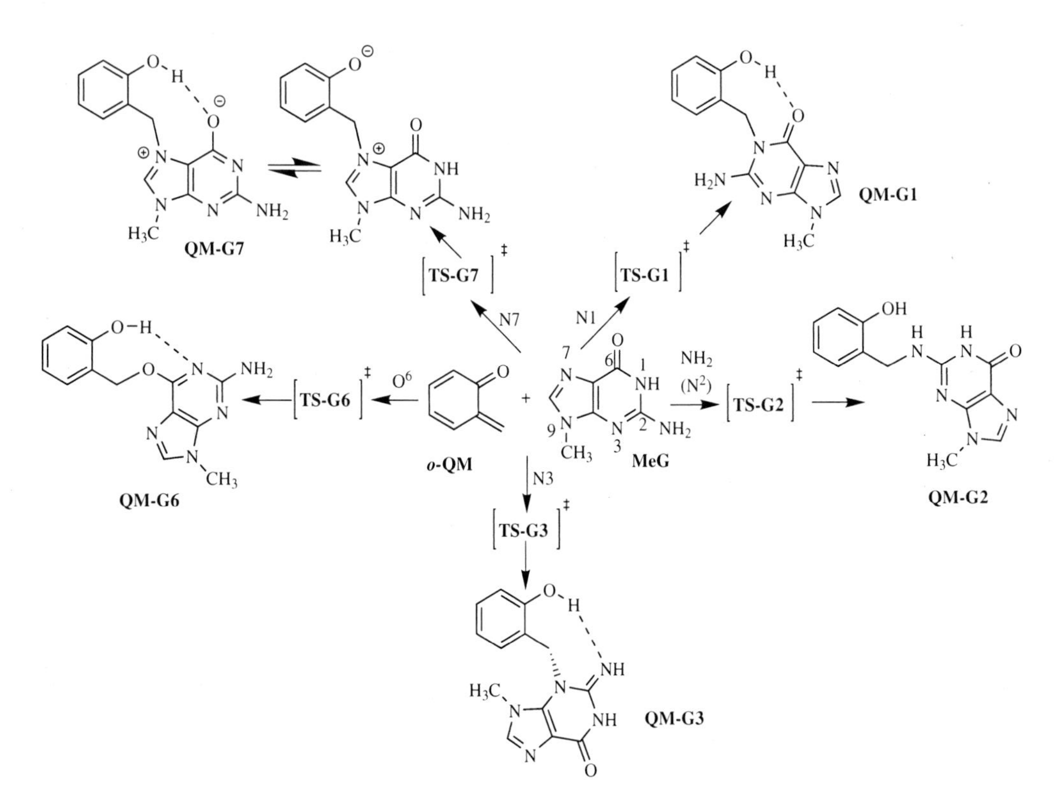

SCHEME 2.6 Reaction pathways for 9-methylguanine alkylation by *o*-QM, investigated by DFT.

TABLE 2.2 9-Methyladenine (MeA) and 9-Methylguanine (MeG) Nucleophilicity Scale in the Gas Phase and in Aqueous Solution, Based on the Activation Energies (in Parentheses, in kcal/mol) for the Alkylation Processes,[a] Through the TSs TS-A1, TS-A3, TS-A6, and TS-A7 for MeA and TS-G1, TS-G2, TS-G3, TS-G6, and TS-G7 for MeG

Purine	Gas Phase	Aqueous Solution
MeA	N1(10.3) > N7(14.9) > NH_2(21.7) ≥ N3 (23.8)	N1(14.5) > N7(16.7) > N3(19.4) > NH_2(22.4)
MeG	O(5.2) > N1(12.3) ~ N7(12.9) ≫ N3(18.3) > NH_2(20.5)	O(16.6) ≥ N7(17.7) > NH_2(19.8) ≫ N3(21.9) ≥ N1(22.9)

Data taken from Ref. [15].

and methyl vinyl ketone (**MVK**) have been investigated at the B3LYP/6-31G(d,p) level of theory,[16] including solvation effects computed by PCM method as implemented in the Gaussian 03 package.[26b] The comparison of the activation enthalpies ($\Delta H^{\ddagger}$) for the *ortho* and *meta* reaction pathways (Table 2.4) reveals a remarkable regioselectivity with the *ortho* attack mode exhibiting transition structures always more stable than the *meta* ones at least by 12.0 kcal/mol for electron-rich alkenes such as **MVE** and 3.8 kcal/mol for electron-deficient dienophiles (i.e., **MVK**). The reactivity, the *ortho* selectivity (see data in Table 2.4), and the asynchronicity [measured by the difference, Δd (Å), of the forming bond length in the TSs in Fig. 2.3] are strongly enhanced by the electron-donor character of the substituent on the

TABLE 2.3 Stability of the QM-9-Methyladenine and QM-9-Methylguanine Adducts (QM-A and QM-G, Respectively) Relatively to Reactants in the Gas Phase (ΔG_{gas}) and in Aqueous Solution (ΔG_{aq}). Activation Energies for Their Decomposition into Free QM and Cytosine ($\Delta G^{\ddagger}_{Rev\text{-}aq}$)

Adduct	ΔG_{gas} [a,b]	$\Delta G_{acetonitrile}$ [b,c]	ΔG_{DMSO} [b,c]	ΔG_{aq} [b,c]	$\Delta G^{\ddagger}_{Rev\text{-}aq}$ [b,c]
QM-A					
QM-A1 (N1)	−7.6	—	—	−5.2	19.7
QM-A3 (N3)	+15.5	—	—	+12.3	—
QM-A6 (NH_2)	−11.6	—	—	−10.7	33.1
QM-A7 (N7)	+8.1	—	—	+3.4	—
QM-G					
QM-G1 (N1)	−14.5	−9.7	−8.7	−7.4	30.3
QM-G2 (NH_2)	−11.4	−11.3	−10.7	−11.7	31.5
QM-G3 (N3)	−3.3	—	—	+0.8	—
QM-G6 (O)	−9.9	—	—	+1.4	—
QM-G7 (N7)	−4.7	—	—	−2.8	20.5

Source: Data taken from Ref. [15].

[a]B3LYP/6-31 + G(d,p).

[b]Gibbs energy in kcal/mol.

[c]C-PCM B3LYP/6-31 + G(d,p).

TABLE 2.4 Activation Enthalpies $\Delta H^{\ddagger}$ (kcal/mol) for the *Ortho*/*Meta* and *Endo*/*Exo* Pathways for the Diels–Alder Reactions Between *o*-QM and MVE, Styrene, and MVK in the Gas Phase[a]

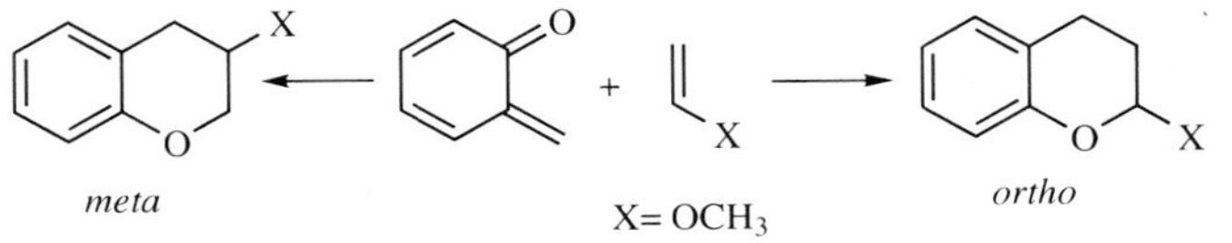

meta ← *o*-QM + CH_2=CH–X → *ortho*

X = OCH_3, Ph, $COCH_3$

Dienophile	Regioselectivity	*Endo*/*Exo*	$\Delta H^{\ddagger}$
MVE	*ortho*	*endo*	6.0
		exo	5.8
	meta	*endo*	18.3
		exo	19.7
Styrene	*ortho*	*endo*	9.8
		exo	10.0
	meta	*endo*	14.7
		exo	15.3
MVK	*ortho*	*endo*	7.8
		exo	8.2
	meta	*endo*	11.7
		exo	12.1

Data taken from Ref. [16].

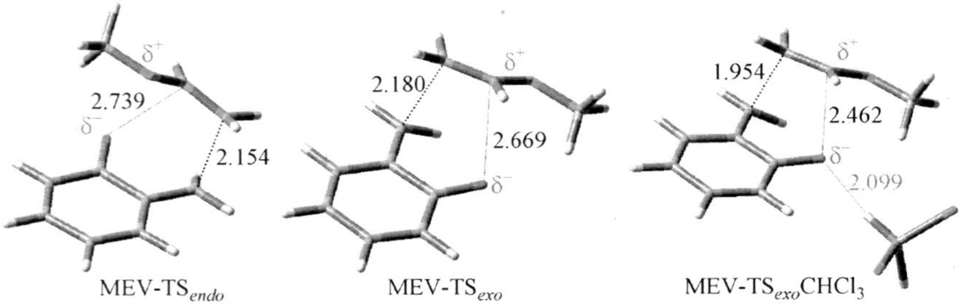

FIGURE 2.3 *Endo* and *exo* TS geometries of the Diels–Alder reaction between *o*-QM and **MVE**, also in the presence of an ancillary $CHCl_3$ molecule (**MEV-TS$_{exo}$CHCl$_3$**), at B3LYP/6-31G(d,p) level, in the gas phase. Forming bond lengths are in Å. (See the color version of this figure in Color Plates section.)

TABLE 2.5 Solvent Effect on the Asynchronicity (Δd, in Å) and Energies ($\Delta E^{\ddagger}$, in kcal/mol) of MEV-TSexo Transition Structure at B3LYP/6-31G(d,p) Level[a]

Solvent	Cy	Toluene	Et_2O	CH_3OH	$CHCl_3$ Bulk	$CHCl_3$ Bulk + Specific
$\Delta E^{\ddagger}$	7.8	5.6	5.3	3.8	4.0	3.4
Δd (Å)	0.51	0.51	0.57	0.65	0.67	0.51

Data taken from Ref. [16].

dienophiles. The asynchronicity for the [4 + 2] cycloaddition with **MVE** and the charge transfer to *o*-QM (0.25 e) in the *endo–exo* TSs are very pronounced; therefore, they may be described as a TSs with a zwitterionic character. Data in Table 2.4 also suggest that *o*-QMs dò not exhibit pronounced *endo–exo* diastereoselectivity. In fact, the *exo* approach is slightly favored over the *endo* for **MVE**, and the opposite is predicted for styrene and **MVK**.

The energy and geometry data listed in Table 2.5 show that the effect of solvent bulk (computed for cyclohexane, toluene, diethylether chloroform, THF, and methanol by PCM model), decreases the activation energy, increasing asynchronicity for the [4 + 2] cycloaddition reactions.

Specific solvation effect has also been modeled adding an ancillary $CHCl_3$ molecule H-bonded to the QM oxygen atom. The authors claimed that the activation energies were decreased in the cycloaddition with styrene (5.1–4.1 kcal/mol), **MEV** (4.0–3.4 kcal/mol), and **MVK** (4.8–3.9 kcal/mol) as a result of HB interactions,[16] but the effect is very small and it could likely be a computational artifact resulting from basis set superposition error (BSSE). These results indicate that the HB of chloroform on *o*-QM reactivity is negligible and that the solvation effects of $CHCl_3$ can be modeled taking into account only bulk effects.

2.4 TAUTOMERIZATIONS INVOLVING QUINONES AND QUINONE METHIDES

The oxidative polymerization of 5,6-dihydroxyindole (**1**) and related tyrosine-derived metabolites is a central, most elusive process in the biosynthesis of eumelanins, which are the characteristic pigments responsible for the dark color of human skin, hair, and eyes. Despite the intense experimental research for more than a century,[36] the eumelanin structure remains uncharacterized because of the lack of defined physicochemical properties and the low solubility, which often prevents successful investigations by modern spectroscopic techniques. The starting step of the oxidative process is a one-electron oxidation of 5,6-dihydroxyindole generating the semiquinone **1-SQ** (Scheme 2.7).

Nevertheless, the further mechanistic steps leading to indole dimerization is not defined and a computational investigation could suggest feasible reaction pathways, providing important anticipation about IR and UV absorption spectra, which could be very useful for the assignment of the intermediates involved. It has been experimentally proposed that the semiquinone **1-SQ** may decay via disproportionation to

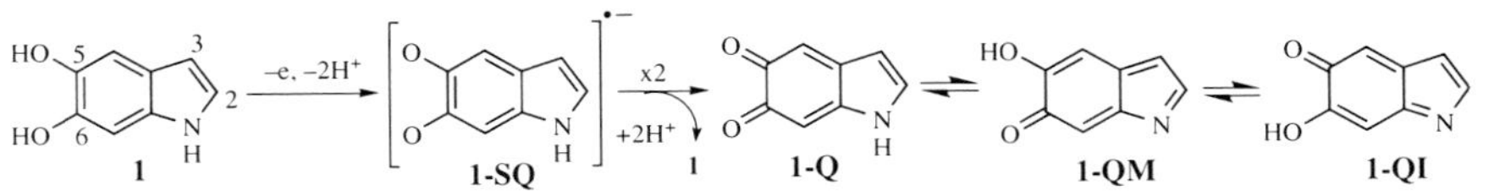

SCHEME 2.7 Generation of conjugated *p*-quinone methide structures by 5,6-dihydroxyindole oxidation.

give a mixture of the *o*-quinone **1-Q**, the quinone methide **1-QM**, and the quinonimine **1-QI** (Scheme 2.7), but the experimental evidences supporting such an hypothesis have been elusive.

Quantum mechanical calculation by methods in the frame of DFT such as B3LYP and PBE0 predicted that 5,6-indolequinone in the gas phase should consist of a mixture of two tautomers, the quinone (**1-Q**) and the quinone methide (**1-QM**).[20] PBE0, also referred as PBE1PBE, is a functional obtained by combining a predetermined amount of exact exchange with the Perdew–Burke–Ernzerhof exchange and correlation functionals.[37] According to B3LYP/6-311 + G(2d,p) data, **1-QM** was destabilized only by 0.6 kcal/mol. This would mean that about 25% of 5,6-indolequinone would exist in the gas phase as quinone methide tautomer. Due to the lower dipole moment of **1-QM** ($\mu = 2.4$ D) in comparison to that of **1-Q** ($\mu = 8.3$ D), the energy difference between these two tautomers increases in aqueous solution with **1-QM** being destabilized by 8 kcal/mol. Therefore, concentration of quinone methide **1-QM** should be negligible in aqueous solution and other polar solvents. Solvation effects have been computed by polarizable conductor model C-PCM[25a]. The quinonimine **1-QI** was always relatively unstable in comparison to **1-Q** ($>$6.4 kcal/mol) both in vacuo and in water, and it is therefore unlikely to contribute to the measured absorption properties or reactivity for the putative quinone.

Vertical excitation energies for the tautomers **1-Q**, **1-QM**, and **1-QI** were obtained by combining time-dependent density functional theory (TD-DFT) technique[38] with the SCRF-CPCM calculation, with both the B3LYP and PBE0 methods in the gas phase and in water. For the five lowest excited states, the excitation energies and oscillator strengths predicted with the PBE0 model were generally larger than those obtained with the B3LYP method, but the differences were relatively small ($<$0.15 eV for the energies and roughly 10% for the intensities). An interesting finding of this study is the prediction of reasonably strong absorption in near-IR for the three tautomers **1-Q**, **1-QM**, and **1-QI**, centered at 603, 700, and $>$800 nm, respectively.

Both B3LYP and PBE0 results suggested that the most stable form **1-Q** exhibits very strong solvatochromism because of the reversal of the energy order for the two lowest excited states.

In contrast, no state reversal was predicted for the quinone methide **1-QM**. The lowest excited state of this molecule had practically identical energy and oscillator strength in water and in the gas phase.

To assess performance of the selected DFT techniques in predicting electronic absorption spectra of quinones, the authors computed excitation energies of *o*-benzoquinone at the same level of theory and compared them to available experimental data.[39] The PBE0 method placed the lowest energy transition of *o*-benzoquinone in $CHCl_3$ at 643 nm. The B3LYP excitation energies were slightly smaller for all states considered. The PBE0 results were in fairly reasonable agreement with a measured spectrum of *o*-benzoquinone in $CHCl_3$ that showed a maxima at 590 nm.

Such results seem to be rather indicative of the opportunity to exploit TD-DFT and the PBE0 functional in predicting spectroscopic properties of mixtures containing the three tautomers **1-Q**, **1-QM**, and **1-QI** in aqueous solution. This approach should be very useful for future experimental mechanistic investigations clarifying the complex mechanisms of dihydroxyindole oxidation.

2.4.1 QM Versus Quinone Stability: Substituent Effects

The relative stability of the **QM** form in comparison to the quinone (**Q**) and quinine imine (**Q-I**) tautomers is not only function of the medium but is also strongly affected by substituents at C-2 and C-3 indole ring (see Scheme 2.7 for numbering). In fact, the tautomeric equilibria involving 3-iodo-5,6-indolequinone (**I**) has been very recently investigated optimizing the geometries of the resulting tautomers in vacuo at the PBE0/6-31 + G(d,p) level of theory.[20] To account for the influence of the aqueous environment, all structures have also been optimized using the PCM solvation model[40] in its UAHF parametrization.[27] The 3-iodo-QM (**I-QM**, in Scheme 2.8) is favored over the *o*-quinone (**I-Q**) by ca. 2.8 kcal/mol, due to the strong (~7 kcal/mol) intramolecular OH–O hydrogen bond stabilization, which is absent in **I-Q**. However, solvation selectively stabilizes the *o*-quinone tautomer (**I-Q**) significantly more than the quinone methide (**I-QM**), so that in water the relative energy of the latter with respect to the *o*-quinone is 6.0 kcal/mol. The present results would therefore indicate that in vacuum or in low polar solvent the 3-iodo substituent stabilizes the quinone methide tautomer **I-QM** relative to the *o*-quinone **I-Q**, but the latter still is much more stable in aqueous solution.

Despite the importance of the oxidative polymerization of 5,6-dihydroxyindole, in the biosynthesis of pigments, little experimental data are known on the oxidation chemistry of the oligomers of **1.** For such reasons, three major dimers of **1**, such as **2-4** (Scheme 2.9), have been computationally investigated at PBE0/6-31 + G(d,p) level of theory both in gas and in aqueous solution (by PCM solvation model) to clarify the quinone methide/*o*-quinone tautomeric distribution.

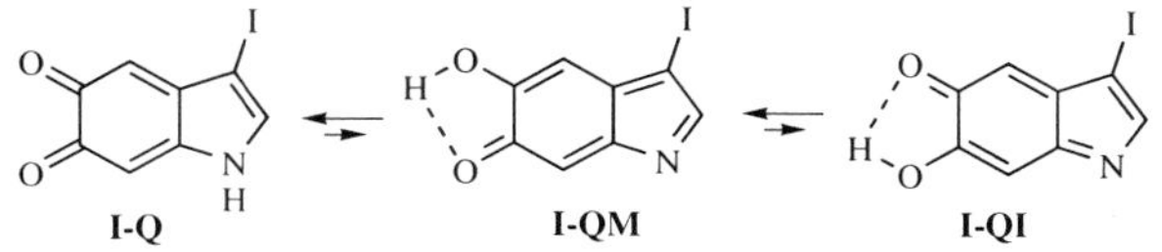

I-Q I-QM I-QI

SCHEME 2.8

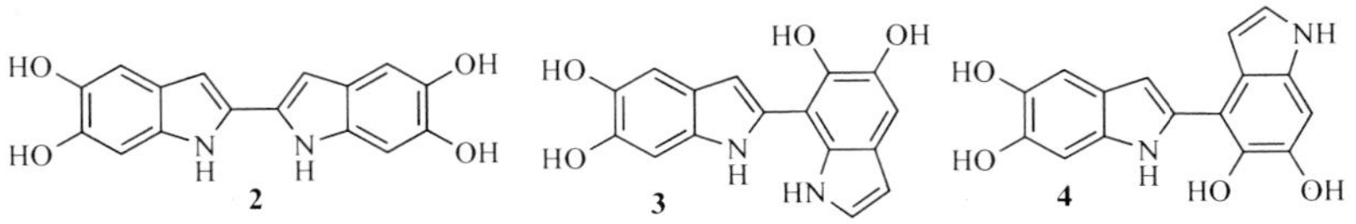

SCHEME 2.9

DFT results suggested that the delocalized QMs **2a–4a** (Scheme 2.10) and their related E isomers (**2b–4b**) are by far the most stable tautomers in the series.[19] The structures that come next, the less delocalized p-QMs **2c–4c** in vacuum, and the quinones **2d–4d** (**Qs**) are less stable by several kcal per mole in aqueous solution.

These DFT data provide a consistent picture for the tautomerization equilibria involving the dimers **2–4**, which highlight the extended quinone methides **2a**, **3a**, and **4a** as the most stable tautomers for all biindolyl quinones investigated.

The tautomeric product distribution has been a prerequisite for a further investigation aimed at predicting absorption properties of the transient semiquinones and quinones generated by pulse radiolytic oxidation of **2–4**. The simulation of electronic absorption spectra has been computed using the TD-DFT approach both in vacuum and in aqueous solution, using the large 6-311 + + G(2d,2p) basis set.[19]

2.5 *o*-QUINONE METHIDE METAL COMPLEXES

As mentioned in the introduction, the prototype *o*-QM is a highly reactive intermediates in organic reactions, including cycloaddition chemistry and DNA covalent modification, due to the high electrophilicity at the exocylic methylene carbon (see its dipolar representation in Scheme 2.11).

In striking contrast, *o*-QM cyclopentadienyl Ir and Rh complexes (Cp*Ir and Cp*Rh) such as **Ir-iPr$_2$**[41] and **RhMe$_2$**[42] (Scheme 2.12) have been isolated, characterized by X-ray diffraction, and have shown nucleophilic reactivity at the exocyclic carbon.

2.5.1 Geometries and Reactivity as Function of the Metal and the Structural Features

To explain the QM "umpolung" in these metal complexes and to rationalize their reactivity as a function of both metal fragment and ligand substitution pattern, a series of DFT calculations on Ir, Rh, Co, and Ru congeners have been performed at B3LYP level using LANL2DZ basis set.[17] This basis set uses the effective core potential of Hay and Wadt[43] to describe the core electrons of the metal atom, Co (341/311/41), Ru (341/321/31), Rh (341/321/31), and Ir (341/321/21). The valence electrons of the

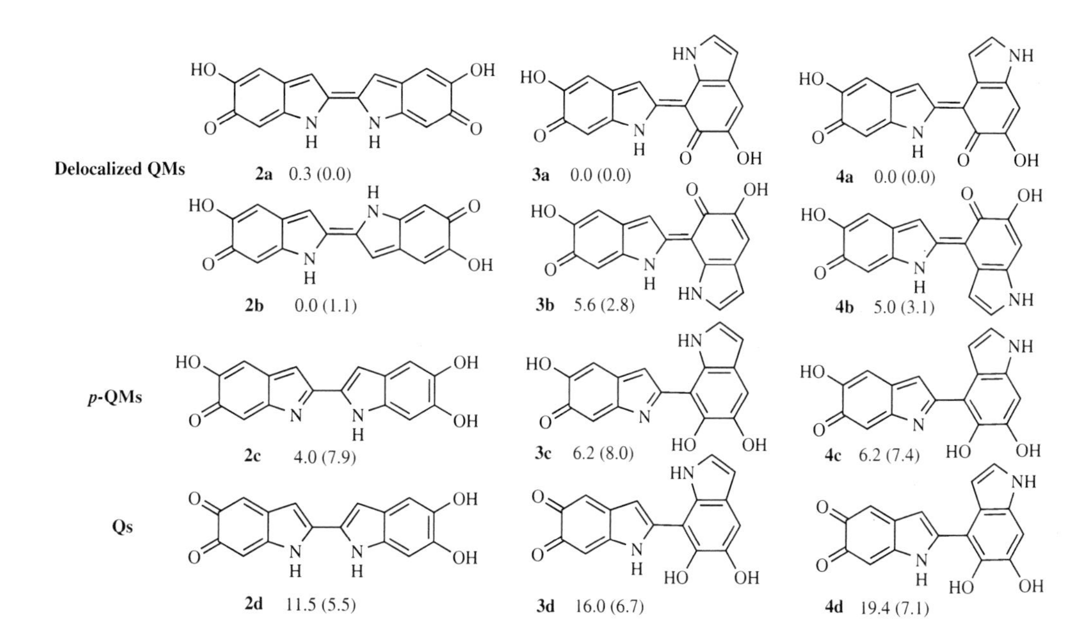

SCHEME 2.10 Structures and relative energies (kcal/mol) of several tautomers/conformers resulting from the bielectronic oxidation of **2–4**, computed at the PBE0/6-31 + G(d,p) level in vacuum and in aqueous solution (in parenthesis, by PCM solvation model, data from Ref. [19]).

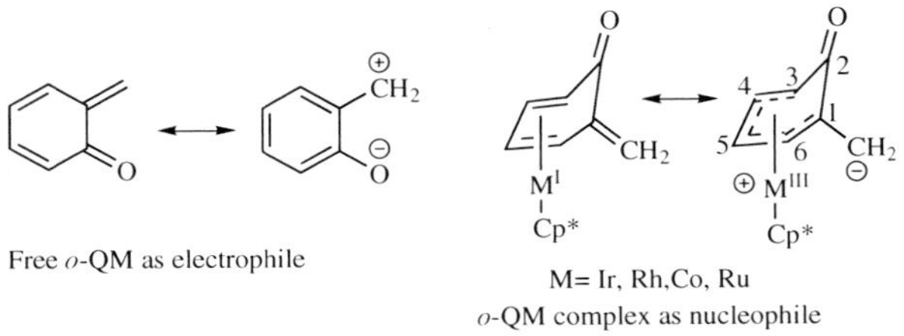

Free *o*-QM as electrophile

M= Ir, Rh,Co, Ru
o-QM complex as nucleophile

SCHEME 2.11

metal atom in addition to those of H, C, and O were treated by the Dunning–Huzinaga double-ζ basis sets (D95).[44] These basis sets were augmented by the addition of a single polarization function to all H, C, and O atoms [designated as LANL2DZ** or LANL2DZ(d,p)]. To examine basis set effects on QM-metal complex geometries, D95 have been replaced by the 6-31G, TZV,[45] and 6-311G basis sets for H, C, N, and O atoms, leaving the LANL2DZ basis set on the transition metal intact. Validation of the computational method used came from the comparison of the computed QM-complexed structures with the known solid-state geometries of **Ir-*i*Pr$_2$** and **Rh-Me$_2$** complexes,[41,42] which displays a fairly good agreement between experiments and calculations without any remarkable basis set dependence. The geometry of the *o*-QM ligand is mostly invariant when the metal center is changed. However, the hinge angle, defined as the deviation of the carbonyl/alkene portion of the *o*-QM from planarity with the coordinated diene moiety [which is measured by the dihedral angle across C(3)–C(6) atoms], varies rather dramatically as a function of the metal fragment.[17] In more detail, the hinge angle is 25° for QM-Co, 27° for QM-Rh, 36° for QM-Ir, and 25° for QM-Ru complexes. Such a variation is the result of back donation from the metal to the aromatic QM-system,[46] which is predicted to increase from Co to Rh and Ir, paralleling the trend of the hinge angle. The electronic back donation is well supported by frontier molecular orbital analysis (considering the HOMO of the complexes, see Fig. 2.4). In fact, the HOMO of the complexes reveals strong π back-bonding between a filled metal "d" orbital and a π* orbital of the *o*-QM, which is particularly localized on the QM-diene subunit (in red, Fig. 2.4).

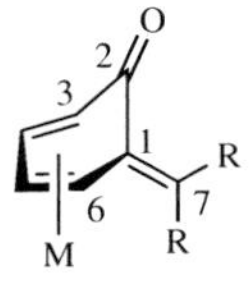

Ir-iPr$_2$ M= Cp*Ir, R=*i*Pr
Rh-Me$_2$ M= Cp*Rh, R=CH$_3$

SCHEME 2.12 *o*-QM metal complexes isolated and characterized by X-ray diffraction, which structures have been used as a validation for DFT calculations.

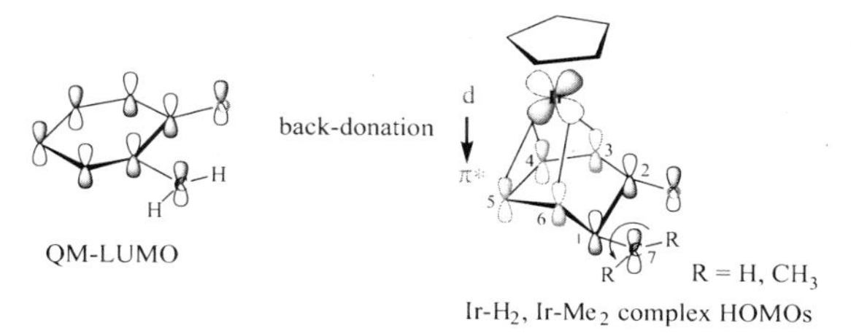

FIGURE 2.4 Schematic representation of the back donation from the metal to the QM-system, in the complexes **Ir-H$_2$** and **Ir-Me$_2$**. (See the color version of this figure in Color Plates section.)

The nucleophilic character of the exocyclic methylene carbon C7 has been described by the partial charges derived from Mulliken and natural orbital analyses. The charge on C7 is affected by both the metal and the substituents at the exocyclic CH_2 group. Concerning the latter aspect, methyl substituents on the exocyclic C7 carbon reduce the extended delocalization and the negative charge by torsion of the $C_2C_1C_7R$ dihedral angle (curved arrow in Fig. 2.4), improving complex stability. The partial negative charge by NPA (natural population analysis) is higher (−0.38) in the prototype complexes (R=H) with Ru and Co than in the Ir complex. The charge data suggest that the QM in the former complexes should be more reactive and nucleophilic than the latter one. Therefore, the computational analysis suggests that the Ru and Co complexes of the prototype *o*-QM may be useful for future synthetic studies, since they should exhibit high reactivity as nucleophiles at the exocyclic methylene group.[17]

2.6 GENERATION OF *o*-QM

2.6.1 Generation of *o*-QM Tethered to Naphthalene Diimides by Mild Activation

Naphthalene diimides (NDIs) tethered to quaternary ammonium salts of Mannich bases (Scheme 2.13) have been shown experimentally to be activatable precursors of QMs (QMPs) under mild conditions.[47] Such a structural merging between an activatable and reactive electrophile such as *o*-QM and an extended aromatic core with DNA binding properties have recently suggested promising diagnostic and therapeutic applications. The QMP activation is induced by both base catalysis and single-electron reduction (triggered by a mild reducing agent such as sodium dithionite), and the resulting transient QMs have been trapped by several nucleophiles (Scheme 2.13) and ethyl vinyl ether.[47]

The base catalysis and the monoelectronic reductive activation processes have been described by a computational investigation at the R(U)B3LYP/6-31 + G(d,p) level of theory for the model imide **NI** (Scheme 2.14),[47] both in the gas phase and in aqueous solution, using PCM solvation model.[40]

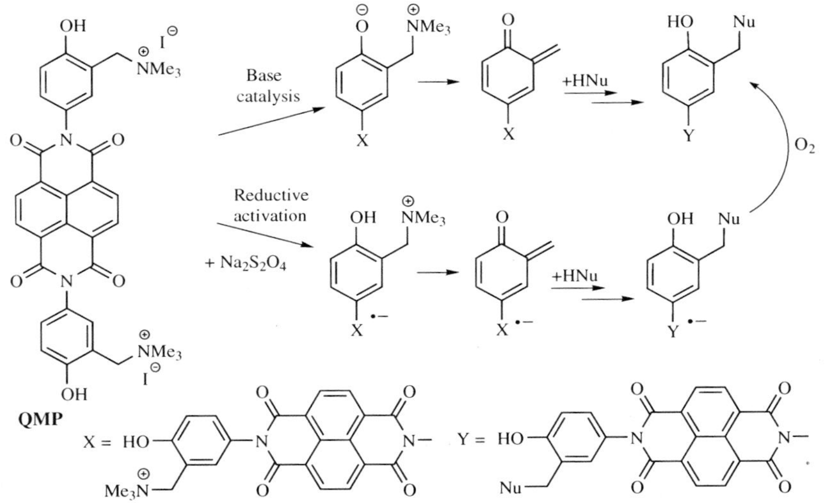

SCHEME 2.13 Activation of **QMPs** (quinone methide precursors) by base catalysis and single-electron reduction (reproduced from Ref. [47] with permission from American Chemical Society).

The activation free energies computed both in the gas phase and in aqueous solution (Table 2.6) suggest that the generation of an alkylating QM (**QM-NI**) becomes a much easier process, passing from the protonated quaternary ammonium salt **NI** to its zwitterionic form **NI**$^-$.

The location of the stationary points, including transition structures **TS-NI** and **TS-NI**$^-$, in the solvent bulk suggests a strong specific base catalysis on the QM generation process, since the activation free energy in aqueous solution from the zwitterionic **NI**$^-$ is reduced by 32 kcal/mol in comparison to the QM generation starting from the protonated precursor **NI**. The chemical reduction of the naphthalimide moiety to its radical anion **NI**$^{\bullet -}$, which generates a QM tethered to the imide radical anion moiety **QM-NI**$^{\bullet -}$, also induces a remarkable activation, lowering the barrier by 30.7 and 29.5 kcal/mol in gas phase and in aqueous solution, respectively. These computational data have supported the experimental findings of a new activation protocol for the mild generation of QMs by single-electron reduction.[47]

2.6.2 Thermal Generation of *o*-QM in Oxidative Processes in the Gas Phase

The partial combustion of toluene, with the generation of the intermediate 2-methylphenyl radical (**MP**, in Scheme 2.15) leading to the prototype quinone methide, has recently been investigated by high-level post-HF and DFT theoretical studies.[22]

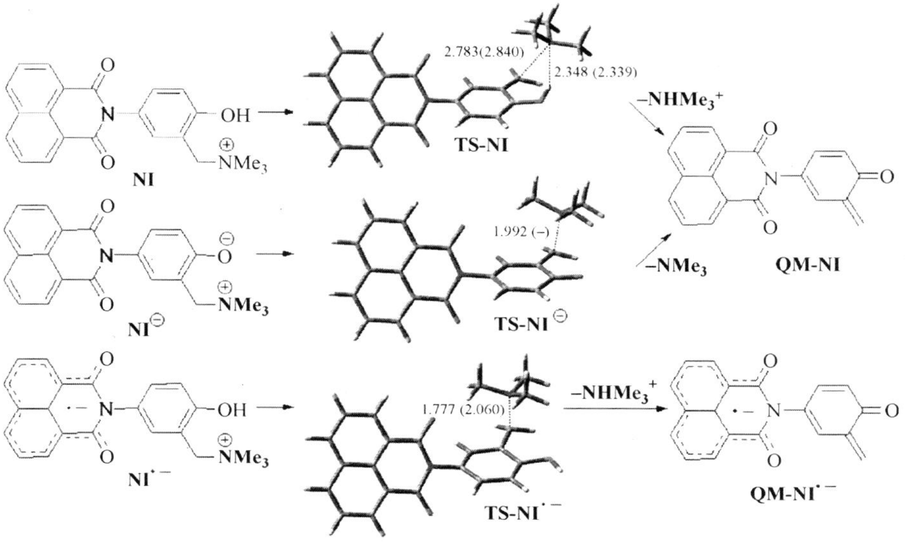

SCHEME 2.14 Uncatalyzed, base-catalyzed, and reductive generation of **QMs** tethered to a naphthalene imide core, through the TSs **TS–NI**, **TS–NI$^-$**, and **TS–NI$^{\bullet-}$**, respectively [bond lengths are in Å; data in parentheses are for full R(U)B3LYP/6-31 + G(d,p) optimization in aqueous solution] (reproduced from Ref. [47] with permission from American Chemical Society). (See the color version of this scheme in Color Plates section.)

A detailed reaction mechanism has been presented for the 2-methylphenyl radical + O_2 reacting system, which generates the 2-methylphenylperoxy radicals (**MPP**). The **MPP** radical, depicted in the general scheme, is the key intermediate, lying 48.7 kcal/mol below the reactants **MP** + O_2. The peroxy radical **MPP**, which is

TABLE 2.6 Activation Gibbs Energy in the Gas Phase and in Aqueous Solution at R(U)B3LYP/6-31 + G(d,p) Level of Theory for the Generation of the QMs Starting from NI, Its Anion NI$^-$, and Its Radical Anion NI$^{\bullet-}$

QMP	ΔE^a	$\Delta G_{gas}{}^b$	ΔG_{aq}
NI	37.0	32.1	46.5[c], 41.1[d], 41.4[e]
NI$^-$	3.7	1.0	—[f], 8.5[d], 8.5[e]
NI$^{\bullet-}$	3.8	1.4	17.0[c], 6.3[d], 6.7[e]

Source: Reproduced from Ref. [47] with permission from American Chemical Society.

[a]Activation electronic energies in kcal/mol.

[b]Activation Gibbs energies in the gas phase.

[c]Stationary points optimized in aqueous solution at R(U)B3LYP/6-31 + G(d,p) using PCM (UA0 radii) solvation model.

[d]Single-point calculation at B3LYP/6-31 + G(d,p) on gas phase geometries (UA0 radii).

[e]Single-point calculation at B3LYP/6-31 + G(d,p) on gas phase geometries (UAHF radii).

[f]**TS-NI$^-$** was not located in aqueous solution.

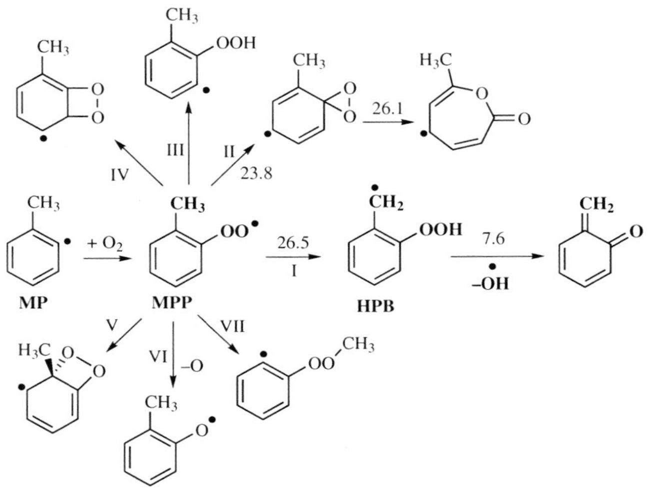

SCHEME 2.15 Reaction pathways for the generation of the prototype quinone methide through the oxidation of the 2-methylphenyl radical, investigated by post-HF and DFT methods (data have been taken from Ref. [22]).

generated with almost no barrier through a very loose TS, may react following seven different reaction pathways according to Scheme 2.15.

All the intermediates and transition structures connecting the 2-methylphenyl radical + O_2 reacting system to *o*-QM and other final products along the seven reaction pathways have been optimized at MP2/6-31G(d) and B3LYP/6-31G(d) level of theory. Enthalpies of formation for these stationary points have been calculated using series of 2–4 isodesmic reactions with the G3[48] and the G3B3[49] methods.[22] G3 and G3B3 are both composite theoretical methods. G3 involves an initial HF/6-31G(d) level geometry optimization and frequency calculation, followed by a further geometry optimization at the MP2/6-31G(d) level of theory. In the G3B3 method, these initial steps are replaced by a B3LYP/6-31G(d) geometry optimization and frequency calculation. In order to accurately determine stationary print energies, the final MP2 and B3LYP geometries have been subjected to a series of quadratic configuration interactions with single and double excitations and triple excitations added perturbatively [QCISD (T)[50]]. Unfortunately, reactions involving radical species, such as those in Scheme 2.15, suffer from spin contamination errors.[51] Spin contamination results in having a wave function that appears to be the desired spin state, but contains a bit of some other spin state mixed in. This occasionally results in slightly lowering the computed total energy due to having more variational freedom. More often, the result is to slightly raise the total energy since a higher energy state is being mixed in. However, this change is an artifact of an incorrect wave function. Since the spin contamination problem is much more severe for HF methods, these errors are greatest for the G3 approach. Comparatively, the B3LYP density functional, which is utilized

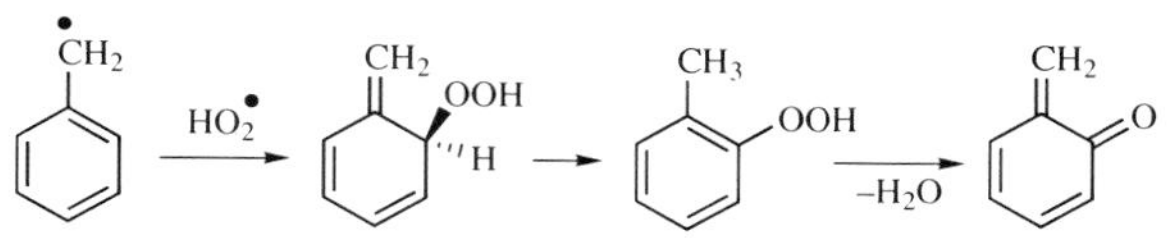

SCHEME 2.16 Additional reaction pathway for the generation of the quinone methide in the gas phase oxidation of 2-methylphenyl radical, investigated by the hybrid functional MPW1K (reproduced from Ref. [23] with permission from American Chemical Society).

in the G3B3 method, is much less susceptible to spin contamination than HF methods.[52] Therefore, according to the authors, it should provide more reliable PESs.

Among the seven reaction pathways, the intramolecular hydrogen abstraction from the methyl group in the **MPP** radical by the radical peroxy oxygen atom leading to 2-hydroperoxybenzyl radical (**HPB**, pathways I, Scheme 2.15) is one of the kinetically most favored since its activation enthalpy is only 26.5 kcal/mol. This benzylic radical is fairly unstable and rapidly generate the *o*-QM by OH radical loss through a very low activation barrier (7.6 kcal/mol).

In addition, another computational study in the frame of DFT, using the hybrid functional MPW1K,[53] had suggested that *o*-QM may be an intermediate in the reaction of the peroxy radical ($HO_2^{\bullet}$) with the benzyl radical at the *ortho*-position (Scheme 2.16),[54] which should be significant in atmospheric processes and low-temperature combustion systems ($T < 1500$ K).

2.7 THERMAL DECOMPOSITION OF *o*-QM IN THE GAS PHASE

The computational modeling of both (i) the *o*-QM isomerization to tropone (Scheme 2.17) and (ii) the QM unimolecular dissociation reactions to fulvene through a decarbonylation process (Scheme 2.18) in the gas phase have been investigated by CBS-QB3 composite theoretical method.[23] The CBS-QB3 method[55] requires an initial B3LYP/6-311G(2d,d,p) geometry optimization and frequency calculation and then uses the resulting stationary point geometries for higher level energy corrections with the CCSD(T)[56] and MP4 theoretical methods, followed by an

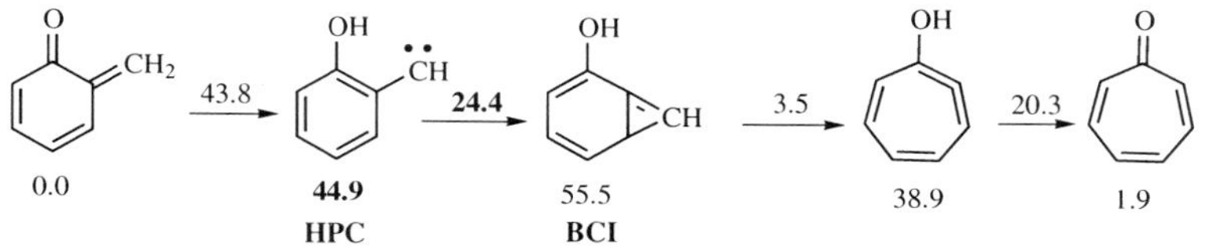

SCHEME 2.17 Generation of tropone from *o*-QM via hydroxyphenylcarbene (**HPC**). Enthalpies of formation for the intermediates and activation enthalpies (data above the arrows) are reported in kcal/mol. Enthalpies of formation and activation enthalpy for the rate-determining step are in bold (data have been taken from Ref. [23]).

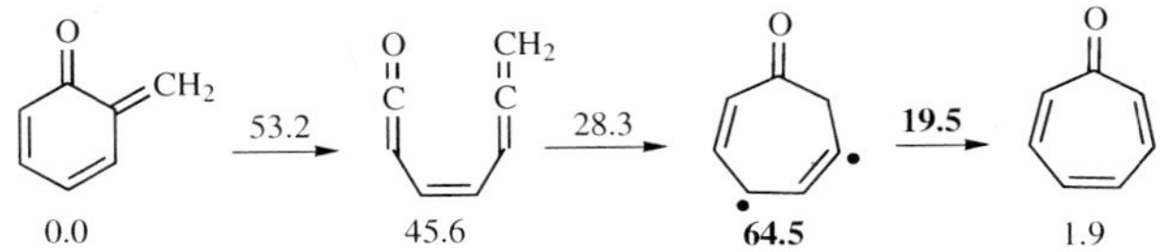

SCHEME 2.18 Generation of tropone from *o*-QM via a ring-opening process. Enthalpies of formation for the intermediates and activation enthalpies (data above the arrows) are reported in kcal/mol. Enthalpies of formation and activation enthalpy for the rate-determining step are in bold (data have been taken from Ref. [23]).

extrapolation to the complete basis set (CBS) limit. Enthalpies of formation at 298 K have been calculated with the CBS-QB3 theoretical method for all the identified stationary points, including the transition states. Activation energies have also been calculated using the functional BB1K[57] within the DFT frame and 6-31 + G(2d,p) basis set. The DFT hybrid functional BB1K has been used since it has been specifically developed by Truhlar and coworkers for accurate thermochemical kinetics.[57] Two reaction pathways have been identified leading to tropone, one pathway via an initial hydrogen shift reaction producing 2-hydroxyphenylcarbene (**HPC**, Scheme 2.17) as intermediate and the other pathway via quinone methide initial ring opening (Scheme 2.18).

Concerning the first reaction pathway reported in Scheme 2.17, *o*-QM undergoes an intramolecular hydrogen shift to produce **HPC**.

The carbene **HPC** follows a further intramolecular addition to the aromatic ring at the *ortho*-position, forming the bicyclic intermediate (**BCI**). The latter is the rate-determining step of this reaction channel, since the TS connecting **HPC** to **BCI** is the stationary point on the PES at higher energy (+ 69.3 and 67.1 kcal/mol, from CBS-QB3 and BB1K methods, respectively), relatively to *o*-QM. The bicyclic intermediate opens to 1,2,4,6-cycloheptatetraen-1-ol and finally intramolecular hydrogen shift occurs from the OH group generating tropone. This type of rearrangement is not a novelty since it is well known that other phenylcarbenes also give seven-membered ring products.[58]

The second reaction pathway investigated was a *o*-QM decomposition initiated by a ring-opening process, generating a conjugated ketenes as intermediate, as shown in Scheme 2.18.

The rate-determining step of this second reaction mechanism is the isomerization of the cyclic diradical to tropone, since the TS connecting the intermediate to the final product lies 84.0 kcal/mol (from CBS-QB3 method) above the starting *o*-QM.

The authors also considered two additional reaction pathways for the *o*-QM decarbonylation to fulvene, both via ring opening (Scheme 2.19).[23]

Kinetic analysis shows that the formation of tropone through a hydroxyphenylcarbene intermediate (which exhibits the lowest activation energy 69.3 kcal/mol) dominates *o*-QM decomposition process up to 1500 K, with fulvene + CO formation becoming competitive at higher temperatures. In fact, the latter decomposition mode although disfavored by its higher activation enthalpy (75.4 versus 69.3 kcal/mol) becomes competitive due to its more positive activation entropy.

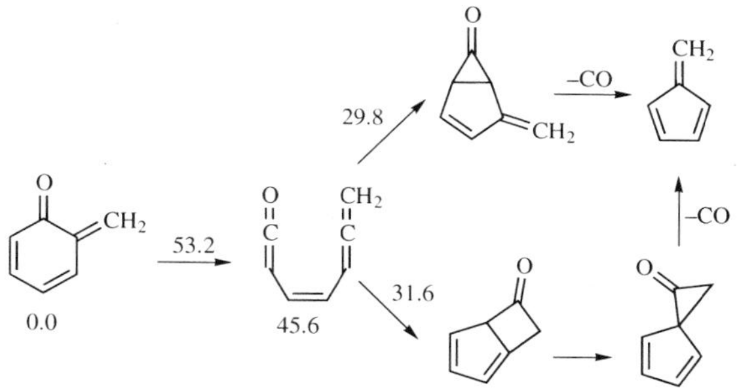

SCHEME 2.19 Generation of fulvene from the decarbonylation of *o*-QM, via two different pathways. Enthalpies of formation for the intermediates and activation enthalpies (data above the arrows) are reported in kcal/mol (data have been taken from Ref. [23]).

The calculated overall rate constant for *o*-QM decomposition is in relatively good agreement with the experimental measurements (67.1 kcal/mol).[59] CBS-QB3-predicted rate constants are slightly slower than the experimental results. The BB1K/6-31 + G(2d,p) data provide a considerably better comparison to the experimental reaction rate. In fact, the calculated total rate constant using the BB1K barrier heights has been fitted to the Arrhenius equation for $T = 800–2400$ K, yielding an observed activation energy of 71.3 kcal/mol.[23]

2.8 QM GENERATION IN LIGNIN FORMATION

Lignin polymerization is a natural process initiated by the enzymatic oxidation of hydroxycinnamyl alcohols such as *p*-coumaryl (**CM**), coniferyl (**CF**), and sinapyl alcohols (**SN**), which are known as monolignols (Scheme 2.20).[60]

The oxidation generates highly delocalized phenoxy radicals (**PhO**$^{\cdot}$, Scheme 2.21), which may initiate (i) a radical polymerization process, trapping the reactant (**CF**) to give a benzyl radical intermediate (**QMR**), or it may (ii) follow a radical coupling to produce the *p*-QM **β-O-QM**, which being a reactive electrophile could undergo cationic polymerization.

Considering coniferyl alcohol as model reagent, these two mechanisms have been investigated in the frame of DFT to clarify which could be responsible for the initiation step of the polymeric lignin growth. In more detail, all calculations were carried out using the B3LYP[24] hybrid density functional as implemented in the Gaussian 98 program.[26a] Restricted (RB3LYP) and unrestricted (UB3LYP)

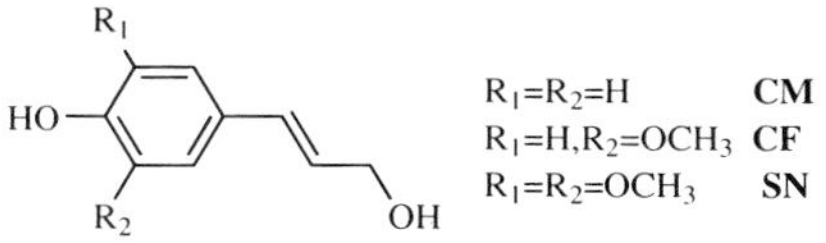

SCHEME 2.20 Hydroxycinnamyl alcohols involved in lignin polymerization.

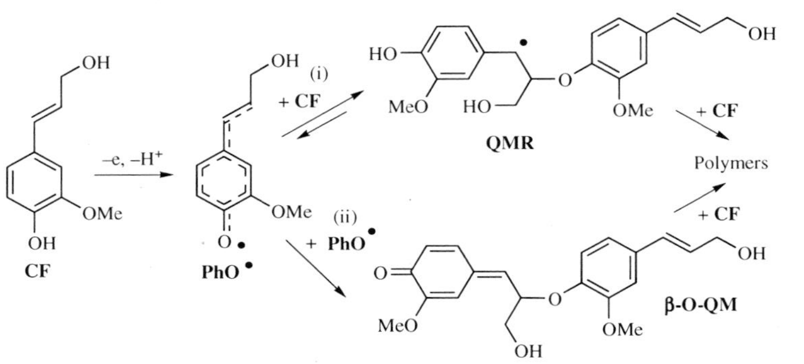

SCHEME 2.21 Two possible reaction mechanisms for the initiation step of the lignin polymerization, investigated at PCM-B3LYP/6-311G(2df,p)//B3LYP/6-31G(d,p) level of theory.

formalisms were used for closed-shell (reactant **CF** and **β-O-QM**) and open-shell systems (delocalized phenoxy radical, **PhO** and **QMR**), respectively. Stationary point (reactants, intermediates, and transition structures) geometries were optimized with the 6-31G(d,p) basis set. Energies were computed by single-point calculations using polarization and diffuse functions [6-311G(2df,p) and 6-311 + + G(2df,p) basis sets]. The B3LYP/6-311G(2df,p) single-point energy calculations were also performed in conjunction with the Gaussian 98 implementation of the PCM.[40]

The initial coupling of two coniferyl alcohol radicals (**PhO**$^{\cdot}$) forming a *p*-QM (**β-O-QM**) proceeds by a very low energy barrier of ~3–5 kcal/mol and the reaction resulted to be strongly exothermic by 23 kcal/mol in a solution with a dielectric constant $\varepsilon = 4$. Solvation effect on the reaction energetic was modeled by PCM solvation model at B3LYP/6-311G(2df,p) level of theory. The coupling of a coniferyl alcohol radical (**PhO**$^{\cdot}$) to coniferyl alcohol to give the benzyl radical **QMR** exhibits a higher activation energy (10.6 kcal/mol) and it is slightly endothermic (+0.4 kcal/mol) at the same level of theory. In other words, the radical addition to **CF** is a reversible process. The computed energetics clearly showed that latter reaction (pathway i) is less favorable than the formation of a *p*-QM (pathway ii). Such an energy difference cannot be easily overcome by the reactant concentration, which should favor the pathway (i).

The reaction mechanism for the water addition that converts the QM **β-O-QM** into a guaiacylglycerol-β-coniferyl ether dilignol (**GGE**) have also been investigated (Scheme 2.22).[18]

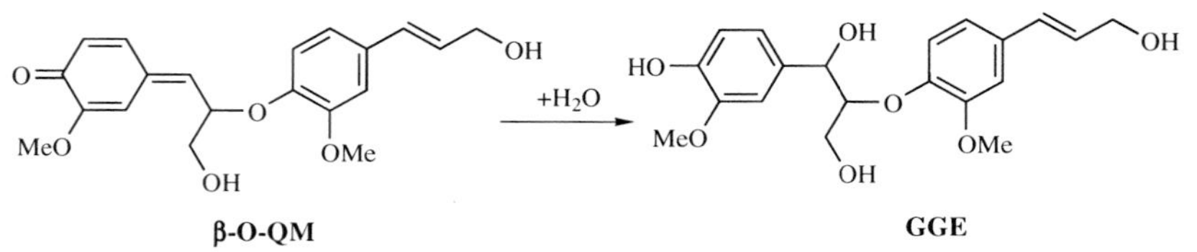

SCHEME 2.22

Employing a model system that includes three water molecules, it was shown that the initial step has a substantial energy barrier ($\geq$30 kcal/mol) that prevents the addition of water from taking place under mild conditions.

To account for the effects of specific acid catalysis, the calculations were also carried out in the presence of an additional proton. The resulting potential energy surface at PCM-B3LYP/6-311G(2df,p)//B3LYP/6-31G(d,p) level of theory suggested that the β-O-4-linked quinone methide (**β-O-QM**) is a fairly stable species and its conversion to a guaiacylglycerol-β-coniferyl ether dilignol has to be an acid-catalyzed process.[18]

2.9 CONCLUSION AND PERSPECTIVE

The general picture of the QM properties and reactivity modeled by DFT methods using B3LYP, PBE0 (PBE1PBE), BB1K, and MPW1K functionals, both in the gas phase and in solvents, appears to be reliable, since it has been possible to reproduce experimental kinetic behavior of both QMs and their adducts. Particularly informative are the computed kinetic data of QMs and their conjugate adducts with the DNA bases in aqueous solution, which are in excellent agreement to the available experimental data. This result has been achieved coupling the DFT approach to PCMs to describe bulk solvation effects, neglecting specific solvation of the solvent.

The predicted reversibility of the alkylation process at the 1-methylcytosine N3, 9-methyladenine N1, and 9-methylguanine oxygen atom foresaw the possibility of exploiting the QM-conjugated nucleosides as mild carriers of alkylating and cross-linking agents. In addition, time-dependent DFT calculations using PBE0 functional also provided fairly reliable UV-visible absorption spectra of conjugated QM-structures and their quinone tautomers. Such computed spectroscopic data could be useful to experimentalist in diagnostic and mechanistic investigations.

The reliability and fairly low demanding computational cost of DFT methods opens the way for *in silico* investigations of the QM generation catalyzed by DNA itself, starting from metastable precursors, which are capable of oligonucleotide recognition.

REFERENCES

1. (a) Qiao, G. G.; Lenghaus, K.; Solomon, D. H.; Reisinger, A.; Bytheway, I.; Wentrup, C. 4,6-Dimethyl-*o*-quinone methide and 4,6-dimethylbenzoxete. *J. Org. Chem.* 1998, 63, 9806–9811. (b) Tomioka, H. Matrix isolation study of reactive *o*-quinoid compounds: generation, detection and reactions. *Pure Appl. Chem.* 1997, 69, 837–840. (c) Tomioka, H.; Matsushita, T. Benzoxetene. Direct observation and theoretical studies. *Chem. Lett.* 1997, 26, 399–400.
2. (a) Wan, P.; Barker, B.; Diao, L.; Fisher, M.; Shi, Y.; Yang, C. 1995 Merck Frosst Award Lecture. Quinone methides: relevant intermediates in organic chemistry. *Can. J. Chem.*

1996, 74, 465–475. (b) Diao, L.; Cheng, Y.; Wan, P. Quinone methide intermediates from the photolysis of hydroxybenzyl alcohols in aqueous solution. *J. Am. Chem. Soc.* 1995, 117, 5369–5370. (c) Brousmiche, D.; Wan, P. Photogeneration of an *o*-quinone methide from pyridoxine (vitamin B6) in aqueous solution. *Chem. Commun.* 1998, 491–492. (d) Chiang, Y. A.; Kresge, J.; Zhu, Y. Flash photolytic generation and study of *p*-quinone methide in aqueous solution. An estimate of rate and equilibrium constants for heterolysis of the carbon—bromine bond in *p*-hydroxybenzyl bromide. *J. Am. Chem. Soc.* 2002, 123, 6349–6356. (e) Chiang, Y. A.; Kresge, J.; Zhu, Y. Flash photolytic generation of *ortho*-quinone methide in aqueous solution and study of its chemistry in that medium. *J. Am. Chem. Soc.* 2001, 123, 8089–8094.

3. (a) Modica, E.; Zanaletti, R.; Freccero, M.; Mella, M. Alkylation of amino acids and glutathione in water by *o*-quinone methide. Reactivity and selectivity. *J. Org. Chem.* 2001, 66, 41–52. (b) Zanaletti, R.; Freccero, M. Synthesis, spectroscopic characterization and chemical reactions of stable *o*-QM on solid phase. *Chem. Commun.* 2002, 1908–1909.
4. Weinert, E. E.; Dondi, R.; Colloredo-Mels, S.; Frankenfield, K. N.; Mitchell, C. H.; Freccero, M.; Rokita, S. E. Substituents on quinone methides strongly modulate formation and stability of their nucleophilic adducts. *J. Am. Chem. Soc.* 2006, 128, 11940–11947.
5. Chiang, Y. A.; Kresge, J.; Zhu, Y. Kinetics and mechanisms of hydration of *o*-quinone methides in aqueous solution. *J. Am. Chem. Soc.* 2000, 122, 9854–9855.
6. (a) Bolton, J. L.; Turnipseed, S. B.; Thompson, J. A.; *Chem.-Biol. Interact.* 1997, 107, 185–200. (b) Bolton, J. L.; Valerio, L. G.; Thompson, J. A. The enzymic formation and chemical reactivity of quinone methides correlate with alkylphenol-induced toxicity in rat hepatocytes. *Chem. Res. Toxicol.* 1992, 5, 816–822. (c) Thompson, D. C.; Perera, K.; Krol, E. S.; Bolton, J. L. *o*-Methoxy-4-alkylphenols that form quinone methides of intermediate reactivity are the most toxic in rat liver slices. *Chem. Res. Toxicol.* 1995, 8, 323–327. (d) Lewis, M. A.; Graff Yoerg, D.; Bolton, J. L.; Thompson, J. A. Alkylation of 2′-deoxynucleosides and DNA by quinone methides derived from 2,6-di-*tert*-butyl-4-methylphenol. *Chem. Res. Toxicol.* 1996, 9, 1368–1374.
7. Chiang, Y. A.; Kresge, J.; Zhu, Y. Flash photolytic generation of *o*-quinone α-phenylmethide and *o*-quinone α-(*p*-anisyl)methide in aqueous solution and investigation of their reactions in that medium. Saturation of acid-catalyzed hydration. *J. Am. Chem. Soc.* 2002, 123, 717–722.
8. McCracken, P. G.; Bolton, J. L.; Thatcher, G. R. J. Covalent modification of proteins and peptides by the quinone methide from 2-*tert*-butyl-4,6-dimethylphenol: selectivity and reactivity with respect to competitive hydration. *J. Org. Chem.* 1997, 62, 1820–1825.
9. (a) Pande, P.; Shearer, J.; Yang, J.; Greenberg, W. A.; Rokita, S. E. Alkylation of nucleic acids by a model quinone methide. *J. Am. Chem. Soc.* 1999, 121, 6773–6779. (b) Veldhuyzen, W. F.; Shallop, A. J.; Jones, R. A.; Rokita, S. E. Thermodynamic versus kinetic products of DNA alkylation as modeled by reaction of deoxyadenosine. *J. Am. Chem. Soc.* 2001, 123, 11126–11132.
10. (a) Zhou, Q.; Turnbull, K. D. Phosphodiester alkylation with a quinone methide. *J. Org. Chem.* 1999, 64, 2847–2851. (b) Zhou, Q.; Turnbull, K. D. Trapping phosphodiester–quinone methide adducts through *in situ* lactonization. *J. Org. Chem.* 2000, 65, 2022–2029.

11. (a) Musil, L.; Koutek, B.; Pisova, M.; Soucek, M. Quinone methides and fuchsones. XIX. Delocalization and stability of *o*-and *p*-quinone methides: a HMO study. *Collect. Czech. Chem. Commun.* 1981, 46, 1148–1159. (b) Krupicka, J.; Koutek, B.; Musil, L.; Pavlickova, L.; Soucek, M. Quinone methides and fuchsones. XVII. Half-wave potentials of quinone methides in dimethylformamide: substituent effects. *Collect. Czech. Chem. Commun.* 1981, 46, 861–872.

12. Musil, L.; Koutek, B.; Velek, J.; Krupicka, J.; Soucek, M. Quinone methides and fuchsones. XXIX. CNDO/S MO study of the fuchsone derivatives with a sterically crowded exocyclic double bond. *Collect. Czech. Chem. Commun.* 1983, 48, 2825–2839.

13. Di Valentin, C.; Freccero, M.; Zanaletti, C.; Sarzi-Amadè, M. *o*-Quinone methide as alkylating agent of nitrogen, oxygen, and sulfur nucleophiles. The role of H-bonding and solvent effects on the reactivity through a DFT computational study. *J. Am. Chem. Soc.* 2001, 123, 8366–8377.

14. Freccero, M.; Di Valentin, C.; Sarzi-Amadè, M. Modeling H-bonding and solvent effects in the alkylation of pyrimidine bases by a prototype quinone methide: a DFT study. *J. Am. Chem. Soc.* 2003, 125, 3544–3553.

15. Freccero, M.; Gandolfi, R.; Sarzi-Amadè, M. Selectivity of purine alkylation by a quinone methide. Kinetic or thermodynamic control? *J. Org. Chem.* 2003, 68, 6411–6423.

16. Wang, H.; Wang, Y.; Han, K.-L.; Peng, X.-J. A DFT study of Diels–Alder reactions of *o*-quinone methides and various substituted ethenes: selectivity and reaction mechanism. *J. Org. Chem.* 2005, 70, 4910–4917.

17. Lev, D. A.; Grotjahn, D. B.; Amouri, H. Reversal of reactivity in diene-complexed *o*-quinone methide complexes: insights and explanations from *ab initio* density functional theory calculations. *Organometallics* 2005, 24, 4232–4240.

18. Durbeej, B.; Eriksson, L. A. Formation of β-*O*-4 lignin models: a theoretical study. *Holzforschung* 2003, 57, 466–478.

19. Pezzella, A.; Panzella, L.; Crescenzi, O.; Napolitano, A.; Navaratman, S.; Edge, R.; Land, E. J.; Barone, V.; d'Ischia, M. Short-lived quinonoid species from 5,6-dihydroxyindole dimers en route to eumelanin polymers: integrated chemical, pulse radiolytic, and quantum mechanical investigation. *J. Am. Chem. Soc.* 2006, 128, 15490–15498.

20. Pezzella, A.; Crescenzi, O.; Natangelo, A.; Panzella, L.; Napolitano, A.; Navaratman, S.; Edge, R.; Land, E. J.; Barone, V.; d'Ischia, M. Chemical, pulse radiolysis and density functional studies of a new, labile 5,6-indolequinone and its semiquinone. *J. Org. Chem.* 2007, 72, 1595–1603.

21. Il'ichev, Y. V.; Simon, J. D. Building blocks of eumelanin: relative stability and excitation energies of tautomers of 5,6-dihydroxyindole and 5,6-indolequinone. *J. Phys. Chem. B* 2003, 107, 7162–7171.

22. Da Silva, G.; Chen, C.-C.; Bozzelli, J. W. Toluene combustion: reaction paths, thermochemical properties, and kinetic analysis for the methylphenyl radical + O_2 reaction. *J. Phys. Chem. A* 2007, 111, 8663–8676.

23. Da Silva, G.; Chen, C.-C.; Bozzelli, J. W. Quantum chemical study of the thermal decomposition of *o*-quinone methide (6-methylene-2,4-cyclohexadien-1-one). *J. Phys. Chem. A* 2007, 111, 7987–7994.

24. (a) Becke, A. D. A new mixing of Hartree–Fock and local density-functional theories. *J. Chem. Phys.* 1993, 98, 1372–1377. (b) Lee, C.; Yang, W.; Parr, R. G. Development of

the Colle–Salvetti correlation-energy formula into a functional of the electron density. *Phys. Rev. B* 1988, 37, 785–789.

25. (a) Barone, V.; Cossi, M. Quantum calculation of molecular energies and energy gradients in solution by a conductor solvent model. *J. Phys. Chem. A* 1998, 102, 1995–2001.(b) Klamt, A.; Schüürmann, G. COSMO: a new approach to dielectric screening in solvents with explicit expressions for the screening energy and its gradient. *J. Chem. Soc., Perkins Trans.* 1993, 799–805. (c) Klamt, A.; Jonas, V.; Burger, T.; Lohrenz, J. C. W. Refinement and parametrization of COSMO-RS. *J. Phys. Chem. A* 1998, 102, 5074–5085. (d) For a more comprehensive treatment of solvation models, see Cramer, C. J.; Truhlar, D. G. Implicit solvation models: equilibria, structure, spectra, and dynamics. *Chem. Rev.* 1999, 99, 2161–2200.

26. (a) Frisch, M. J.; Trucks, G. W.; Schlegel, H. B.; Scuseria, G. E.; Robb, M. A.; Cheeseman, J. R.; Zakrzewski, V. G.; Montgomery, J. A., Jr; Stratmann, R. E.; Burant, J. C.; Dapprich, S.; Millam, J. M.; Daniels, A. D.; Kudin, K. N.; Strain, M. C.; Farkas, O.; Tomasi, J.; Barone, V.; Cossi, M.; Cammi, R.; Mennucci, B.; Pomelli, C.; Adamo, C.; Clifford, S.; Ochterski, J.; Petersson, G. A.; Ayala, P. Y.; Cui, Q.; Morokuma, K.; Malick, D. K.; Rabuck, A. D.; Raghavachari, K.; Foresman, J. B.; Cioslowski, J.; Ortiz, J. V.; Baboul, A. G.; Stefanov, B. B.; Liu, G.; Liashenko, A.; Piskorz, P.; Komaromi, I.; Gomperts, R.; Martin, R. L.; Fox, D. J.; Keith, T.; Al-Laham, M. A.; Peng, C. Y.; Nanayakkara, A.; Gonzalez, C.; Challacombe, M.; Gill, P. M. W.; Johnson, B.; Chen, W.; Wong, M. W.; Andres, J. L.; Gonzalez, C.; Head-Gordon, M.; Replogle, E. S.; Pople, J. A.; *Gaussian 98, Revision A.7;* Gaussian, Inc: Pittsburgh, PA, 1998. (b) Frisch, M. J.; Trucks, G. W.; Schlegel, H. B.; Scuseria, G. E.; Robb, M. A.; Cheeseman, J. R.; Montgomery, J. A., Jr; Vreven, T.; Kudin, K. N.; Burant, J. C.; Millam, J. M.; Iyengar, S. S.; Tomasi, J.; Barone, V.; Mennucci, B.; Cossi, M.; Scalmani, G.; Rega, N.; Petersson, G. A.; Nakatsuji, H.; Hada, M.; Ehara, M.; Toyota, K.; Fukuda, R.; Hasegawa, J.; Ishida, M.; Nakajima, T.; Honda, Y.; Kitao, O.; Nakai, H.; Klene, M.; Li, X.; Knox, J. E.; Hratchian, H. P.; Cross, J. B.; Adamo, C.; Jaramillo, J.; Gomperts, R.; Stratmann, R. E.; Yazyev, O.; Austin, A. J.; Cammi, R.; Pomelli, C.; Ochterski, J. W.; Ayala, P. Y.; Morokuma, K.; Voth, G. A.; Salvador, P.; Dannenberg, J. J.; Zakrzewski, V. G.; Dapprich, S.; Daniels, A. D.; Strain, M. C.; Farkas, O.; Malick, D. K.; Rabuck, A. D.; Raghavachari, K.; Foresman, J. B.; Ortiz, J. V.; Cui, Q.; Baboul, A. G.; Clifford, S.; Cioslowski, J.; Stefanov, B. B.; Liu, G.; Liashenko, A.; Piskorz, P.; Komaromi, I.; Martin, R. L.; Fox, D. J.; Keith, T.; Al-Laham, M. A.; Peng, C. Y.; Nanayakkara, A.; Challacombe, M.; Gill, P. M. W.; Johnson, B.; Chen, W.; Wong, M. W.; Gonzalez, C.; Pople, J. A.; *Gaussian 03, Revision D.01;* Gaussian, Inc: Wallingford, CT, 2004.

27. Barone, V.; Cossi, M.; Tomasi, J. A new definition of cavities for the computation of solvation free energies by the polarizable continuum model. *J. Chem. Phys.* 1997, 107, 3210–3221.

28. Guo, H.; Sirois, S.; Proynov, E. I.; Salaub, D. R.In *Theoretical Treatment of Hydrogen Bonding;* Hadzy, D., Ed.; Wiley: New York. 1997.

29. Rablen, P. R.; Lockman, J. W.; Jorgensen, W. L. *Ab initio* study of hydrogen-bonded complexes of small organic molecules with water. *J. Phys. Chem. A* 1998, 102, 3782–3797.

30. (a) Del Bene, J. E.; Person, W. B.; Szczepaniak, K. Properties of hydrogen-bonded complexes obtained from the B3LYP functional with 6-31G(d,p) and 6-31 + G(d,p) basis sets: comparison with MP2/6-31 + G(d,p) results and experimental data. *J. Phys. Chem.*

1995, 99, 10705–10707. (b) Kim, K.; Jordan, K. D. Comparison of density functional and MP2 calculations on the water monomer and dimer. *J. Phys. Chem.* 1994, 98, 10089–10094.

31. (a) Arnaud, R.; Juvin, P.; Vallee, Y. Density functional theory study of the dimerization of the sulfine H_2CSO. *J. Org. Chem.* 1999, 64, 8880–8886. (b) Rutting, P. J.; Burgers, P. C.; Francis, J. T.; Terlouw, J. K. Sulfine, $CH_2{=}S{=}O$: determination of its heat of formation, basicity, and bond strengths by quantum chemistry. *J. Phys. Chem.* 1996, 100, 9694–9697.
32. Kendall, R. A.; Dunning, T. H., Jr; Harrison, R. J. Electron affinities of the first-row atoms revisited. Systematic basis sets and wave functions. *J. Chem. Phys.* 1992, 96, 6769–6806.
33. Rokita, S. E.; Yang, J.; Pande, P.; Greenberg, W. A. Quinone methide alkylation of deoxycytidine. *J. Org. Chem.* 1997, 62, 3010–3012.
34. Weinert, E. E.; Frankenfield, K. N.; Rokita, S. E. Time-dependent evolution of adducts formed between deoxynucleosides and a model quinone methide. *Chem. Res. Toxicol.* 2005, 18, 1364–1370.
35. Wang, H.; Wahi, M. S.; Rokita, S. E. Immortalizing a transient electrophile for DNA cross-linking. *Angew. Chem. Int. Ed.* 2008, 47, 1291–1293.
36. Prota, G. The chemistry of melanins and melanogenesis. *Fortschr. Chem. Org. Naturst.* 1995, 64, 93–148.
37. Adamo, C.; Barone, V. Toward reliable density functional methods without adjustable parameters: the PBE0 model. *J. Chem. Phys.* 1999, 110, 6158–6170.
38. (a) Casida, M. E.; Jamorski, C.; Casida, K. C.; Salahub, D. R. Molecular excitation energies to high-lying bound states from time-dependent density-functional response theory: characterization and correction of the time-dependent local density approximation ionization threshold. *J. Chem. Phys.* 1998, 108, 4439–4449. (b) Stratmann, R. E.; Scuseria, G. E.; Frisch, M. J. An efficient implementation of time-dependent density-functional theory for the calculation of excitation energies of large molecules. *J. Chem. Phys.* 1998, 109, 8218–8224. (c) Adamo, C.; Scuseria, G. E.; Barone, V. Accurate excitation energies from time-dependent density functional theory: assessing the PBE0 model. *J. Chem. Phys.* 1999, 111, 2889–2899.
39. Engelhard, M.; Luettke, W. The σ-homologs of *o*-benzoquinone. *Chem. Ber.* 1977, 110, 3759–3769.
40. (a) Cossi, M.; Scalmani, G.; Rega, N.; Barone, V. New developments in the polarizable continuum model for quantum mechanical and classical calculations on molecules in solution. *J. Chem. Phys.* 2002, 117, 43–54. (b) Scalmani, G.; Barone, V.; Kudin, K. N.; Pomelli, C. S.; Scuseria, G. E.; Frisch, M. J. Achieving linear-scaling computational cost for the polarizable continuum model of solvation. *Theor. Chem. Acc.* 2004, 111, 90–100.
41. Amouri, H.; Besace, Y.; Bras, J. L.; Vaissermann, J. General synthesis, first crystal structure, and reactivity of stable *o*-quinone methide complexes of Cp*Ir. *J. Am. Chem. Soc.* 1998, 120, 6171–6172.
42. Amouri, H.; Vaissermann, J.; Rager, M. N.; Grotjahn, D. B. Rhodium-stabilized *o*-quinone methides: synthesis, structure, and comparative study with their iridium congeners. *Organometallics* 2000, 19, 5143–5148.
43. Hay, P. J.; Wadt, W. R. *Ab initio* effective core potentials for molecular calculations. Potentials for the transition metal atoms Sc to Hg. *J. Chem. Phys.* 1985, 82, 270–283.

44. Dunning, T. H., Jr; Hay, P. J. In *Modern Theoretical Chemistry;* Schaefer, H. F., III, Ed.; Plenum: New York, 1976; Vol. 3, p 1.
45. Schaefer, A.; Huber, C.; Ahlrichs, R. Fully optimized contracted Gaussian basis sets of triple zeta valence quality for atoms Li to Kr. *J. Chem. Phys.* 1994, 100, 5829–5835.
46. Amouri, H.; Le Bras, J. Taming reactive phenol tautomers and *o*-quinone methides with transition metals: a structure–reactivity relationship. *Acc. Chem. Res.* 2002, 35, 501–510.
47. Di Antonio, M.; Doria, F.; Mella, M.; Merli, D.; Profumo, A.; Freccero, M. Novel naphthalene diimides as activatable precursors of bisalkylating agents, by reduction and base catalysis. *J. Org. Chem.* 2007, 72, 8354–8360.
48. (a) Curtiss, L. A.; Raghavachari, K.; Redfern, P. C.; Rassolov, V.; Pople, J. A. Gaussian-3 (G3) theory for molecules containing first and second-row atoms. *J. Chem. Phys.* 1998, 109, 7764–7776. (b) Curtiss, L. A.; Raghavachari, K.; Redfern, P. C.; Pople, J. A. Assessment of Gaussian-3 and density functional theories for a larger experimental test set. *J. Chem. Phys.* 2000, 112, 7374–7383.
49. Baboul, A. G.; Curtiss, L. A.; Redfern, P. C.; Raghavachari, K. Gaussian-3 theory using density functional geometries and zero-point energies. *J. Chem. Phys.* 1999, 110, 7650–7657.
50. Curtiss, L. A.; Raghavachari, K.; Pople, J. A. Gaussian-2 theory: use of higher level correlation methods, quadratic configuration interaction geometries, and second-order Møller–Plesset zero-point energies. *J. Chem. Phys.* 1995, 103, 4192–4120.
51. (a) Montoya, A.; Truong, T. N.; Sarofim, A. F. Spin contamination in Hartree–Fock and density functional theory wavefunctions in modeling of adsorption on graphite. *J. Phys. Chem. A* 2000, 104, 6108–6110. (b) Cioslowski, J.; Liu, G.; Martinov, M.; Piskorz, P.; Moncrief, D. Energetics and site specificity of the homolytic C–H bond cleavage in benzenoid hydrocarbons: an *ab initio* electronic structure study. *J. Am Chem. Soc.* 1996, 118, 5261–5254.
52. (a) Bally, T.; Borden, W. T. In *Reviews in Computational Chemistry;* Lipkowitz, K. B.; Boyd, D. B.,Eds.; VCH-Wiley: New York, 1999; Vol. 13, p 1. (b) Wong, M. W.; Radom, L. Radical addition to alkenes: an assessment of theoretical procedures. *J. Phys. Chem.* 1995, 99, 8582–8588.
53. Lynch, B. J.; Fast, P. L.; Harris, M.; Truhlar, D. G. Adiabatic connection for kinetics. *J. Phys. Chem. A* 2000, 104, 4811–4815.
54. Skokov, S.; Kazakov, A.; Dryer, F. L. A theoretical study of oxidation of phenoxy and benzyl radicals by HO_2. *Fourth Joint Meeting of the U.S. Sections of the Combustion Institute, Philadelphia, PA, March 20–23*, 2005.
55. Montgomery, J. A., Jr; Frisch, M. J.; Ochterski, J. W.; Petersson, G. A. A complete basis set model chemistry. VII. Use of the minimum population localization method. *J. Chem. Phys.* 2000, 112, 6532–6542.
56. Pople, J. A.; Head-Gordon, M.; Raghavachari, K. Quadratic configuration interaction. A general technique for determining electron correlation energies. *J. Chem. Phys.* 1987, 87, 5968–5975.
57. Zhao, Y.; Lynch, B. J.; Truhlar, D. G. Development and assessment of a new hybrid density functional model for thermochemical kinetics. *J. Phys. Chem. A* 2004, 14, 2715–2719.

58. Gaspar, P. P.; Hsu, J.-P.; Chari, S. The phenylcarbene rearrangement revisited. *Tetrahedron* 1985, 41, 1479–1507.

59. Dorrestijn, E.; Pugin, R.; Ciriano Nogales, M. V.; Mulder, P. Thermal decomposition of chroman. Reactivity of *o*-quinone methide. *J. Org. Chem.* 1997, 62, 4804–4810.

60. Freudenberg, K. Lignin. Its constitution and formation from *p*-hydroxycinnamyl alcohols. *Science* 1965, 148, 595–600.

3

QUINONE METHIDE STABILIZATION BY METAL COMPLEXATION

ELENA POVERENOV AND DAVID MILSTEIN

Department of Organic Chemistry, The Weizmann Institute of Science, Rehovot 76100, Israel

3.1 INTRODUCTION

Coordination of reactive and/or unstable molecules to metal centers is a useful approach for their stabilization,[1] and it presents unique opportunities for their characterization by spectroscopic methods and for elucidation of their structure. Moreover, under appropriate conditions the coordinated species can be chemically modified. In addition, displacement of the coordinated compound from the metal and its trapping in solution by reactions with suitable substrates can form the basis for useful synthetic methodology.

Quinone methides (QMs), especially the simple ones (those not having substituents at the exocyclic methylene group), are very unstable compounds. Their isolation is very difficult and normally requires very dilute solutions and low temperatures.[2] Due to the aromatic zwitterionic form (Scheme 3.1), quinone methides react very rapidly with both electrophiles and nucleophiles, with the medium, or in self-condensation reactions.

Since quinone methides can be viewed as electron-deficient conjugated olefins, their coordination to electron-rich, low-valent metal centers is expected to be very favorable, resulting in stabilization of high energy d orbitals of the transition metal.[3] Therefore, such metal–olefin complexation might be strong enough to allow stabilization of QM species even at the expense of gaining aromaticity. Generation of the complexed quinone methide moiety by transformation of another ligand in the coordination sphere of the metal (rather than trapping of a highly unstable free QM) is the best approach for the formation of QM complexes.

Quinone Methides, Edited by Steven E. Rokita

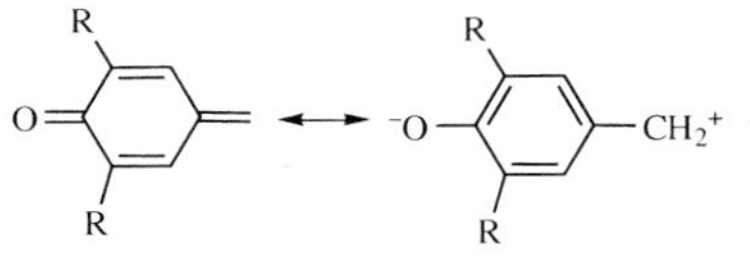

SCHEME 3.1

3.2 QM-BASED PINCER COMPLEXES

3.2.1 Formation

Complexes of quinone methides with a nonsubstituted exocyclic methylene carbon were obtained by Milstein and coworkers. In the course of studies of the reactivity of aromatic pincer-type PCP complexes,[4] a dearomatization process, which led to formation of the *para*-quinone methide complex **1**, was observed (Scheme 3.2).[5,6]

In this process, a phenolic pincer Rh(III) methyl chloride complex, obtained by C—C activation[7] of the corresponding pincer ligand, was converted upon heating into a Rh(I) QM complex **1**. This unusual reactivity indicates that when a late transition metal in high oxidation state is coordinated to a *p*-oxybenzyl group, intramolecular charge transfer might take place, resulting in a two-electron reduction of the metal center at the expense of aromaticity of the corresponding ligand. The observation of this reaction was followed by intensive investigation of the quinonoid chemistry of transition metal complexes, including exploration of the driving forces behind the dearomatization processes. Some of these studies are illustrated below.

3.2.2 Reactivity and Modifications

Unlike free quinone methides that react rapidly even with relatively weak nucleophiles such as water or alcohols to give 1,6-addition products, the complexed *p*-QM moiety shows remarkable stability toward nucleophilic attack even upon moderate heating, demonstrating complete blockage of the methylene group reactive site. In addition, the whole complex is very stable. In contrast to other Rh(I)-olefin complexes that are normally very reactive in oxidative addition of compounds such as iodomethane, dihydrogen, or hydrogen chloride,[8] the QM Rh(I) complex did not react with these compounds even under forcing conditions.[5]

Moreover, the stability of the QM pincer complexes allows selective modifications of both the metal center and the carbonyl part of the quinone methide moiety, still with

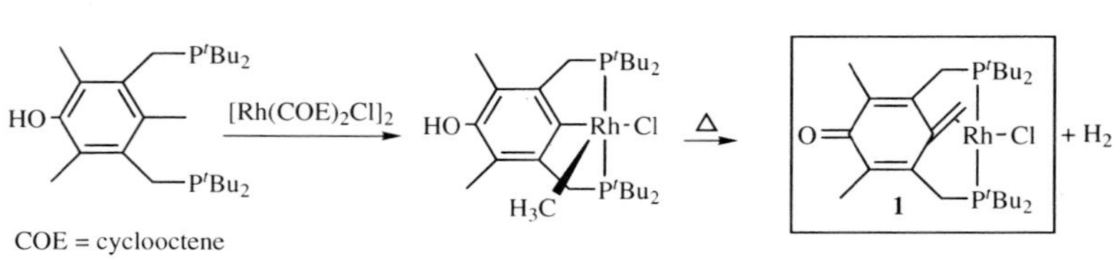

COE = cyclooctene

SCHEME 3.2

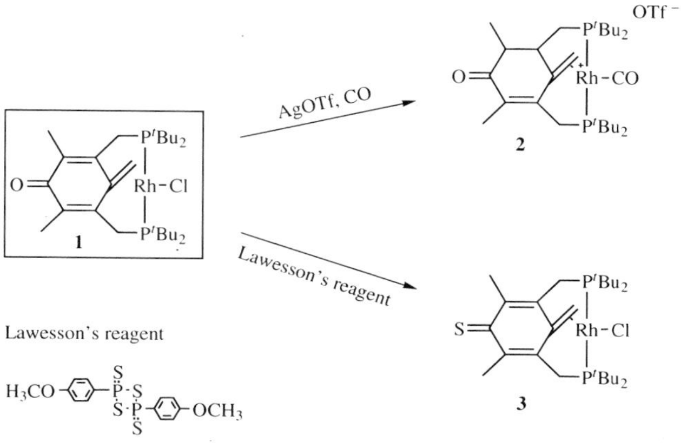

SCHEME 3.3

no aromatization taking place (Scheme 3.3).[5] For instance, abstraction of the chloride ligand with AgOTf followed by addition of CO proceeded smoothly, in analogy to the reactivity of simple aromatic PCP-based Rh complexes, forming complex **2**. Interestingly, conversion of the QM carbonyl group to a thiocarbonyl group was easily accomplished using Lawesson's reagent, forming the thioquinone methide complex **3**.[6] It is noteworthy that organic thioquinone methides are very rare. They are extremely reactive and even less stable than quinone methides.[9]

Interestingly, reaction of the QM complex **1** with strong electrophiles, such as HOTf or Me_3SiOTf, did not lead to the expected rearomatization. Rather, the first stable methylene arenium complexes **4** were formed (Scheme 3.4).[10]

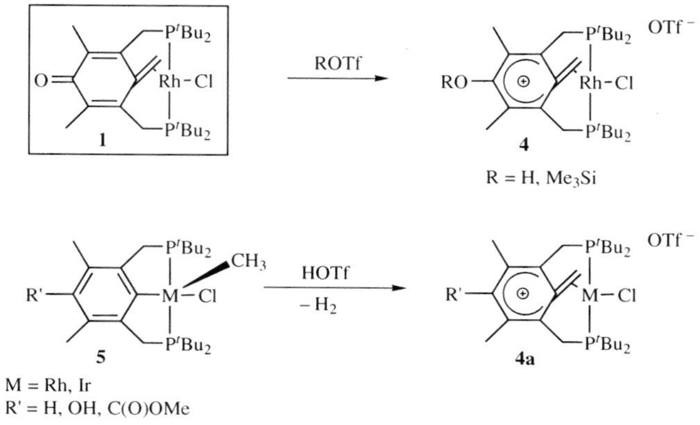

SCHEME 3.4

Following this observation, a general approach for the synthesis of pincer-type methylene arenium compounds was developed (Scheme 3.4). Upon reaction of the methyl rhodium (or iridium) complexes **5** with a slight excess of triflic acid, dihydrogen (not methane!) was evolved to form the methylene arenium complexes **4a**.[11] Thus, the methylene arenium form is clearly preferred over the benzylic M(III) form, in which the positive charge is localized at the metal center.*

An additional notable mode of the reactivity of these quinone methide complexes is formation of metal stabilized *p*- and *o*-xylylenes.[10]

Upon reaction of complex **1** with two equivalents of MeLi, the first equivalent attacked the metal center and substituted the chloride ligand, while the second equivalent attacked the QM carbonyl group. Competing 1,2- and 1,4-elimination of LiOH resulted in formation of the *p*- and *o*-xylylenes **6a** and **6b**, respectively (Scheme 3.5). Free xylylenes are highly reactive species that undergo spontaneous polymerization even at very low temperatures.[12] It has been calculated that the energy difference between the ground state (dimethylene cyclohexadiene) and the transition state (biradical of dimethylene benzene) is less than 6 kcal/mol, making the isolation of these compounds under normal conditions impossible.[13] In both the *p*-xylylene complex **6a** and the *o*-xylylene complex **6b**, the metal center is coordinated in an η^2 fashion to only *one* of the exocyclic double bonds, while the phosphine chelation effect and the distortion of complex geometry from planarity seem to be the most important stabilization factors of these unique complexes.[10]

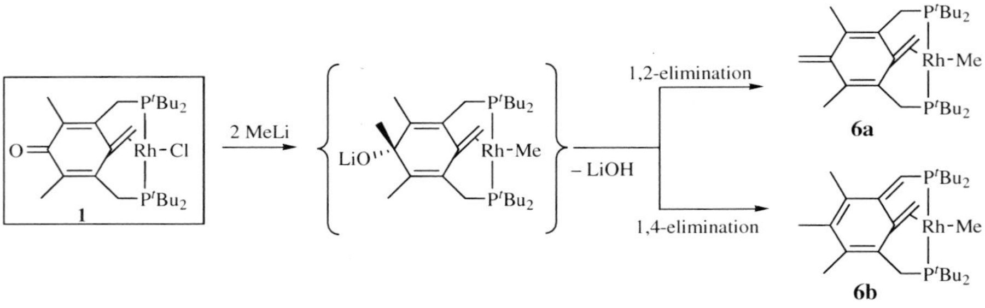

SCHEME 3.5

3.2.3 Os-Based, *p*-QM Complexes

A similar approach to the one described above was utilized for the formation of quinone methide derivatives of osmium.[14] Reaction of $OsCl_2(PPh_3)_3$ with a phenolic diphosphine ligand in the presence of Et_3N resulted in phosphine exchange followed by C–H activation and deprotonation by the base to form the two isomeric QM

* In addition to the methylene arenium case, in which a coordinatively unsaturated positively charged metal center is stabilized by transfer of positive charge to the aromatic ring, stabilization can be accomplished by η^2-C–H or η^2-C–C agostic interactions with the aromatic system (see Ref. [5]).

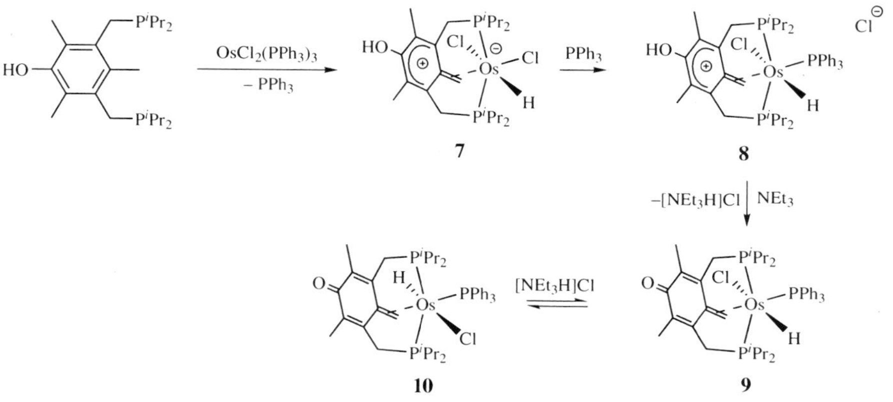

SCHEME 3.6

complexes **9** and **10** (Scheme 3.6). In agreement with chemistry of the corresponding rhodium QM complexes, compounds **9** and **10** exhibit good stability, even in an alcoholic medium. Interestingly, in this work the methylene arenium intermediates **7** and **8** were isolated, shedding light on the mechanism of the unusual dearomatization process that eventually leads to the quinone methide complexes.

3.3 ONE-SITE COORDINATED QM COMPLEXES

3.3.1 η^2-*ortho*-QM Complexes

3.3.1.1. Formation The first example of an isolated QM complex was reported by Harman and coworkers.[15]

o-QM complexes were formed by condensation of aldehydes with an Os(II) complex of an η^2-coordinated phenol ligand. Water elimination from the initially formed aldol intermediate led to a series of substituted η^2-*o*-QM complexes **11** with a variety of R substituents.[16] A similar approach was used by the same group to form a W-based η^2-*o*-QM complex **11a**[17] (Scheme 3.7).

3.3.1.2. Release and Reactivity of η^2-o-QMs Although the η^2-*o*-QM Os complexes **11** are stable when exposed to air or dissolved in water, the quinone methide moiety can be released upon oxidation (Scheme 3.8).[16] For example, reaction of the Os-based *o*-QM **12** with 1.5 equivalents of CAN (ceric ammonium nitrate) in the presence of an excess of 3,4-dihydropyran led to elimination of free *o*-QM and its immediate trapping as the Diels–Alder product tetrahydropyranochromene, **14**. Notably, in the absence of the oxidizing agent, complex **12** is completely unreactive with both electron-rich (dihydropyran) and electron-deficient (*N*-methylmaleimide) dienes.

Oxidation of the *o*-QM complex **13** (formed by treating the phenol complex with (*R*)-citronellal and pyridine) with CAN resulted in an intramolecular Diels–Alder reaction to form the benzo[*c*]chromene **15** (Scheme 3.8).

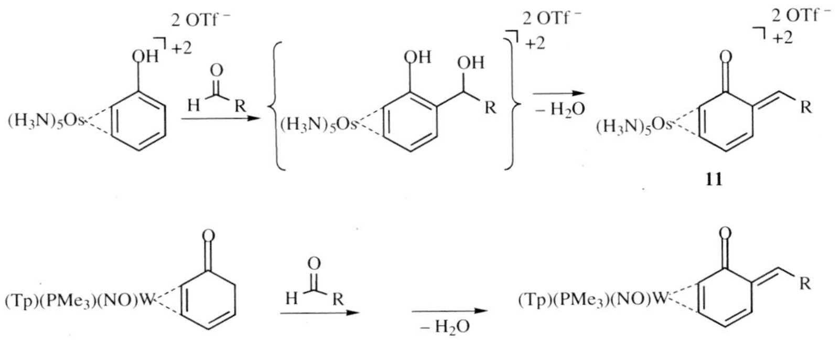

Tp = hydridotris(pyrazolyl)borate

SCHEME 3.7

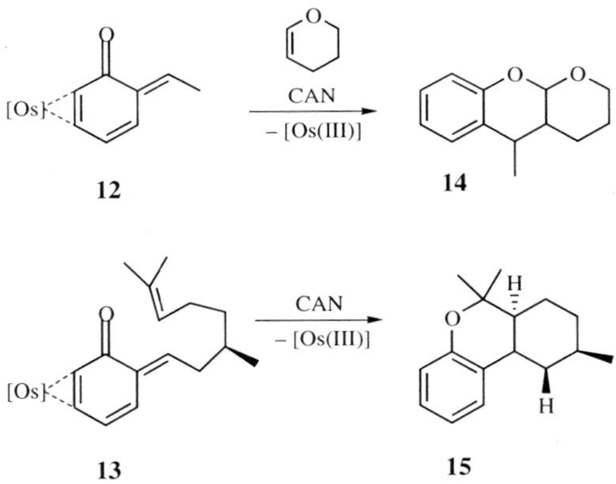

[Os] = $[Os(NH_3)_5](OTf)_2$
CAN = ceric ammonium nitrate

SCHEME 3.8

3.3.2 η^2-*p*-QM Complexes

3.3.2.1. Formation An approach to form QM complexes in which the QM ligand is coordinated only via the exocyclic C=C bond and can be released from the metal was developed by Milstein.[18,19] The synthetic strategy is based on the idea that a zwitterionic η^1-methylene-*p*-phenoxy metal complex may undergo charge transfer to the metal, resulting in an η^2-methylene *p*-QM complex of the reduced metal center (Scheme 3.9).

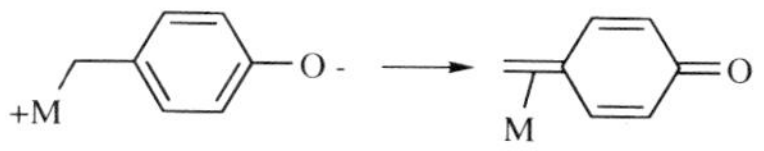

SCHEME 3.9

Oxidative addition of a silyl-protected 4-(bromomethyl)phenol precursor to (tmeda)Pd(II)Me_2 (tmeda = tetramethylethylenediamine), followed by ethane reductive elimination, resulted in formation of the benzylic complex **16** (Scheme 3.10). Exchange of tmeda for a diphosphine ligand (which is better suited for stabilizing the ultimate Pd(0) QM complex), followed by removal of the protecting silyl group with fluoride anion, resulted in the expected *p*-QM Pd(0) complex, **17**, via intermediacy of the zwitterionic Pd(II) benzyl complex. In this way a stable complex of *p*-BHT-QM, **17b**, the very important metabolite of the widely used food antioxidant BHT[20] (BHT = butylated hydroxytoluene) was prepared. Similarly, a Pd(0) complex of the elusive, simplest *p*-QM, **17a**, was obtained (Scheme 3.10).

The strong backbonding from the chelated diphosphine Pd(0) metal center to the electron-poor exocylic C=C bond of the QM moieties results in remarkable stability of the complex, with the QM ligand remaining unaffected in water or alcohol, even upon heating.

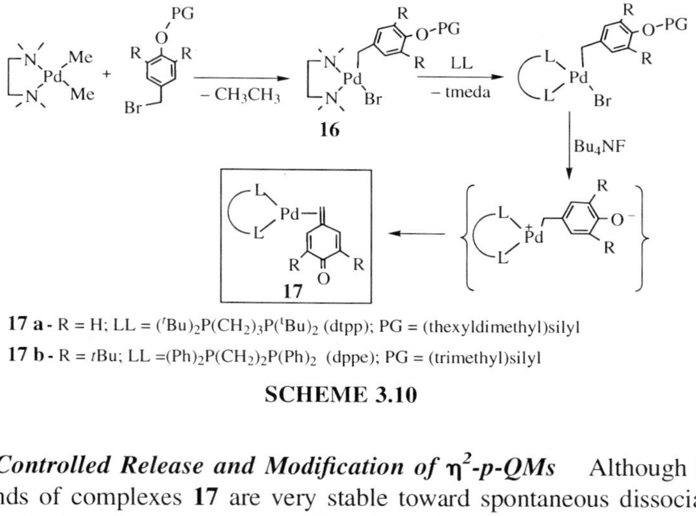

17 a - R = H; LL = $(^tBu)_2P(CH_2)_3P(^tBu)_2$ (dtpp); PG = (thexyldimethyl)silyl

17 b - R = *t*Bu; LL =$(Ph)_2P(CH_2)_2P(Ph)_2$ (dppe); PG = (trimethyl)silyl

SCHEME 3.10

3.3.2.2. Controlled Release and Modification of η^2-p-QMs Although the η^2-*p*-QM ligands of complexes **17** are very stable toward spontaneous dissociation, the quinone methide moiety can be released by substitution with electron-deficient alkenes (e.g., dibenzylideneacetone, DBA) or with diphenylacetylene (Scheme 3.11). The unstable free QM was trapped in the solution. With methanol, the expected 2,6-di-*tert*-butyl-4-methoxyphenol, the 1,6-Michael-type reaction product, was formed.[18,19]

Interestingly, the zwitterionic palladium intermediate postulated in Section 3.3.2.1 was trapped using a platinum system (Scheme 3.12). When the chelating imino-pyridine-based Pt(IV) complex **18** bearing a silyl-protected oxy-benzyl group was

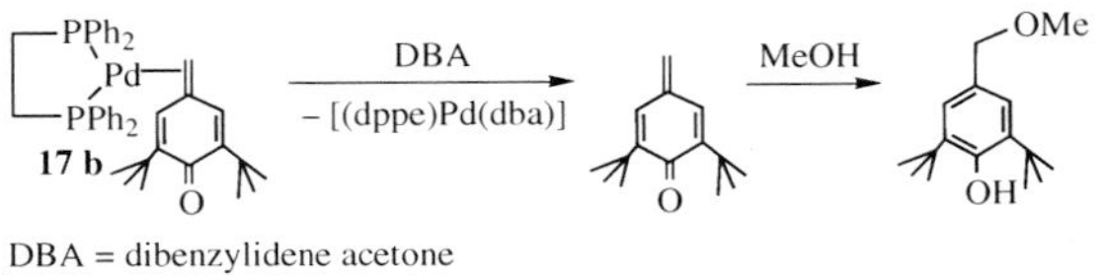

DBA = dibenzylidene acetone

SCHEME 3.11

reacted with fluoride anion in a polar solvent (acetone), the η^1-methylene-*p*-phenoxy-Pt(IV) intermediate **19** was obtained and characterized at −30 °C. Upon warming of this compound to room temperature, free BHT-QM was released into the solution. Apparently, the unobserved QM complex is unstable, as a result of the Pt(II) metal center not being sufficiently electron rich to stabilize the coordinated electron-deficient quinonoid moiety, in contrast to the lower valent Pd(0) case. In nonpolar solvents, which cannot stabilize the zwitterionic intermediate (for instance, benzene), reaction with fluoride led to immediate rearrangement and release of BHT-QM, even at −30 °C. In methanol, the 1,6-Michael-type adduct was trapped (Scheme 3.12).[21]

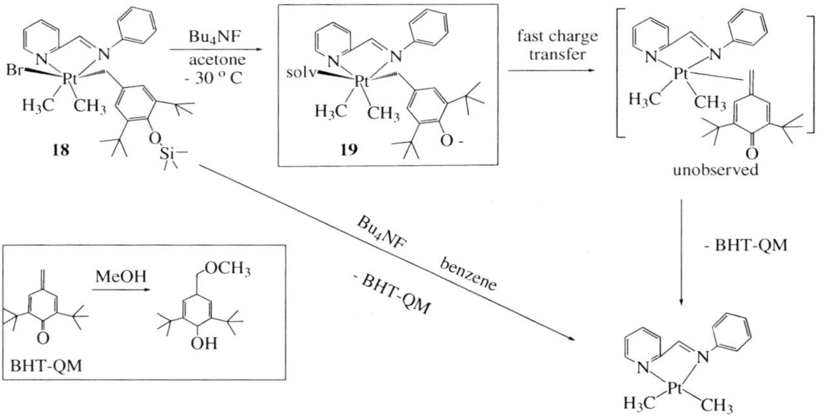

SCHEME 3.12

As observed with the pincer-type quinone methide complexes (Section 3.2), the one-site coordinated QMs can also undergo chemical transformation to other quinonoid compounds. For instance, reaction of complex **20** with MeOTf resulted in the methylene arenium complex **21** (Scheme 3.13).[22]

Experimental observations,[23] supported by high-level *ab initio* calculations, [24] indicate that two extreme resonance forms contribute to the general energy of the benzyl cation: the aromatic form **A,** in which the positive charge is concentrated at the methylene group, and the *nonaromatic*, "*methylene arenium*" form **B** with a sp^2 *ipso*-carbon atom and ring-localized charge (Scheme 3.13). Unlike benzyl cations of the form **A**, which were isolated and studied, especially by Olah and coworkers,[23] compounds represented by the form **B** remained elusive. Thus, metal complexation

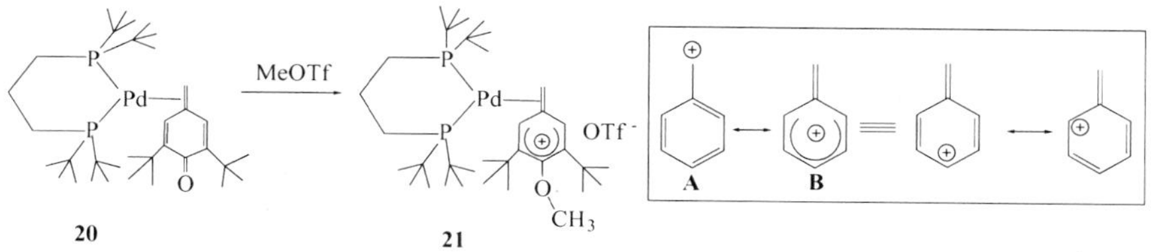

SCHEME 3.13

(via the QM complex) allowed for the first time isolation, structural characterization, reactivity investigation, and controlled release[22] of a compound with the methylene arenium form of the benzyl cation.[10,22]

3.4 η^4-COORDINATED QM COMPLEXES

3.4.1 Formation of η^4-Coordinated QM Complexes

An alternative route for stabilization of quinone methides by metal coordination involves deprotonation of a η^5-coordinated oxo-dienyl ligand. This approach was introduced by Amouri and coworkers, who showed that treatment of the [Cp*Ir(oxo-η^5-dienyl)]$^+$ BF_4^- **22** with a base (*t*-BuOK was the most effective) resulted in formation of stable Cp*Ir(η^4-*o*-QM) complexes **23** (Scheme 3.14).[25] Using the same approach, a series of η^4-*o*-QM complexes of rhodium was prepared (Scheme 3.14).[26] Structural data of these complexes and a comparison of their reactivity indicated that the *o*-QM ligand is more stabilized by iridium than by rhodium.

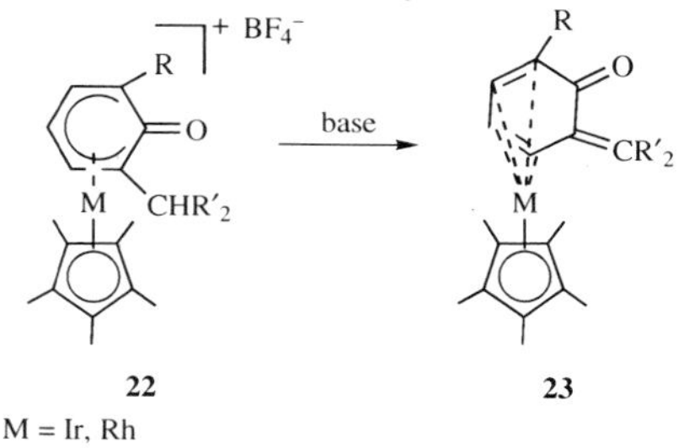

SCHEME 3.14

Although deprotonation at the benzylic position of arenes coordinated to ruthenium and chromium was reported,[27] in the case of the coordinated oxo-η^5-dienyl unit, nucleophilic attack at one terminus of the complexed η^5-dienyl ligand, rather than deprotonation, was expected.[28] The reason for the successful deprotonation (even at relatively hindered isopropyl sites) is, according to the authors, the cationic nature of the Cp*M fragment. In addition, the transition state for the deprotonation might involve stabilization by the metal (Scheme 3.15).

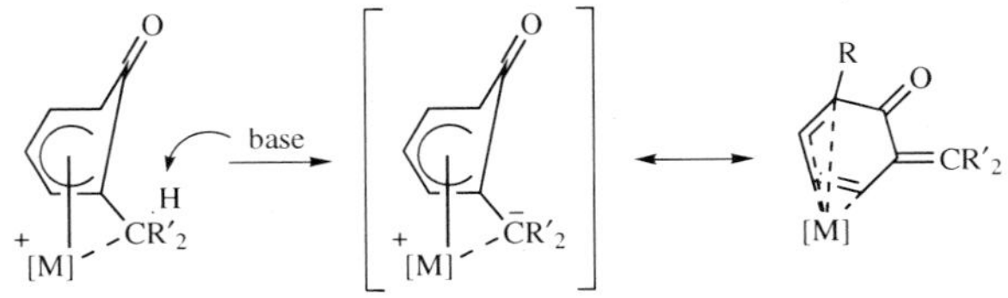

[M] = Cp*Ir, Cp*Rh

SCHEME 3.15

3.4.2 Reactivity of η^4-Coordinated QM Complexes

As mentioned earlier, metal complexation not only allows isolation of the QM derivatives but can also dramatically modify their reactivity patterns.[29] *o*-QMs are important intermediates in numerous synthetic[30] and biological[31,32] processes, in which the exocyclic carbon exhibits an *electrophilic* character.[30–33] In contrast, a metal-stabilized *o*-QM can react as a base or nucleophile (Scheme 3.16).[29] For instance, protonation of the Ir-η^4-QM complex **24** by one equivalent of HBF_4 gave the initial oxo-dienyl complex **25**, while in the presence of an excess of acid the dicationic complex **26** was obtained. Reaction of **24** with I_2 led to the formation of new oxo-dienyl complex **27**, instead of the expected oxidation of the complex and elimination of the free *o*-QM. Such reactivity of the exocyclic methylene group can be compared with the reactivity of electron-rich enol acetates or enol silyl ethers, which undergo electrophilic iodination.[34]

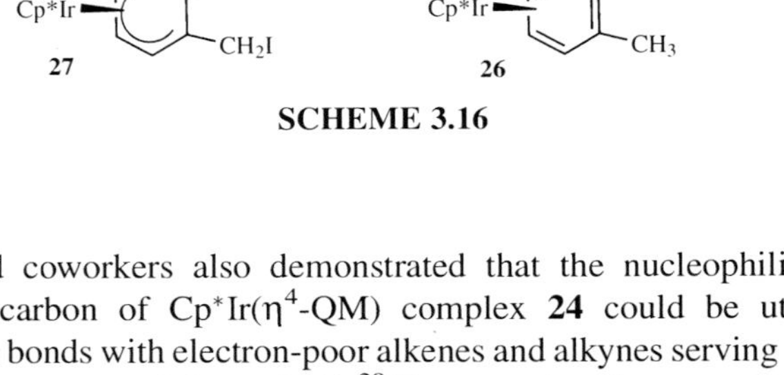

SCHEME 3.16

Amouri and coworkers also demonstrated that the nucleophilic reactivity of the exocyclic carbon of Cp*Ir(η^4-QM) complex **24** could be utilized to form carbon–carbon bonds with electron-poor alkenes and alkynes serving as electrophiles or cycloaddition partners (Scheme 3.17).[29] For example, when complex **24** was treated with the electron-poor methyl propynoate, a new *o*-quinone methide complex **28** was formed. The authors suggest that the reaction could be initiated by nucleophilic attack of the terminal carbon of the exocyclic methylene group on the terminal carbon of the alkyne, generating a zwitterionic oxo-dienyl intermediate, followed by proton transfer

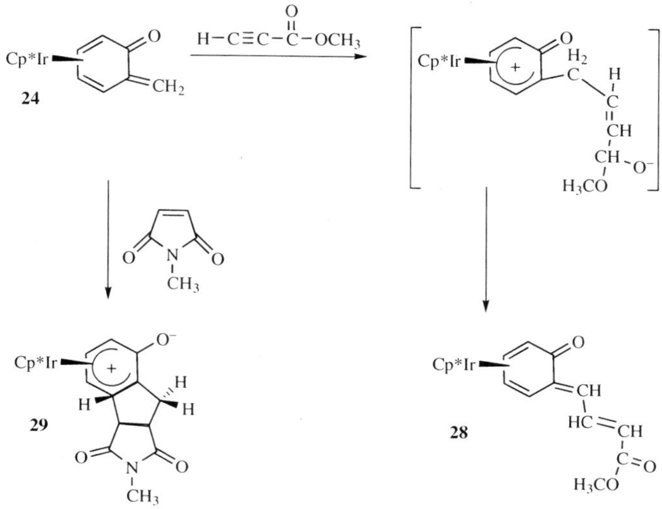

Cp*Ir
24
O
CH$_2$
H−C≡C−C(=O)−OCH$_3$
Cp*Ir
+
H$_2$
C
H
C
CH
CH
O$^-$
H$_3$CO
O
N
CH$_3$
Cp*Ir
+
O$^-$
H
H
H
29
O
N
O
CH$_3$
Cp*Ir
O
28
CH
HC
CH
C=O
H$_3$CO

SCHEME 3.17

from the side chain of the oxo-dienyl cation to the enolate to give the observed product (Scheme 3.17). Many nucleophilic additions to electrophilic alkynes take place in a similar manner.[35]

The coordinated quinone methide π-system of complex **24** can also undergo cycloaddition (Scheme 3.17). When **24** was reacted with *N*-methylmaleimide, a [3 + 2] cycloaddition took place to give the tricyclic iridium complex **29**. The closest example to this unprecedented reactivity pattern is a formal [3 + 2] cycloaddition of *p*-quinone methides with alkenes catalyzed by Lewis acids, although in that reaction the QMs serve as electron-poor reagents.[36]

These reactions clearly indicate that the exocyclic carbon of the complexed QM in these systems is nucleophilic in character, in contrast to its electrophilic nature in free *o*-quinone methides. The Cp*Ir metal center stabilizes the mesomeric form in which the exocyclic carbon experiences high electron density (Scheme 3.18).[29]

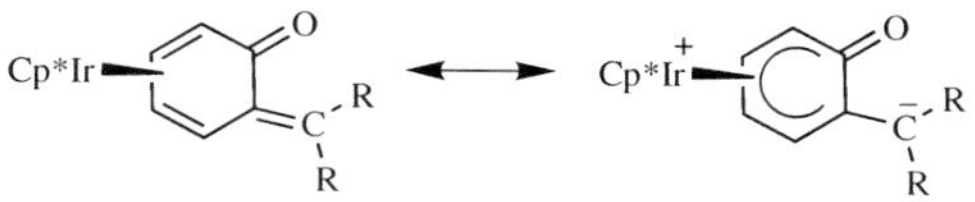

Cp*Ir
O
C
R
R
Cp*Ir$^+$
O
C$^-$
R
R

SCHEME 3.18

3.4.3 η^4-Coordinated QM Complexes of Mn

In a similar approach, double deprotonation of η^6-coordinated *ortho*- and *para*-cresols, **30** and **31,** with *t*-BuOK led to formation of stable η^4-coordinated *p*- and *o*-quinone methide complexes of manganese, **32** and **33** (Scheme 3.19).[37]

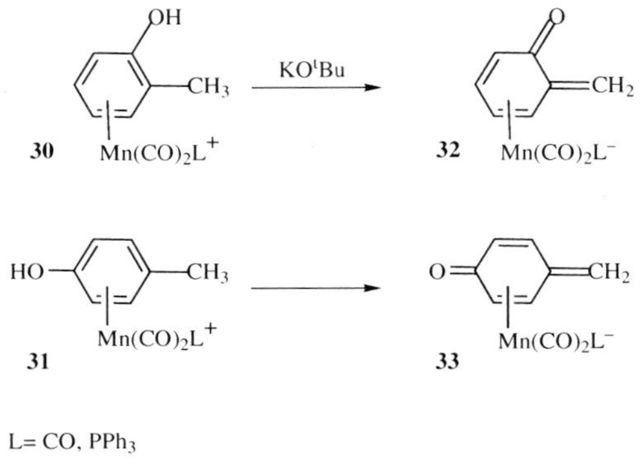

L= CO, PPh_3

SCHEME 3.19

3.5 CHARACTERIZATION OF QM COMPLEXES

Quinone methide complexes were characterized by standard spectroscopic techniques. It is possible to elucidate the structural and electronic properties of the QM complexes and also to estimate the reactivity tendencies of the coordinated QM moiety based on IR, NMR, and X-ray diffraction spectroscopes.

3.5.1 IR

The IR spectrum of the reported metal-coordinated QM compounds depends on the nature of the metal fragment. For example, pincer-type *p*-QMs exhibit a carbonyl stretch at about 1595 cm^{-1} for Rh complexes[6] and 1629 cm^{-1} for Os complexes.[14] η^2-Coordinated *p*-QMs of Pd give rise to signals at 1587[19] and 1598[18] cm^{-1}, while the C=O stretches of η^4-bent *o*-QMs coordinated to Ir and Rh appear at 1631–1643[25] cm^{-1}.

3.5.2 ^1H and $^{13}C\{^1H\}$ NMR

Obviously, the NMR pattern of the QM moiety depends on its coordination mode. As expected, in case of ring coordination, such as η^4-coordination[25,26] or η^2-coordination,[16,17] the ring signals are affected, while in the case of exocyclic bond coordination, the signals of the exocyclic bond are influenced.[6,14,18,19]

For all QM complexes, the ring signals in both ^1H and $^{13}C\{^1H\}$ NMR spectra indicate the lack of aromaticity. The carbonyl group appears in the $^{13}C\{^1H\}$ NMR spectrum in the range of 184[18]–201[16] ppm, which is the region observed for 2,5-cyclohexadienones and quinones.[38]

In the case of η^4-coordinated QMs, all ring atoms (carbons and protons) are strongly affected. The carbons of the exocyclic double bond resonate in the range of 103–128 ppm (exocyclic carbon $=CR_2$) and 130–142 ppm (ring carbons).[25]

In cases of complexes bearing an exocyclic double bond directly coordinated to the metal center, the carbons of the double bond usually exhibit coupling with NMR-active metal centers and/or auxiliary ligands.[6,14,18,19] The chemical shifts of the quaternary carbon atom vary from 66.97[6] to 82.28[18] ppm, while the methylene group gives rise to signals at 29.16,[14] 41.91,[6] or 51.34[18] ppm in the $^{13}C\{^{1}H\}$ NMR spectra. As one can see, the chemical shift variation is relatively broad and significantly affected by the nature of the metal center.

3.5.3 X-Ray

The quinonoid (as compared to aromatic) character of the ligand is strongly reflected in the inequality of the ring bond distances, two of the bonds being substantially shorter than the other C—C bonds of the ring (see Figs. 3.1–3.3). The carbonyl C—O distance is usually in the range of reported carbonyl double bonds (1.23–1.25 Å).[6,14,18,25]

In the case of η^2-coordination of the exocyclic C=C bond, it becomes substantially elongated compared with the double bond of free alkenes, as a result of back donation from the metal to the π^* orbitals of the double bond. For instance, in complex **17b** the coordinated bond length is 1.437 Å (see Fig. 3.2).[18] This is also reflected in the loss of planarity around the quaternary exocyclic carbon, the methylenic carbon being bent out of the ring plane by 10.78°.[18] Similar structural features were also observed with other P_2Pd conjugated olefin complexes.[39]

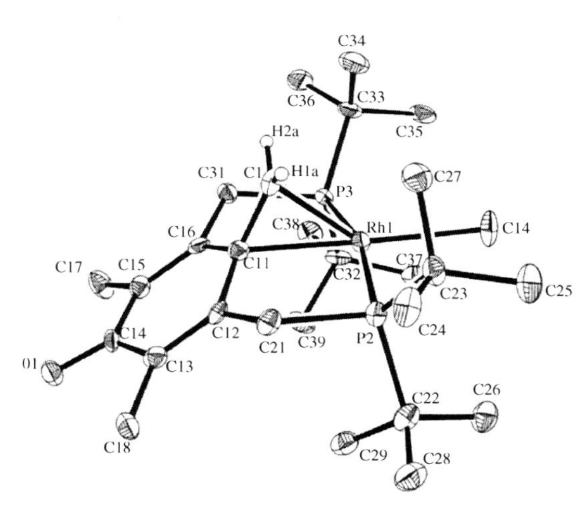

FIGURE 3.1 ORTEP view of complex **1** at 50% probability level. The hydrogen atoms (except H(1a) and H(2a)) are omitted for clarity. Selective bond distances (Å) and angles (deg): Rh(1)–C(1), 2.052(6); Rh(1)–C(11), 2.183(5); C(1)–C(11), 1.441(8); O(1)–C(14), 1.239(6); C(12)–C(13), 1.349(8); C(13)–C(14), 1.477(8); C(16)–C(11)–C(12), 118.7(5). Reproduced by permission from the American Chemical Society.

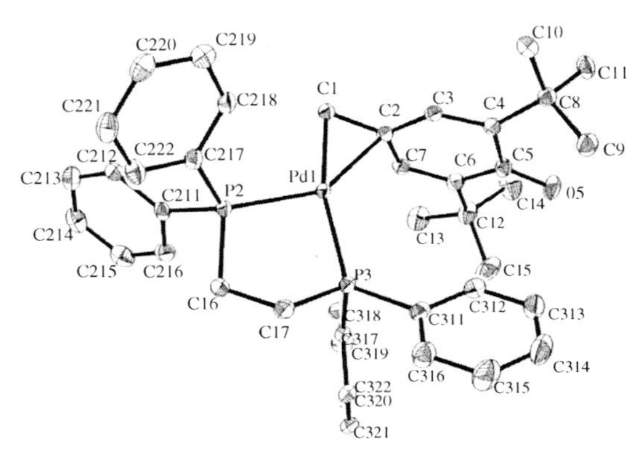

FIGURE 3.2 ORTEP view of complex **17b** at 50% probability level. Selected bond distances (Å) and angles (deg): Pd(1)–C(1), 2.088(2); Pd(1)–C(2), 2.208(2); C(1)–C(2), 1.437(2); C(5)–O(5), 1.251(2); C(2)–C(3), 1.443(3); C(3)–C(4) 1.362(2); C(4)–C(5) 1.474(3); C(5)–C(6) 1.474(3); C(6)–C(7) 1.364(3); C(2)–C(7) 1.441(2); P(2)–Pd(1)–P(3) 86.73(2); Pd(1)–C(1)–C(2) 75.06(10); Pd(1)–C(2)–C(1) 65.97(10). Reproduced by permission from the American Chemical Society

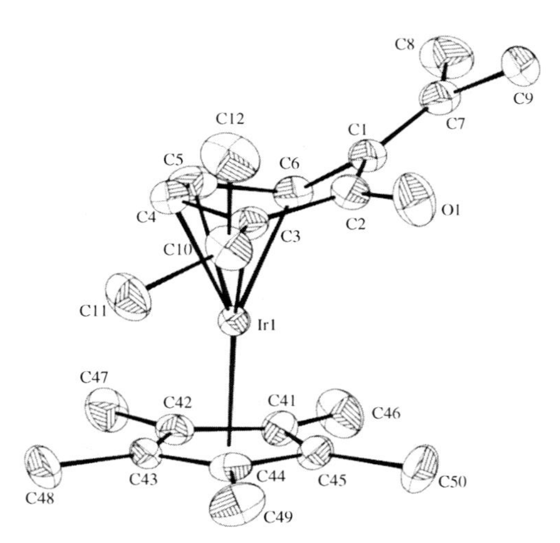

FIGURE 3.3 ORTEP view of complex **23** at 30% probability; one of the two independent molecules is being depicted. The hydrogen atoms are omitted for clarity. Selective bond distances (Å) and angles (deg): C(1)–C(2), 1.49(1); C(2)–C(3), 1.48(1); C(3)–C(4), 1.44(1); C(4)–C(5), 1.43(1); C(5)–C(6), 1.42(1); C(1)–C(6), 1.46(1); C(1)–C(7), 1.34(1); C(2)–O(1), 1.23(1); Ir(1)–C(3), 2.193(8); Ir(1)–C(4), 2.143(9); Ir(1)–C(5), 2.122(9); Ir(1)–C(6), 2.169(8); Ir(1)–C(41), 2.179(8); Ir(1)–C(42), 2.220(7); Ir(1)–C(43), 2.246(7); Ir(1)–C(44), 2.208(7); Ir(1)–C(45), 2.180(8); C(2)–C(1)–C(6), 109.7(7); C(1)–C(2)–C(3), 114.7(7); C(2)–C(3)–C(4), 121.4(8); C(1)–C(6)–C(5), 120.6(8). Reproduced by permission from the American Chemical Society.

In η^4-coordinated *o*-QM complexes, the distance from the metal to the π-bonded carbon center is 1.76 Å for the quinonoid ligand.[25] The uncoordinated part of the *o*-QM ligand is bent away from the metal, giving a large (33° across C3—C6) dihedral angle (see Fig. 3.3).[25]

3.6 CONCLUSION AND FUTURE APPLICATIONS

Stabilization of quinone methides by coordination to transition metals might have various applications, beyond the important structural and spectroscopic studies of these coordinated compounds. For example, it is possible to manipulate their structure while coordinated to the metal center and affect their controlled release and trapping, leading to new synthetic procedures.

In this chapter, we have described some useful properties of coordinated QM complexes. For instance, remarkably stable η^2-*o*-QM complexes of Os can release the QM moiety upon oxidation, followed by controlled reaction with suitable dienophiles to form the chroman ring system.[15,16] η^2-Coordinated *p*-QM complexes can be easily released upon exposure to an electron-deficient alkene and trapped as the 1,6-Michael addition product.[18,19]

With regard to the reactivity of the QM ligands, it was shown that they are capable of undergoing a number of different chemical transformations to give complexes of otherwise elusive methylene arenium,[10,22] thioquinone methides,[5] and xylylenes.[10] The Cp*Ir-based η^4-*o*-QM moiety exhibits unusual nucleophilic reactivity of the exocyclic carbon. It reacted with electron-deficient $HC{\equiv}CCO_2CH_3$ or *N*-methylmaleimide to give a conjugated *o*-QM complex or [3 + 2] cycloaddition product, respectively.[29] Such reactivity can be utilized in formation of polycyclic organic products.

The remarkable reactivity properties of coordinated QMs and the possibility of their release might be useful in organic synthesis. Moreover, the preparation of QM complexes of metals bearing chiral ligands might provide opportunities for asymmetric synthesis, such as in cycloaddition reactions. In addition, in the future metal–quinone methide complexes might be useful in biological studies and could be relevant to drug delivery. In the cases of biologically active quinone methides that are not isolable and/or are incompatible with biotic media, their metal complexes might be useful as stable quinone methide carriers. In another direction, the pincer-type QM complexes might be of interest regarding nonlinear optical applications, since the energy difference between the ground and transition states of the QM moiety is low, while the dipole moment difference is expected to be high. The first excited state of the QM is zwitterionic in nature, while the solvent-dependent transformations between the quinonoid and zwitterionic platinum complexes[21] indicate that the energy levels of these two forms are close.

Thus, research in the relatively new field of quinone methide complexes is attractive from both fundamental and applied point of views. Advantages of metal complexation include stabilization of the reactive QM moiety, the ability to affect its controlled release, and the opportunity to tune their reactivity, since even small changes in the nature of the metal produce a significant alteration in the properties of the coordinated quinone methides.

ACKNOWLEDGMENTS

Financial support by the Minerva Foundation, the Israel Science Foundation, and the Helen and Martin Kimmel Center for Molecular Design is gratefully acknowledged.

REFERENCES

1. (a) *The Chemistry of the Metal–Carbon Bond. The Structure, Preparation, Thermochemistry and Characterization of Organometallic Compounds;* Hartley, F. R.; Patai, S.,Eds.; John Wiley & Sons: New York, 1982; Vol. 1. (b) For a recent review see Dyke, A. M.; Hester, A. J.; Lloyd-Jones, G. C. Organometallic generation and capture of *ortho*-arynes. *Synthesis* 2006, 24, 4093–4112.
2. Dyall, L. K.; Winstein, S. Nuclear magnetic resonance spectra and characterization of some quinone methides. *J. Am. Chem. Soc.* 1972, 94, 2196–2199.
3. Silvestre, J.; Albright, T. A. Haptotropic rearrangements in polyene–ML*n* complexes. 3. Polyene–ML2 systems. *J. Am. Chem. Soc.* 1985, 107, 6829–6841.
4. (a) For recent reviews on pincer-type complexes, see van der Boom, M. E.; Milstein, D. Cyclometalated phosphine-based pincer complexes: mechanistic insight in catalysis, coordination, and bond activation. *Chem. Rev.* 2003, 103, 1759–1792. (b) Singleton, J. T. The uses of pincer complexes in organic synthesis. *Tetrahedron* 2003, 59, 1837–1857. (c) Albrecht, M.; van Koten, G. Platinum group organometallics based on "pincer" complexes: sensors, switches, and catalysts. *Angew. Chem. Int. Ed.* 2001, 40, 3750–3781.
5. Vigalok, A.; Milstein, D. Advances in metal chemistry of quinonoid compounds: new types of interactions between metals and aromatics. *Acc. Chem. Res.* 2001, 34, 798–807.
6. Vigalok, A.; Milstein, D. Metal-stabilized quinone- and thioquinonemethides. *J. Am. Chem. Soc.* 1997, 119, 7873–7874.
7. Rybtchinski, B.; Milstein, D. Metal insertion into C—C bonds in solution. *Angew. Chem. Int. Ed.* 1999, 38, 871–883.
8. Collman, J. P.; Roper, W. R. Oxidative-addition reactions of d^8 complexes. In *Advances in Organometallic Chemistry;* Stone, F. G. A.; West, R. Eds.; Academic Press: New York, 1968; Vol. 7, pp 53–92.
9. (a) Raasch, M. S. Monothioanthraquinones. *J. Org. Chem.* 1979, 44, 632–633. (b) Itoh, T.; Fujikawa, K.; Kubo, M. *p*-Thioquinone methides: synthesis and reaction. *J. Org. Chem.* 1996, 61, 8329–8331.
10. Vigalok, A.; Shimon, L. J. W.; Milstein, D. Methylene arenium cations via quinone methides and xylylenes stabilized by metal complexation. *J. Am. Chem. Soc.* 1998, 120, 477–483.

11. Vigalok, A.; Rybtchinski, B.; Shimon, L. J. W.; Ben-David, Y.; Milstein, D. Metal-stabilized methylene arenium and σ-arenium compounds: synthesis, structure, reactivity, charge distribution, and interconversion. *Organometallics* 1999, 18, 895–905.

12. Errede, L. A.; Hoyt, J. M. Chemistry of xylylenes. III. Some reactions of *p*-xylylene that occur by free radical intermediates. *J. Am. Chem. Soc.* 1960, 82, 436–439.

13. (a) Errede, L. A.; Szwarc, M. Chemistry of *p*-xylylene, its analogs, and polymers *Quart. Rev.* 1958, 12, 301–320. (b) Coppinger, G. M.; Bauer, R. H. Stability of hetero-*p*-benzoquinones. *J. Phys. Chem.* 1963, 67, 2846–2848.

14. Gauvin, R. M.; Rozenberg, H.; Shimon, L. J. W.; Ben-David, Y.; Milstein, D. Osmium-mediated C–H and C–C bond cleavage of a phenolic substrate: *p*-quinone methide and methylene arenium pincer complexes. *Chem. Eur. J.* 2007, 13, 1382–1393.

15. Kopach, M. E.; Harman, W. D. Novel Michael additions to phenols promoted by osmium (II): convenient stereoselective syntheses of 2,4- and 2,5-cyclohexadienones. *J. Am. Chem. Soc.* 1994, 116, 6581–6592.

16. Stokes, S. M. J.; Ding, F.; Smith, P. L.; Keane, J. M.; Kopach, M. E.; Jervis, R.; Sabat, M.; Harman, W. D. Formation of *o*-quinone methides from η^2-coordinated phenols and their controlled release from a transition metal to generate chromans. *Organometallics* 2003, 22, 4170–4171.

17. Todd, M. A.; Grachan, M. A.; Sabat, M.; Myers, W. H.; Harman, W. D. Common electrophilic addition reactions at the phenol ring: the chemistry of TpW(NO)(PMe_3)(η^2-phenol). *Organometallics* 2006, 25, 3948–3954.

18. Rabin, O.; Vigalok, A.; Milstein, D. Metal-mediated generation, stabilization, and controlled release of a biologically relevant, simple *para*-quinone methide: BHT-QM. *J. Am. Chem. Soc.* 1998, 120, 7119–71120.

19. Rabin, O.; Vigalok, A.; Milstein, D. A novel approach towards intermolecular stabilization of *para*-quinone methides. First complexation of the elusive, simplest quinone methide, 4-methylene-2,5-cyclohexadien-1-one. *Chem. Eur. J.* 2000, 6, 454–462.

20. (a) Guyton, K. Z.; Dolan, P. M.; Kensler, T. W. Quinone methide mediates *in vitro* induction of ornithine decarboxylase by the tumor promoter butylated hydroxytoluene hydroperoxide. *Carcinogenesis* 1994, 15, 817–821. (b) Yamamoto, K.; Kato, S.; Tajima, K.; Mizutani, T. Electronic and structural requirements for metabolic activation of butylated hydroxytoluene analogs to their quinone methides, intermediates responsible for lung toxicity in mice. *Biol. Pharm. Bull.* 1997, 20, 571–573. (c) McCracken, P. G.; Bolton, J. L.; Thatcher, G. R. J. Covalent modification of proteins and peptides by the quinone methide from 2-*tert*-butyl-4,6-dimethylphenol: selectivity and reactivity with respect to competitive hydration. *J. Org. Chem.* 1997, 62, 1820–1825. (d) Reed, M.; Thompson, D. C. Immunochemical visualization and identification of rat liver proteins adducted by 2,6-di-*tert*-butyl-4-methylphenol (BHT). *Chem. Res. Toxicol.* 1997, 10, 1109–1117. (e) Lewis, M. A.; Yoerg, D. G.; Bolton, J. L.; Thompson, J. Alkylation of 2′-deoxynucleosides and DNA by quinone methides derived from 2,6-di-*tert*-butyl-4-methylphenol. *Chem. Res. Toxicol.* 1996, 9, 1368–1374.

21. Poverenov, E.; Shimon, L. J. W.; Milstein, D. Quinone methide generation based on a *cis*-(*N,N*) platinum complex. *Organometallics* 2007, 26, 2178–2182.

22. Poverenov, E.; Leitus, G.; Milstein, D. Synthesis and reactivity of the methylene arenium form of a benzyl cation, stabilized by complexation. *J. Am. Chem. Soc.* 2006, 128, 16450–16451.

23. (a) Olah, G. A.; Porter, R. D.; Jeuell, C. L.; White, A. M. Stable carbocations. CXXV. Proton and carbon-13 magnetic resonance studies of phenylcarbenium ion (benzyl cations). Effect of substituents of the stability of carbocations. *J. Am. Chem. Soc.* 1972, 94, 2044–2052. (b) Van Pelt, P.; Buck, H. M. Proton catalyzed hydride transfer from alkanes to methylated benzyl cations. I. Kinetics. *Recl. Trav. Chim. Pays-Bas.* 1973, 3092, 1057–1066. (c) Bollinger, J. M.; Comisarow, M. B.; Cupas, C. A.; Olah, G. A. Stable carbonium ions. XLV. Benzyl cations. *J. Am. Chem. Soc.* 1967, 89, 5687–5691.
24. Reindl, B.; Clark, T.; Scheleyer, P. v. R. Modern molecular mechanics and *ab initio* calculations on benzylic and cyclic delocalized cations. *J. Phys. Chem. A.* 1998, 102, 8953–8963; and references therein.
25. Amouri, H.; Besace, Y.; Le Bras, J. General synthesis, first crystal structure, and reactivity of stable *o*-quinone methide complexes of Cp*Ir. *J. Am. Chem. Soc.* 1998, 120, 6171–6172.
26. Amouri, H.; Vaissermann, J.; Rager, M. N.; Grotjahn, D. B. Rhodium-stabilized *o*-quinone methides: synthesis, structure, and comparative study with their iridium congeners. *Organometallics* 2000, 19, 5143–5148.
27. (a) Casado, C. M.; Wagner, T.; Astruc, D. Deprotonation of the complexes [Ru(arene) Cp] + PF-6 (arene = C_6Me_6 and fluorene): X-ray crystal structure of [Ru(η^5-$C_6Me_5CH_2$) Cp] and determination of the pK_a values using the iron analogs. *J. Organomet. Chem.* 1995, 502, 143–145. (b) Schmalz, H.-G.; Volk, T.; Bernicke, D.; Huneck, S. On the deprotonation of η^6-1,3-dimethoxybenzene–$Cr(CO)_3$ derivatives: influence of the reaction conditions on the regioselectivity. *Tetrahedron* 1997, 53, 9219–1932.
28. Le Bras, J.; Rager, M. N.; Besace, Y.; Vaissermann, J.; Amouri, H. Activation and regioselective *ortho*-functionalization of the A-ring of β-estradiol promoted by "CpIr": an efficient organometallic procedure for the synthesis of 2-methoxyestradiol. *Organometallics* 1997, 16, 1765–1771.
29. (a) Amouri, H.; Vaissermann, J.; Rager, M. N.; Grotjahn, D. B. Stable *o*-quinone methide complexes of iridium: synthesis, structure, and reversed reactivity imparted by metal complexation. *Organometallics* 2000, 19, 1740–1748. (b) Amouri, H.; Le Bras, J. Taming reactive phenol tautomers and *o*-quinone methides with transition metals: a structure–reactivity relationship. *Acc. Chem. Res.* 2002, 35, 501–510.
30. (a) Padwa, A.; Lee, G. A. Photochemical transformation of 2,2-disubstituted chromenes. *J. Chem. Soc., Chem. Commun.* 1972, 795–796. (b) Padwa, A.; Au, A.; Lee, G. A.; Owens, W. Photochemical transformations of small ring heterocyclic systems. LX. Photochemical ring-opening reactions of substituted chromenes and isochromenes. *J. Org. Chem.* 1975, 40, 1142–1149. (c) Padwa, A.; Lee, G. A.; Owens, W. Carbonyl group photochemistry via the enol form. Photoisomerization of 4-substituted 3-chromanones. *J. Am. Chem. Soc.* 1976, 98, 3555–3564. (d) Karabelas, K.; Moore, H. W. Trimethylsilylmethyl-1,4-benzoquinones. Generation and trapping of *o*-quinone methides. *J. Am. Chem. Soc.* 1990, 112, 5372–5373. (e) Huang, C.-G.; Beveridge, K. A.; Wan, P. Photocyclization of 2-(2′-hydroxyphenyl)benzyl alcohol and derivatives via *o*-quinonemethide type intermediates. *J. Am. Chem. Soc.* 1991, 113, 7676–7684. (f) Huang, C.-G.; Shukla, D.; Wan, P. Mechanism of photoisomerization of xanthene to 6*H*-dibenzo[*b,d*]pyran in aqueous solution. *J. Org. Chem.* 1991, 56, 5437–5442. (g) Katritzky, A. R.; Zhang, Z.; Lan, X.; Lang, H. *O*-(α-Benzotriazolylalkyl)phenols: novel precursors for the preparation of *ortho*-substituted phenols via intermediate *o*-quinone methides. *J. Org. Chem.* 1994, 59, 1900–1903.

31. (a) Egberston, M.; Danishefsky, S. J. Modeling of the electrophilic activation of mitomycins: chemical evidence for the intermediacy of a mitosene semiquinone as the active electrophile. *J. Am. Chem. Soc.* 1987, 109, 2204–2205. (b) Tomasz, A. K.; Chawla, A. K.; Lipman, R. Mechanism of monofunctional and bifunctional alkylation of DNA by mitomycin C. *Biochemistry* 1988, 27, 3182–3187. (c) Ouyang, A.; Skibo, E. B. Design of a cyclopropyl quinone methide reductive alkylating agent. *J. Org. Chem.* 1998, 63, 1893–1900. (d) Lin, A. J.; Pradini, R. S.; Cosby, L. A.; Lillis, B. J.; Shansky, C. W.; Sartorelli, A. C. Potential bioreductive alkylating agents. 2. Antitumor effect and biochemical studies of naphthoquinone derivatives. *J. Med. Chem.* 1973, 16, 1268–1271. (e) Antonini, I.; Lin, T.-S.; Cosby, L. A.; Dai, Y.-R.; Sartorelli, A. C. 2- and 6-Methyl-1,4-naphthoquinone derivatives as potential bioreductive alkylating agents. *J. Med. Chem.* 1982, 25, 730–735.

32. Akylations of DNA by related quinone methides: Chatterjee, M.; Rokita, S. E. Inducible alkylation of DNA using an oligonucleotide–quinone conjugate. *J. Am. Chem. Soc.* 1990, 112, 6397–6399.Chatterjee, M.; Rokita, S. E. Sequence-specific alkylation of DNA activated by an enzymatic signal. *J. Am. Chem. Soc.* 1991, 113, 5116–5117. Chatterjee, M.; Rokita, S. E. The role of a quinone methide in the sequence specific alkylation of DNA. *J. Am. Chem. Soc.* 1994, 116, 1690–1697.

33. (a) Wagner, H.-U.; Gompper, R.;In *The Chemistry of the Quinonoid Compounds;* Patai, S. Ed.; Wiley: New York, 1974; Part 2, Chapter 18.(b) Berson, J.; In *The Chemistry of the Quinonoid Compounds;* Patai, S.; Rappoport, Z. Eds.; Wiley: Chichester. 1988; Vol. 2, Part 1, pp 455. (c) Thompson, D. C.; Thompson, J. A.; Sugumaran, M.; Moldeus, P. Biological and toxicological consequences of quinone methide formation. *Chem.-Biol. Interact.* 1993, 86, 129–162.

34. (a) Cambie, R. C.; Hayward, R. C.; Jurlina, J. L.; Rutledge, P. S.; Woodgate, P. D. Reactions of enol acetates with thallium(I) acetate–iodine. *J. Chem. Soc., Perkin Trans. 1* 1978, 2, 126–130. (b) Rubottom, G. M.; Mott, R. C. Reaction of enol silyl ethers with silver acetate–iodine. Synthesis of α-iodo carbonyl compounds. *J. Org. Chem.* 1979, 44, 1731–1734.

35. Lavallée, J.-F.; Berthiaume, G.; Deslongchamps, P.; Grein, F. Intramolecular Michael addition of cyclic β-keto esters onto conjugated acetylenic ketones. *Tetrahedron Lett.* 1986, 27, 5455–5458.

36. Angle, S. R.; Arnaiz, D. O. Formal [3 + 2]cycloaddition of benzylic cations with alkenes. *J. Org. Chem.* 1992, 57, 5937–5947.

37. Reingold, J. A.; Son, S. U.; Kim, S. B.; Dullaghan, C. A.; Oh, M.; Frake, P. C.; Carpenter, G. B.; Sweigart, D. A. Pi-bonded quinonoid transition-metal complexes. *J. Chem. Soc., Dalton Trans.* 2006, 20, 2385–2398.

38. Pouchert, C. J.; Behnke, J. *The Aldrich Library of* ^{13}C *and* ^{1}H *FT NMR Spectra;* Aldrich Chemical Co., Inc.: Milwaukee, WI. 1993; Vol. 1.

39. For example, in the structurally related $(PMe_3)_2Pd(\eta^2\text{-}CH_2{=}C_5Me_4)$, the exocyclic double bond distance of the coordinated fulvene molecule is 1.424(2) Å and it is at an 10.81° angle with the fulvene ring: Werner, H.; Crisp, G. T.; Jolly, P. W.; Kraus, H.-J.; Kruger, C. Synthesis of (1,2,3,4-tetramethylfulvene) palladium(0) complexes from η^5-pentamethylcyclopentadienyl)palladium(II) precursors. The crystal structure of [Pd $(PMe_3)2(\eta^2\text{-}CH_2{:}C_5Me_4)$]. *Organometallics* 1983, 2, 1369–1377. (b) Benn, R.; Betz, P.; Goddard, R.; Jolly, P. W.; Kokel, N.; Kruger, C.; Topalovic, I. Z. Intermediates in the palladium-catalyzed reaction of 1,3-dienes. Part 6. A solid-state NMR spectroscopic

investigation of butadiene complexes of nickel, palladium and platinum. *Z. Naturforsch.* 1991, 46B, 1395–1405. (c) Herrmann, W. A.; Thiel, W. R.; Brossmer, C.; Öfele, K.; Priermeier, T. (Dihalomethyl)palladium(II) complexes from dibenzylideneacetone palladium(0) precursors: synthesis, structure, and reactivity. *J. Organomet. Chem.* 1993, 461, 51–60. (d) Kranenburg, M.; Delis, J. G. P.; Kamer, P. C. J.; van Leeuwen, P. W. N. M.; Vrieze, K.; Veldman, N.; Spek, A. L.; Goubitz, K.; Fraanje, J. Palladium(0)-tetracyanoethylene complexes of diphosphines and a dipyridine with large bite angles, and their crystal structures. *J. Chem. Soc., Dalton Trans.* 1997, 1839–1849.

4

INTERMOLECULAR APPLICATIONS OF *o*-QUINONE METHIDES (*o*-QMs) ANIONICALLY GENERATED AT LOW TEMPERATURES: KINETIC CONDITIONS

THOMAS PETTUS[1] AND LIPING PETTUS[2]

[1] *Department of Chemistry and Biochemistry, University of California at Santa Barbara, Santa Barbara, CA 93106, USA*

[2] *Department of Chemical Research and Discovery, Amgen, Inc., Thousand Oaks, CA 91320, USA*

4.1 INTRODUCTION TO *o*-QMs

There exist many fundamental and philosophical truths in the universe. Among these is the notion that the value of an item is determined by its inherent characteristics and its surrounding environment. For example, a fishing pole near the sea is a useful tool, but when relocated to the desert its usefulness decreases.

o-Quinone methides (*o*-QMs) are highly reactive species with short lifetimes because of the thermodynamic drive to undergo rearomatization. However, the above philosophical truth applies to these highly electrophilic intermediates. In the absence of stabilizing residues, these transient species prove so reactive that the setting in which they are produced dictates the range and scope of their subsequent application. Figure 4.1 shows some of the accepted canonical representation for *o*-QMs. These depictions should give readers some appreciation regarding the reactivity of *o*-QMs and impart some credence to the notion that the olefin geometry in an *o*-QM

Quinone Methides, Edited by Steven E. Rokita

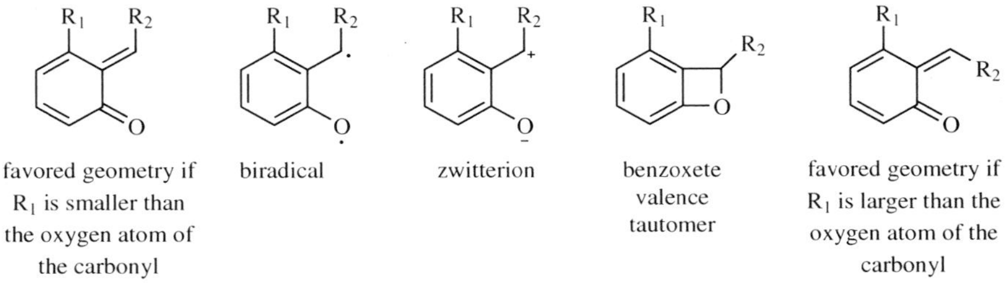

favored geometry if R_1 is smaller than the oxygen atom of the carbonyl

biradical

zwitterion

benzoxete valence tautomer

favored geometry if R_1 is larger than the oxygen atom of the carbonyl

FIGURE 4.1 Some plausible canonical representation of *o*-QMs.

is fluxional, being determined by the energy difference between competing steric effects (R_2 and R_1) versus (R_2 and O).

Given their extraordinary reactivity, one might assume that *o*-QMs offer plentiful applications as electrophiles in synthetic chemistry. However, unlike their more stable *para*-quinone methide (*p*-QM) cousin, the potential of *o*-QMs remains largely untapped. The reason resides with the propensity of these species to participate in undesired addition of the closest available nucleophile, which can be solvent or the *o*-QM itself. Methods for *o*-QM generation have therefore required a combination of low concentrations and high temperatures to mitigate and reverse undesired pathways and enable the redistribution into thermodynamically preferred and desired products. Hence, the principal uses for *o*-QMs have been as electrophilic heterodienes either in *intramolecular* cycloaddition reactions with nucleophilic alkenes under thermodynamic control or in *intermolecular* reactions under thermodynamic control where a large excess of a reactive nucleophile thwarts unwanted side reactions by its sheer vast presence.

4.2 THERMAL GENERATION CONDITIONS

Therefore, most of the nonoxidative generation methods that have evolved can be viewed as a "crossover" reaction of sorts whereby one *o*-QM product is exchanged for another by application of heat. The stereochemistry accruing in the products of these procedures is expectedly subject to thermodynamic control. For example, while exploring a synthetic approach for nomofungin (Fig. 4.2), Funk recently showed that

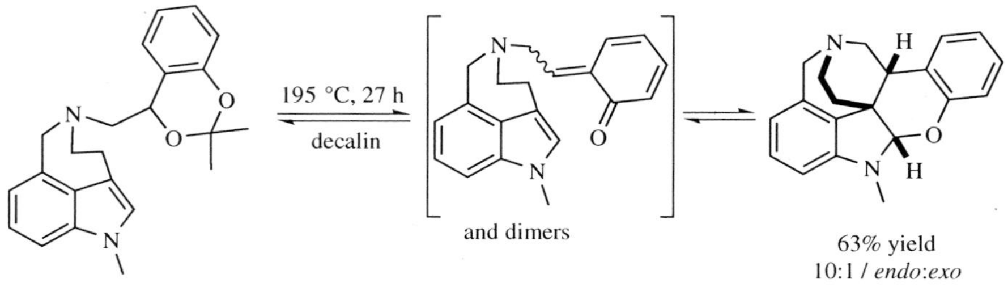

195 °C, 27 h

decalin

and dimers

63% yield

10:1 / *endo*:*exo*

FIGURE 4.2 The Funk crossover strategy for the hexacyclic core of nomofungin.

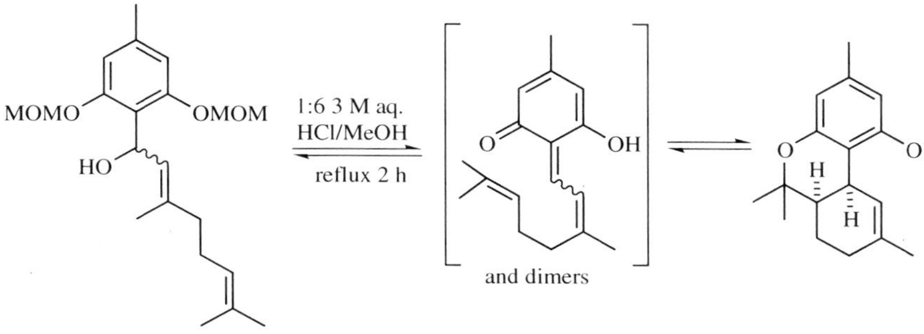

FIGURE 4.3 The Snider strategy for the tetracyclic core of the bisabosquals.

heating 2,2-dimethyl-1,3-benzodioxin to 195 °C caused expulsion of acetone resulting in the formation of an *o*-QM intermediate, which subsequently succumbed to a [4 + 2] reaction with the attached indole.[1] Whether or not the intermediate *o*-QM undergoes cycloadditions with itself or some other nucleophile or recombines with acetone proves rather inconsequential as the ultimate product and the stereochemical outcome (10:1 *endo*/*exo*) are decided by relative thermodynamic stabilities. Moreover, under thermal conditions the *o*-QM intermediate can even undergo reaction with extremely unreactive nucleophiles, such as the indole.

Another elegant example of the thermal generation and subsequent intramolecular cycloaddition of an *o*-QM can be found in Snider's biomimetic synthesis of the tetracyclic core of bisabosquals.[2] Treatment of the starting material with acid causes the MOM ethers to cleave from the phenol core (Fig. 4.3). Under thermal conditions, a proton transfer ensues from one of the phenols to its neighboring benzylic alcohol residue. Upon expulsion of water, an *o*-QM forms. The *E* or *Z* geometry of the *o*-QM intermediate and its propensity toward interception by formaldehyde, water, or itself, again prove inconsequential as the outcome is decided by the relative thermodynamic stabilities among accessible products.

Recently, Ohwada has employed benzoxazines as the starting precursor for various β,β-unsubstituted *o*-QMs (Fig. 4.4).[3] In this protocol, the benzoxazine is gently warmed; in some cases, 50 °C proves sufficient to expel methyl cyanoformate and

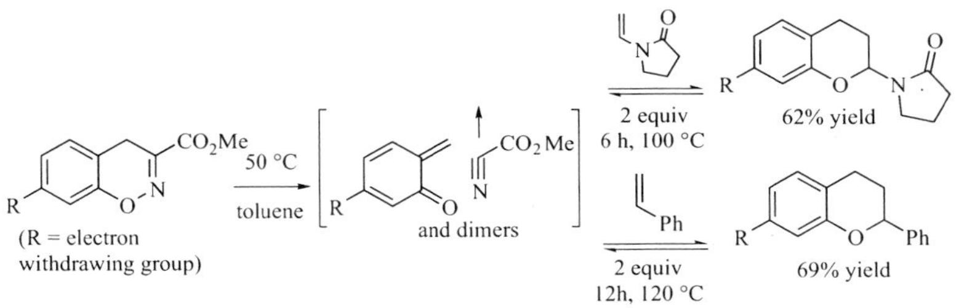

FIGURE 4.4 The Ohwada method of a benzoxazine crossover for *o*-QM generation.

generate the *o*-QM intermediate. The initial expulsion is irreversible because methyl cyanoformate is a poor nucleophile. Electron-withdrawing groups within the *o*-QM species appear to facilitate reaction with less reactive nucleophiles. However, the choice of subsequent conditions depends on the overall reactivity of the nucleophile. Long reaction times and high temperatures (>100 °C) are necessary in many examples, such as the cycloaddition between an *o*-QM and nonnucleophilic alkenes. These facts suggest that the product distribution in these high-temperature examples remains under thermodynamic control and enables the products of undesired *o*-QM side reactions, such as dimerization, to revert and channel into the preferred product. In addition, the conspicuous absence of benzopyrans afforded from *o*-QMs with β-substituents may suggest that these reactions afford *syn*/*anti* ratios as the result of a thermodynamic distribution among isomers.

For the remaining discussion, we focus on low-temperature methods for generation of *o*-QMs (< 25 °C) as reported in the literature since 2001 as well as their subsequent synthetic applications. Surprisingly, only three general procedures adhere to this stringent criterion. All of the methods can be considered as examples of anionic generation of *o*-QMs. In our opinion, these three procedures are unique because any *o*-QM intermediate generated in a nonoxidative fashion at low temperature can then be utilized in reactions under kinetic control. For past several decades, kinetically controlled reactions have largely supplanted thermodynamic regimes in synthetic applications because of likelihood of better stereocontrol and greater precision.

4.3 LOW-TEMPERATURE KINETIC GENERATION OF *o*-QMs

4.3.1 Formation of the *o*-QMs Triggered by Fluoride Ion

There are only a few examples of low-temperature conditions reported to lead to a species behaving as an *o*-QM. All of these, except for our *O*-acyl transfer methods that will be discussed later, use a fluoride ion to trigger the formation of the *o*-QM in an almost instantaneous manner. In these examples, a high concentration of the intended nucleophile is necessary to prevent any side reactions with the *o*-QM, because given the low-temperature conditions its formation is usually irreversible.

The newest protocol, which was reported by Yoshida in 2004, is a testimony to these issues (Fig. 4.5).[4] Treatment of a TMS-aryl compound displaying an *ortho*-OTf residue with a fluoride ion at 0 °C causes the formation of a symmetric benzyne, which succumbs to a [2 + 2] cycloaddition with the carbonyl of various benzaldehydes to form a benzoxete. This four-membered ring then undergoes immediate valence isomerization to its corresponding *o*-QM, which as expected proves highly reactive and indiscriminate in its search for a nucleophile. It could undergo dimerization with itself or the addition of another weak nucleophile. However, since some of the reactive benzyne still remains, it undergoes an immediate [4 + 2] cycloaddition at low temperature to restore aromaticity and affords the corresponding 9-aryl-xanthene. The authors are able to push the reaction to completion by providing three equivalents

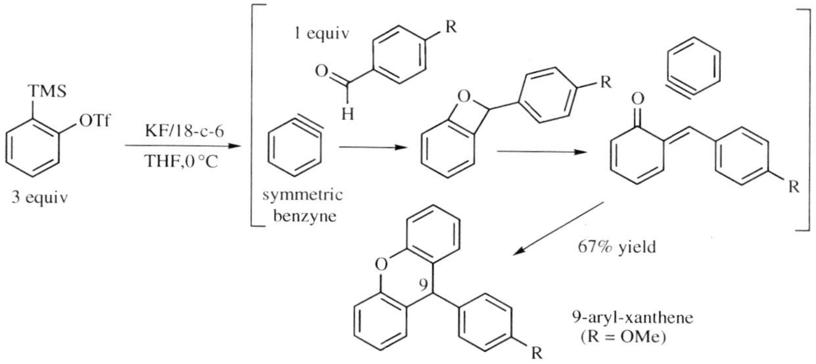

FIGURE 4.5 Yoshida fluoride-triggered [2 + 2] benzoxetane formation and rearrangement.

of the benzyne precursor. Thorough experimentation shows that in order for the reaction to proceed in good yield, the aryl aldehyde must contain an electron-donating group (such as R = OMe) to ensure the rapid and complete formation of the benzoxete precedes subsequent formation of the *o*-QM intermediate, which is most likely the rate-determining step in the cascade.

The first of the few low-temperature methods for the formation of an *o*-QM was a method developed by Rokita.[5] It is principally used for reversible DNA alkylation. However, it has recently begun to find its way into some synthetic applications. It utilizes a silylated phenol, which proves vastly more manageable as an *o*-QM precursor than the corresponding *o*-hydroxyl benzyl halide (Fig. 4.6). In this kinetically controlled process, expulsion of a benzylic leaving group is triggered at low temperature by treatment with a fluoride ion, which causes a β-elimination.

The most comprehensive examination of the Rokita kinetic procedure from a synthetic standpoint was carried out by Barrero and coworkers.[6] They examined the effects of various leaving groups, solvents, nucleophiles, and their equivalents on subsequent [4 + 2] cycloadditions. A vast excess of the intended nucleophile (50–100 equiv) must be employed, because the fluoride triggered β-elimination proves nearly instantaneous at room temperature resulting in a high concentration of a species that is prone to undergo dimerization and other undesired side reactions that are irreversible at these low temperatures (Fig. 4.7). Use of fewer equivalents of the intended nucleophile led to a rapid drop off in yield. For example, 5–10 equivalents of ethoxyvinyl ether (EVE) affords only a 5–10% yield of the desired benzopyran

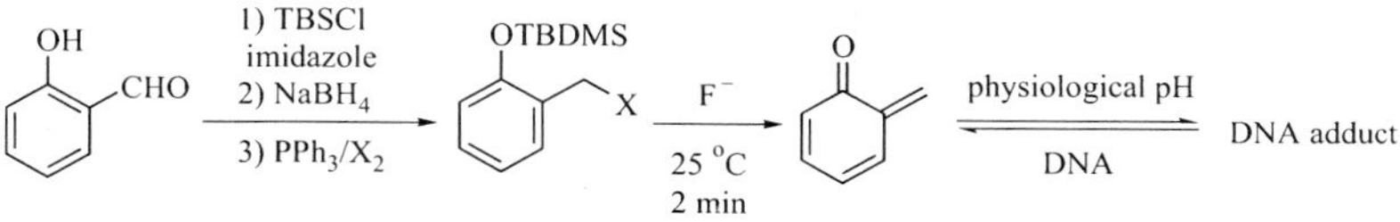

FIGURE 4.6 The Rokita fluoride-triggered expulsion for generation of an *o*-QM.

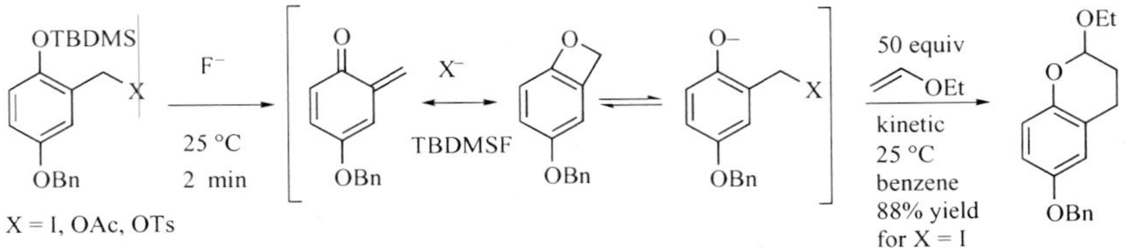

FIGURE 4.7 Barrero investigation of synthetic parameters surrounding the Rokita method.

adduct. Methylene chloride was reported to be nearly useless as a solvent, whereas less polar solvents are more productive. Interestingly, acetate and tosylate precursors succumbed to degradation rather than undergo cycloaddition. This finding is not altogether surprising given the pK_a of HCl, HBr, and HI versus HOAc or HOTs, but it may also suggest that the leaving groups provide some modicum of reversibility and possibly stabilizes the *o*-QM intermediate.

Scheidt reports using an amended Rokita method in conjunction with an umpolung derivative of several aryl aldehydes for the synthesis of α-aryl ketones.[7] The procedure is indeed useful for the synthesis of 2-arylated benzofurans as shown by the synthesis of demethylmoracin I (Fig. 4.8).

Some time ago, McLoughlin[8] and later Mitchell[9] observed the reductions of various aryl ketones and aldehydes displaying an *ortho*-*O*-acyl residue (Fig. 4.9). The transformation appeared to involve the transfer of two hydrides and the researchers speculated the formation of an *o*-QM intermediate that subsequently underwent 1,4-reduction.

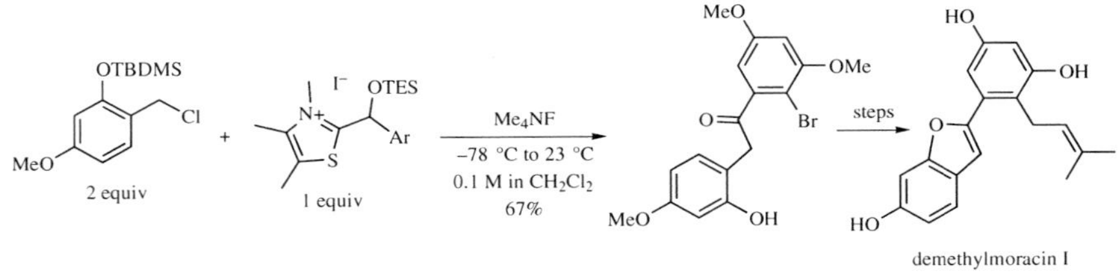

FIGURE 4.8 The Scheidt synthesis of demethylmoracin I using the Rokita method.

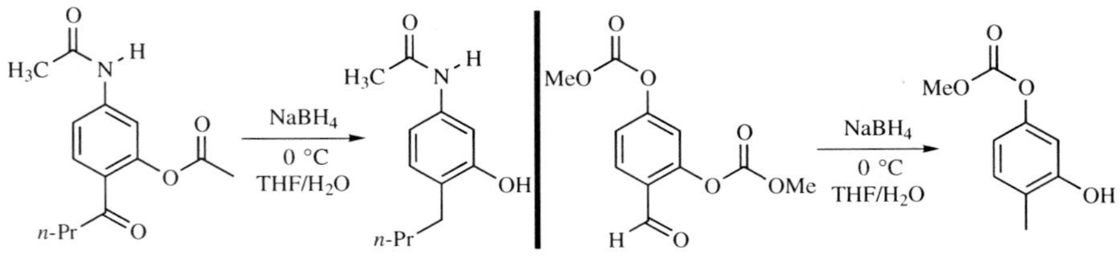

FIGURE 4.9 McLoughlin and Mitchell reductive processes presumed to involve *o*-QMs.

4.3.2 Stepwise Formation of *o*-QMs

While engaged in several synthetic projects, we required access to an assortment of differentially protected 4- and 5-alkylated resorcinol and hydroquinone derivatives. The usual synthetic methods failed to provide these aromatic materials in an expedient manner. We speculated that nucleophiles other than [H^-] could be used in the McLoughlin–Mitchell cascade. After extensive investigation, members of the Pettus group found a variety of *ortho-O*-acyl analogues that participated in analogous reductions (–OBoc, –OAc, –OPiv). However, the procedure would have been useless had we not accidentally stumbled upon a series of procedures that enabled the gradual and stepwise formation of the *o*-QM intermediate.[10] After much experimentation, we chose to focus upon *ortho*-OBoc derivatives because they were less susceptible to direct reaction with organolithium and organomagnesium reagents or their corresponding alkoxides. We found that an assortment of aryl ketones, aldehydes, and alcohols could be elaborated in this one-pot three-component process that was more efficient and mild than competing tactics such as directed *ortho* lithiation (D*o*M), aryl Claisen rearrangement, and electrophilic substitution (A–E, Fig. 4.10).

The examples shown in Figs. 4.11–24 should give readers an indication as to the breadth of the compounds accessible from this process. Ryan Jones and fellow graduate students mapped out some of the unusual reactivity of these systems and the effects of solvent and temperature.[11] Two undergraduate students, Tuttle and Rodriguez, subsequently adapted method A for a short synthesis of (±)-mimosifoliol (**4**).[12] The synthesis begins by the selective removal of the two of the three Me ethers in compound **1** followed by protection of the resulting phenol with the bis-Boc to afford compound **2** (Fig. 4.11). The sequential addition of phenyl lithium and vinyl Grignard to aldehyde **2** gives the phenol **3** in 63% yield. It is difficult to imagine a more straightforward synthesis of this deceptively simple differentially protected polyhydroxylated phenol.

Mejorado investigated the asymmetric addition of various organometallic nucleophiles using method A, but the reaction could not be catalyzed. The intermediates proved to be far too reactive. However, he established that the addition of a stoichiometric amount of a preformed chiral complex [an admixture of Taddol (*trans*-α,α′-(dimethyl-1,3-dioxolane-4,5-diyl)bis(diphenylmethanol)) and EtMgBr] to **5** affords some *enantiomeric excess* in the resulting phenol product **6** (Fig. 4.12).[13]

Jacobi reports using a variant of method A to access the A,B,E-ring system of wortmannin.[14] The sequential addition of methyl lithium and acetylenic Grignard reagent followed by triflation proceeds from **7** to the corresponding triflate **8** in 74% yield (Fig. 4.13). Subsequent carbonylation of the alkyne and the phenol produces the acyl oxazole **9**, which is smoothly converted into the furanolactone **10** over three more steps.

Without a means to easily control the chirality of benzylic stereocenter, Pettus group members decided to focus on the development of procedures for the construction of adducts without chiral centers. These researchers found that organomagnesium reagents can be used to both generate and consume the *o*-QM from the corresponding

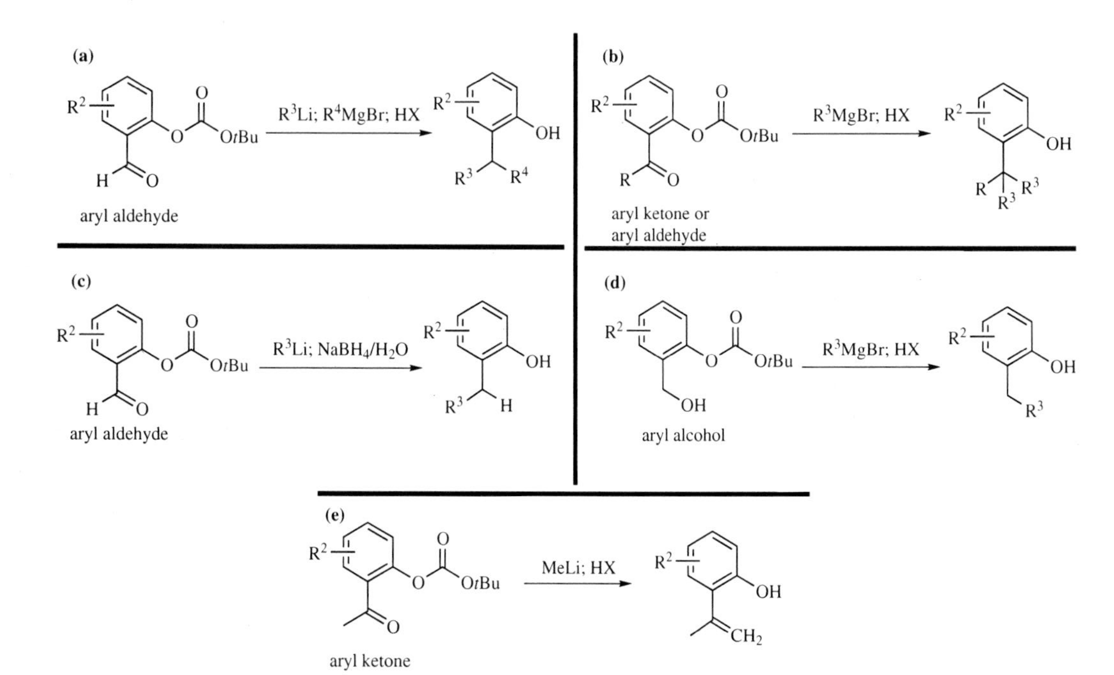

FIGURE 4.10 Pettus one-pot methods for alkylation presumed to involve *o*-QMs.

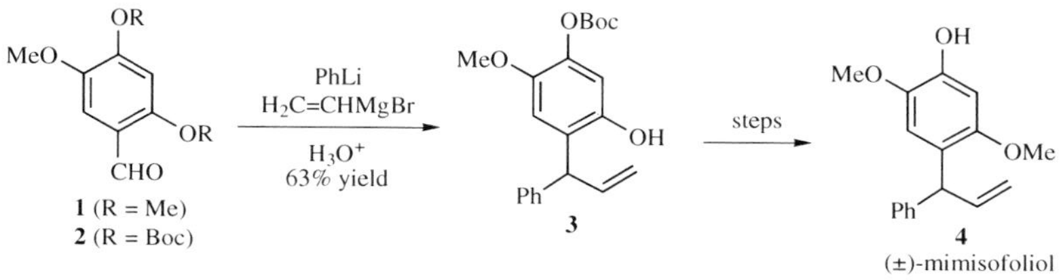

FIGURE 4.11 Method A: sequential addition of RLi/RMgBr for synthesis of (±)-**4**.

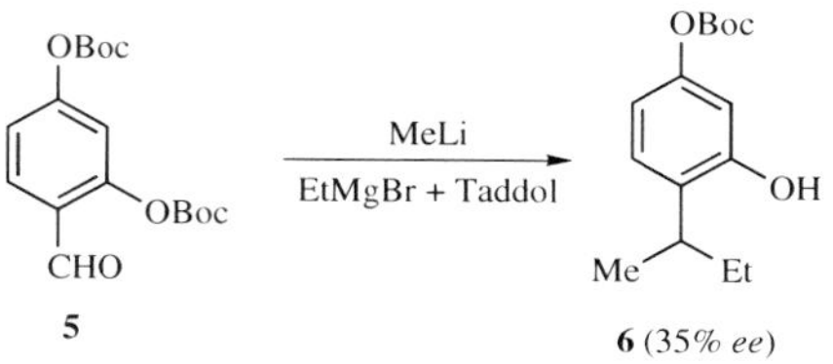

FIGURE 4.12 Stoichiometric amounts of chiral additives afford low %*ee*.

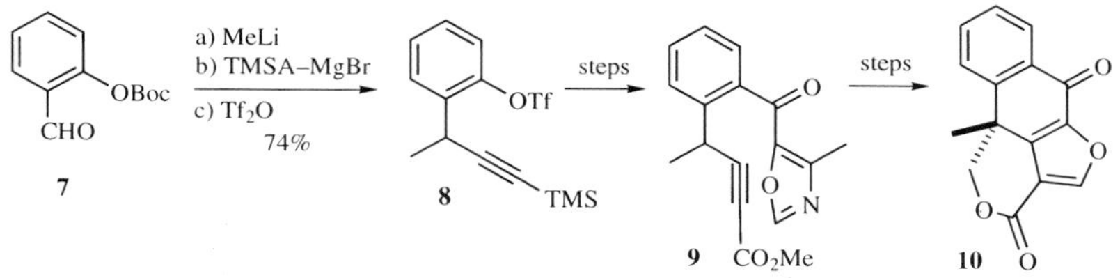

FIGURE 4.13 Jacobi's use of method A in a strategy for wortmannin.

o-OBoc benzaldehydes and acetophenones (method B, Fig. 4.14). Indeed, it appeared that magnesium had an almost magical property, because without Mg^{2+} the reactions fail to provide any identifiable product for the corresponding lithium reagents. For example, upon addition of two equivalents of a Grignard reagent such as methyl magnesium bromide, the *o*-OBoc aldehyde **7** proceeds to the corresponding *o*-isopropyl phenol **11**, a bis-adduct, in 86% yield, whereas the corresponding lithium reagent produced nothing identifiable.[10]

Group members further devised two additional methods (methods C and D, Fig. 4.15) leading to monoalkylation products. In the first, an alkyl lithium reagent is added to the aldehyde **5** (0.1 M in Et_2O, −78 °C). Upon disappearance of starting

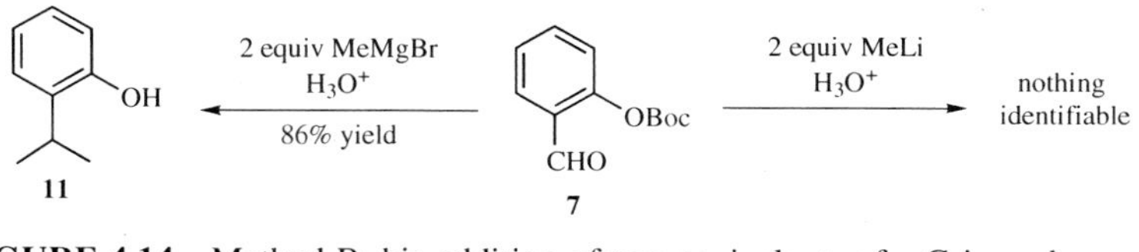

FIGURE 4.14 Method B: bis-addition of two equivalents of a Grignard reagent.

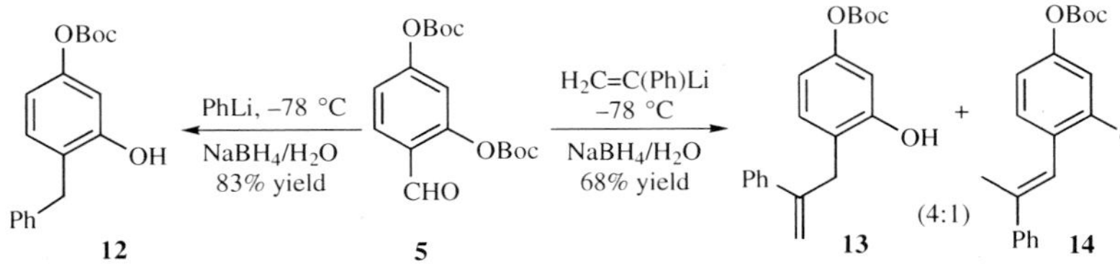

FIGURE 4.15 Method C: RLi addition followed by $NaBH_4$reduction.

material as observed by TLC, $NaBH_4$ (4 H^- equiv) and water (10 equiv) are added, and the reaction is permitted to warm to room temperature whereupon the alkoxide is quenched with 0.5 M HCl to facilitate isolation of the phenol. As Mitchell and McLoughlin previously reported, the addition of water to these borohydride reductions is critical. Without the addition of water, the yields are very poor. In the case of 2,4-bis-OBoc-benzylaldehyde **5**, the addition of phenyl lithium and sodium borohydride/water affords an 83% yield of the expected phenol **12**. However, method C is not without restrictions. The process does not work well for the addition of vinyl anions. For example, in the case of the α-styryl phenyl lithium reagent, a mixture of adducts **13**:**14** forms in a nearly inseparable 4:1 ratio mixture that is the result of 1,4- and 1,6-reduction.[10]

To overcome this dilemma and permit the addition of less reactive nucleophiles by a slight modification to the overall process, a new method was developed (method D, Fig. 4.16). The aldehyde must be first reduced to the corresponding benzyl alcohol. Short exposure (<10 min) to sodium borohydride proved effective. However, overreduction often occurred without careful monitoring. We subsequently found that borane–dimethyl sulfide complex readily reduces the corresponding o-OBoc aldehyde without risk of overreduction, and unlike the Baldwin process for reduction of the corresponding o-OAcetates,[15] we found that the aryl OBoc residue does not transfer to the newly formed benzyloxyanion. Van De Water then showed that the benzyl alcohol undergoes reaction with two equivalents of the Grignard reagent

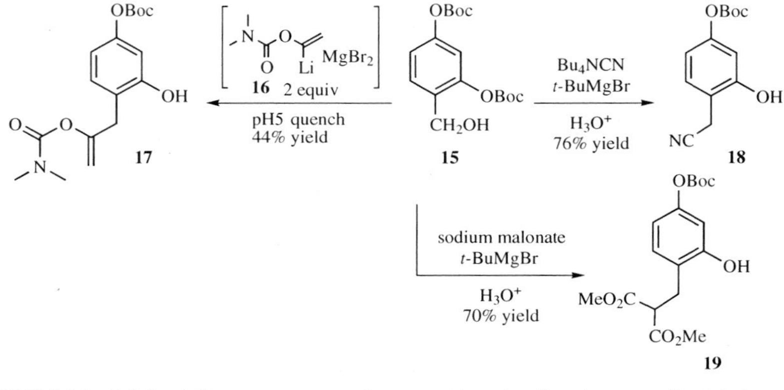

FIGURE 4.16 Method D: organomagnesium reactions leading to monoalkylated products.

and affords the corresponding *ortho*-functionalized phenol.[16] For example, the combination of the 2,4-bis-*o*-OBoc benzyl alcohol **15** with two equivalents of the lithiated carbamyl enol ether **16** and magnesium bromide affords the expected enol ether **17** in a moderate 44% yield. Higher yields are observed when adding less acid labile functionality. By a slight modification to method D, softer nucleophiles can be introduced as well. Hoarau reported that upon the sequential combination of the benzyl alcohol **15**, *t*-butyl-ammonium cyanide and *t*-butyl Grignard, the expected nitrile **18** was produced in 76% yield.[17] Similarly, sequential combination of the benzyl alcohol **15**, sodium malonate and *t*-butyl Grignard affords the desired diester **19** in 70% yield after acid workup.[1]

Magdziak reported that an *o*-OBoc acetophenone undergoes addition of two equivalents of a Grignard reagent in a manner that seems similar to method B for the corresponding *o*-OBoc benzaldehydes (Fig. 4.17).[10] For example, 2,5-bis-OBoc-acetophenone **20** (0.1 M in Et_2O, −78 °C) undergoes reaction with methyl Grignard upon warming to room temperature and affords the corresponding *o*-*t*-butyl-phenol **21** in 78% yield after workup.

Very surprisingly, lithium reagents also prove productive in their reaction with *o*-OBoc acetophenones, but the reaction unexpectedly proceeds to styrene derivatives. For example, addition of two equivalents of methyl lithium to 2,5-bis-OBoc-acetophenone **20** affords the styrenyl adduct **22** in 69% yield almost instantaneously even at low temperature (−78 °C).[10] This protocol, which we designate as method E, is in stark contrast from the corresponding reaction employing similar two equivalents of lithium reagent with the analogous *o*-Boc benzaldehyde that had resulted in the slow decomposition of starting material (Fig. 4.14). This divergent behavior between magnesium and lithium reagents has always been troubling for us from a mechanistic perspective. In subsequent sections, we will speculate as to the cause for this discrepancy between metals, as well as the unusual role that Grignard reagents and magnesium bromide may play in these reactions.

From a perspective of utility and convenience, methods C and D are proving most synthetically useful (Figs. 4.18–24). Lindsey elaborated the 2,5-bis-OBoc-benzyl alcohol **23** into the corresponding prenylated phenol **24** in 65% yield using method D for use in subsequent syntheses of the *N*-SMase inhibitors F11263 and F11334B1.[18] Hoarau demonstrated that this prenylation method was useful for most phenol nuclei, excepting the 2,6-bis-OBoc-benzyl alcohol **25**, which affords only a 25% yield of the desired phenol **26**.[19] This result was attributed to problems regarding unfavorable steric interactions for the 1,4-addition and the surprising instability of the benzyl alcohol **25**.

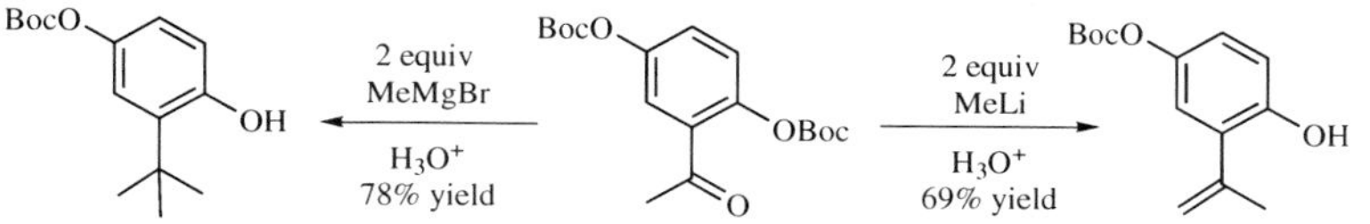

BocO, OH, **21** ← 2 equiv MeMgBr / H_3O^+, 78% yield — BocO, OBoc, O, **20** — 2 equiv MeLi / H_3O^+, 69% yield → BocO, OH, **22**

FIGURE 4.17 Methods B and E: disparate reactivity of *o*-OBoc acetophenones with organometallic reagents.

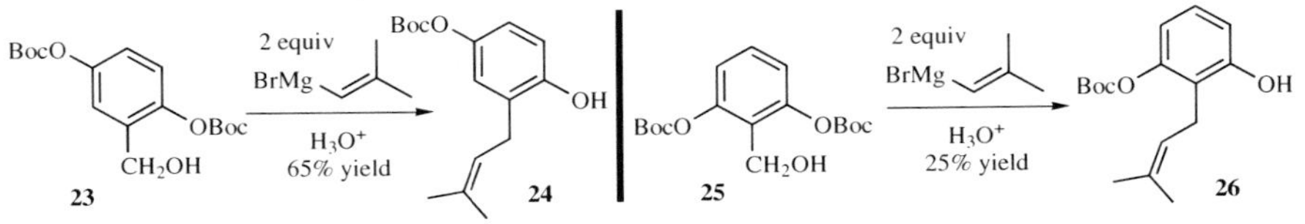

FIGURE 4.18 Synthesis of *o*-prenyl phenols by method D.

While pursuing quinone and cyclohexenone natural products **29** and **30** (Fig. 4.19) isolated from *Tapirira guianensis*, Hoarau developed several modifications for method C that enabled the use and recovery of more precious nucleophiles.[20] For example, addition of a lithiated sulfone to the 2,4-bis-OBoc-3-bromobenzaldehyde **27** proceeds to the phenol **28** in 75% yield. However, unlike the corresponding Grignard reagent, the sulfone moiety enables any unreacted nucleophile to be reisolated and reutilized in subsequent coupling reactions. Although magnesium bromide is usually not needed for the sodium borohydride prompted formation and 1,4-reduction of the *o*-QM intermediate, in this case it results in higher yields and shorter reaction times for the one-pot procedure.

Wang used method D to fashion a key intermediate for the synthesis of rishirilide B (Fig. 4.20).[21] The 2,4-bis-OBoc-3-methyl-benzyl alcohol (**31**) undergoes the addition of two equivalents of corresponding Grignard reagent to afford phenol **32** in 75% yield (Fig. 4.20). This material was subsequently elaborated by Mejorado in three steps (61% yield) to the corresponding 2,5-chiral cyclohexadienone **33**, which was ultimately transformed into (+)-rishirilide B (**34**).[22]

o-QMs generated via modified version of method D were also useful in the synthesis of several unique spironitronates.[23] For example, deprotonation of the 2-OBoc-4-methoxy-benzyl-alcohol **35** in the presence of methyl nitroacetate by addition of two equivalents of sodium *t*-butoxide gives the desired phenol **36** in 74% yield. Subsequent phenol oxidation affords a spironitronate ester **37** (Fig. 4.21).

Shenlin Huang has implemented method C with 2,4-bis-OBoc-3-bromobenzyaldehyde **27** (Fig. 4.22) and 2-(trimethylsilyl)ethoxy]methyl-lithium **38** at −78 °C in THF.[24] Surprisingly, lithium–halogen exchange does not happen; and the intermediate benzyl alcohol undergoes reduction with sodium borohydride in the same pot to afford the desired bromophenol **39** in 68% yield. This material

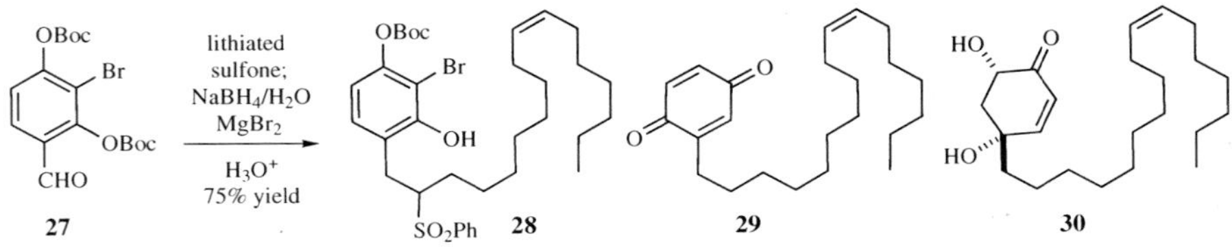

FIGURE 4.19 Synthesis of assigned structures isolated from *Tapirira guianensis*.

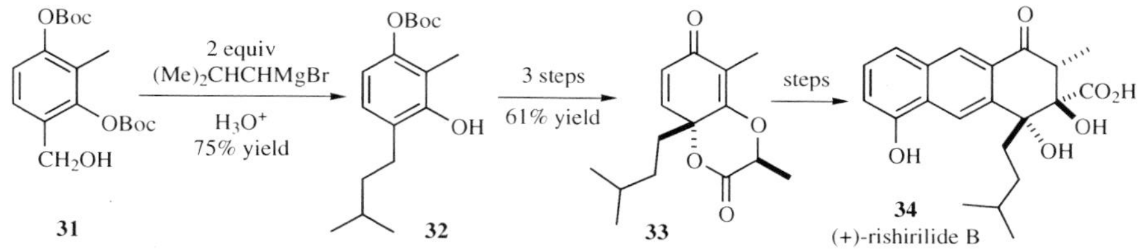

FIGURE 4.20 Synthesis of (+)-rishirilide B using method D.

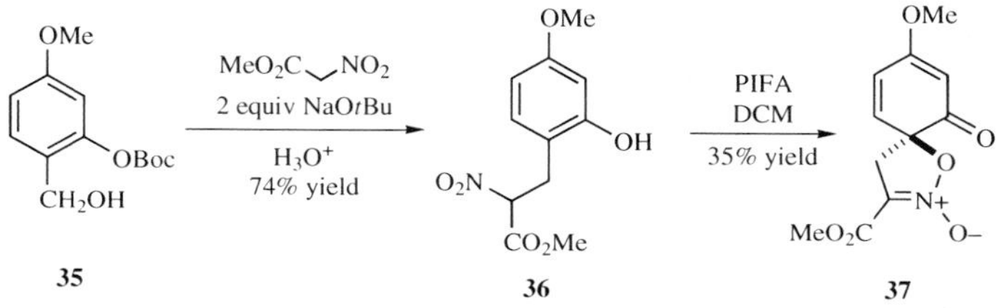

FIGURE 4.21 Synthesis of spironitronates using method D.

transforms in three pots to the corresponding *nonracemic* 2,5-cyclohexadieneone **40**, which readily converts into epi-cleroindicin D (**41**) and cleroindicin F (**42**).

By modification of method A, Jones has transformed 2,4-bis-OBoc-benzyaldehyde **5** into the 3-carbomethoxy dihydrocoumarin **43** in 68% yield (Fig. 4.23).[11a] The reaction proceeds by the addition of phenyl Grignard followed by addition of a preformed mixture of methyl malonate and sodium hydride and warming to room temperature. This particular example obviates the need for prior initiation by an organolithium reagent.

John Ward has functionalized an indane using method D in route to tetrapetalone A (**46**) (Fig. 4.24).[25] The *o*-OBoc benzyl alcohol **44** undergoes addition with two equivalents of Grignard and affords after acidic workup the phenolic indane **45** in 73% yield. Because of steric effects, only one diastereomer is observed after hydrolysis of the enol ether and thermodynamic equilibration of the

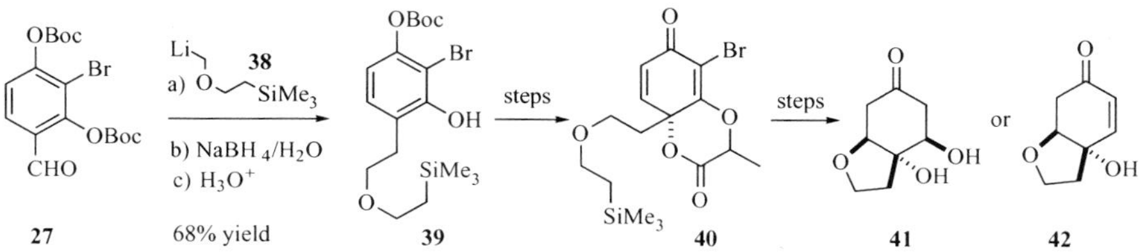

FIGURE 4.22 Synthesis of some cleroindicins using method C.

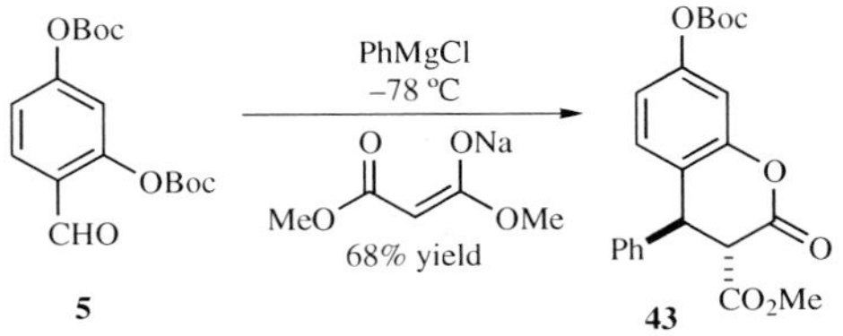

FIGURE 4.23 Synthesis of dihydrocoumarins via method A.

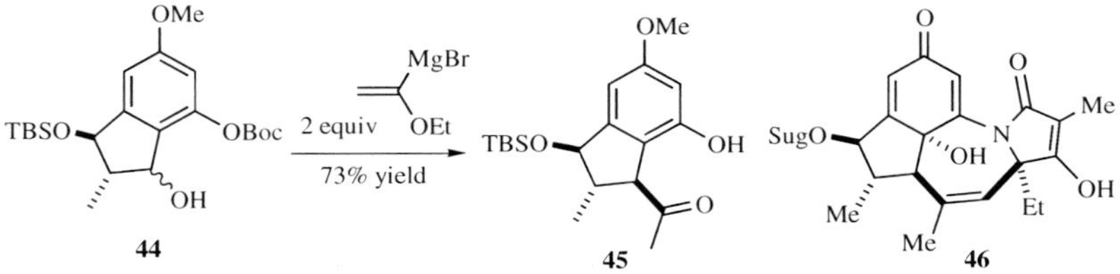

FIGURE 4.24 Strategy for tetrapetalone A (**46**) using method D.

methyl ketone. However, a 1:1 ratio is apparent in crude ^{1}H-NMR spectra taken of the initial reaction mixture that emerges from addition of the vinyl anion to the *o*-QM intermediate.

A modification to method C devised by Jake Cha has enabled the synthesis of 2-substituted benzodihydrofurans, while in route to phenolic constituents from *Dalbergia cochinchinensis* such as **49**.[26] For example, addition of phenyl lithium to 2-OBoc-4,5-bis-methoxybenzaldehyde **47** followed by the addition of iodomethyl Grignard proceeds to the benzodihydrofuran **48** in a 60% yield (Fig. 4.25).

4.3.3 Kinetically Controlled Cycloadditions

The Pettus group has also developed three methods (F–H, Fig. 4.26) enabling low-temperature, inverse demand cycloadditions of *o*-QM intermediates. Jones and Selenski began by investigating the reactions of styrenes with *o*-OBoc benzalcohols

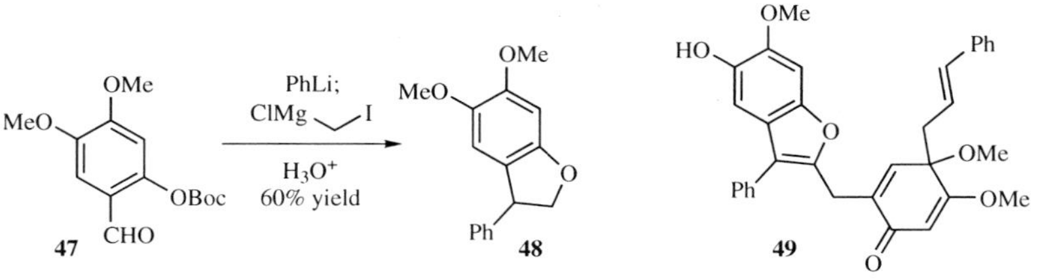

FIGURE 4.25 Strategy for phenol **49** from *Dalbergia cochinchinensis*.

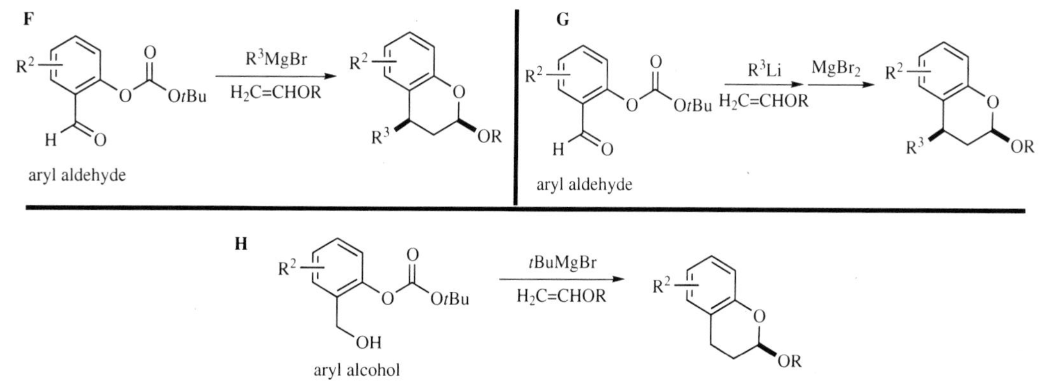

FIGURE 4.26 Pettus one-pot methods; kinetically controlled cycloaddition of *o*-QMs.

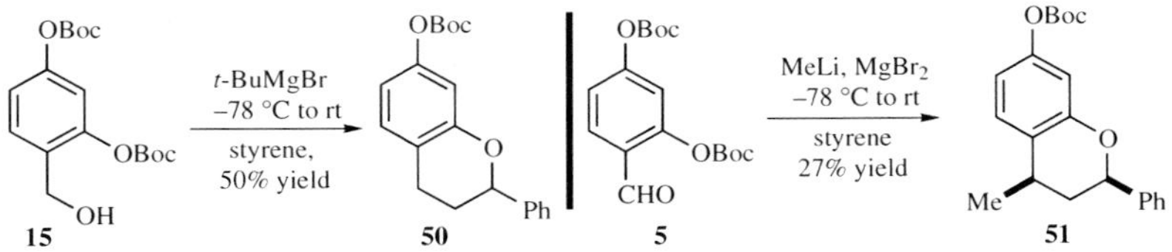

FIGURE 4.27 Some examples of low-temperature cycloadditions with styrene.

and *o*-OBoc benzaldehydes.[27] Unlike other *o*-QM cycloadditions conducted under thermal conditions, they observed a large preference for *cis* stereochemistry when an organometallic reagent was used to initiate the reaction from an *o*-OBoc benzaldehyde (methods F and G) suggesting an *endo* transition state between the styrene and *o*-QM.

The low yields, which are observed among styrenyl adducts, reflect a combination of the poor reactivity of the styrene at the low temperature of the reaction. For example, the combination of *t*-butyl Grignard with the 2,4-bis-OBoc-benzyl alcohol **15** affords the corresponding benzopyran **50** in only 50% yield even when carried out in the presence of 5–10 equivalents of the styrene (method H, Fig. 4.27).[27] Yields for substituted benzopyran styrene adducts are still lower (method G, Fig. 4.27). For example, addition of methyl lithium to 2,4-bis-OBoc-benzylaldehyde **5** followed by the addition of the dienophile and magnesium bromide affords benzopyran **51** in a paltry 27% yield. Method F is entirely ineffective in these cases, because the methyl Grignard reagent competes with the enol ether and with styrene 1,4-addition of methyl supercedes cycloaddition.

Better yields result from more nucleophilic styrene dienophiles. For example, method F proves successful with the benzaldehyde **5** and α-methoxystyrene to afford the benzopyran **52** in a 55% yield (Fig. 4.28).[27] The preferred diastereomer reflects an *endo* orientation with the more reactive moiety, which in this case is the vinyl ether portion of the dienophile. However, the diastereoselectivity for this and other 1,1-substituted alkenes is less than that for the corresponding mono-substituted systems.

Selenski pioneered further modifications to method G by initiating the formation of the *o*-QM intermediate through the reduction of the starting aldehyde (Fig. 4.29).[28] For example, the desired benzopyran **54** is formed in a 45% yield from the aldehyde **53** and

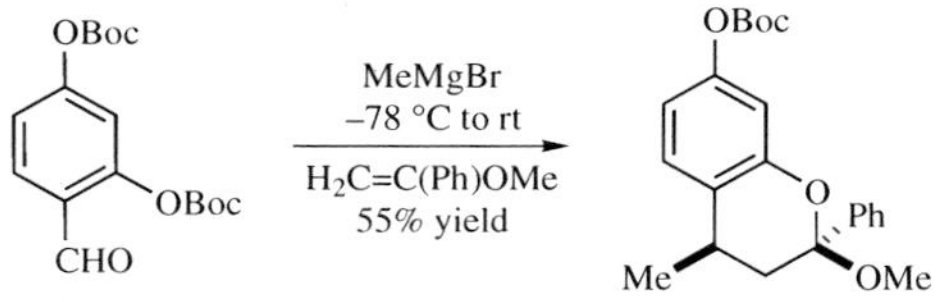

FIGURE 4.28 Cycloaddition with an activated styrene.

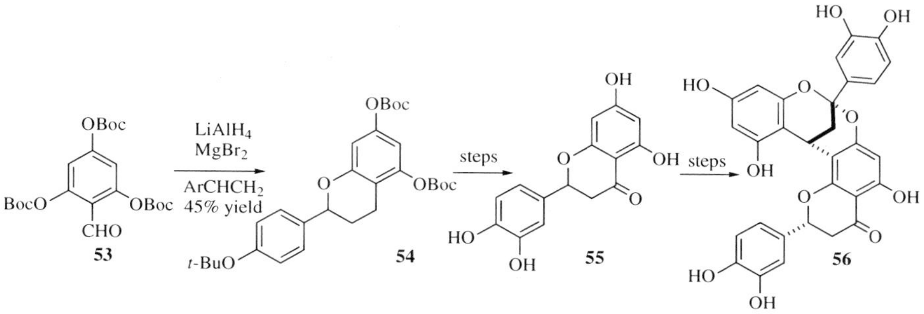

FIGURE 4.29 Selenski synthesis of eriodictyol (**55**) and diinsininone (**56**).

p-O*t*Bu-styrene. The reaction begins by the addition of lithium aluminum hydride to a solution containing the 2,4,6-tris-OBoc-benzaldehyde **53** and *para-tert*-butoxy-styrene. Next, magnesium bromide is added to cause the formation of the *o*-QM that results in the subsequent [4 + 2]-cycloaddition. Whether or not this hydride modification will work on other benzaldehydes that display only one flanking *o*-OBoc residue remains to be tested. Nevertheless, Selenski transformed the benzopyran adduct **54** into the flavanone known as eriodictyol (**55**) and further elaborated that structure into an unusual proanthocyanidin known as (±)-diinsininone (**56**).[28]

Jones and Selenski subsequently examined the reactivity of enol ethers in similar cycloaddition processes.[27] The yields are significantly higher than for styrenyl examples. For instance, implementation of method F by the addition of vinyl Grignard to 2,4-bis-OBoc-benzyaldehyde **5** at −78 °C in the presence of ethoxyvinyl ether (EVE) affords the corresponding chroman ketal **57** in a 70% yield after warming to room temperature and workup with acid (Fig. 4.30). Interestingly, the reaction works well even if an excess of the initiating Grignard reagent is added, which suggests that the cycloaddition with EVE is faster than the conjugate addition with the residual Grignard reagent.

Method G is used to introduce the alkyl fragment when less reactive alkenes are employed or for cases where functionality within the dienophilic alkene undergoes reaction with the Grignard reagent. Following this procedure, a lithium anion is first added to the aldehyde **5** at −78 °C.[27] After consumption of the aldehyde has been determined by TLC, the dienophile is added and magnesium bromide is introduced. The cycloaddition occurs as the reaction warms to room temperature. In the case of

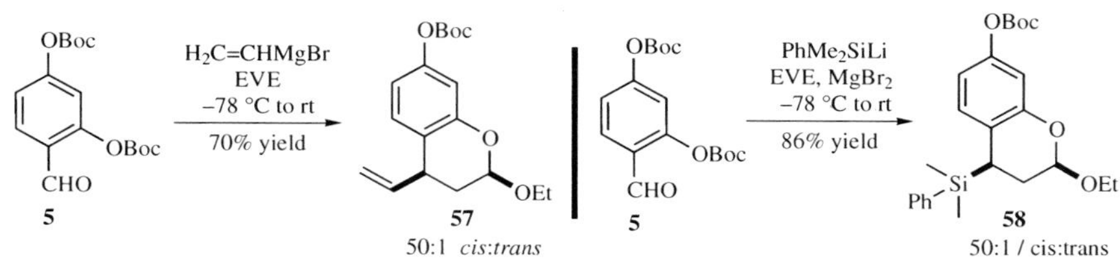

FIGURE 4.30 Cycloaddition with EVE, method F (left) and method G (right).

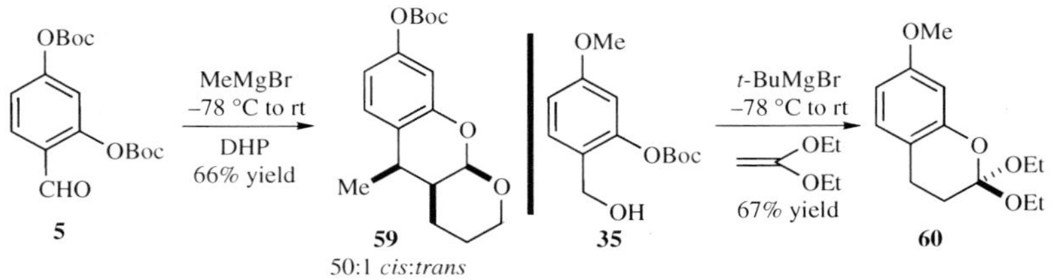

FIGURE 4.31 Reactions with some other varieties of enol ethers.

$PhMe_2SiLi$, the corresponding procedure yields the benzopyran **58** in 86% yield from benzaldehyde **5** with better than 50:1 diastereoselectivity.

As expected, other enol ethers work well in these procedures. For example, Jones and Selenski find that implementation of method F, which occurs by addition of MeMgBr to benzaldehyde **5** in the presence of dihydropyran (DHP) at −78 °C affords a 66% yield of the corresponding tricyclic ketal **59** with better than 50:1 *endo* diastereoselectivity (Fig. 4.31).[27] On the contrary, Lindsey reports use of method H with the benzyl alcohol **35** and diethylketene acetal. The cycloaddition reaction occurs almost instantaneously upon deprotonation of the benzyl alcohol **35** by *t*-butylmagnesium bromide in the presence of the ketene acetal and yields the corresponding benzopyran *ortho* ester **60** in a 67% yield.[29]

Methods F and H have also been employed with aromatic nuclei serving as the corresponding dienophile. Jones and Selenski found that the addition of methyl magnesium bromide to 2,4-bis-OBoc-benzaldehyde **5** in the presence of four equivalents of furan affords a 76% yield of the expected cycloadduct **61** (Fig. 4.32).[27] The regiochemistry reflects the respective orbital coefficients in the HOMO of furan which serves as a dienophile for this reaction. Lindsey has observed the corresponding cycloaddition with 2-amino-oxazoles to produce **62**.[30] The cycloaddition is triggered by the addition of a slight excess of *t*-butyl-magnesium bromide to the benzyl alcohol **35** stirring at −78 °C with four equivalents of the corresponding 2-amino-oxazole. However, not all heteroaromatic nuclei participate in these protocols. Simpler oxazoles and nitrogen substituted pyrroles and indoles, such as the one participating in the Funk procedure shown in Fig. 4.2, fail to undergo cycloaddition. Presumably, at these low operational temperatures, the resonance stabilization for these heterocyclic

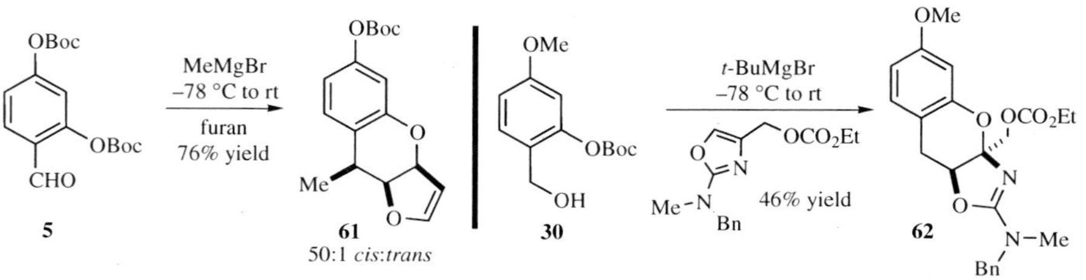

FIGURE 4.32 Some examples of aromatic compounds as dienophiles.

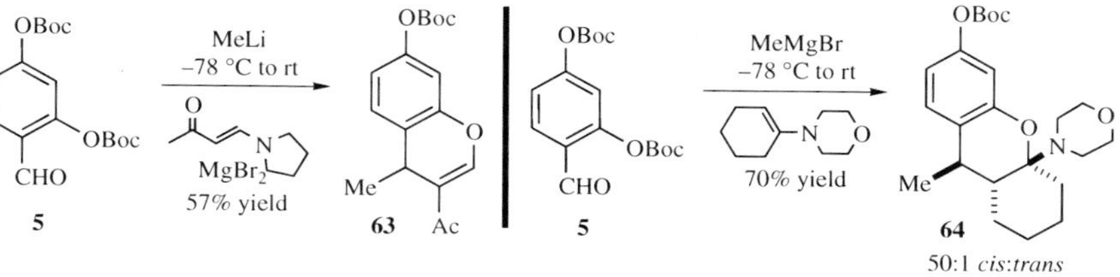

FIGURE 4.33 Some examples of enamine dienophiles.

aromatic dienophiles is significantly larger than the energy released by the rearomatization caused by the reaction with the *o*-QM.

Several enamines also participate in these cycloaddition reactions. For example, the addition of methyl lithium to benzaldehyde **5** and the sequential introduction of the vinylogous amide and magnesium bromide results in the cycloaddition elimination product chromene **63** (method G, Fig. 4.33).[27] The introduction of methyl magnesium bromide to a solution of the benzaldehyde **5** and two equivalents of the morpholine enamine produces the cycloadduct **64** in 70% yield with better than 50:1 diastereoselectivity (method F). Less reactive enamides, such as that used by Ohwada in Fig. 4.4, however, fail to participate in these conditions.

Imines, on the contrary, proved particularly reactive under these conditions (Fig. 4.34). For example, Jones and Selenski report that the introduction of one equivalent of methyl magnesium bromide to benzaldehyde **5** stirring at −78 °C in the presence of one and half equivalents of the imine that is derived from the condensation of benzyl amine and benzaldehyde proceeds immediately to the aminal **65** in 94% yield.[27] Only the *trans* isomer is observed from this low-temperature cycloaddition. While the relative stereochemistry appears to be result of an *exo* transition state, we suspect that initial *cis* adduct from and *endo* addition may epimerize under these conditions.

Unlike their thermal counterparts, these low-temperature procedures result in good diastereoselectivity, presumably because the reactions proceed under kinetic control. For example, while investigating a now defunct strategy aimed at heliquinomycin, Lindsey reports an exclusive cycloadduct **66** forms in 31% yield when the alcohol **35** and the oxazole are mixed together with *t*-butyl Grignard

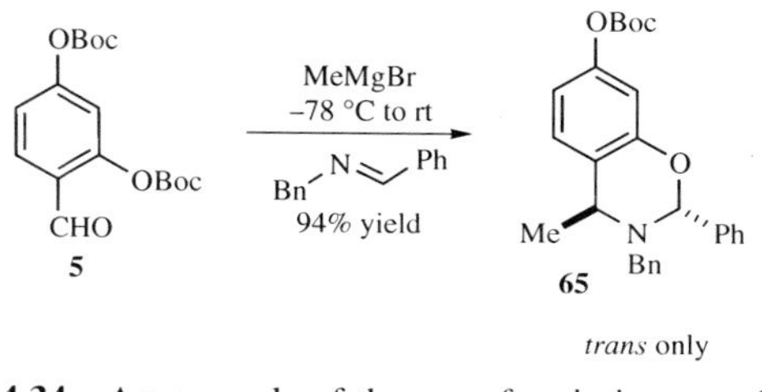

FIGURE 4.34 An example of the use of an imine as a dienophile.

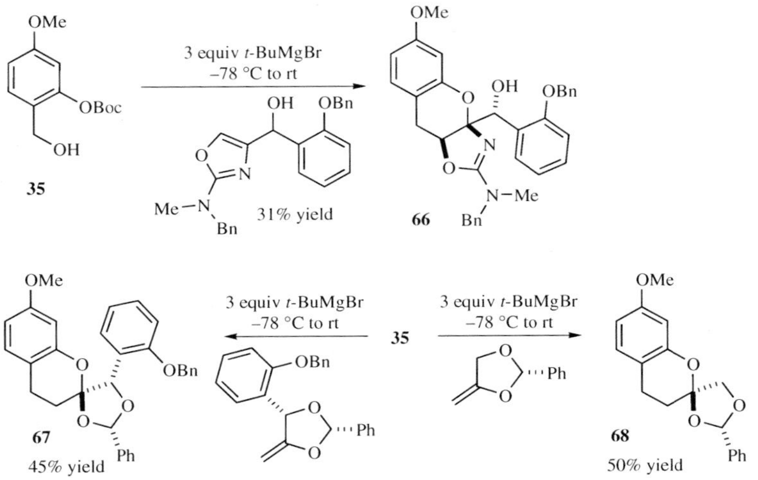

FIGURE 4.35 Some diastereoselective examples using chiral dienophiles.

(method H, Fig. 4.35).[30] He also observed that alcohol **35** undergoes diastereoselective reactions with various enol ethers derived from chiral 1,3-dioxolanes (for example, spirocycles **67** and **68** are formed in respective yields of 45% and 50%, diastereoselective ratio >20:1).[31]

Yaodong Huang, while pursuing the synthesis of (+)-berkelic acid (**69**), reported a diastereoselective cycloaddition using method H that leads to another type of 5,6-aryloxy spiroketals (Fig. 4.36).[32] For example, addition of three equivalents of *t*-butyl magnesium bromide to alcohol **70** in the presence of the exocyclic enol ether **71** proceeds in a 72% yield to the spiroketal **72** with a 4.5:1 selectivity favoring the *endo* approach (Fig. 4.36). Additional experiments suggest the bromine atom decreases the HOMO–LUMO band gap and improves diastereoselectivity.

Selenski investigated the use of chiral enol ether auxiliaries in order to adapt method F–H for enantioselective syntheses. After surveying a variety of substituted and unsubstituted enol ethers derived from a vast assortment of readily available chiral alcohols, she chose to employ enol ethers derived from *trans*-1,2-phenylcyclohexanol such as **73** and **74** (Fig. 4.37). These derivatives were found to undergo highly diastereoselective cycloadditions resulting in the formation of **75** and **76** in respective

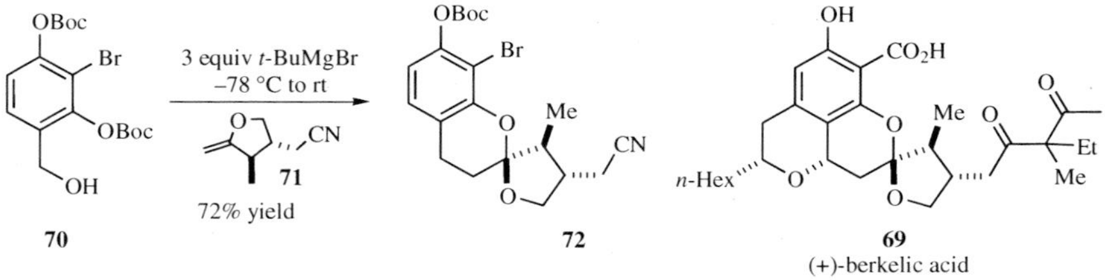

FIGURE 4.36 A diastereoselective cycloaddition strategy for (+)-berkelic acid 69.

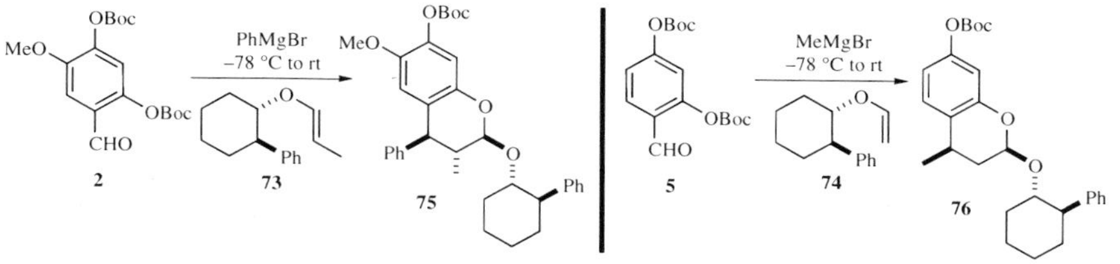

FIGURE 4.37 Some examples of diastereoselective cycloadditions using chiral auxiliaries.

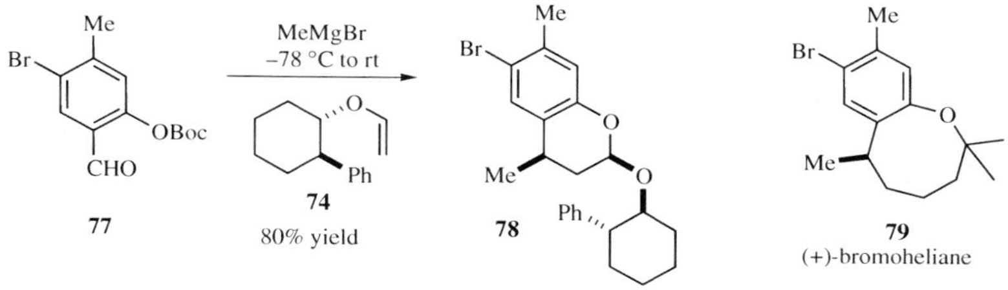

FIGURE 4.38 A diastereoselective cycloaddition to the synthesis of (+)-bromoheliane.

yields of 60% and 75%, both in >95% *de*. Compound **75** was subsequently used in an enantioselective synthesis of (+)-mimosifoliol.[33]

Jason Green has successfully applied the Selenski method to the synthesis of (+)-bromoheliane (**79**, Fig. 4.38).[34] In this example, two equivalents of the chiral enol ether are added to the benzaldehyde **77** in diethyl ether (0.1 M) and cooled to −78 °C. Methyl Grignard is then added. The cycloaddition occurs while the reaction warms to room temperature. The benzopyran adduct **78** forms in 80% yield with 50:1 diastereoselectivity. DFT calculations and experiments suggest that the diastereoselectivity depends on the magnitude of the HOMO–LUMO band gap. In this instance, the LUMO of the supposed *o*-QM intermediate is computed to be −2.6 eV, whereas the HOMO of the enol ether is −5.9 eV. A 50:1 selectivity is recorded for resulting 3.3 eV gap. For reactions of 2,5-bis-OBoc-4-methyl-benzaldehyde, where the HOMO–LUMO gap is larger (3.6 eV), a 20:1 ratio of diastereomers is observed.

4.4 MECHANISTIC INVESTIGATIONS

From the synthetic investigations that have been described in the previous schemes, an appreciation of the mechanism for these reactions (methods A–H) has emerged in our group. However, the characteristics and exact nature of the *o*-quinone methide intermediate are still debated. Our past observations clearly indicate the cascade leading to the reactive species that behaves as an *o*-quinone methide should behave is

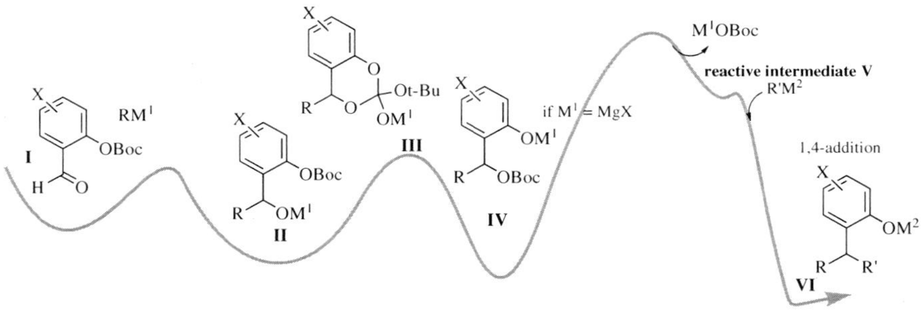

FIGURE 4.39 Early speculation as to the cascade.

initiated by the formation of a benzyl metal alkoxide **II**, as caused by deprotonation (methods D and H), or by the addition of nucleophile R^3M to the benzaldehyde (methods A, B, C, F, G), or by the addition of nucleophile R^3M to the corresponding acetophenone (methods B and E). The alkoxide then undergoes reaction with the adjacent phenolic acyl residue to form the tetrahedral transition state **III** that collapses to yield the more stable phenoxide anion **IV**, and in doing so transfers the –OBoc protecting group to the benzyl alcohol (Fig. 4.39). This first half of the cascade leading to intermediate **IV** seems energetically reasonable given the energy of the entering organometallic species as compared with the intermediate metal alkoxide **II** and ending phenoxide **IV**. The existence of intermediate **IV** was corroborated by acylation of the lithium phenoxide **IV** by acetyl chloride (**81**, Fig. 4.40). In the absence of acetyl chloride, however, the subsequent fate of lithium species **IV** remains something of an enigma because it fails to behave as expected and undergoes decomposition. On the contrary, we have been unable to intercept intermediate **IV** for reactions initiated by magnesium reagents, even in the presence of HMPA. It would seem that in the case of magnesium intermediate **IV** the subsequent reaction is almost instantaneous. For example, at −78 °C, addition of one-half of an equivalent of methyl magnesium bromide to the aldehyde **5** affords a 1:1 ratio of the bis-addition product **82** along with the starting material **5** (Fig. 4.40).

Assuming the pathway is similar for both lithium and Grignard reagents, one interpretation of these observations is that the rate-limiting step is the formation of the reactive intermediate **V** from **IV**; magnesium(II) facilitates a transformation into an *o*-quinone methide, whereas it proves impossible for the corresponding lithium(I) species. Other stark examples of contrast between the lithium species **IV** and its magnesium counterpart support this notion (Fig. 4.40). Among these was the observation that "magnesium reactivity" could be reintroduced by the subsequent introduction of a magnesium reagent to the lithium species **80** as shown by the formation of compounds **82** and **83** from **80**. Without the addition of the Grignard reagent, compound **80** simply decomposes. Similar findings were observed for acetate **81**, which was reintroduced to the cascade by the addition of the respective lithium or magnesium species. We initially presumed that these divergent results were caused by a simple metal exchange with magnesium bromide that is available from the Schlenk

OBoc
OBoc
5
H O
a) MeLi
−78 °C
2 MeMgBr ⇌ $Me_2Mg + MgBr_2$
OBoc
2 equiv
MeMgBr
O
83
OEt
OEt
−78 °C to rt
Et_2O
OBoc
OLi
80
Me OBoc
MeMgBr
−78 °C to rt
Et_2O
OBoc
OH
Me Me 82
10 equiv
OEt
−78 °C to rt
MeMgBr
b) AcCl
−78 °C to rt
EtO 2–10 equiv
MeLi; $MgBr_2$
−78 °C to rt
EtO
OBoc
OAc
O
81
O Ot-Bu
MeLi
EtO
2–10 equiv
−78 °C to rt
Et_2O
starting
material
decomposition

FIGURE 4.40 Mechanistic studies on transformation of lithium species **80**.

equilibrium via the corresponding organomagnesium reagent. However, the fact that the lithium species **80** failed to undergo cycloaddition or 1,4-additions, or produce any product for that matter except for acetophenone examples subsequently discussed (Fig. 4.43), remained quite troubling if the reactive intermediate were simply an *o*-quinone methide species.

Based on these findings, we originally speculated that in the presence of $MgBr_2$ the corresponding magnesium phenoxide undergoes a β-elimination of BrMgOBoc, to form yet another lower energy alkoxide and the *o*-QM (Fig. 4.41). Once the

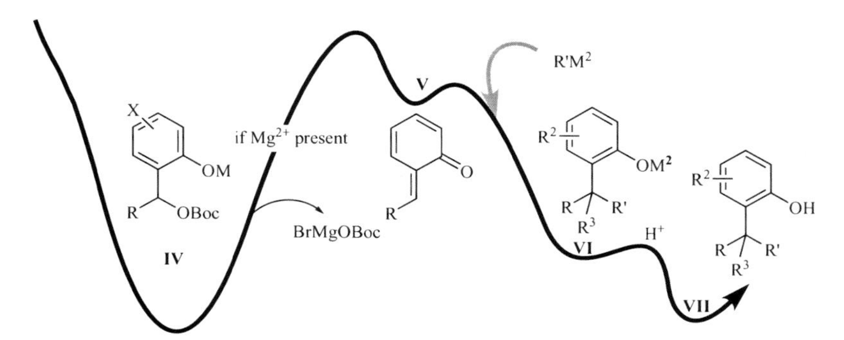

FIGURE 4.41 Speculation as to the second half of the cascade.

o-QM species **V** formed from this elimination, it undergoes immediate reaction with the most reactive nucleophile it encounters. From extensive studies, the β-elimination step appears to be both temperature and structure dependent, as congested systems (R = *i*Pr) and conjugated systems (R = Ph) are slow to undergo subsequent reactions.

While not fully understood, the transformation is significant for several reasons. First, the generation and subsequent kinetic consumption occur in a single pot only if magnesium(II) is present. This requirement provides a low-temperature trigger to access the reactive intermediate **V**. Second, the method appears to enable formation of low concentrations of the *o*-QM species for application in subsequent reactions under kinetic control, and this fact seems to enable use of a limited quantity of the subsequent nucleophile. Prior applications of *o*-QMs intended for intermolecular reactions have proven tedious. They have required the preparation of a stable precursor that was subsequently derivatives to the *o*-QM intermediate by high temperatures and low pressure. If not conducted at low pressure, the temperature had to be high enough to reverse unwanted side products back into the synthetic stream by leading to the most stable product. In the case of the anionic Rokita kinetic conditions, a vast quantity of the subsequent nucleophile is required to prevent unwanted side reactions. For these reasons, our methods A–H are able to deliver a far greater range of adducts than the closest competing strategies. Moreover our procedures prove diastereoselective and scalable.

However, the differences in the fates of lithium and magnesium phenoxides **IV** remain quite troubling, as do the energy requirements if the process involves a simple *o*-QM intermediate as **V**. A β-elimination leading to the *o*-QM **V** can be viewed as a very energetically demanding dearomatization. Therefore, it is difficult to imagine the difference in energy between the phenoxide **IV** and OBoc alkoxides **II** as proving sufficient to facilitate the formation of **V** merely because of the difference in Lewis acidities between magnesium and lithium cations. Moreover, chelation causes the –OBoc leaving to be orthogonal with the π system of the phenoxide enolate and would discourage rather than encourage the requisite β-elimination (Fig. 4.42).

Further peculiarities (Fig. 4.43) have surfaced among the reaction products of the corresponding acetophenones, which have caused us to reconsider the fate of

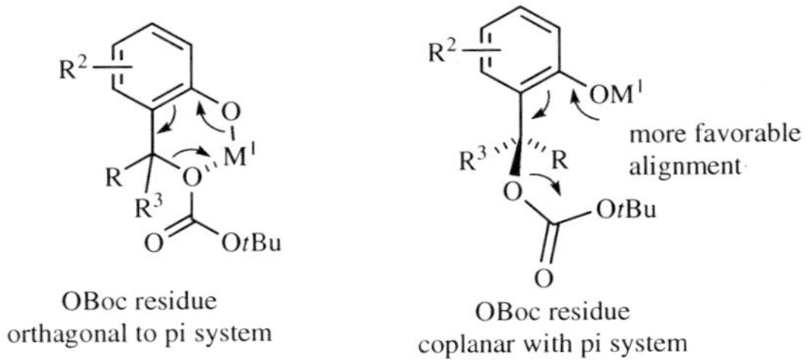

OBoc residue
orthagonal to pi system

more favorable
alignment

OBoc residue
coplanar with pi system

FIGURE 4.42 Questionable π alignment for β-elimination.

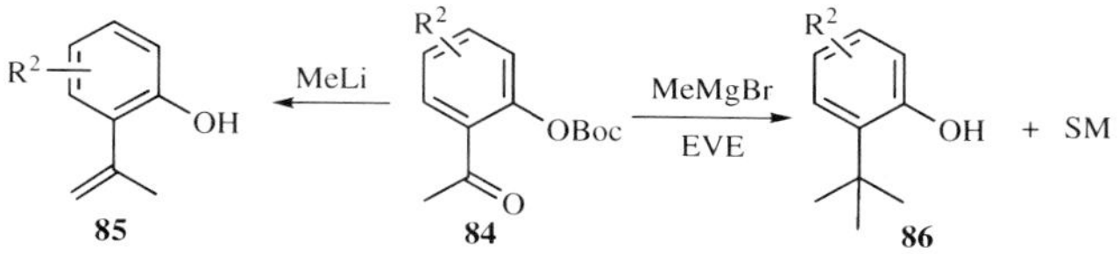

FIGURE 4.43 Is *o*-QM the reactive species at all?

intermediate **IV** and reevaluate its *o*-QM character. For example, addition of one equivalent of methyl lithium to the acetophenone **84** affords the styrene **85** in good yield.[16] The most reasonable explanation for its formation is an energetically favorable 1,5-sigmatropic shift in an *o*-QM intermediate, yet lithium reagents are not supposed to be able to access the *o*-QM intermediate. On the contrary, addition of one equivalent of methyl Grignard to the corresponding acetophenone both in the presence and outside the presence of ethoxyvinyl ether (EVE) affords the corresponding *t*-butyl phenol **86** and starting material in a 1:1 ratio. It is interesting to note that EVE fails to trap intermediate **IV** and that 1,4-addition of a methyl anion proves faster than cycloaddition or the corresponding sigmatropic 1,5-rearrangement. This result suggests that perhaps the reactive intermediate is not an *o*-QM species at all, but perhaps a biradical intermediate.

Therefore, in Fig. 4.44 we have presented a different interpretation of our many findings and results. Perhaps in cases of lithium reagents, β-elimination can and does occur via a coplanar elimination (Fig. 4.42) from the monodentate lithium phenoxide resulting in a very high energy *o*-QM intermediate (**V**). Once formed it undergoes polymerization ($R' = H$) or reaction with the first nucleophile it finds. We would expect that sigmatropic rearrangement, in the case of β,β-disubstituted *o*-QMs would supercede decomposition, as we have observed. On the contrary, in the presence of magnesium bromide, perhaps a remarkable magnesium species **V** forms, which serves to further stabilize the system. It may resemble after some fashion a biradical canonical representation of an *o*-QM. This unusual species undergoes homocoupling with additional Grignard reagent to form sterically demanding products such as the *o*-*t*-butyl-phenol **86**, rather than participate in the sigmatropic rearrangement. This notion would explain the tail–tail dimer, such as **88**, which we have isolated on occasion and some of the unusual chroman products arising from unreactive alkenes such as **87**, which is observed when vinyl Grignard reagents are used to initiate the cascade from **23**.

4.5 LONG-TERM PROSPECTS

Irrespective of the exact mechanism, the recent advent of low-temperature methods clearly rank among the most powerful methods for constructing compounds from intermediates resembling *o*-QMs, and these processes have largely domesticated these previously untamed and highly reactive species. As discussed in the preceding section,

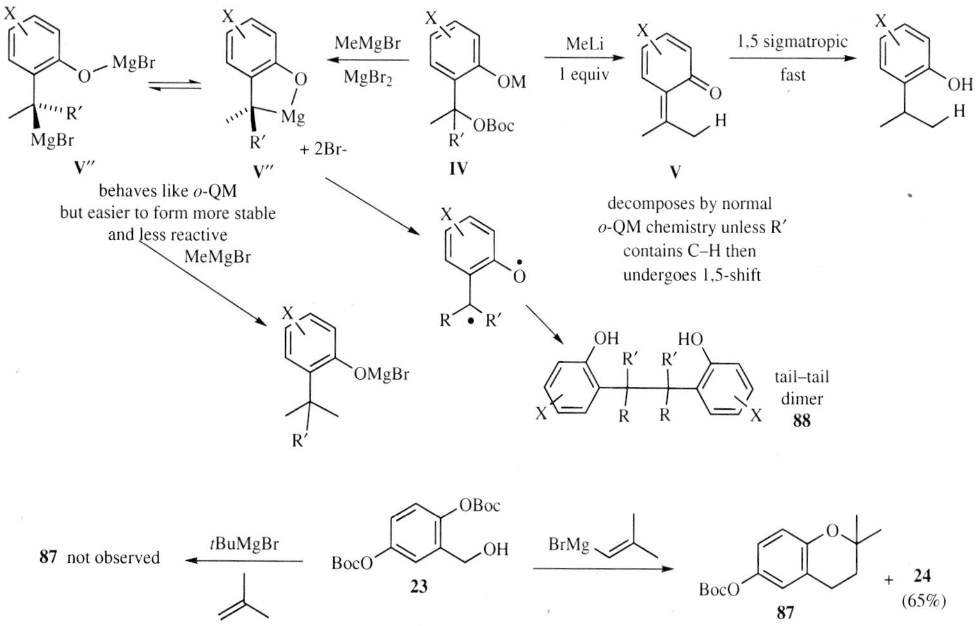

MeMgBr
$MgBr_2$
MeLi
1 equiv
1,5 sigmatropic
fast
V″
V″
+ 2Br-
IV
V
behaves like *o*-QM
but easier to form more stable
and less reactive
MeMgBr
decomposes by normal
o-QM chemistry unless R′
contains C–H then
undergoes 1,5-shift
tail–tail
dimer
88
87 not observed
*t*BuMgBr
23
BrMg
87
+ **24**
(65%)

FIGURE 4.44 Magnesium intermediates may account some unusual reactivity.

these new methods have enabled a gradual and controlled formation and consumption of an entity behaving as an *o*-QM should and have led to product distributions that reflect a reaction under kinetic control. These new processes should prove to be expedient stratagems in many future total syntheses.

REFERENCES

1. Crawley, S. L.; Funk, R. L. A synthetic approach to nomofungin/communesin B. *Org. Lett.* 2003, 5, 3169–3171.
2. Snider, B. B.; Lobera, M. Synthesis of the tetracyclic core of the bisabosquals. *Tetrahedron Lett.* 2004, 45, 5015–5018.
3. Sugimoto, H.; Nakamura, S.; Ohwada, T. Generation and application of *o*-quinone methides bearing various substituents on the benzene ring. *Adv. Synth. Catal.* 2007, 349, 669–679.
4. Yoshida, H.; Watanabe, M.; Fukushima, H.; Ohshita, J.; Kunai, A. A 2:1 coupling reaction of arynes with aldehydes via *o*-QM: straightforward synthesis of 9-arylxanthenes. *Org. Lett.* 2004, 6, 4049–4045.
5. Pande, P.; Shearer, J.; Yang, J.; Greenberg, W. A.; Rokita, S. E. Quinone methide alkylation of deoxycytidine. *J. Am. Chem. Soc.* 1999, 121, 6773–6779.
6. Barrero, A. F.; Moral, J. F. Q.; Herrador, M. M.; Arteaga, P.; Cortés, M.; Benites, J.; Rosellón, A. Mild and rapid method for the generation of *o*-QM intermediates. Synthesis of puupehedione analogues. *Tetrahedron* 2006, 62, 6012–6017.
7. Mattson, A. E.; Scheidt, K. A. Nucleophilic acylation of *o*-QM: an umpolung strategy for the synthesis of aryl ketones and benzofurans. *J. Am. Chem. Soc.* 2007, 129, 4508–4509.
8. McLoughlin, B. J. A novel reduction of carbonyl to methylene by the action of sodium borohydride. *J. Chem. Soc. D* 1969, 10, 540–541.
9. Mitchell, D.; Doecke, C. W.; Hay, L. A.; Koenig, T. M.; Wirth, D. D. *Ortho*-hydroxyl assisted deoxygenation of phenones. Regiochemical control in the synthesis of mono-protected resorcinols and related polyphenolic hydroxyl systems. *Terahedron Lett.* 1995, 36, 5335–5338.
10. Van de Water, R. W.; Magdziak, D. J.; Chau, J. N.; Pettus, T. R. R. New construction of *ortho* ring-alkylated phenols via generation and reaction of assorted *o*-QM. *J. Am. Chem. Soc.* 2000, 122, 6502–6503.
11. (a) Jones, R. M.; Van de Water, R. W.; Lindsey, C. C.; Hoarau, C.; Ung, T.; Pettus, T. R. R. A mild anionic method for generating *o*-QM: facile preparations of *ortho*-functionalized phenols. *J. Org. Chem.* 2001, 66, 3435–3441. (b) Van de Water, R. W.; Pettus, T. R. R. *o*-Quinone methides: intermediates underdeveloped and underutilized in organic synthesis. *Tetrahedron Lett.* 2002, 58, 5367–5405.
12. Tuttle, K.; Rodriguez, A. A.; Pettus, T. R. R. An expeditious synthesis of (±)-mimosifoliol utilizing a cascade involving an *o*-QM intermediate. *Synlett* 2003, 14, 2234–2236.
13. Mejorado, L. H. Ph.D. Thesis, Development and application of nonracemic cyclohexa-2,5-dienones in the total synthesis of (+)-epoxysorbicillinal and (+)-rishirilide B. University of California, Santa Barbara, 2006.

14. Sessions, E. H.; O'Connor, R. T., Jr.; Jacobi, P. A. Furanosteroid studies. Stereoselective synthesis of the A, B, E-ring core of wortmannin. *Org. Lett.* 2007, 9, 221–3224.
15. Rodriguez, R.; Adlington, R. M.; Moses, J. E.; Cowley, A.; Baldwin, J. E. A new and efficient method for *o*-quinone methide intermediate generation: application to the biomimetic synthesis of (±)-alboatrin. *Org. Lett.* 2004, 6, 3617–3619.
16. Van De Water, R. W. Ph.D. Thesis, Alkylation and dearomatization of resorcinal derivatives: A synthetic route towards scyphostatin. University of California, Santa Barbara, 2003.
17. Mejorado, L. H.; Hoarau, C.; Pettus, T. R. R. Diastereoselective dearomatization of resorcinols directed by a lactic acid tether: unprecedented enantioselective access to *p*-quinols. *Org. Lett.* 2004, 6, 1535–1538.
18. Lindsey, C. C.; Gomez-Diaz, C.; Villalba, J. M.; Pettus, T. R. R. Synthesis of the F11334's from *o*-prenylated phenols: muM inhibitors of neutral sphingomyelinase (*N*-SMase). *Tetrahedron* 2004, 58, 4559–4565.
19. Hoarau, C.; Pettus, T. R. R. Strategies for the preparation of differentially protected *ortho*-prenylated phenols. *Synlett* 2003, 1, 127–137.
20. (a) Hoarau, C.Ph.D. Thesis, Synthetic approaches for natural products isolated from Tapirira guianensis. University of California, Santa Barbara, 2005. (b) Hoarau, C.; Pettus, T. R. R. General synthesis for chiral 4-alkyl-4-hydroxycyclohexenones. *Org. Lett.* 2006, 8, 2843–2846.
21. Wang, J.; Pettus, L. H.; Pettus, T. R. R. Cycloadditions of *o*-quinone dimethides with *p*-quinol derivatives: regiocontrolled formation of anthracyclic ring systems. *Tetrahedron Lett.* 2004, 45, 1793–1796.
22. Mejorado, L. H.; Pettus, T. R. R. Total synthesis of (+)-rishirilide B: development and application of general processes for enantioselective oxidative dearomatization of resorcinol derivatives. *J. Am. Chem. Soc.* 2006, 128, 15625–15631.
23. Marsini, M.; Huang, Y.; Van De Water, R. W.; Pettus, T. R. R. Synthesis and reactions of spironitronates. *Org. Lett.* 2007, 9, 3229–3232.
24. Huang, S. University of California, Santa Barbara, CA. Private communication, 2008.
25. Ward, J. University of California, Santa Barbara, CA. Private communication, 2008.
26. Cha, J. University of California, Santa Barbara, CA. Private communication, 2008.
27. Jones, R. M.; Selenski, C.; Pettus, T. R. R. Rapid syntheses of benzopyrans from *o*-OBOC salicylaldehydes and salicyl alcohols: a three-component reaction. *J. Org. Chem.* 2002, 67, 6911–6915.
28. Selenski, C.; Pettus, T. R. R. (±)-Diinsininone: made nature's way. *Tetrahedron* 2006, 62, 5298–5307.
29. Lindsey, C.C. Ph.D. Thesis, Synthetic approach for models subromycin and related natural products. University of California, Santa Barbara, 2005.
30. Lindsey, C. C.; Hoarau, C.; Ung, T.; Pettus, T. R. R. Unusual cycloadditions of *o*-QM with oxazoles. *Tetrahedron Lett.* 2006, 47, 201–204.
31. Marsini, M. A.; Huang, Y.; Lindsey, C. C.; Wu, K.-L.; Pettus, T. R. R. Diastereoselective syntheses of chroman spiroketals via [4 + 2] cycloaddition of enol ethers and *o*-quinone methides. *Org. Lett.* 2008, 10, 1477–1480.

32. Huang, Y.; Pettus, T. R. R. A cycloaddition strategy for use toward berkelic acid, a MMP inhibitor and potent anticancer agent displaying a unique chroman spiroketal motif. *Synlett* 2008, 1353–1356.
33. Selenski, C.; Pettus, T. R. R. Enantioselective [4 + 2] cycloadditions of *o*-quinone methides: total synthesis of (+)-mimosifoliol and formal synthesis of (+)-tolterodine. *J. Org. Chem.* 2004, 69, 9196–9203.
34. Green, J. University of California, Santa Barbara, CA. Private communication, 2008.

5

SELF-IMMOLATIVE DENDRIMERS BASED ON QUINONE METHIDES

Rotem Erez and Doron Shabat

Department of Organic Chemistry, School of Chemistry, Raymond and Beverly Sackler Faculty of Exact Sciences, Tel Aviv University, Tel Aviv 69978, Israel

5.1 INTRODUCTION

It was in 1981 when the group of Katzenellenbogen reported a novel chemical linkage for solving certain problems in prodrug design.[1] The report suggested to link a drug to an enzyme substrate through a self-immolative linker and thereby generating a tripartite prodrug (Fig. 5.1). The prodrug is stable as long as the enzymatic substrate stays attached. However, cleavage of the substrate by the enzyme generates an intermediate that rapidly releases the free active drug.

Compound **1** was synthesized as a model prodrug molecule (Fig. 5.2). In this model, 4-nitroaniline is simulating a drug and a Boc-lysine amino acid is applied as an enzymatic substrate for trypsin. 4-Amino-benzylalcohol is used to link between the two as a self-immolative linker. Cleavage of the Boc-lysine by trypsin generated intermediate molecule **2** that spontaneously disassembled through 1,6-elimination and decarboxylation to release the model drug. This elimination is accompanied by the generation of the azaquinone methide **3** as a by-product that usually trapped by a water molecule to give 4-amino-benzylalcohol. Since that report by Katzenellenbogen, hundreds of publications that appeared in the scientific literature use this technique in prodrug design. Similarly to 4-amino-benzylalcohol, 4-hydroxy-benzylalcohol was also applied as a self-immolative linker that spontaneously eliminates through quinone methide rearrangement.

Quinone Methides, Steven E. Rokita

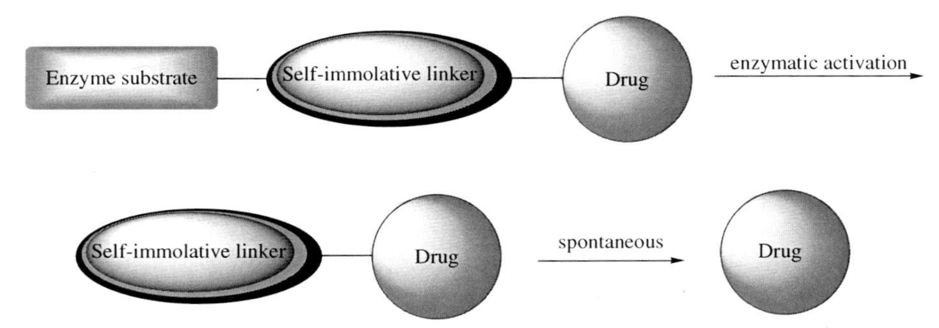

FIGURE 5.1 Graphical structure of a tripartite prodrug with self-immolative linker. (See the color version of this figure in Color Plates section.)

Recently, we realized that the azaquinone methide rearrangement could take place through three benzylic substituents (two *ortho* and one *para*) of the corresponded aniline derivative.[2] Figure 5.3 shows the triple azaquinone methide elimination mechanism, which is initiated by cleavage of an enzymatic substrate in the AB_3 dendritic adaptor **4**, resulting in the release of the three reporter groups. This observation led to the discovery of a new kind of molecules that were termed as self-immolative dendrimers. In fact, three groups (including ours) almost

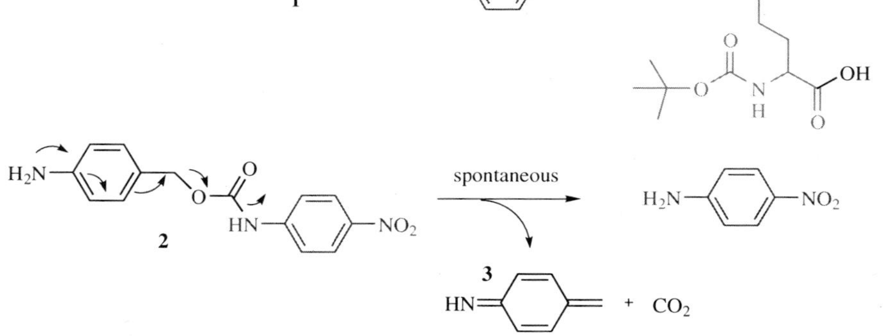

FIGURE 5.2 Activation mechanism of a tripartite prodrug through enzymatic cleavage and elimination of an azaquinone methide. (See the color version of this figure in Color Plates section.)

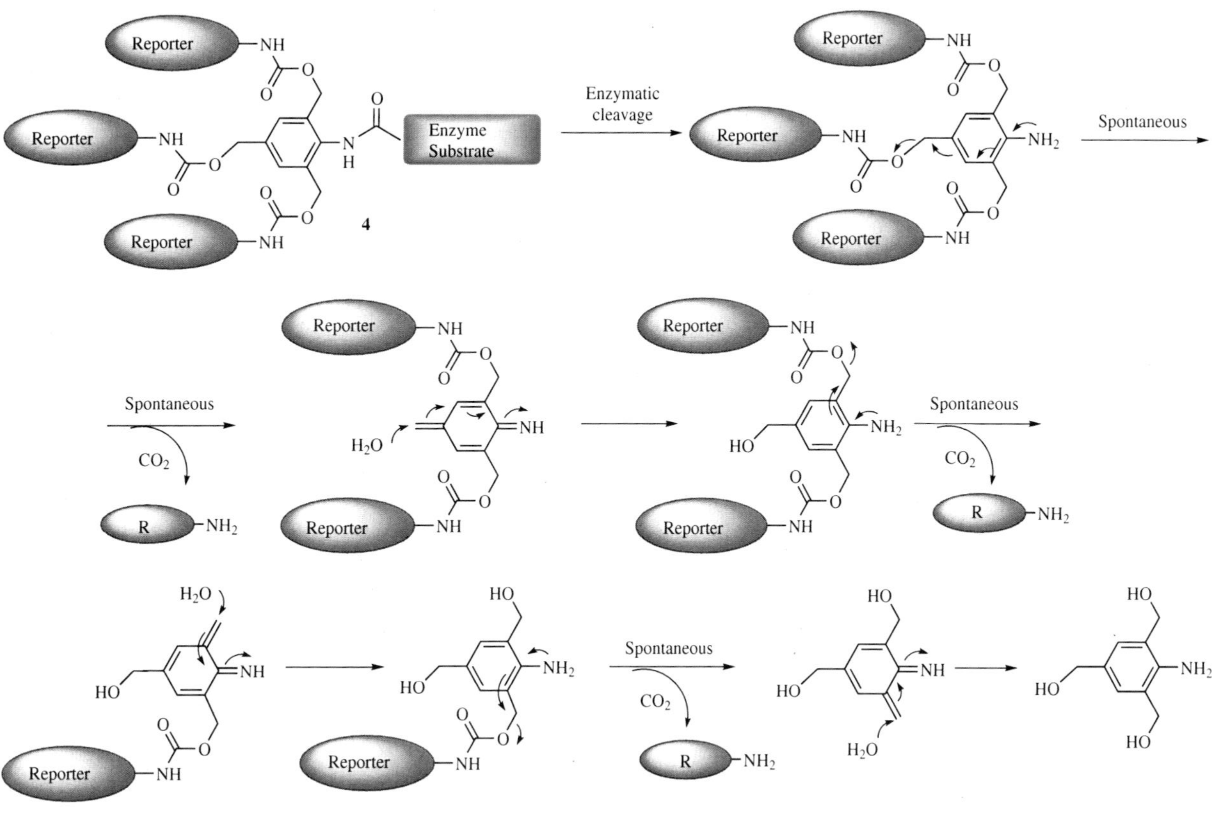

FIGURE 5.3 Disassembly mechanism of a self-immolative AB_3 dendritic adaptor.

simultaneously reported the design and synthesis of novel dendritic structures with triggers that initiate the fragmentation of the dendrimer molecule into its building blocks.[3–5] As a consequence of the self-immolative fragmentation, the tail-group units are released. In all three designs, the dendrimer skeleton is constructed such that it disintegrates into known molecular fragments once the disintegration process has been initiated. The dendritic branching unit that was used in the dendrimers' design disassembles through double or triple quinone methide/azaquinone methide type elimination reactions, in a similar mechanism as presented in Fig. 5.3.

We were able to translate this concept into a novel, self-immolative, dendritic prodrug system. Drug molecules are incorporated as the tail units and an enzyme substrate is used as the trigger, generating a multiple-prodrug unit that is activated with a single enzymatic cleavage. We showed that self-immolative, dendritic prodrugs have a significant advantage in tumor cell-growth inhibition compared with classical monomeric prodrugs.[6] Furthermore, we synthesized a single-triggered heterodimeric prodrug with the anticancer agents doxorubicin and camptothecin as tail units.[7] For the first time, it was possible to release two different chemotherapeutic drugs simultaneously at the same location. We have also designed and synthesized fully biodegradable dendrimers that are disassembled through multienzymatic triggering followed by self-immolative chain fragmentation. The concept of multi-triggered, self-immolative dendrons was recently applied to synthesis of a prodrug activated through a molecular "OR" logic trigger (a dual trigger activated by either one of the two different enzymes).[8,9] In this chapter, we summarized how the quinone methide chemistry can be applied to shape unique dendritic molecules with self-immolative disassembly mechanism.

5.2 SUBSTITUENT-DEPENDENT DISASSEMBLY OF DENDRIMERS

Like in other controlled-release systems, we sought to fine-tune the disassembly rate of the dendritic platform. The deceleration or acceleration could be achieved through a substituent effect[10] on the aromatic building block.[11] In this example, we evaluated dendrons with *p*-nitroaniline (PNA) end groups that can be released by chemical activation. As a trigger we used *t*-butylcarbamate (Boc) group that is removed under acidic conditions. The disassembly mechanism of the self-immolative dendrons is presented in Fig. 5.4. When dendron **5** is subjected to the acidic environment of TFA (trifluoroacetic acid), amine **6** is formed. The latter is cyclized to generate phenol **7** and a dimethylurea derivative. The phenol undergoes double quinone methide rearrangement and decarboxylation to release the two molecules of PNA (compound **8** is formed after the reactive quinone methide is trapped with methanol).

It was rationalized that one could gain control of the disassembly rate by changing the substituent R on the aromatic ring. Electron-withdrawing substituents should accelerate the cyclization step since phenol **7** will become a better leaving group in the acyl substitution of **6** to **7**. Four dendrons were selected for this study (Fig. 5.5): two (**9** and **11**) with a methyl substituent and two others

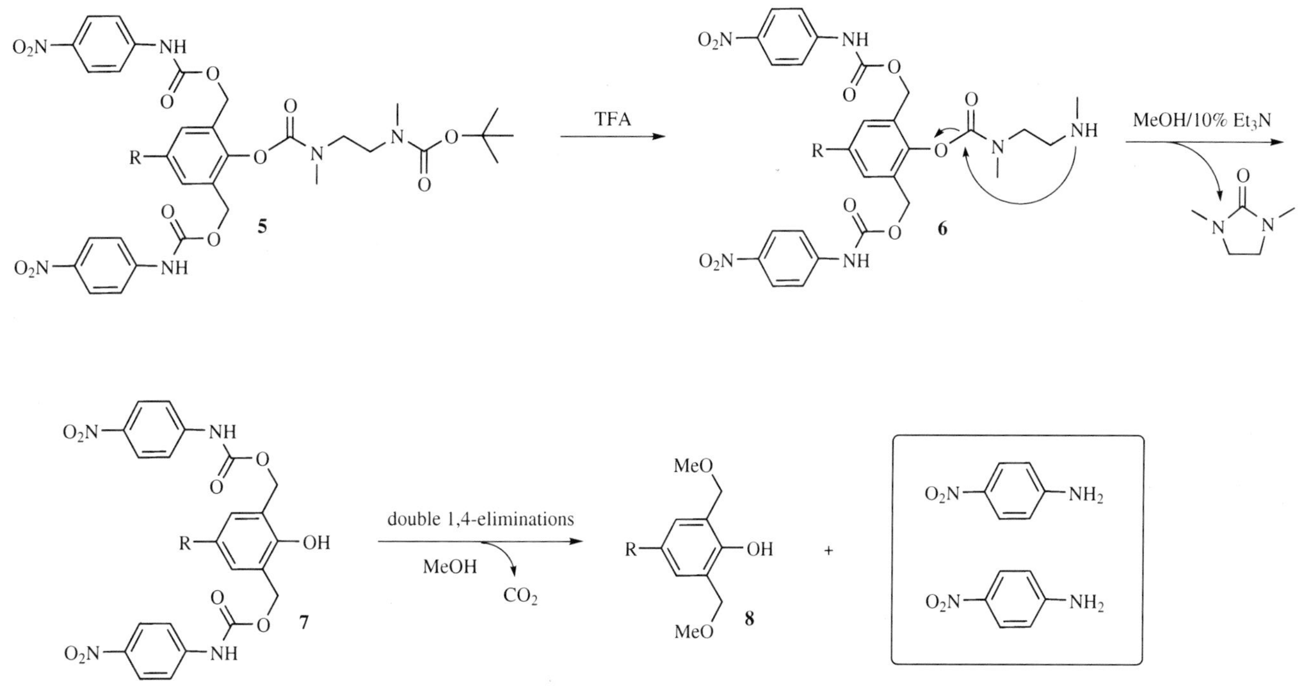

FIGURE 5.4 Disassembly mechanism of first-generation self-immolative dendron.

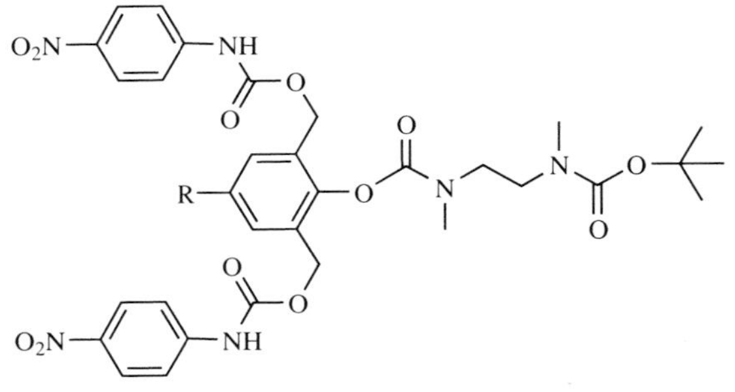

9 R = Methyl

10 R = CO_2Et

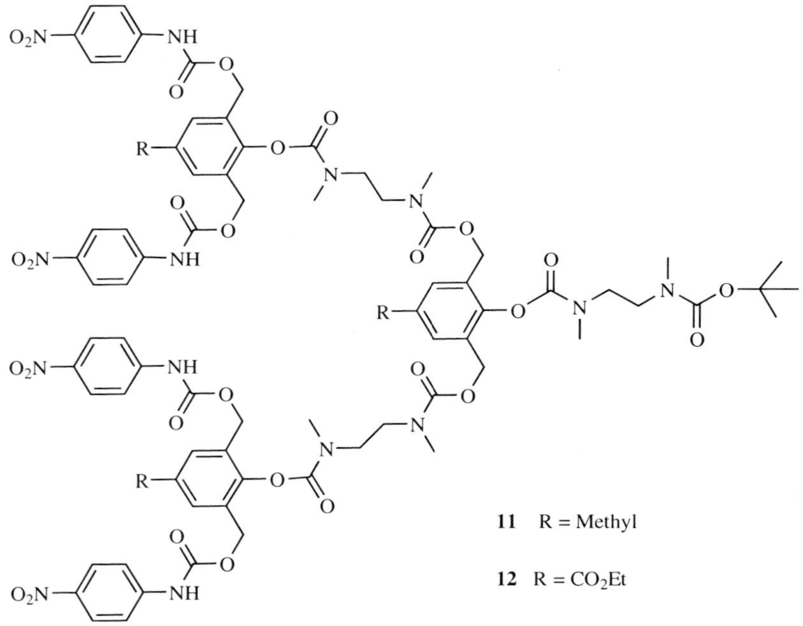

11 R = Methyl

12 R = CO_2Et

FIGURE 5.5 Chemical structures of first- and second-generation self-immolative dendrons.

with an ethylcarboxy ester substituent (**10** and **12**). All dendrons had the same end groups (PNA) and activating triggers (Boc).

The dendrons were initially incubated with TFA to afford their amine salts, which dissolved in dimethylsulfoxide (DMSO). Then, the solutions were diluted into methanol with 10% triethylamine. The sequential fragmentation of compounds

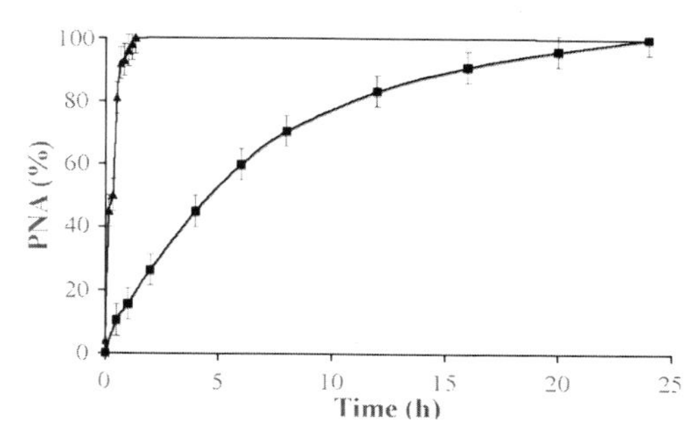

FIGURE 5.6 Release of PNA from dendrons **9**(▲) and **10** (■) after chemical activation. Starting concentration: 500 μM; $\lambda = 348$ nm; 90% MeOH/10% Et_3N; room temperature.

9–12 was monitored by reverse-phase analytical HPLC, using C-18 analytical column, at a wavelength of 348 nm. The release of PNA from dendrons **9** and **10** is presented in Fig. 5.6. The disassembly of dendron **10** occurred within a little bit more than 1 h, while 25 h were needed to complete the disassembly of dendron **9**. Similarly, Fig. 5.7 shows that second-generation dendron **12** disassembled much faster than dendron **11**; 50 h for dendron **11** and only 3 h for dendron **12**. (The disassembly rate was quantified based on the pick area of the HPLC chromatograms. The data were normalized to percent units where the sum of all picks is equal to 100%. We observed 100% conversion of the starting material to PNA.)

The acceleration of dendrons **10** and **12** disassembly can be explained by the stabilization of the phenolate species, which is released upon the intracyclization of amine intermediate **6** (Fig. 5.4). While the ester substituent (an electron-withdrawing group) oriented *para* to the phenol stabilizes the negative charge on the phenolic oxygen, the methyl substituent destabilizes it through an electron-donating inductive effect. The kinetic measurements indicate that dendrons **10** and **12** disassembled approximately 30 times faster than dendrons **9** and **11**. This result is further supported by the difference in pK_a values of cresol and 4-hydroxybenzoic acid, ethyl ester (10.26 and 8.34, respectively). The increase of the disassembly rate by one and a half orders of magnitude is proportional to difference in the pK_a values.

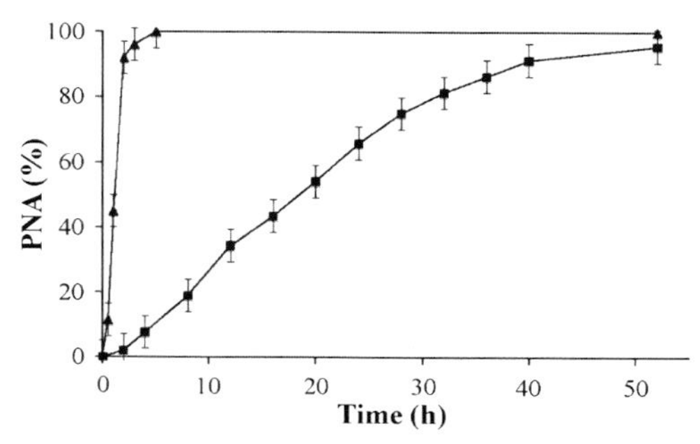

FIGURE 5.7 Release of PNA from dendrons **11**(▲) and **12**(■) after chemical activation. Starting concentration: 500 µM; $\lambda = 348$ nm; 90% MeOH/10% Et_3N; room temperature.

5.3 ELIMINATION-BASED AB_3 SELF-IMMOLATIVE DENDRITIC AMPLIFIER

Rapid release of tail-drug units from the dendritic platform is essential in order to achieve maximal drug concentration at a specific location. Therefore, self-immolative dendrons with a fast disassembly mechanism should have a significant advantage in a dendritic prodrug system.[2] AB_3 prodrug system **13** releases three molecules of active drug upon single cleavage by PGA (Fig. 5.8). The disassembly pathway is initiated by removal of phenylacetic acid, elimination of azaquinone methide, and decarboxylation to generate amine intermediate **14**. The latter cyclizes to release dimethylurea derivative and phenol **15**, which rapidly undergoes triple azaquinone methide elimination to release the three drug units. The rate-determining step was found to be the cyclization of **14** to **15**.

In order to enhance the release rate of the peripheral drug units, we sought to design a system that would disassemble without the slow cyclization step. Taking into consideration this objective, we designed AB_3 dendritic molecule **16**. The first step in the reaction is catalytic cleavage of phenylacetic acid by PGA, followed by elimination of azaquinone methide and decarboxylation to release amine intermediate **17**. This amine intermediate further disassembles through triple elimination to release the three drug units (Fig. 5.9). The disassembly of this molecule after the enzymatic cleavage is solely based on azaquinone methide elimination reactions and therefore is expected to occur very rapidly.

To compare the disassembly rate of the above dendritic systems, we synthesized AB_3 molecules **18** and **19** (Fig. 5.10). Both the molecules are designed for activation by PGA and have three units of tryptophan as a model drug. Tryptophan was used for initial evaluation as it contains a strong UV chromophore, allowing us to monitor the disassembly reaction.

The disassembly rate of dendritic molecules **18** and **19** was evaluated in phosphate buffered saline (PBS, pH 7.4) in the presence and absence of PGA. The release of tryptophan was monitored by a reverse-phase HPLC at a wavelength of 320 nm. The results are presented in Figs. 5.11 and 5.12. No disassembly of either system was observed in the buffer without PGA (data not shown). In the presence of PGA, dendritic molecule **18** disassembled to release tryptophan within approximately 4 days (Fig. 5.11), whereas dendritic molecule **19** released its tryptophan tail units within 40 min (Fig. 5.12).

Under the experiment conditions, the enzymatic cleavage occurs within seconds. Therefore, the observed release time of the tryptophan is also the actual disappearance time of the intermediate forms after the enzymatic cleavage. This dramatic enhancement of tail-unit release with the elimination-based system (dendritic molecule **19**) compared to the cyclization-based system (dendritic molecule **18**) is best viewed by superimposition of the graphs (Fig. 5.13).

We decided to apply the elimination-based dendritic system to the synthesis of an anticancer prodrug and to evaluate it in a tumor cell cytotoxicity assay. Dendritic prodrugs **20** and **21** were synthesized with the chemotherapeutic drug melphalan as a tail unit and a trigger that is activated by PGA (Fig. 5.14).

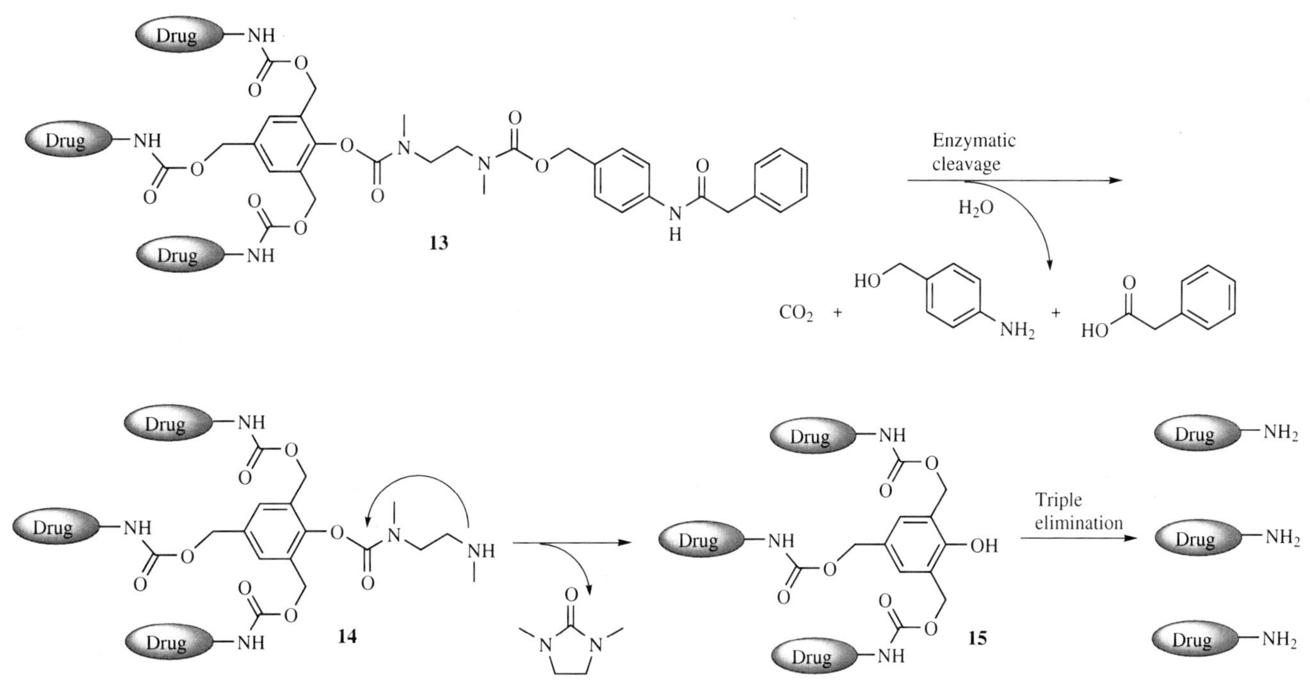

FIGURE 5.8 Disassembly mechanism of AB_3 self-immolative dendritic molecule **13**.

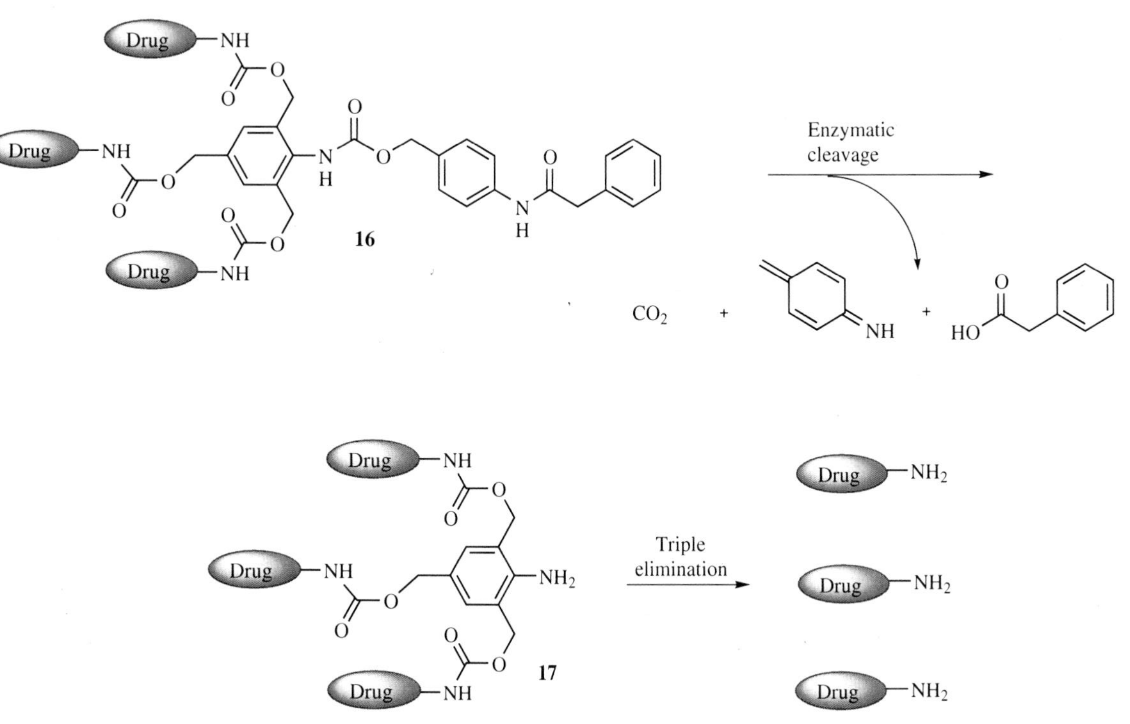

FIGURE 5.9 Disassembly mechanism of AB_3 self-immolative dendritic molecule **16**.

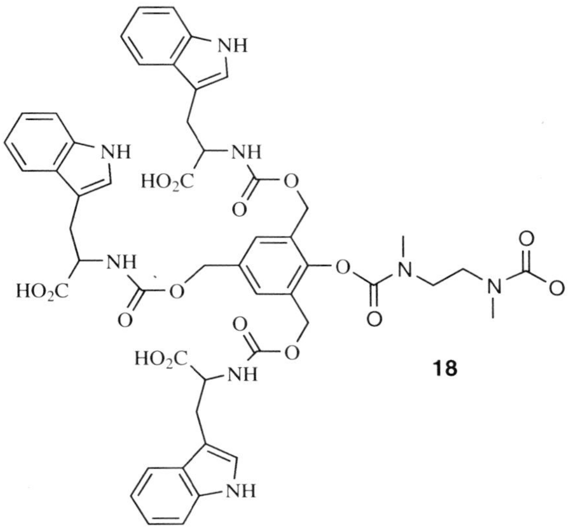

18

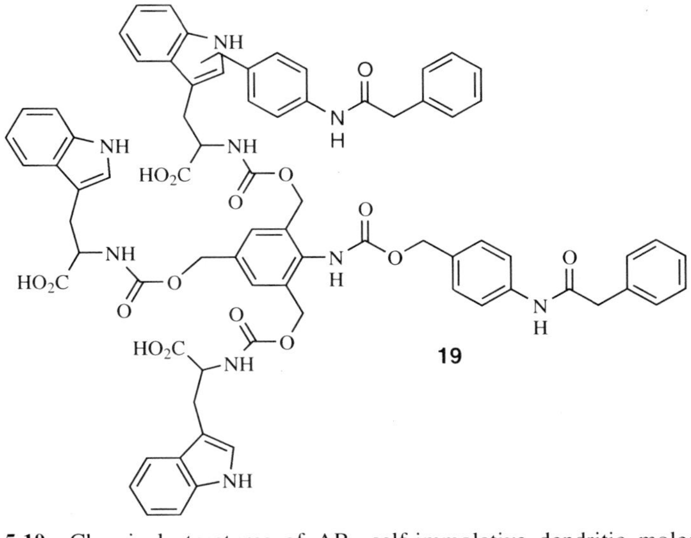

19

FIGURE 5.10 Chemical structures of AB_3 self-immolative dendritic molecules with tryptophan tail units and a trigger that is activated by PGA.

In order to evaluate the *in vitro* antitumor activity of the prodrugs, compounds **20** and **21** were incubated at varied concentrations with human T-lineage acute lymphoblastic leukemia MOLT-3 cells in the presence or absence of 1 mM of PGA. The data from the cell proliferation assays are presented in Fig. 5.15.

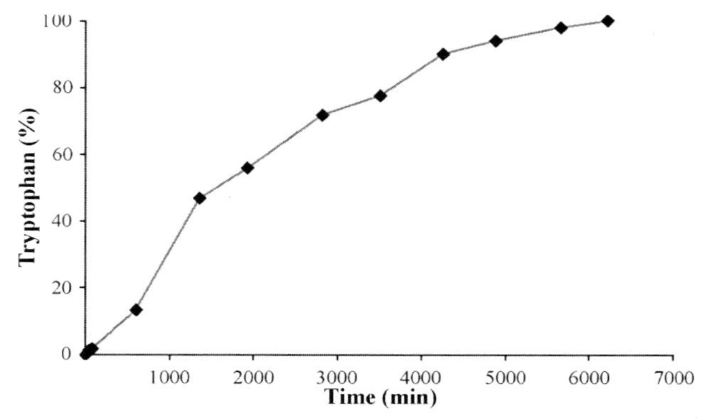

FIGURE 5.11 PGA-catalyzed release of tryptophan from dendritic compound **18** (compound **18** (500 mM) in PBS, PGA (1 mg/mL)).

A colorometric assay based on the tetrazolium salt XTT was used to evaluate the cytotoxity of the compounds.

Melphalan prodrugs **20** and **21** exhibited significantly (about 100-fold) reduced toxicity than free melphalan in the absence of PGA. XTT cytotoxicity assays showed a decrease of 17-fold in IC_{50} for prodrug **20** (100 μM versus 6 μM with PGA) and 200-fold for prodrug **21** (100 μM versus 0.5 μM with PGA). In the presence of PGA, prodrug **20** showed some increased cytotoxicity but still notably less than that of free melphalan (6 and 0.3 μM, respectively). However, when prodrug **21** was activated by PGA, the cytotoxicity was almost identical to that of free melphalan (IC50 values of free melphalan and prodrug **21** in the presence of PGA were 0.3 and 0.5 μM, respectively). Interestingly, in the absence of PGA both prodrugs were relatively not toxic even at very high doses of 100 μM. No toxicity was observed for the platform building blocks within the activity range of melphlane or melphlane prodrug.

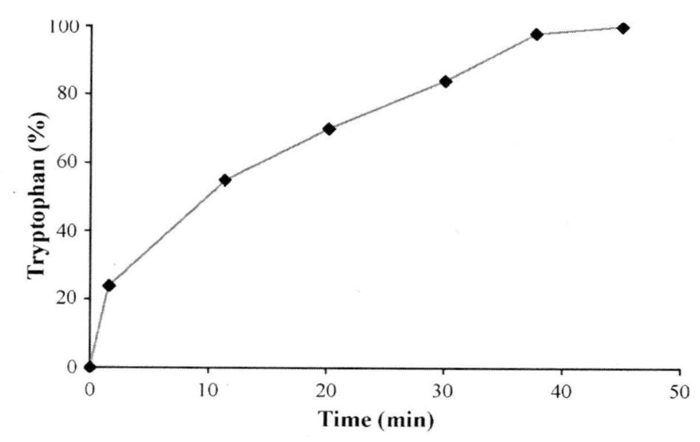

FIGURE 5.12 PGA-catalyzed release of tryptophan from dendritic compound **19** (compound **19** (500 mM) in PBS, PGA (1 mg/mL)).

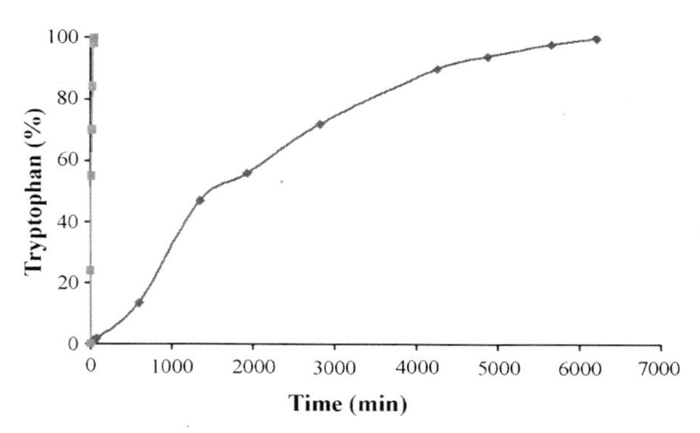

FIGURE 5.13 PGA-catalyzed release of tryptophan from dendritic compound **18** (purple) with $t_{1/2}$ of 1400 min versus release from dendritic compound **19** (red) with $t_{1/2}$ of 10 min.

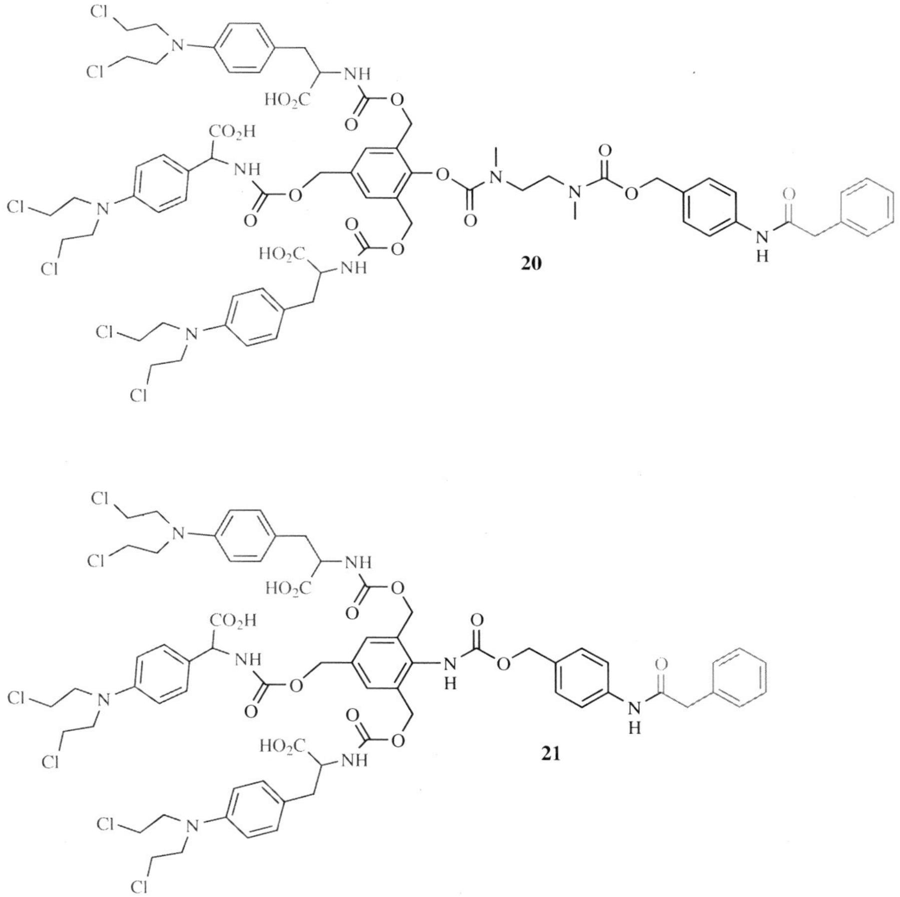

20

21

FIGURE 5.14 Chemical structures of AB_3 self-immolative dendritic prodrugs with melphalan tail units and a trigger that is activated by PGA. (See the color version of this figure in Color Plates section.)

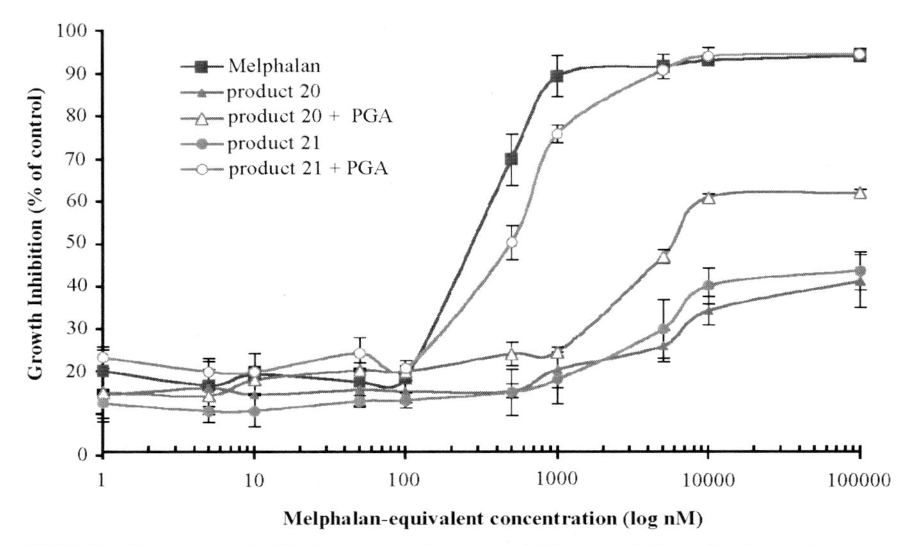

FIGURE 5.15 Growth inhibition assay of leukemia MOLT-3 cell line with dendritic prodrug **20** or **21** in the presence or absence of PGA. Cells were incubated for 72 h. Full squares represent melphalan, full triangles prodrug **20**, empty triangles prodrug **20** with PGA, full circles prodrug **21**, and empty circles prodrug **21** with PGA. Symbols represent mean ± SD.

The elimination-based AB_3 dendritic prodrug showed significant enhancement of drug release in comparison to a cyclization-based AB_3 dendritic prodrug. This difference was noticeably reflected in a cytotoxicity assay. This new trimeric prodrug system could offer significant advantages in inhibition of tumor growth relative to regular monomeric prodrugs, especially if the targeted or secreted enzyme exists at relatively low levels in the malignant tissue.

5.4 CONTROLLED SELF-ASSEMBLY OF PEPTIDE NANOTUBES

The aromatic dipeptide nanotubes (ADNT) represent a unique class of organic nanostructures. These bioinspired structures are formed by the self-assembly of the core recognition motif of the β-amyloid polypeptide into hollow tubes of remarkable persistence length and rigidity. A limiting factor in the utilization of this system, however, is the ability to temporally control the assembly process. Therefore, we decided to explore the ability of a self-immolative dendritic system to serve as a platform for the controlled assembly of peptide nanotubes.[12] The extremely short length of the peptide building blocks and their ability to self-assemble make these units ideal candidates for controlled-assembly applications. We sought to evaluate the ability of the elimination-based AB_3 self-immolative dendritic system to serve as a carrier platform for controlled assembly of diphenylalanine. Compound **22** (Fig. 5.16) was prepared with three units of diphenylalanine and a trigger that is activated by PGA.

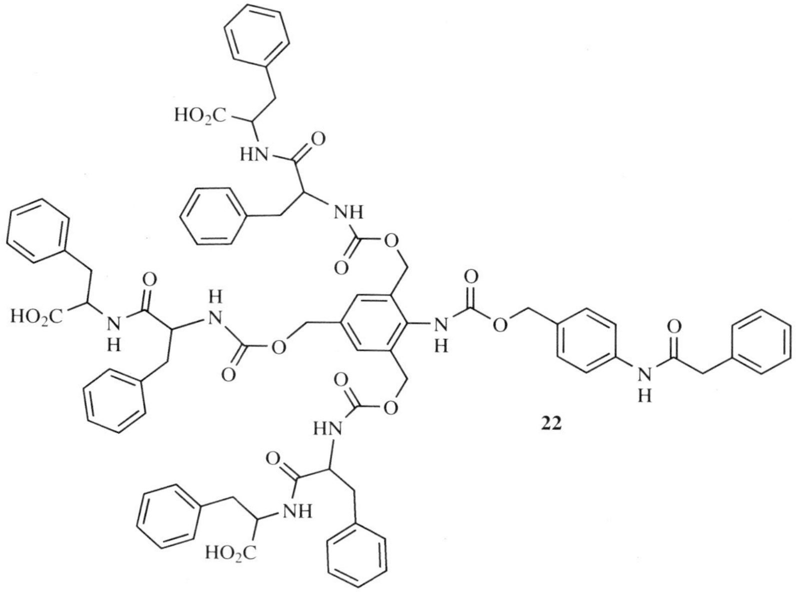

FIGURE 5.16 AB_3 self-immolative dendritic system with diphenylalanine end units and a trigger designed for activation by PGA.

In order to evaluate the release and self-assembly of the diphenylalanine end groups, dendron **22** was incubated in PBS, pH 7.4, without PGA at a concentration of 1.5 mM. Transmission electron microscopy (TEM) analysis revealed that attachment of the diphenylalanine to the dendritic platform prevented self-assembly, and therefore, no organized structures were observed (Fig. 5.17a). Then, dendron

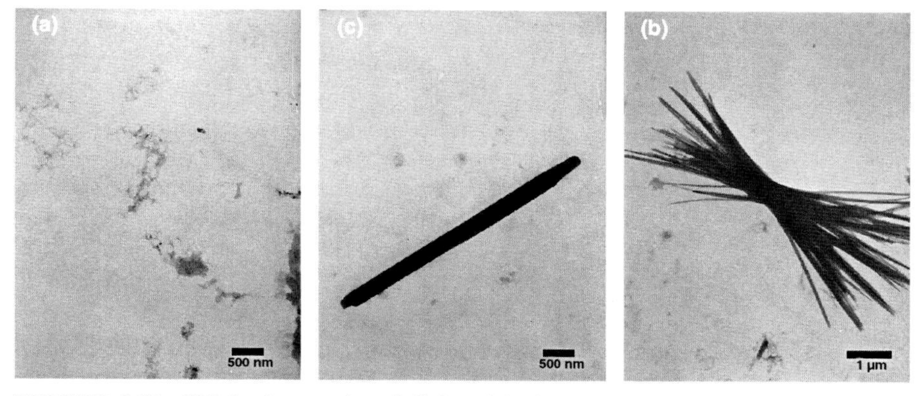

FIGURE 5.17 TEM micrographs of diphenylalanine peptide nanotubes self-assembled after the enzymatic cleavage. (a) TEM images of **22** prior to the enzymatic cleavage. (b and c) TEM images of **22** after the enzymatic cleavage.

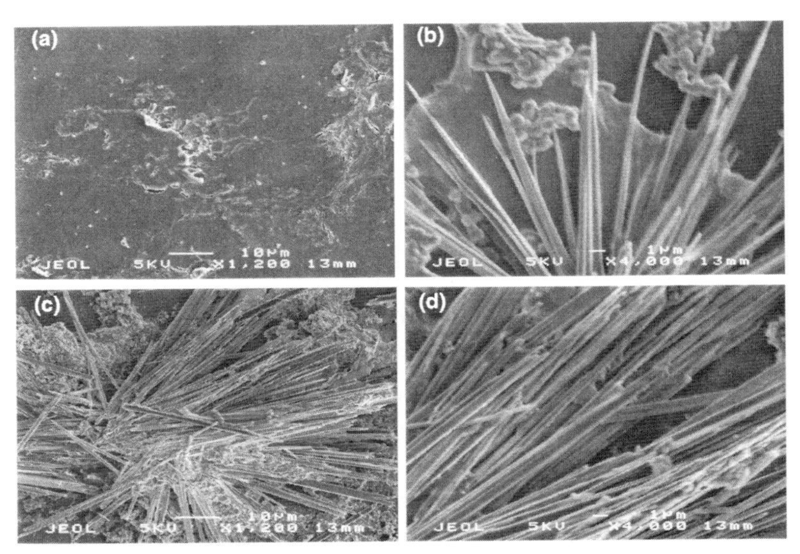

FIGURE 5.18 SEM micrographs of diphenylalanine peptide nanotubes self-assembled after the enzymatic cleavage. SEM images of **22** incubated at room temperature (a) prior to the enzymatic cleavage, (b and c) 2 h after the enzymatic cleavage, and (d) 5 days after the enzymatic cleavage.

22 was incubated in the PBS buffer in the presence of PGA (1.5 mM of **22** and 1 mg/mL of PGA) and after 1 h a sample was analyzed by TEM. Formation of peptide nanotubes was clearly observed (Fig. 5.17b and c). Scanning electron microscopy (SEM) analysis confirmed the self-assembly into organized structures. No structures were observed in the absence of PGA (Fig. 5.18a), but nanotubes were clearly present 2 h (Fig. 5.18b and c) and 5 days (Fig. 5.18d) after PGA cleavage.

AB_3 dendritic platform **23** functioned as an efficient carrier for the controlled formation of diphenylalanine nanotubes (Fig. 5.19). First, it prevented the formation of any organized structures when the peptides were attached. Second, the three units of diphenylalanine were rapidly released upon cleavage of the trigger. Finally, the platform allowed control of the release of the end units through a variety of triggering agents.

The current example is a significant addition to the attempts to control bioorganic assembly at the nanoscale and should serve as an important step toward the

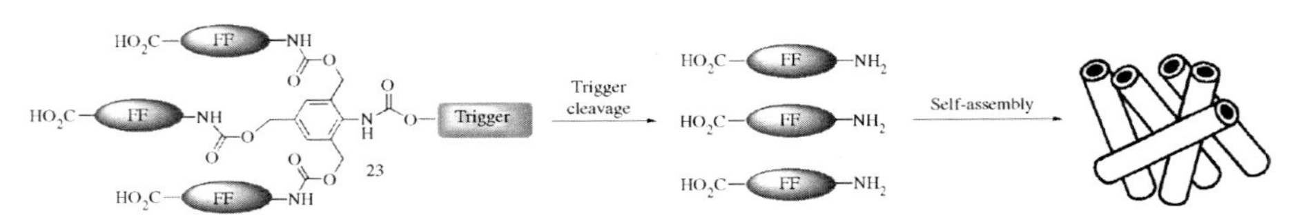

FIGURE 5.19 Controlled formation of diphenylalanine peptide nanotubes.

development of nanobiotechnological applications. The ability to control the self-assembly process may also allow the use of ADNT in microelectronic and micro-electromechanics (MEMS) processes where physical and chemical signals could be used for the precise fabrication. Since the peptide nanostructures are compatible with high temperatures and various organic solvents, the dendritic platform could be applied in lithographic processes that will include both organic and inorganic materials. Furthermore, this novel method may have implications for design of inhibitors for self-assembly of peptides. These inhibitors could be used as drugs for preventing the formation of protein or peptide aggregates in Alzheimer's and Parkinson's diseases, two major disorders that are associated with the self-assembly of neurotoxic nanostructures.

5.5 AB_6 SELF-IMMOLATIVE DENDRITIC AMPLIFIER

The basic building block of a first-generation dendrimer is referred as an AB_n unit, where A is the head and B is the tail. Figure 5.20 shows the general molecular structure of AB_1 (**24**), AB_2 (**25**), and AB_3 (**26**) self-immolative dendritic adaptors. The release of an active reporter molecule from compound **24** is initiated upon the cleavage of the trigger, followed by removal of the cyclic dimethylurea derivative and 1,6-quinone methide rearrangement. Dendron **25** similarly undergoes double 1,4-quinone methide rearrangement to release two reporter units, whereas dendron **26** undergoes 1,6- and double 1,4-quinone methide rearrangement to release three reporter units.

Elimination of reporter groups by the quinone methide rearrangement was showed to be an efficient reaction for designing of AB_n self-immolative dendritic adaptors. This elimination takes place when there is a good leaving group *para* or *ortho* of the phenolic oxygen. Therefore, the maximal number of reporters linked to one benzene ring is limited to three, since there are only two *ortho* and one *para* positions available. We recently developed a novel AB_6 self-immolative dendritic molecule, in which the number of substituents that are conjugated with the phenolic oxygen has been doubled.[13] In this AB_6 dendron, a single cleavage event releases six reporter units. In order to generate an AB_6 self-immolative dendritic adaptor from a benzene ring, we have designed dendron **27** (Fig. 5.21). The number of the substituents that are conjugated with the phenolic oxygen was doubled through a short split extension. This molecular design allows amplification of a single cleavage reaction into release of six active reporters.

The disassembly mechanism of dendron **27** is illustrated in Fig. 5.22. Cleavage of the trigger initiates the cyclization of a dimethylurea derivative to release phenol **28**. The latter can undergo 1,8-elimination followed by decarboxylation to release one reporter unit and to generate quinone methide **29**. In the next step, a nucleophile (most likely a solvent molecule) presumably reacts with the highly electrophilic quinone methide to generate the phenol **30**. Similarly, we hypothesize that one more 1,8-elimination and four 1,6-eliminations take place to lead to the release of all six reporters.

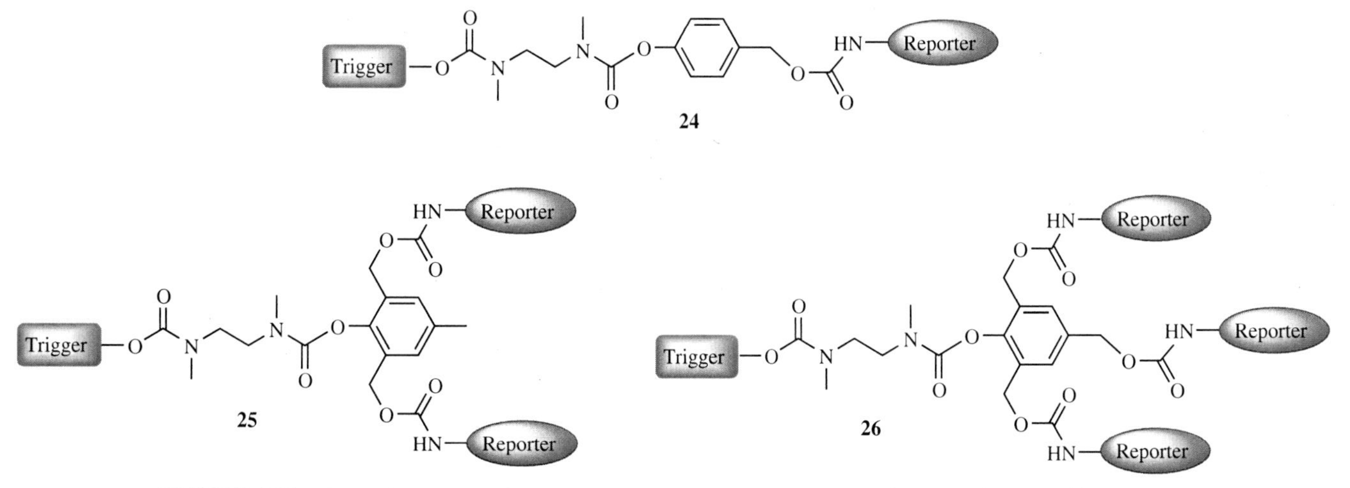

FIGURE 5.20 General structure of AB_1, AB_2, and AB_3 single-triggered self-immolative dendritic adaptors.

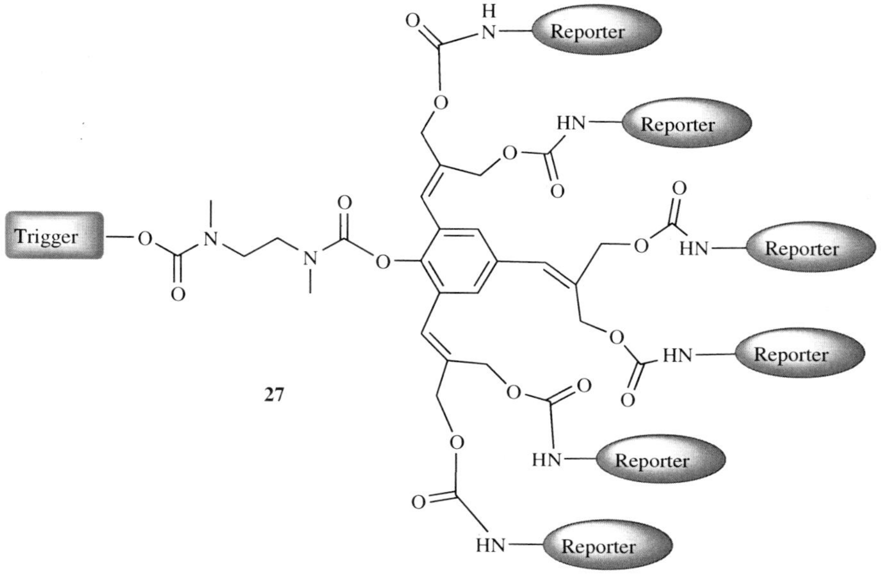

FIGURE 5.21 Molecular structure of a single-triggered AB_6 self-immolative dendritic adaptor.

Then, we synthesized AB_6 dendron **31** with six molecules of aminomethyl-pyrene as reporter units and the *t*-butyloxycarbonyl (Boc) protecting group as a trigger (Fig. 5.23). The dendron was designed to release its six tail units upon removal of the protecting group.

To evaluate the disassembly activity of dendron **31**, the Boc trigger was removed with trifluoroacetic acid to generate the corresponding ammonium salt; this was incubated in 1:1 MeOH/DMSO in a 2% aqueous solution of tetrabutylammonium hydroxide (Bu_4NOH). As a control, dendron **31** with intact trigger was incubated under identical conditions. The release of free aminomethyl-pyrene was monitored by reverse-phase HPLC. Figure 5.24 shows that the pyrene tail units were completely released within 6 h. Importantly, no release was observed in the control solution. In the HPLC chromatogram, the peak corresponding to the aminomethyl-pyrene was observed directly from disassembly of the ammonium salt of dendron **31** without any other intermediates. This observation suggests that the rate-determining step of the disassembly is the cyclization of the amine to release dimethylurea derivative and phenol **28** (Fig. 5.22). The latter has a short lifetime and the six elimination reactions to release the observed aminomethyl-pyrene are probably fast.

Additional support for this disassembly mechanism was obtained by monitoring the release of the pyrene tail units by fluorescence spectroscopy. The confined proximity of the pyrene units in the dendritic molecule results in formation of excimers. The excimer fluorescence generates a broad band at a wavelength of 470 nm in the emission spectrum of dendron **31** (Fig. 5.25). Upon the release of the pyrene units from the dendritic platform, the 470 nm band disappeared from

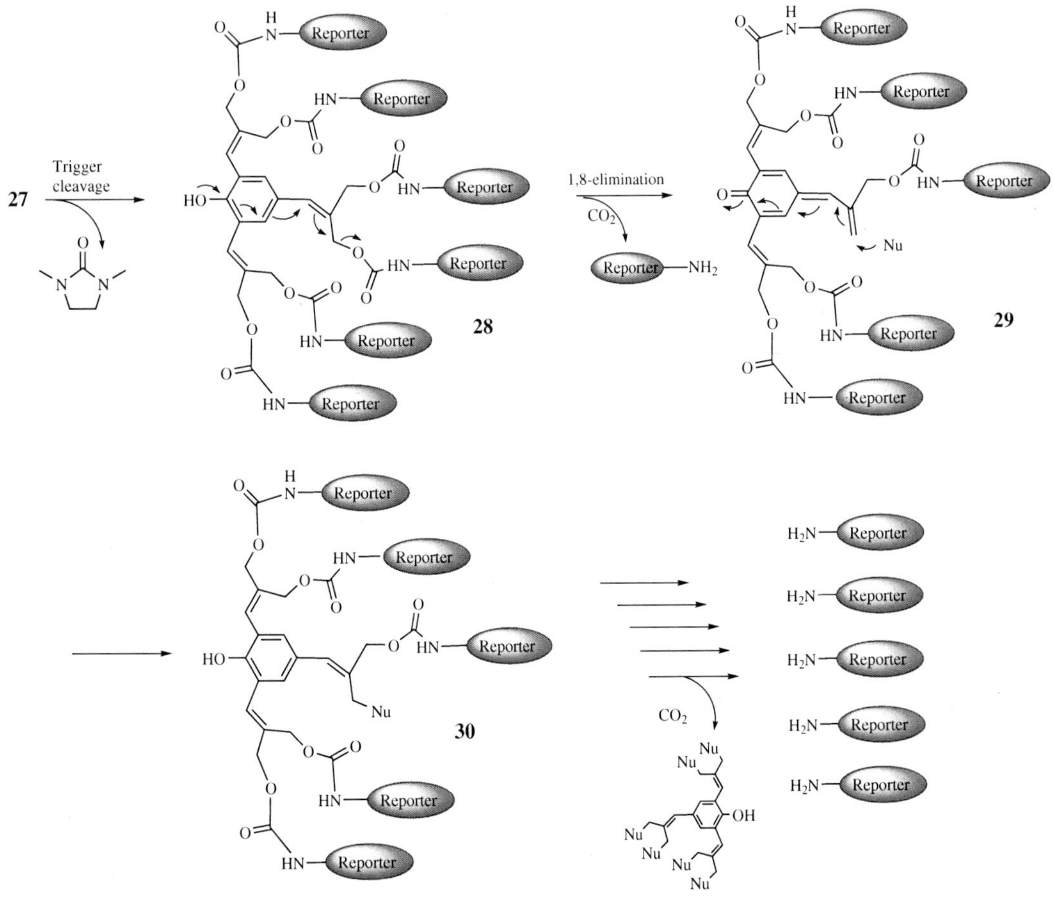

FIGURE 5.22 Disassembly mechanism of AB_6 self-immolative dendron **27**.

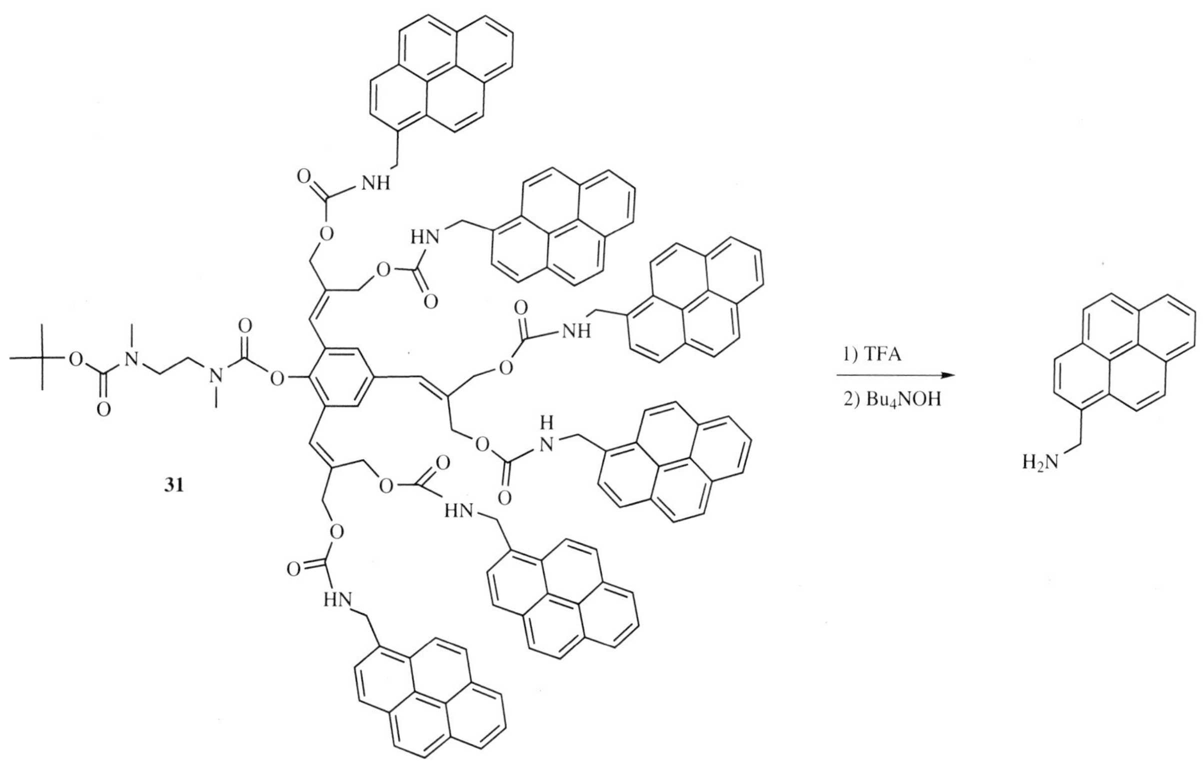

FIGURE 5.23 Molecular structure of a self-immolative AB_6 dendron with aminomethyl-pyrene reporter units and a Boc protecting group as a trigger.

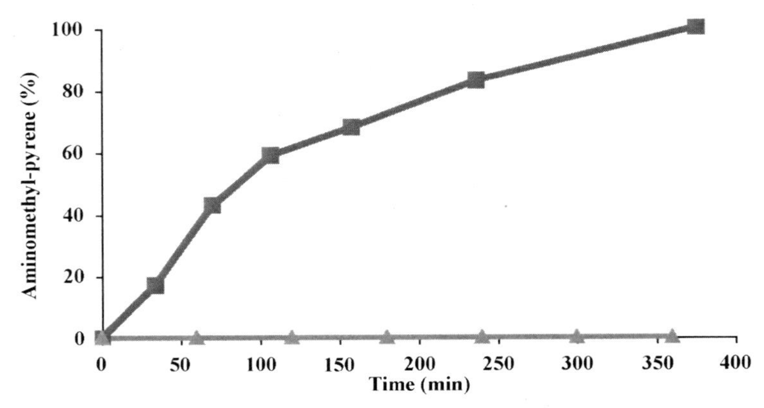

FIGURE 5.24 Release of aminomethyl-pyrene from dendron **31** upon removal of the Boc protecting group trigger (1:1 MeOH/DMSO mixture with 2% aqueous solution of Bu_4NOH) (■). Control reaction under similar conditions with dendron **31** with the Boc trigger intact (▲).

the spectrum and the band at 408 nm that corresponds to free aminomethyl-pyrene increased. Similarly as obtained in the HPLC assay, no decrease in excimer band was observed for the control reaction with dendron **31** with Boc trigger intact (data not shown).

We also wanted to evaluate the disassembly of our dendritic system under physiological conditions. Thus, we synthesized a self-immolative AB_6 dendron **32** with water-soluble tryptophan tail units and a phenylacetamide head as a trigger (Fig. 5.26) to evaluate disassembly in aqueous conditions. The phenylacetamide is selectively cleaved by the bacterial enzyme penicillin G amidase (PGA). The trigger was designed to disassemble through azaquinone methide rearrangement and cyclic dimethylurea elimination to release a phenol intermediate that will undergo six quinone methide elimination reactions to release the tryptophan tail units.

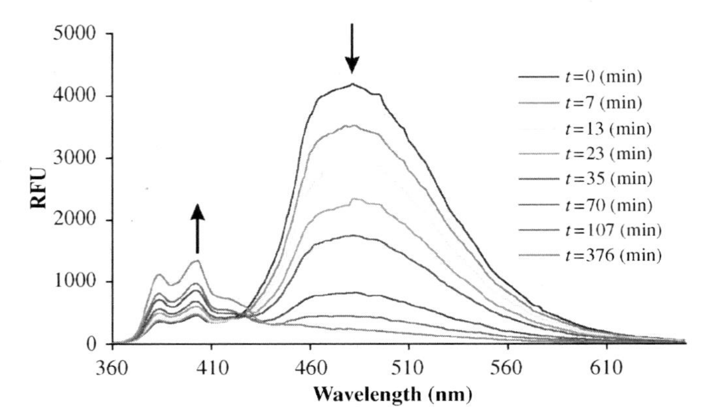

FIGURE 5.25 Emission fluorescence spectra (l_{ex} = 340 nm) of dendron **31** (50 mM) upon removal of the Boc trigger upon incubation in 1:1 MeOH/DMSO mixture with 2% aqueous solution of Bu_4NOH. (See the color version of this figure in Color Plates section.)

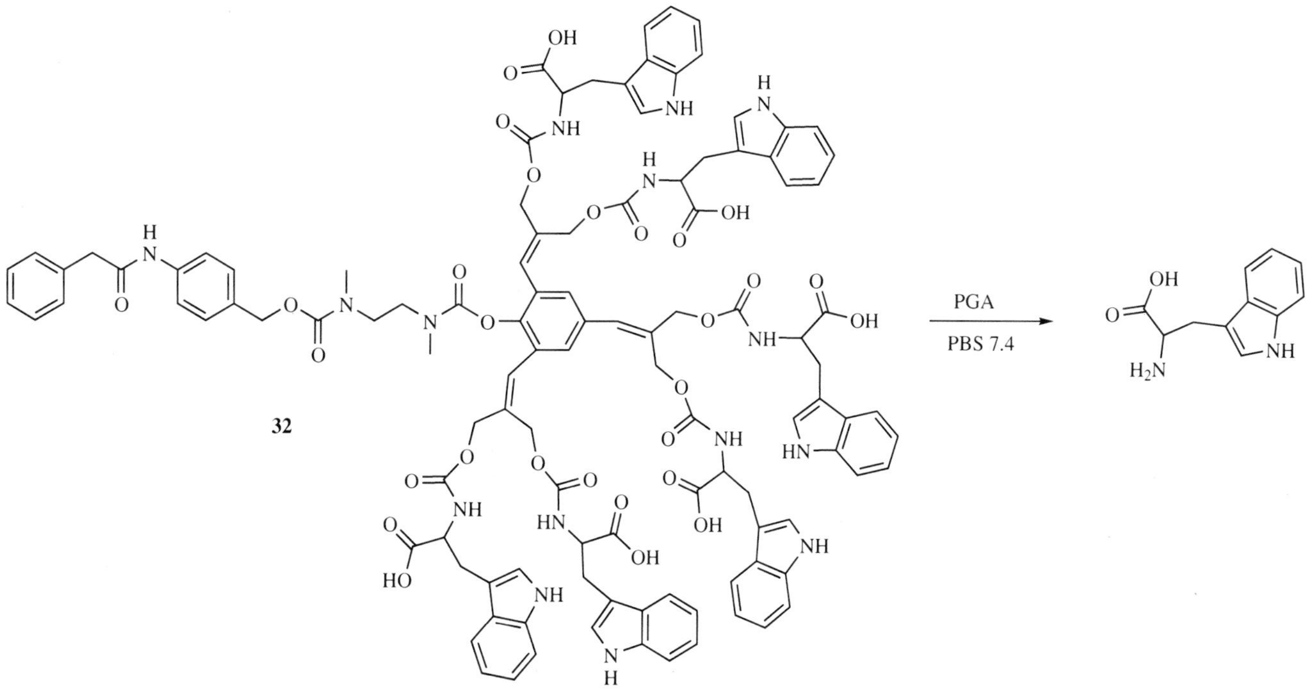

FIGURE 5.26 Molecular structure of a self-immolative AB_6 dendron with tryptophan tail units and phenylacetamide group as a trigger.

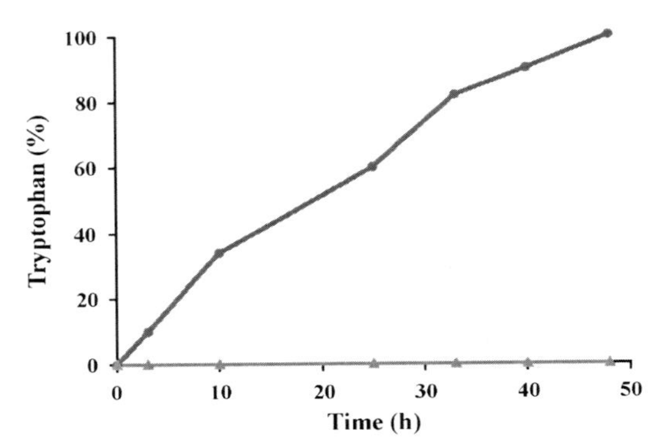

FIGURE 5.27 Release of tryptophan from dendron **32** upon incubation with PGA. Conditions: [dendron **32**], 50 mM; [PGA], 0.1 mg/mL (Ω). Control reaction under similar conditions without PGA (▲).

Dendron **32** was incubated in phosphate buffer saline, pH 7.4, in the presence and in the absence of PGA. The progress of the disassembly was monitored by RP-HPLC, and the results are presented in Fig. 5.27. Tryptophan was gradually released from dendron **32** upon incubation with PGA. The release was completed within 48 h in the presence of PGA; the control reaction without PGA showed no release at all. Although the disassembly of this dendron occurred more slowly under physiological conditions than dendron **31** in the MeOH/DMSO environment (Fig. 5.24), PGA cleaved its phenylacetamide substrate from dendron **32** and the resulting amine intermediate was disassembled to release the total six molecules of tryptophan.

Since tryptophan is also a fluorescence molecule, we monitored the disassembly through spectroscopic measurements. It was found that within the AB_6 dendron (**32**), the tryptophan fluorescence was significantly quenched due to the confined proximity, which is forced by the dendritic skeleton. Upon the disassembly, we observed a gradual increase of two bands on 400 and 760 nm that correspond to free tryptophan molecules (Fig. 5.28).

A single AB_6 dendritic subunit is able to multiply a triggering event sixfold, whereas previous AB_2 and AB_3 self-immolative dendritic subunits multiplied a triggering event only twofold or threefold, respectively. Thus, AB_6-based self-immolative dendrimers achieve higher degrees of signal amplification at lower dendrimer generations. Since higher generation dendrimers are typically more difficult to synthesize, the AB_6-based dendrimer may be a more efficient dendritic amplification subunit.

The generated quinone methide intermediates, during the disassembly, are highly reactive electrophiles and rapidly react with any available nucleophile (methanol or tetrabutylammonium hydroxide under organic solvent conditions). We could not isolate any significant amount of material that derived from the core molecule, probably due to generation of a mixture of compounds by the addition of different nucleophiles to the quinone methide. This molecule acts as an amplifier of a cleavage

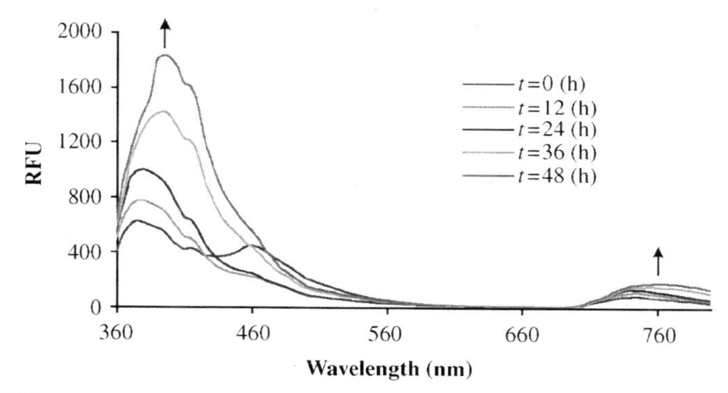

FIGURE 5.28 Emission fluorescence spectra (l_{ex} = 260 nm) of dendron **32** (12 mM) upon incubation in PBS, 37 °C with PGA (0.1 mg/mL). (See the color version of this figure in Color Plates section.)

reaction. Thus, a single cleavage event in the dendron focal point is translated into a release of six tail units.

5.6 ENZYMATIC ACTIVATION OF SECOND-GENERATION SELF-IMMOLATIVE DENDRIMERS

Two options for the triggering/activation of self-immolative dendrimers were proposed. One approach is based on chemical triggering whereas the other is based on enzymatic bioactivation. Chemical and enzymatic activation of first- and second-generation self-immolative dendrimers was easily achieved. However, when second-generation self-immolative dendrimer was tested for enzymatic activation, no fragmentation was observed at all. We assumed that the hydrophobic structure of the dendritic prodrugs generates aggregation under aqueous conditions, which prevented the enzyme from accessing the triggering substrate. Enzymatic activation of second-generation self-immolative dendrimers was achieved through conjugation of polyethylene glycol (PEG) to the dendritic platform via click chemistry.[14] The PEG tails significantly decreased the hydrophobicity of the dendrimers and thereby prevented aggregate formation. We designed and synthesized second-generation dendritic molecule **33** (Fig. 5.29). The molecule has an enzymatic trigger that is activated by PGA (red), four reporter groups of 4-nitroaniline (blue), and two acetylene functional groups that are ready to be coupled to various PEG-azide units (purple).

Dendritic molecule **33** was conjugated with two equivalents of commercially available PEG400-azide via the copper-catalyzed click reaction to give conjugate **34** (Fig. 5.30). The location of the PEG tails in the dendron's periphery was highly important for aqueous solubility during the dendritic platform fragmentation as presented in Fig. 5.31. The PEG tails remained attached to the dendron fragments until the reporter groups of 4-nitroaniline were released. Cleavage of the phenylacetamide group by PGA triggered the disassembly of dendron **34** through a known self-immolative reaction sequence.

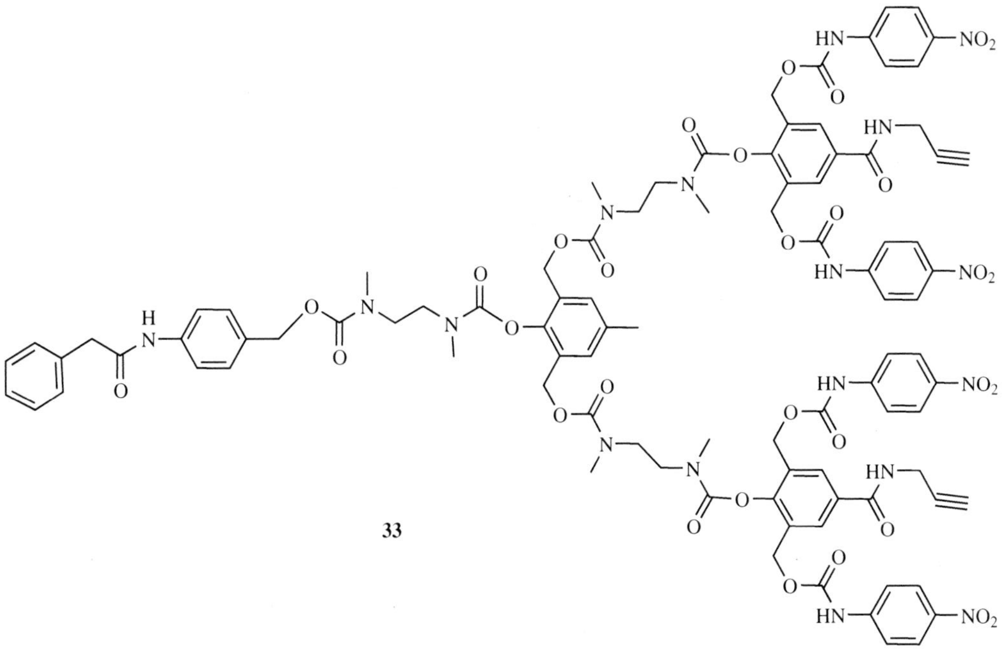

33

FIGURE 5.29 Molecular structure of second-generation self-immolative dendron with a trigger designed for activation by PGA, reporter groups of 4-nitroaniline, and acetylene functional groups for click conjugation.

33 $\xrightarrow[Cu/CuSO_4]{PEG400\text{-}N_3}$

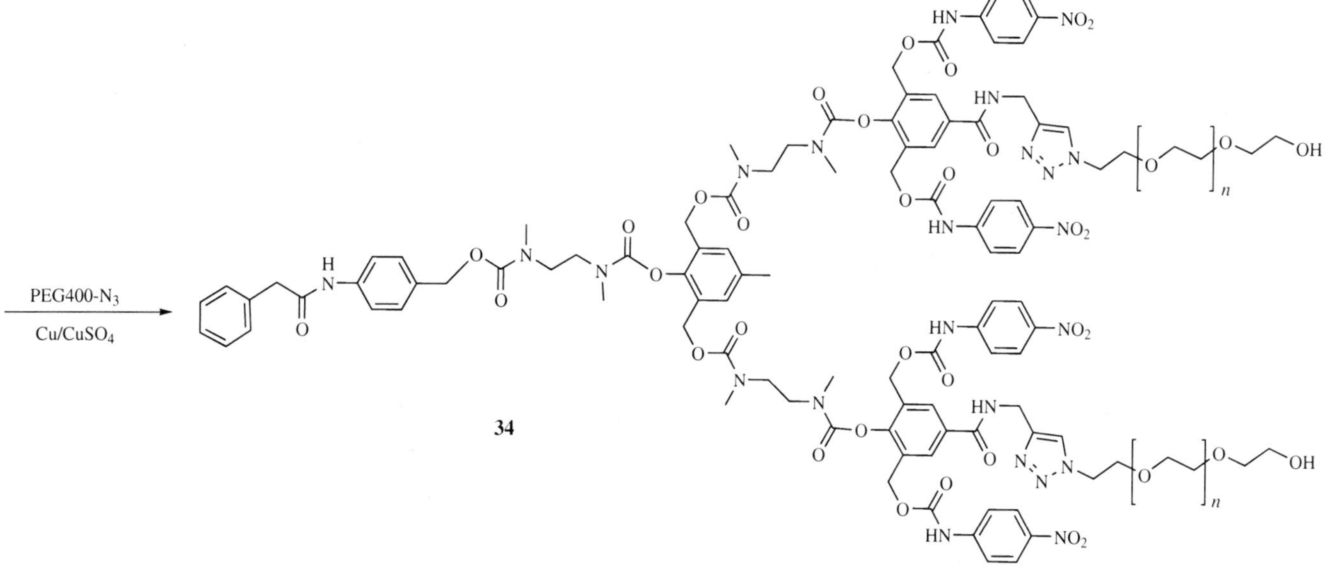

34

FIGURE 5.30 Conjugation of dendritic molecule **33** with PEG400-azide.

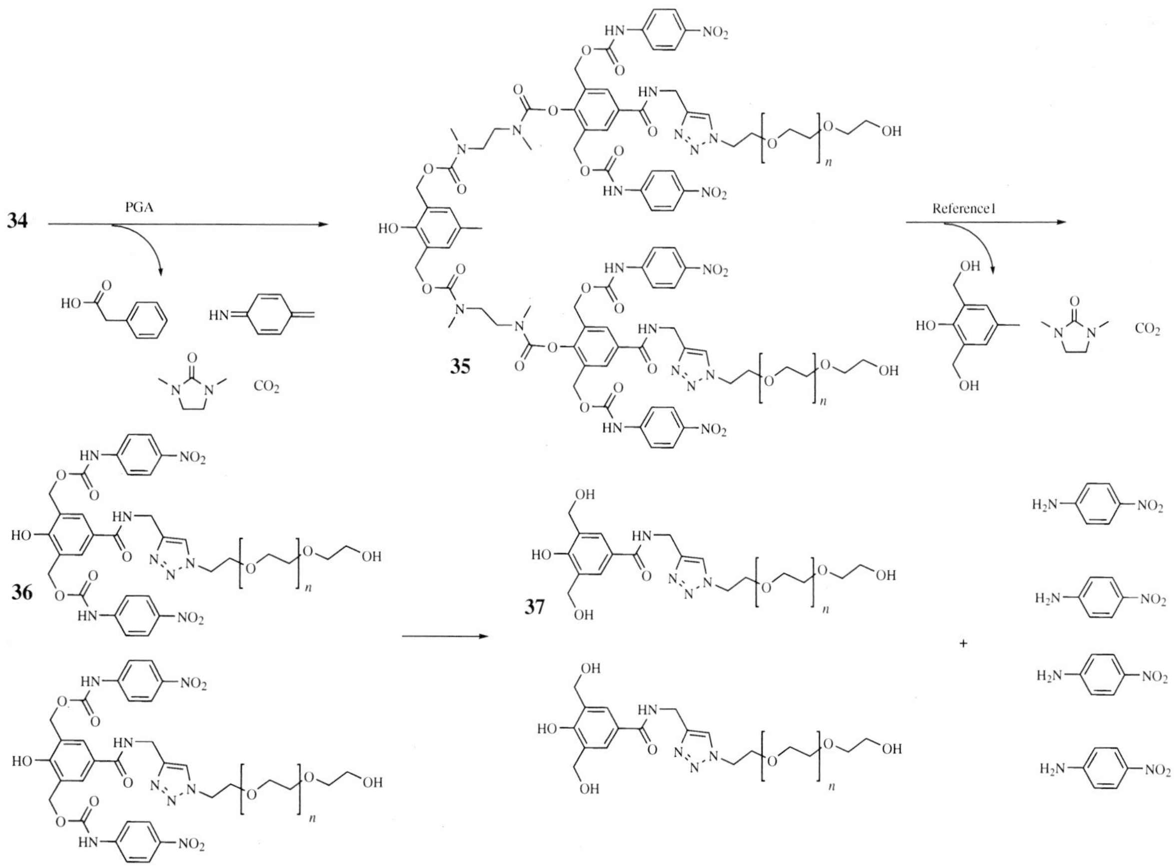

34 → PGA → 35 → Reference1 → 36 → 37 + H_2N–C6H4–NO_2

FIGURE 5.31 Disassembly pathway of second-generation dendritic molecule **34** triggered by enzymatic activation of PGA.

Following the enzymatic cleavage, azaquinone methide was rapidly eliminated and decarboxylation occurred, leading to internal cyclization that released a urea derivative and phenol **35**. The latter was disassembled as previously described to generate two equivalents of phenol **36**, which was further fragmented to release the four reporter groups.

Dendritic molecules **33** and **34** were then incubated with PGA in PBS (pH 7.4) at 37 °C. Control solutions were composed of buffer without the enzyme. The sequential fragmentation illustrated in Fig. 5.31 was monitored by observing the disappearance of dendrons **33** or **34** and the release of 4-nitroaniline by RP-HPLC. As expected, dendron **33** could not be activated by PGA and remained intact for 72 h (data not shown). However, dendron **34** showed clear activation upon incubation with PGA and its corresponding peak completely disappeared from the HPLC chromatogram as 4-nitroaniline appeared (Fig. 5.32). No 4-nitroaniline was observed in the control experiment when dendron **34** was incubated in the buffer without PGA.

Then, we evaluated the enzymatic activation of a second-generation dendritic prodrug with camptothecin (CPT), an anticancer agent, in place of the reporter groups (dendron **38**).

Second-generation CPT dendritic prodrug **38** was prepared with a trigger that is activated by PGA and PEG5000 tails to allow sufficient aqueous solubility (Fig. 5.33). The aqueous solubility of dendritic prodrug **38** was measured and found to be over 11 mg/mL. The bioactivation of the CPT dendritic prodrug **38** by PGA was then evaluated. The prodrug was incubated in PBS (pH 7.4) at 37 °C with and without PGA, and the release of free CPT was monitored by RP-HPLC. As in the model system described above, the PEG-conjugated prodrug **38** was efficiently activated by PGA, whereas its parent prodrug (without the PEG) remained intact. The release of CPT from dendritic prodrug **38** as a function of time is plotted in

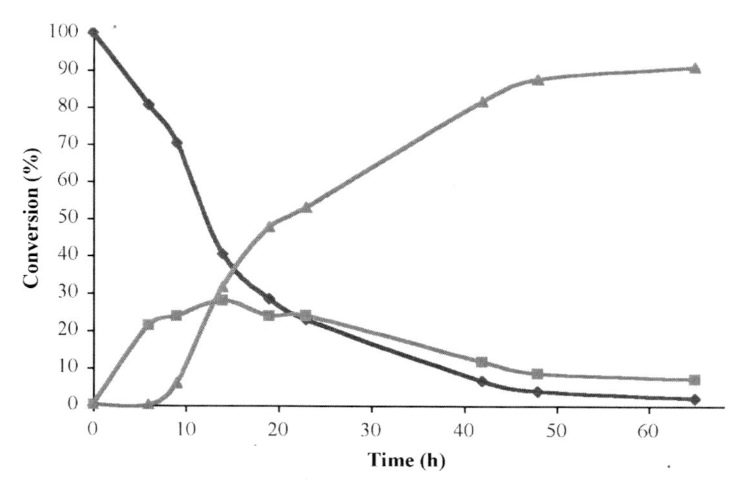

FIGURE 5.32 Bioactivation of dendron **34** by PGA. (◆) Dendron **34**, (■) intermediates, and (▲) 4-nitroaniline.

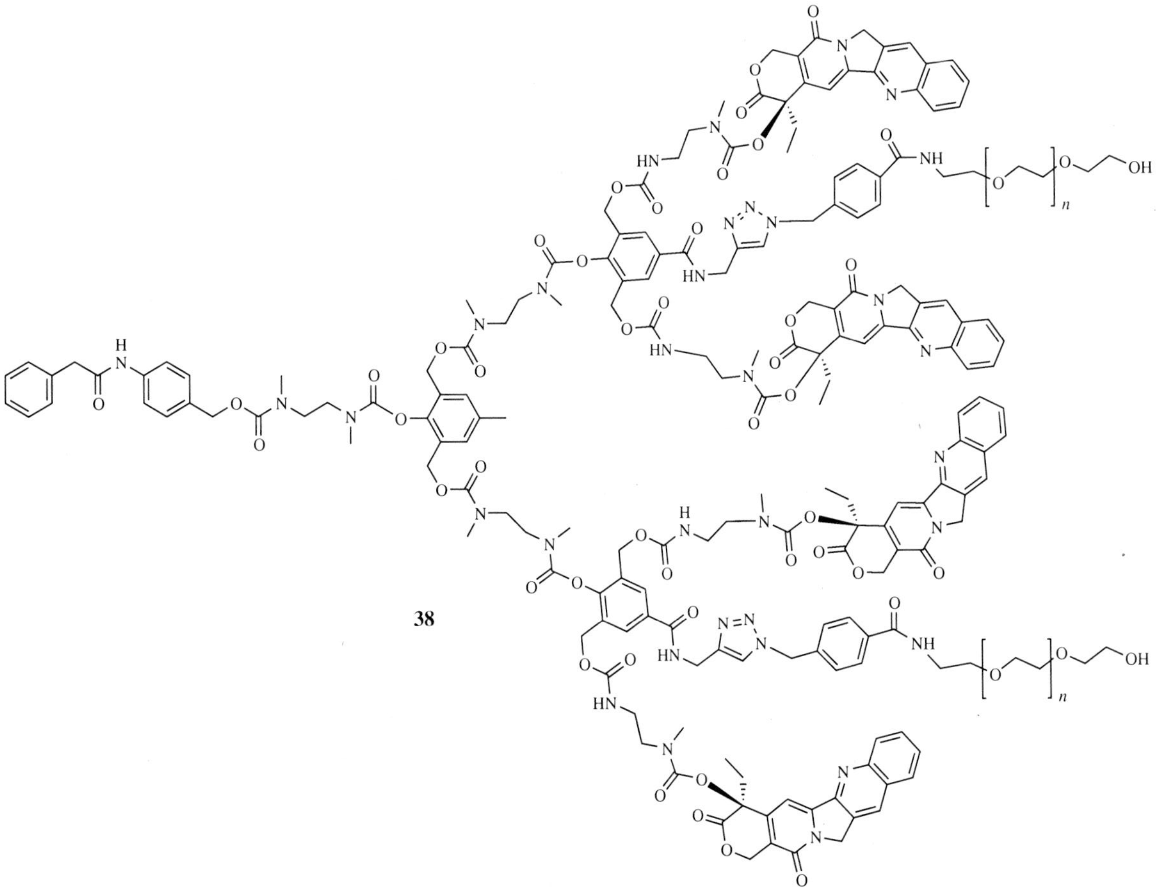

38

FIGURE 5.33 Molecular structure of second-generation, self-immolative, dendritic CPT prodrug with a trigger designed for activation by PGA.

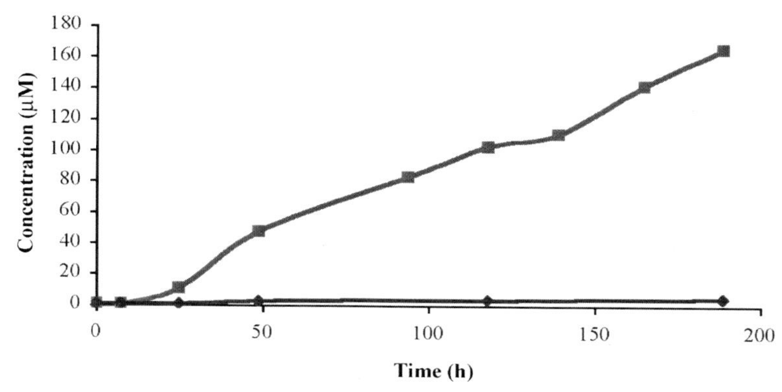

FIGURE 5.34 CPT release from dendritic prodrug **38** with PGA (■) and without PGA (◆).

Fig. 5.34. No release was observed when prodrug **38** was incubated in the buffer without PGA.

Next we evaluated the ability of dendritic prodrug **38** to inhibit cell proliferation in the presence of PGA using three different cell lines: the human T-lineage acute lymphoblastic leukemia cell line MOLT-3, the human leukemia T cell line JURKAT, and the human kidney embryonic HEK-293 cell line. The results are summarized in Table 5.1 and the full data from the cell assays are presented in Fig. 5.35. PGA effectively activated dendritic prodrug **38**, and its toxicity was significantly increased in the cell-growth inhibition assays of three cancerous cell lines. The IC_{50} of prodrug **38** was between 100- and 1000-fold less than free CPT. When PGA was added, the prodrug was activated and its toxicity approached that of free CPT.

In this example even two short PEG400 oligomers were sufficient to permit access of the enzyme to substrate in the dendron's focal point and allow the bioactivation of the system. However, the PEG400 oligomers were not large enough to significantly affect dendrons' aqueous solubility. We assumed that the conjugation of PEG oligomers with higher mass should achieve sufficient hydrophilicity to solublize the dendritic prodrug system in water. Indeed, when PEG5000 was conjugated to a second-generation dendritic prodrug bearing four molecules of the hydrophobic drug CPT, the conjugate had significant aqueous solubility and PGA efficiently activated it.

TABLE 5.1 IC_{50} (nM) Values from Cell-Growth Inhibition Assays

Drug/Prodrug	MOLT-3		JURKAT		HEK-293	
	$IC_{50}{}^{a}$	$IC_{50}{}^{b}$	$IC_{50}{}^{a}$	$IC_{50}{}^{b}$	$IC_{50}{}^{a}$	$IC_{50}{}^{b}$
CPT	2.9	2.9	3.1	3.1	12	12
Pro-CPT 38	4200	41	2000	13	4100	22

[a]Cells were incubated in medium with drug/prodrug.
[b]Cells were incubated in medium with drug/prodrug + 1 mM of PGA.

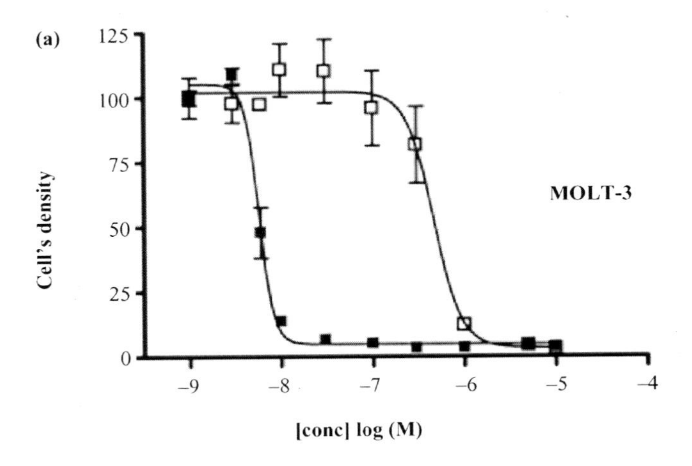

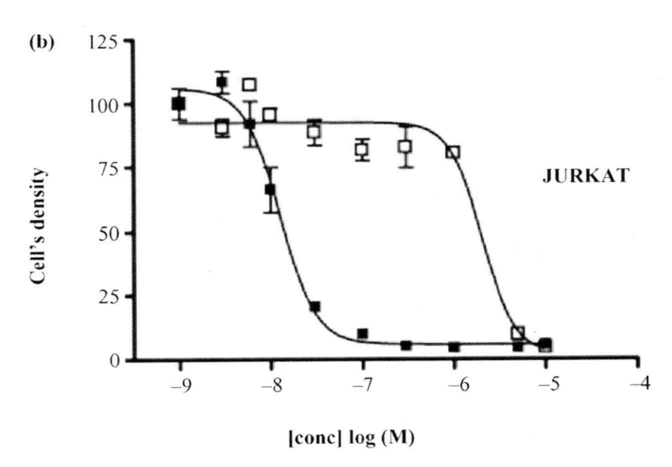

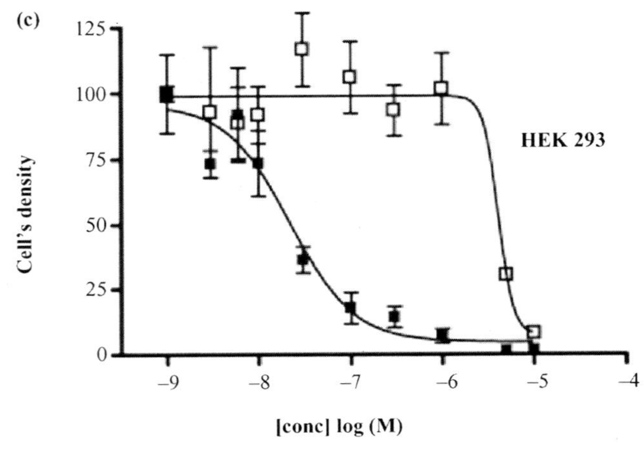

FIGURE 5.35 Growth inhibition assay of three human cancerous cell lines, with dendritic prodrug **38** in the presence (■) and absence (□) of PGA; cells were incubated for 72 h.

5.7 DUAL OUTPUT MOLECULAR PROBE FOR ENZYMATIC ACTIVITY

A molecular probe with dual output signals offers two detection modes allowing use of the same probe in different environments. We have demonstrated how an AB_2 self-immolative dendron with double quinone methide release mechanism can be applied to create a molecular probe with UV–Vis and fluorescence modes for the detection of a specific catalytic activity.[15] The molecular probe is illustrated in Fig. 5.36. The central unit of the probe (the molecular adaptor) is linked to an enzymatic substrate that acts as a trigger and to two different reporter molecules. Cleavage of the enzymatic substrate triggers the release of the two reporters and a consequent activation of their signals.

In order to construct a molecular probe with the required activity, we have used AB_2 self-immolative dendron **39** (Fig. 5.37). In structure **39**, a phenylacetamide group (red) is used as a trigger (cleaved by PGA), and the reporter units are 4-nitrophenol and 6-aminoquinoline. The reporter molecules are attached through stable carbamate linkages that maintain the signals in an OFF position. Upon cleavage of the trigger and release of the reporter units, the signal from each is turned ON. The 4-nitrophenol is detected by visible yellow color and 6-aminoquinoline by fluorescence emission. PEG5000 is also attached to the dendritic adaptor in order provide substantial aqueous solubility.

The disassembly pathway of molecular probe **39** is initiated by catalytic cleavage of phenylacetic acid by PGA, elimination of azaquinone methide, decarboxylation, and cyclization to release dimethylurea derivative and phenol **40** (Fig. 5.38). The latter rapidly undergoes double quinone methide elimination to release the two reporter units and by-product **41**. The output of these cascade

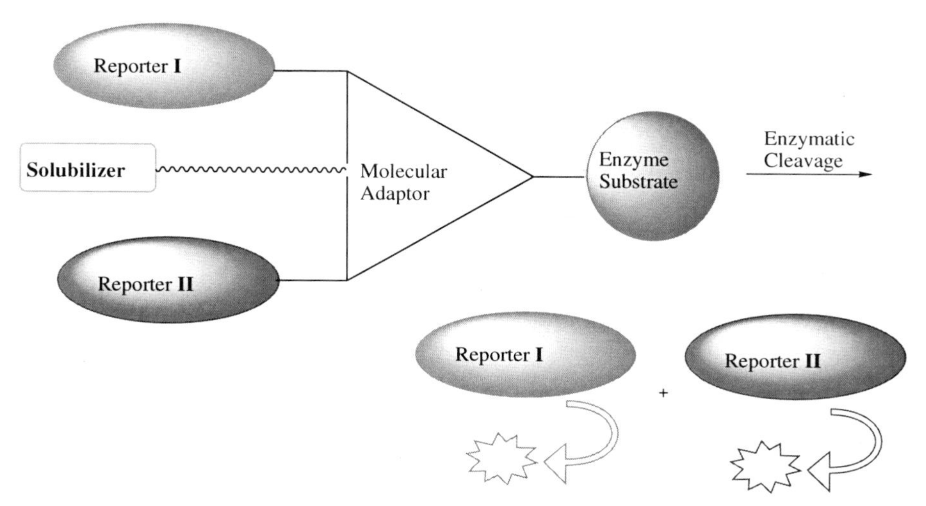

FIGURE 5.36 A graphical illustration of a molecular probe for detection of enzymatic activity with dual output. (See the color version of this figure in Color Plates section.)

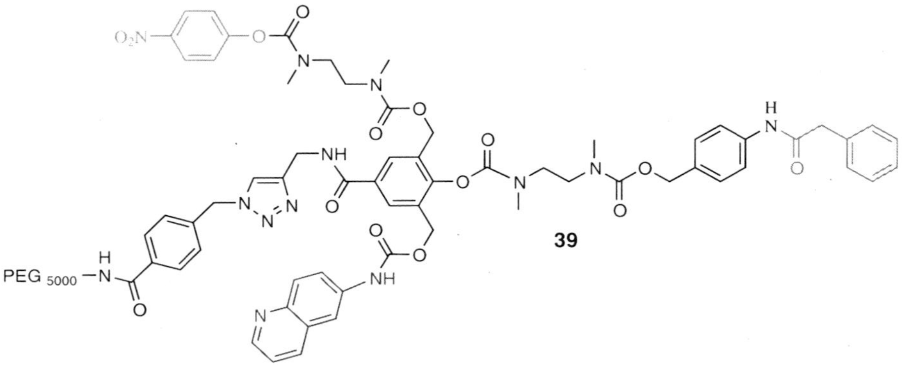

FIGURE 5.37 Chemical structure of a molecular probe with UV–Vis and fluorescence outputs for penicillin G amidase activity. The phenylacetamide group (red) is a substrate for PGA. The reporter units, 4-nitrophenol and 6-aminoquinoline, provide a visible signal and a fluorescence signal, respectively, upon release. (See the color version of this figure in Color Plates section.)

reactions should be expressed by new absorbance and fluorescence signals as a result of the release of 4-nitrophenol and 6-aminoquinoline, respectively.

In order to test our molecular probe, compound **39** was incubated in PBS (pH 7.4) at 37 °C (with and without PGA) and the release of the reporters, 6-aminoquinoline and 4-nitrophenol, was monitored by fluorescence and UV–Vis spectroscopy. The fluorescence spectrum of **39** exhibited a strong emission band at 390 nm that decreased over time as the probe fragmented (Fig. 5.39a). The generation of a new band at 460 nm indicated the formation of free 6-aminoquinoline. In order to evaluate the kinetic behavior of the sequential fragmentation, the maximal intensities of the band at 460 nm were plotted as a function of the time (Fig. 5.39b). The enzymatic cleavage was also monitored in the visible range; upon incubation of **39** in the presence of PGA, the release of 4-nitrophenol resulted in a significant increase in absorbance with a maximum at approximately 405 nM (Fig. 5.39c). This absorbance gradually increased as a function of time in the presence of PGA (Fig. 5.39d). Importantly, no release of 6-aminoquinoline or 4-nitrophenol was observed when compound **39** was incubated in the buffer without PGA (data not shown).

As far as we know, this is the first molecular probe that includes two different types of reporter units activated upon on a specific stimulus. The other option to achieve dual detection would be to use two separate probes. However, in this case there could be a problem of competitive catalysis (circumstances in which the K_m of the two substrate is not identical). In our probe, 6-aminoquinoline and 4-nitrophenol, detected by fluorescence and absorbance spectroscopy, respectively, were used as reporter units. Due to the synthetic flexibility of our approach, other reporter molecules with different types of functional groups, like amine or hydroxyl, can be linked to our molecular probe. The two assays must be orthogonal to each other, in order to prevent disturbances in the detection measurement. Another advantage of the probe is the aqueous solubility

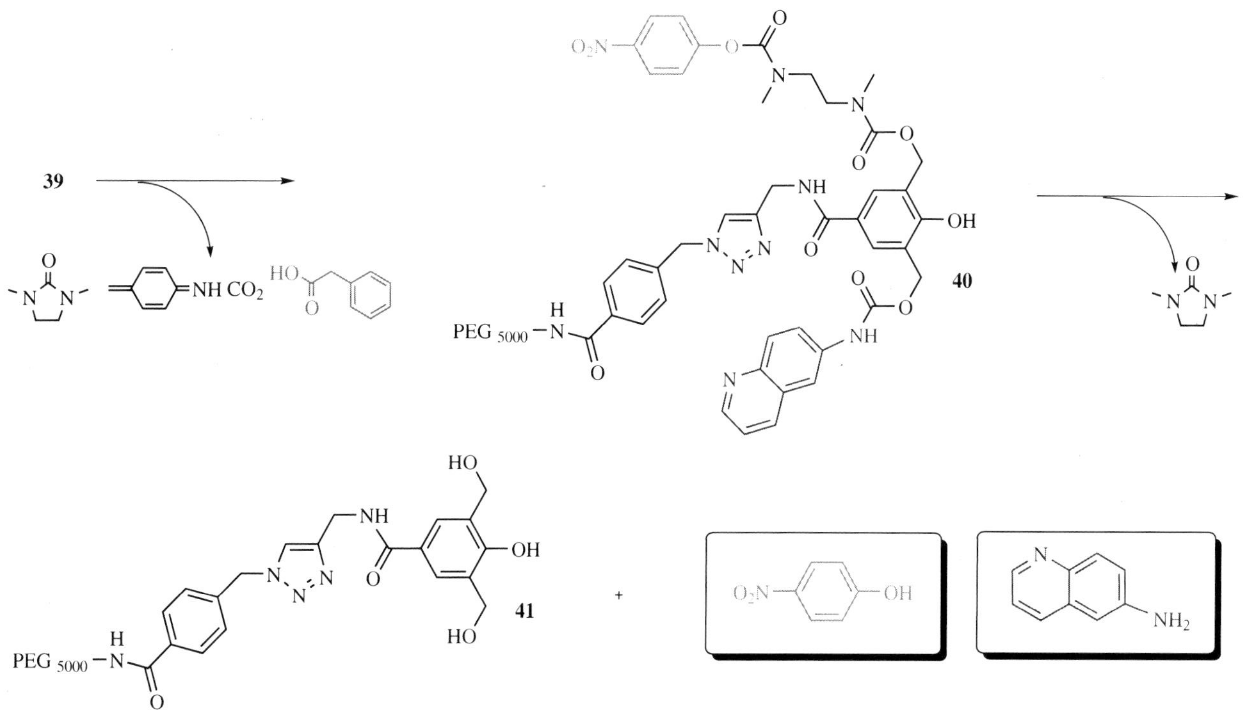

FIGURE 5.38 Disassembly pathway of AB_2 self-immolative dendritic molecule **39**. (See the color version of this figure in Color Plates section.)

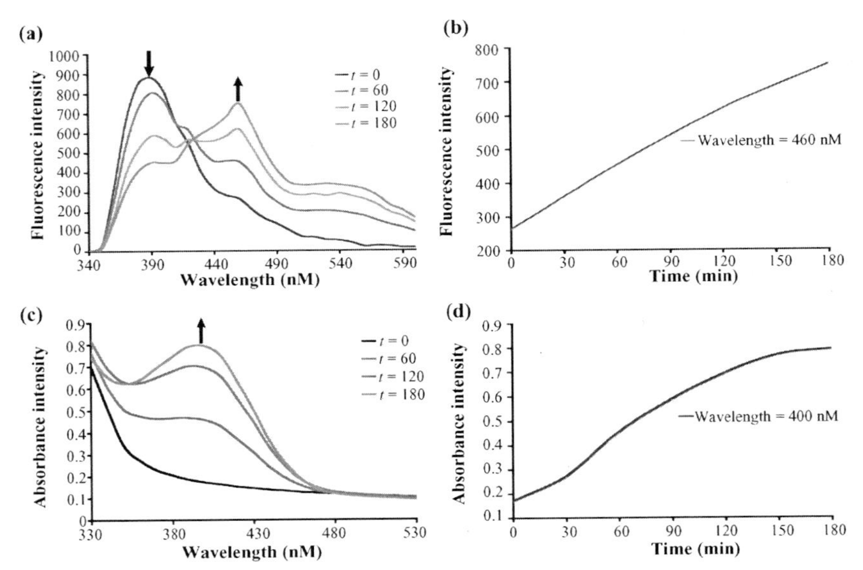

FIGURE 5.39 (a) Emission fluorescence ($l_{ex} = 250$ nm) of **39** (500 mM) upon addition of PGA (1.0 mg/mL). (b) Fluorescence of **39** at 460 nm in the presence of PGA (1.0 mg/mL) as a function of time. (c) UV–Vis absorbance spectra of **39** (500 mM) in the presence of PGA (1.0 mg/mL). (d) Absorbance of **39** at 400 nm after addition of PGA (1.0 mg/mL) as a function of time. (See the color version of this figure in Color Plates section.)

provided by the PEG5000 tail; this tail should guarantee sufficient solubility for most organic reporter molecules in assay media. As demonstrated, this probe functioned under physiological conditions and this design should allow detection of any number of specific bioactivities. We have used phenylacetamide as a triggering substrate for PGA; however, many other substrates can be applied instead. Specifically, this system could be used for detection of proteolytic activity by PGA, which is expressed in bacteria.

5.8 CLEAVAGE SIGNAL CONDUCTION IN SELF-IMMOLATIVE DENDRIMERS

The structural design of nerve cells is a striking example of dendritic architecture, which acts as a signal transduction system. Neurons are known to send out a series of long specialized processes that will either receive electrical signals (dendrites) or transmit these electrical signals (axons) to their target cells (Fig. 5.40).

The dendritic architecture of a neuron inspired us to design dendrimers with a signal transduction pathway related to that of a nerve cell. We have designed and demonstrated a self-immolative dendritic molecule, which is capable of transferring a cleavage signal in a convergent manner to the core and then amplifying it divergently to the periphery.[16] This synthetic system has analogy to the signal transduction pathway of a neuron.

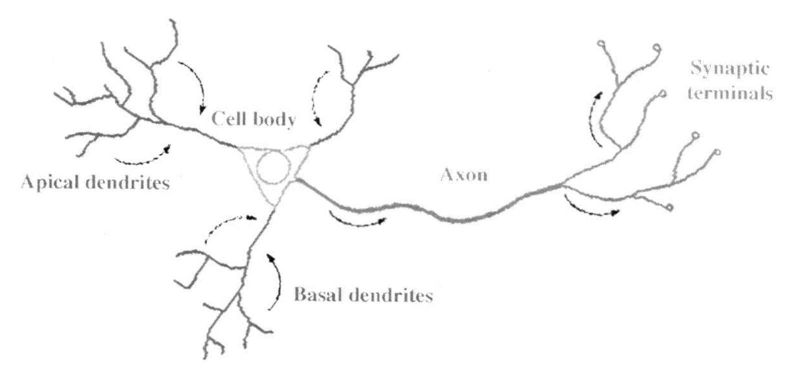

FIGURE 5.40 Schematic of the dendritic architecture of a neuron. The electrical signal is transferred in a convergent manner from the dendrites toward the axon, where it diverges to the synaptic terminals.

We used multitriggered self-immolative dendron as a receiver and linked it through a short spacer to the single-triggered self-immolative dendron that acts as an amplifier. In this design (shown schematically in Fig. 5.41), a signal is received through activation of either one of the triggers. The signal is transferred to the focal point, where it is divergently amplified through the right dendron, and both reporter units are released. During the signal propagation, the dendritic molecule is disassembled into small fragments.

Based on the design illustrated in Fig. 5.41, we synthesized two dendritic molecules: compound **42** (first generation) and compound **43** (second generation) shown in Fig. 5.42. In each, the signal transduction is programmed to initiate through enzymatic cleavage of the phenylacetamide trigger by penicillin G amidase. 6-Aminoquinoline was used as a reporter unit, since its release can be monitored by fluorescence spectroscopy. Upon the release of 6-aminoquinoline from the dendrimer, the conjugation of its amine functional group with the quinoline p-system is increased, and a new band at 460 nm appears in the emitted fluorescence spectrum. PEG400

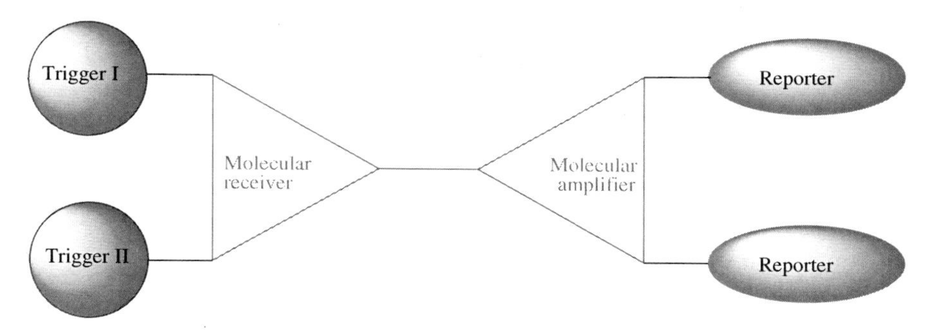

FIGURE 5.41 Graphical structure of a receiver–amplifier dendritic molecule. (See the color version of this figure in Color Plates section.)

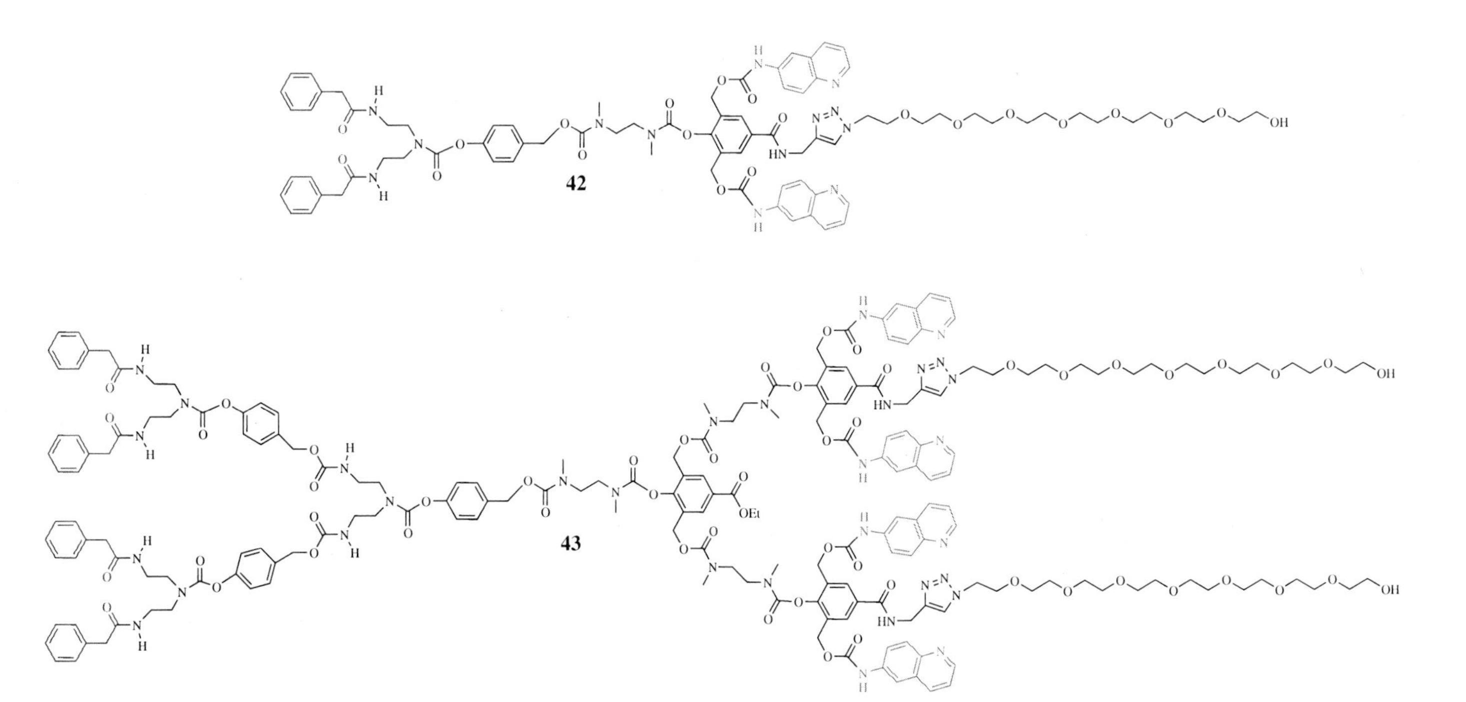

FIGURE 5.42 Chemical structures of first-generation (**42**) and second-generation (**43**) self-immolative, receiver–amplifier dendritic molecules with an enzymatic trigger (blue), cleaved by PGA and 6-aminoquinoline (red) reporter groups. (See the color version of this figure in Color Plates section.)

oligomers were attached to the right side of the dendritic molecules in order to obtain enough aqueous solubility to allow enzymatic activation.

The signal transfer mechanism of the first-generation dendritic molecule (**42**) is illustrated in Fig. 5.43. Enzymatic cleavage of either one of the phenylacetamide groups by PGA exposes free amine intermediate **44**. The latter is cyclized to initiate a series of self-immolative fragmentations that release phenol **45** and several other short fragments. Phenol **45** is disassembled through a double quinone methide type rearrangement to release carbon dioxide, compound **46**, and most importantly, free the two fluorescent molecules of 6-aminoquinoline. The second-generation dendritic molecule **43** disassembles via a mechanism similar to that of molecule **42** (Fig. 5.44). Enzymatic cleavage of one of the four phenylacetamide groups by PGA releases amine intermediate **47**, which initiates the signal transfer through self-immolative fragmentations. The output is expressed in the form of a fluorescence signal as a result of the release of four 6-aminoquinoline molecules.

Dendritic molecules **42** and **43** were incubated with PGA in PBS (pH 7.4) at 37 °C. Control solutions were incubated in the buffer without the enzyme. The sequential fragmentation illustrated in Figs. 5.43 and 5.44 was monitored through the release of 6-aminoquinoline. As shown in Fig. 5.45, free 6-aminoquinoline is generated upon addition of PGA to a solution of **42** or **43**. The fluorescence spectrum of **42** and **43** exhibited one emitting band at 390 nm that disappeared during the dendrimers' fragmentation. The generation of a new band at 460 nm indicated the formation of free 6-aminoquinoline. In order to evaluate the kinetic behavior of the sequential fragmentation, the intensities of the bands at 390 and 460 nm were plotted as a function of the time (Figs. 5.45b and d). The release of 6-aminoquinoline from first-generation dendritic molecule **42** was complete in approximately 4 h, whereas the fragmentation of second-generation dendritic molecule **43** required over 50 h. No release was observed when compounds **42** and **43** were incubated in the buffer without PGA.

This example demonstrates new dendritic molecules that act as a receiver–amplifier device. A cleavage signal received by one side of the dendritic molecule is transferred in a convergent manner to the core and then amplified divergently toward the other side. The signal is propagating through self-immolative sequential fragmentations to release reporter molecules that are visualized by fluorescence. This system has similarities to the dendritic architecture and to the function of neurons and other dendritic transduction pathways in nature.

5.9 FUTURE PROSPECTS

The quinone or azaquinone methide rearrangement was demonstrated to be a powerful efficient tool in the construction of dendritic molecules with spontaneous disassembly mechanism. The circumstances in which a single cleavage reaction in the focal point is translated into multiple release of peripheral molecules result in a unique amplification effect. The system acts as a molecular amplifier capable to amplify one cleavage signal to several cleavages of tail units that generate new visual or therapeutic signals. Self-immolative dendritic prodrugs, activated through a single catalytic reaction by

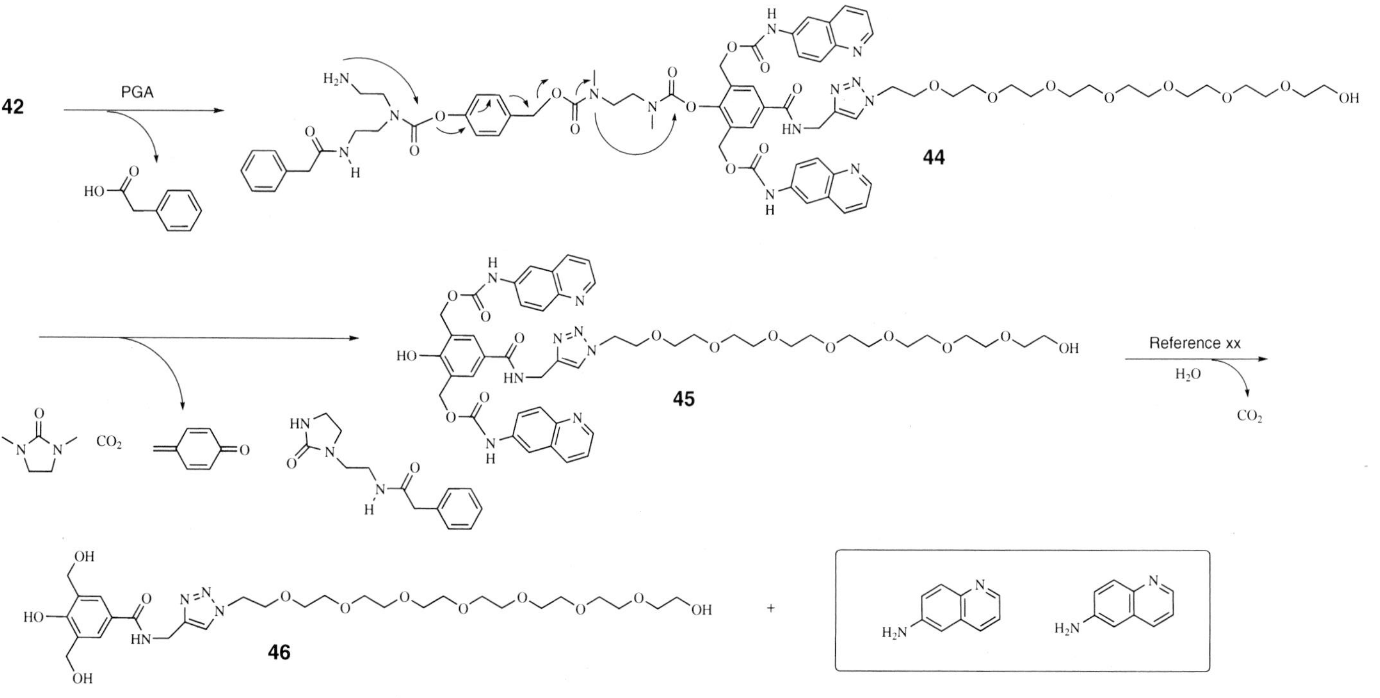

FIGURE 5.43 Signal transduction mechanism for dendritic molecule **42**, through a self-immolative reaction sequence.

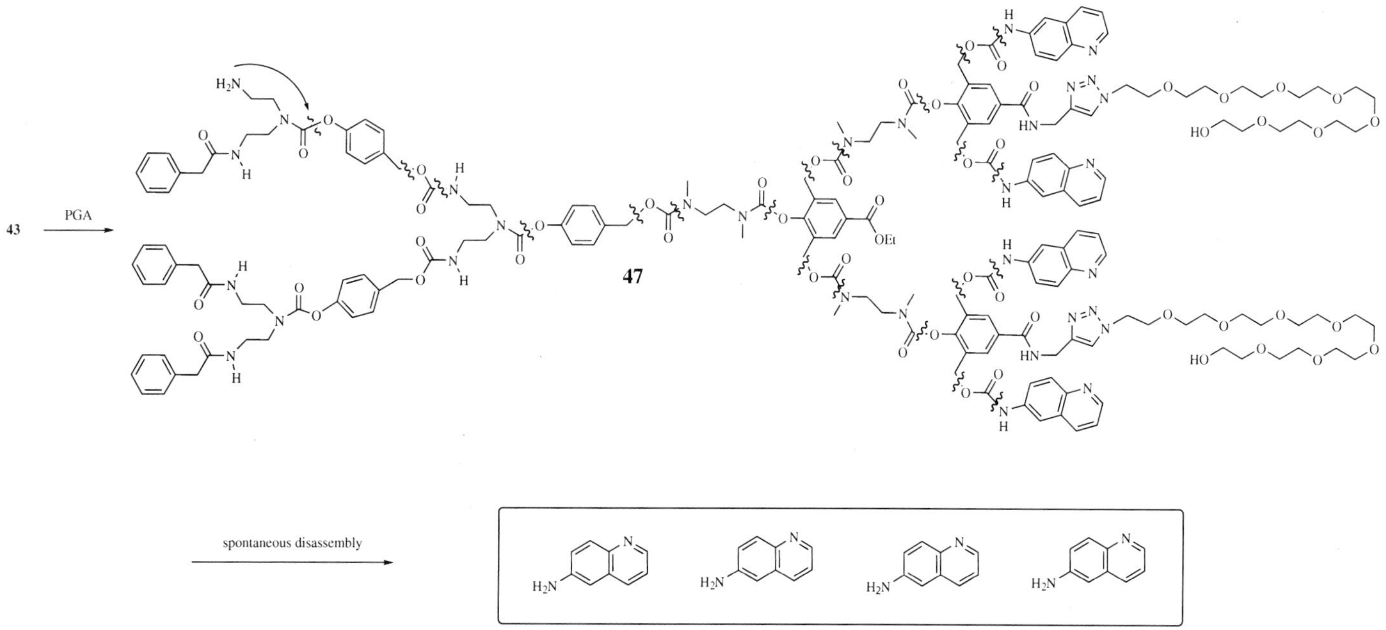

43 $\xrightarrow{\text{PGA}}$ 47 $\xrightarrow{\text{spontaneous disassembly}}$

FIGURE 5.44 Signal transduction pathway for dendritic molecule **43**, through a self-immolative reaction sequence.

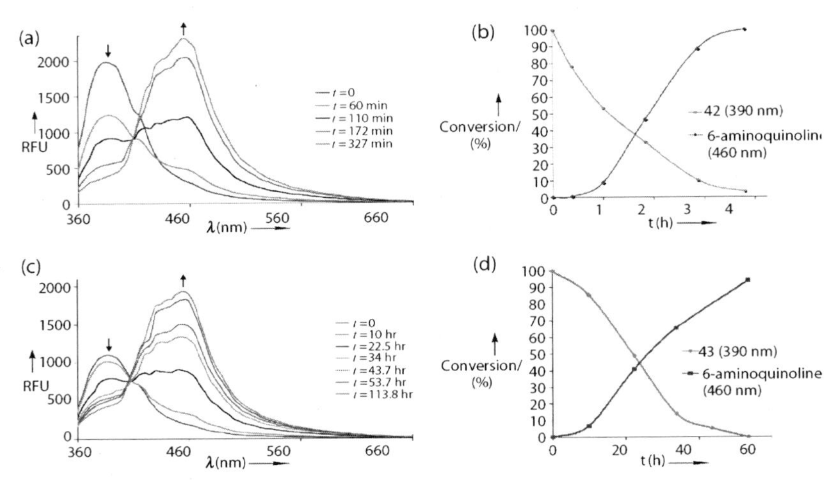

FIGURE 5.45 Emission fluorescence spectra ($\lambda_{ex} = 250$ nm) of **42** (a and b) and **43** (c and d) upon addition of PGA (0.1 mg/mL). The concentrations of dendritic molecules **42** and **43** were 25 and 12 mM, respectively. (See the color version of this figure in Color Plates section.)

a specific enzyme, could offer significant advantages in inhibition of tumor growth, especially if the targeted or secreted enzyme exists at relatively low levels in the malignant tissue. We anticipate that single-triggered dendritic molecules will be further exploited to improve selective chemotherapeutic approaches in cancer therapy and amplification of molecular signal for diagnostic purposes.

REFERENCES

1. Carl, P. L.; Chakravarty, P. K.; Katzenellenbogen, J. A. A novel connector linkage applicable in prodrug design. *J. Med. Chem.* 1981, 24, 479–480.
2. Sagi, A.; Segal, E.; Satchi-Fainaro, R.; Shabat, D.; Remarkable drug-release enhancement with an elimination-based AB_3 self-immolative dendritic amplifier. *Bioorg. Med. Chem.* 2007, 15, 3720–3727.
3. Amir, R. J.; Pessah, N.; Shamis, M.; Shabat, D. Self-immolative dendrimers. *Angew. Chem., Int. Ed. Engl.* 2003, 42, 4494–4499.
4. de Groot, F. M.; Albrecht, C.; Koekkoek, R.; Beusker, P. H.; Scheeren, H. W. Cascade-release dendrimers liberate all end groups upon a single triggering event in the dendritic core. *Angew. Chem., Int. Ed. Engl.* 2003, 42, 4490–4494.
5. Szalai, M. L.; Kevwitch, R. M.; McGrath, D. V. Geometric disassembly of dendrimers: dendritic amplification. *J. Am. Chem. Soc.* 2003, 125, 15688–15689.
6. Haba, K.; Popkov, M.; Shamis, M.; Lerner, R. A.; Barbas, C. F., III; Shabat, D. Single-triggered trimeric prodrugs. *Angew. Chem., Int. Ed. Engl.* 2005, 44, 716–720.
7. Shamis, M.; Lode, H. N.; Shabat, D. Bioactivation of self-immolative dendritic prodrugs by catalytic antibody 38C2. *J. Am. Chem. Soc.* 2004, 126, 1726–1731.

8. Amir, R. J.; Popkov, M.; Lerner, R. A.; Barbas, C. F., III; Shabat, D. Prodrug activation gated by a molecular OR logic trigger. *Angew. Chem., Int. Ed. Engl.* 2005, 44, 4378–4381.
9. Amir, R. J.; Shabat, D. Self-immolative dendrimer biodegradability by multi-enzymatic triggering. *Chem. Commun.* 2004, 1614–1615.
10. Weinert, E. E.; Dondi, R.; Colloredo-Melz, S.; Frankenfield, K. N.; Mitchell, C. H.; Freccero, M.; Rokita, S. E. Substituents on quinone methides strongly modulate formation and stability of their nucleophilic adducts. *J. Am. Chem. Soc.* 2006, 128, 11940–11947.
11. Perry, R.; Amir, R. J.; Shabat, D. Substituent-dependent disassembly of self-immolative dendrimers. *New J. Chem.* 2007, 31, 1307–1312.
12. Adler-Abramovich, L.; Perry, R.; Sagi, A.; Gazit, E.; Shabat, D. Controlled assembly of peptide nanotubes triggered by enzymatic activation of self-immolative dendrimers. *Chem. Biochem.* 2007, 8, 859–862.
13. Shabat, D. Self-immolative dendrimers as novel drug delivery platforms. *J. Polym. Sci. A: Polym. Chem.* 2006, 44, 1569–1578.
14. Gopin, A.; Ebner, S.; Attali, B.; Shabat, D. Enzymatic activation of second-generation dendritic prodrugs: conjugation of self-immolative dendrimers with poly(ethylene glycol) via click chemistry. *Bioconj. Chem.* 2006, 17, 1432–1440.
15. Danieli, E.; Shabat, D. Molecular probe for enzymatic activity with dual output. *Bioorg. Med. Chem.* 2007, 15, 7318–7324.
16. Amir, R. J.; Danieli, E.; Shabat, D. Receiver–amplifier, self-immolative dendritic device. *Chem. Eur. J.* 2007, 13, 812–821.

6

ORTHO-QUINONE METHIDES IN TOCOPHEROL CHEMISTRY

THOMAS ROSENAU AND STEFAN BÖHMDORFER
Department of Chemistry, University of Natural Resources and Applied Life Sciences, Muthgasse 18, A-1190 Vienna, Austria

6.1 INTRODUCTION

Vitamin E, being first reported barely a century ago, is the biologically most important fat-soluble antioxidant and has become a commodity product and bulk chemical in the meantime. Besides its antioxidant function, several nonantioxidant actions of the compound have been recently identified and new ones are still being discovered.[1]

The major, large-scale application of vitamin E is animal nutrition, and many pharmaceutical, health care, and cosmetic products contain the substance. In the consumer notion, "vitamin E" is connected with terms such as antioxidant, radical scavenging, and antiaging. Although usually, and due to the dominance of the α-homologue in all kinds of applications, the term vitamin E is widely—but not correctly—used synonymous with α-tocopherol or even α-tocopheryl acetate, it is in fact a generic descriptor of all tocol and tocotrienol derivatives exhibiting qualitatively the vitamin E activity (as determined by specific biological tests) of α-tocopherol, that is, covering the four tocopherols and the four tocotrienols (α-, β-, γ-, and δ-homologues). These four forms are distinguished by the number and position of methyl groups on the aromatic ring, with α-tocopherol being the "permethylated" congener carrying three methyl groups at positions 5, 7, and 8. The tocotrienols are distinguished from the tocopherols by the presence of double bonds in the isoprenoid side chain.

Quinone Methides, Edited by Steven E. Rokita

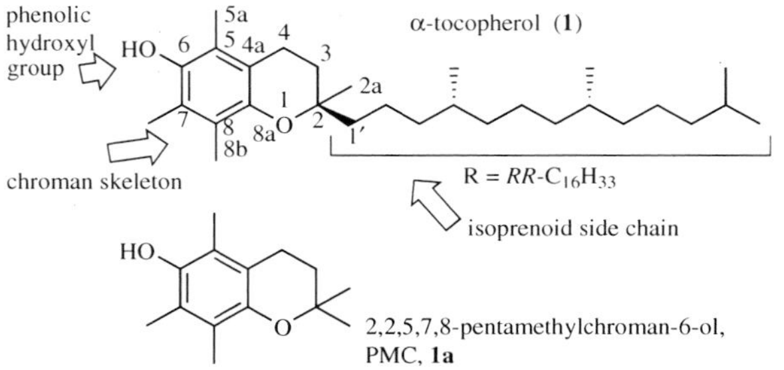

FIGURE 6.1 Chemical structure of α-tocopherol (**1**) and its model compound PMC (**1a**). Here and in the following the R-substituent denotes the *R,R*-configured isoprenoid $C_{16}H_{33}$ side chain of the tocopherol.

Chemical modification of α-tocopherol can in principle occur at three distinguishable regions of the molecule: the phenolic OH group, the chroman skeleton, and the isoprenoid side chain (Fig. 6.1). The isoprenoid side chain is fairly resistant toward all types of chemical attacks. Although the presence of the isoprenoid side chain determines the lipophilicity and the apolarity of the molecule to a great extent, it has only a negligible influence on the general chemical reactivity of the chroman system. This is also the reason why in many reactions 2,2,5,7,8-pentamethylchroman-6-ol (PMC, **1a**), the truncated model compound carrying a methyl group instead of the side chain, has been used in reactions in lieu of the tocopherol itself. The nomenclature and the atom numbering in tocopherols have been regulated by IUPAC,[2,3] and this numbering (Fig. 6.1) is also used throughout this chapter.

Vitamin E has been in the focus of research mainly due to its physiological and medical applications and effects, but it is a very intriguing molecule for mechanistic studies too. This is mainly due to its ability to create an *ortho*-quinone methide (*o*-QM) upon oxidation, a reaction that mainly determines its chemistry. This ability made it an interesting object of study not only for vitamin E chemists and medical scientists but also for the general organic chemists who seek to obtain deeper insight into general phenol and antioxidant chemistry. Within the last several years, the knowledge about the *o*-QM derived from α-tocopherol has been considerably expanded on both fields: the "classical" vitamin E chemistry and the field of general antioxidant and phenol chemistry in which the application of the *o*-QM as a model compound afforded new general insights into formation and reaction mechanisms of *o*-QMs. These different aspects of α-tocopherol and its derived *o*-QM will be discussed in the following section.

6.2 α-TOCOPHEROL AND ITS DERIVED *o*-QM: GENERAL ASPECTS

Oxidation chemistry of α-tocopherol (**1**) generally involves three primary intermediates (**2–4**), which are formed according to the respective reaction conditions used, their

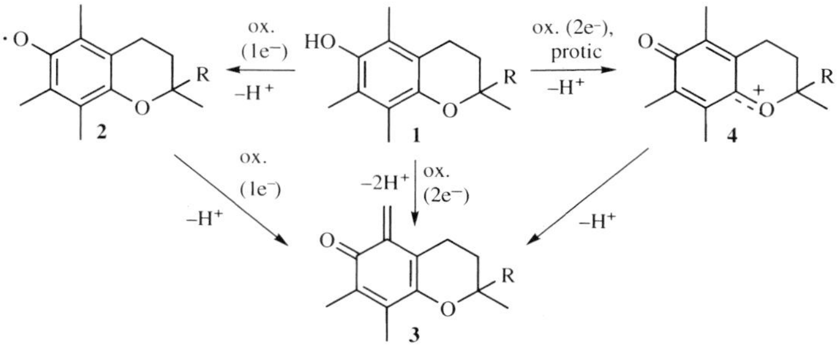

FIGURE 6.2 Primary key intermediates in the oxidation chemistry of α-tocopherol.

intermediacy being especially dependent on the oxidant employed (two-electron or one-electron oxidant) and the solvent chosen (polar or apolar, aprotic or protic, dry or containing traces of water) (see Fig. 6.3). While oxidation in apolar media favors tocopheroxyl radical **2** and the *o*-QM **3**, oxidation in protic or aqueous media favors the chromanoxylium cation **4**; in micellar systems mostly all three intermediates can be observed (Fig. 6.2).

The α-tocopheroxyl radical (**2**) is the primary homolytic (one-electron, radical) oxidation product of α-tocopherol. Its formation and occurrence is comprehensively supported and confirmed by EPR[4–8] and ENDOR[9] experiments. Under physiological conditions, when tocopherol acts as the classical antioxidant, the tocopheroxyl radical is formed directly by H-atom abstraction, and so it is in most *in vitro* model systems. The resulting tocopheroxyl radical is relatively stable and unable to continue the radical chain process, which is the basis of the vitamin's antioxidant action. *In vivo*, the radical is reduced back to the phenol by ascorbate, glutathione, or other reductants in a quite effective recycling system. However, also side reactions occur that deplete the organism of the vitamin. Under special experimental conditions, such as UV-irradiation in inert media, a transient radical cation can also be produced and its occurrence verified by EPR, which in a second step releases the phenolic proton to afford the tocopheroxyl radical. This process is a two-step mechanism and thus distinguished from the usual way of tocopheroxyl radical formation.

The spin density of tocopheroxyl radical **2**, a classical phenoxyl radical, is mainly concentrated at oxygen O-6, which is the major position for coupling with other C-centered radicals, leading to chromanyl ethers **5**. These products are found in the typical lipid peroxidation scenarios. Also at *ortho*- and *para*-positions of the aromatic ring, the spin density is increased. At these carbon atoms, coupling with other radicals, especially O-centered ones, proceeds. Mainly the *para*-position (C-8a) is involved (Fig. 6.3), leading to differently 8a-substituted chromanones **6**.

Heterolytic (two-electron, ionic) oxidation of **1**, or alternatively further one-electron loss from the primary radical **2**, affords chromanoxylium cation **4** with its positive charge mainly localized at C-8a. Cation **4** is stabilized by resonance so that a positive partial charge results also at C-5 and C-7, where nucleophilic attack is

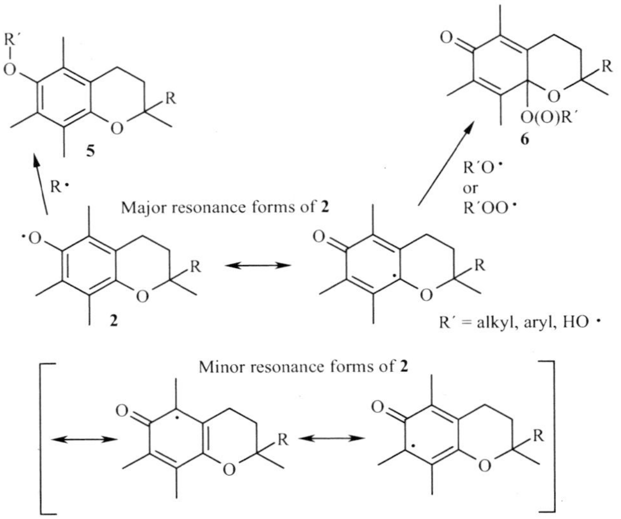

FIGURE 6.3 Resonance forms and general reactions of chromanoxyl radical **2**.

consequently facilitated. The occurrence of chromanoxylium cation **4** as an actual intermediate had been a matter of debate until Webster succeeded in trapping the cation in the presence of non-nucleophilic bulky counterions and determining the crystal structure of the product.[10]

Chromanoxylium cation **4** preferably adds nucleophiles in 8a-position producing 8a-substituted tocopherones **6**, similar in structure to those obtained by radical recombination between C-8a of chromanoxyl **2** and coreacting radicals (Fig. 6.4). Addition of a hydroxyl ion to **4**, for instance, results in a 8a-hydroxy-tocopherone, which in a subsequent step gives the *para*-tocopherylquinone (**7**), the main (and in most cases, the only) product of two-electron oxidation of tocopherol in aqueous media. A second interesting reaction of chromanoxylium cation **4** is the loss of a proton at C-5a, producing the *o*-QM **3**. This reaction is mostly carried out starting from tocopherones **6** or *para*-tocopherylquinone (**7**) under acidic catalysis, so that chromanoxylium **4** is produced in the first step, followed by proton elimination from C-5a. In the overall reaction of a tocopherone **6**, a [1,4]-elimination has occurred. The central species in the oxidation chemistry of α-tocopherol is the *o*-QM **3**, which is discussed in detail subsequently.

The third primary intermediate in the oxidation chemistry of α-tocopherol, and the central species in this chapter, is the *ortho*-quinone methide **3**. In contrast to the other two primary intermediates **2** and **4**, it can be formed by quite different ways (Fig. 6.4), which already might be taken as an indication of the importance of this intermediate in vitamin E chemistry. *o*-QM **3** is formed, as mentioned above, from chromanoxylium cation **4** by proton loss at C-5a, or by a further single-electron oxidation step from radical **2** with concomitant proton loss from C-5a. Its most prominent and most frequently employed formation way is the direct generation from α-tocopherol by two-electron oxidation in inert media. Although in aqueous or protic media, initial

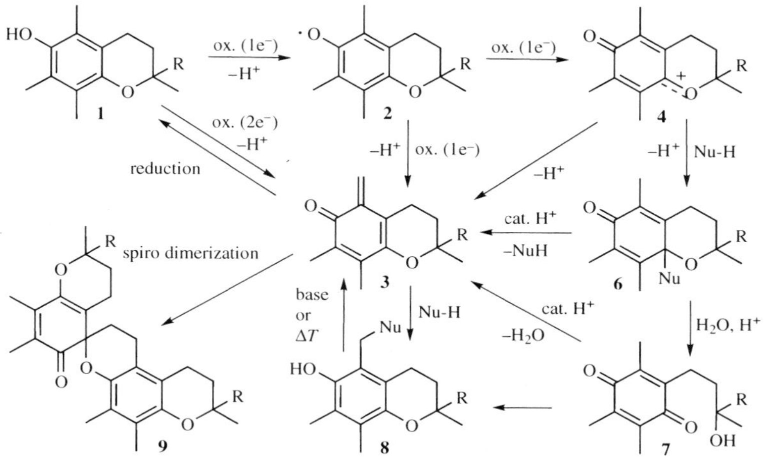

FIGURE 6.4 Reactions and products of the primary oxidation intermediates of α-tocopherol (1), the tocopheroxyl radical 2, *ortho*-quinone methide 3, and chromanoxylium cation 4.

formation of 4 and its trapping products, such as tocopherones 6 and *para*-tocopherylquinone 7, is preferred (which in subsequent reactions might give rise to *o*-QM 3 as well), in apolar media, aprotic media, or media containing small amounts of water, only the *o*-QM 3 is the prominent, if not only, primary oxidation product. An almost equally important formation reaction, although not using α-tocopherol as the starting material, is the degradation of 5a-substituted tocopherols (8), which release the 5a-substituent and the phenolic proton in a [1,4]-elimination process. This requires electronegative moieties (e.g., O-, N-, and S-moieties) at C-5a, which are readily eliminated by base or thermal treatment. This formation pathway is actually the reversal of the formation reaction of such 5a-substituted tocopherols, which involves addition of a nucleophile to the intermediate *o*-QM 3.

In contrast to the α-tocopheroxyl radical (2) and chromanoxylium cation 4 for which the oxidation allows only one structure to form, generation of an *o*-QM from α-tocopherol could proceed, theoretically, involving either of the two methyl groups C-5a or C-7a. The reason for the large selectivity of *o*-QM formation, that is, the nearly exclusive involvement of position 5a, will be discussed in more detail in Section 6.3.1. The overall formation of the *o*-QM from the parent phenol α-tocopherol means a loss of H_2, or more detailed, of two electrons and two protons. In which order and as which species those are released, for example, as protons, H-atoms, or hydride ions, will have major implications on *o*-QM formation and chemistry, which is discussed in Section 6.3.2.

o-QM 3 formed by oxidation from α-tocopherol (1) or from 5a-substituted tocopherols (8) by one of the above pathways, is a highly unstable, and thus reactive, molecule. It undergoes immediate reactions to stable and analytically accessible molecules that are indicators of its intermediacy. The stabilization reactions can be generalized into three categories: (a) reduction back to the parent phenol, (b) 1,4-addition of nucleophiles to C-5a and O-6 forming 5a-substituted tocopherols (**8**), and (c) hetero-Diels–Alder reaction with inverse electron demand, the *o*-QM itself reacting as the electron-deficient diene, the most prominent example being the self-dimerization to spiro dimer **9**. Starting from the *o*-QM's cyclohexadienone structure, all the three reaction types (Fig. 6.4) restore aromaticity, which is the driving force of the reactions. The reactions of tocopherol-derived *o*-QMs are discussed in Section 6.4.

6.3 CHEMO- AND REGIOSELECTIVITY IN THE *o*-QM FORMATION FROM TOCOPHEROL

6.3.1 *o*-QM Versus "5a-Chromanolmethyl" Radicals

In the early days of vitamin E research and up to the late 1980s, the radical reactions of the compounds were given wide attention and all other chemical conversions of the compound were seen as perhaps academically interesting, but as side reactions that do not significantly contribute to the main reactivity picture they were not given enough attention. In the late 1980s, this notion slowly started to change, and in 2007 it was shown that a major part of the reactions commonly assigned to hypothetical tautomers of the tocopheroxyl radical **2** had been in fact wrongly attributed to this primary intermediate and are actually reactions of another primary intermediate, the *o*-QM **3**.[11] This not only changed the perception of basic tocopherol chemistry quite drastically but also promoted the "appreciation" of the *o*-QM **3** as a central species in tocopherol chemistry.

The occurrence of a 5a-C-centered tocopherol-derived radical **10**, often called "chromanol methide radical" or "chromanol methyl radical," had been postulated in literature dating back to the early days of vitamin E research,[12–19] which have been cited or supposedly reconfirmed later (Fig. 6.5).[8,20–22] In some accounts, radical structure **10** has been described in the literature as being a resonance form (canonic structure) of the tocopheroxyl radical, which of course is inaccurate. If indeed existing, radical **10** represents a tautomer of tocopheroxyl radical **2**, being formed by a chemical reaction, namely, a 1,4-shift of one 5a-proton to the 6-oxygen, but not just by a "shift of electrons" as in the case of resonance structures (Fig. 6.5). In all accounts mentioning

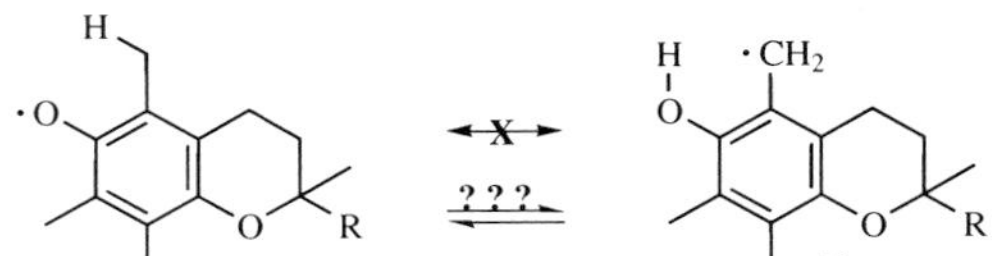

H
·O
O
R
2
???
H
O
·CH2
O
R
10

FIGURE 6.5 The hypothetical chromanol methide radical **10** as a tautomer of α-tocopheroxyl radical **2**.

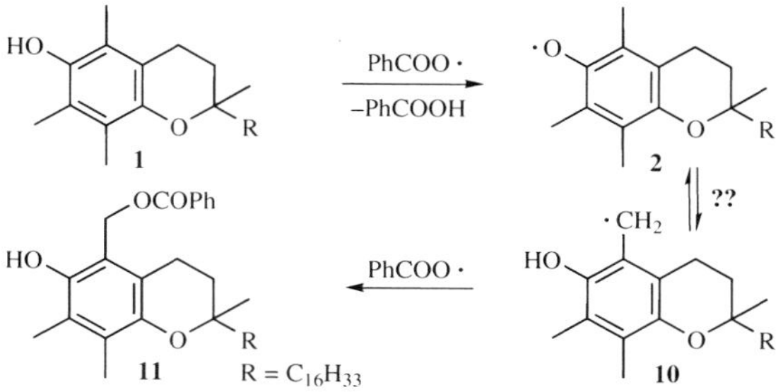

FIGURE 6.6 Hypothetical radical mechanism for the formation of 5a-α-tocopheryl benzoate (**11**) by reaction of α-tocopherol (**1**) with dibenzoyl peroxide.

α-tocopherol-derived C-centered radicals, the spin density was described to be centered at 5a-C, but not at alternative carbons, such as 7a-C or 8b-C. The occurrence of 5a-C-radicals was concluded from experimental observations that seemed to support 5a-C-radicals.

Basically, three reactions were evoked to support the occurrence of 5a-C-centered radicals **10** in tocopherol chemistry. The first one is the formation of 5a-substituted derivatives (**8**) in the reaction of α-tocopherol (**1**) with radicals and radical initiators. The most prominent example here is the reaction of **1** with dibenzoyl peroxide leading to 5a-α-tocopheryl benzoate (**11**) in fair yields,[12] so that a "typical" radical recombination mechanism was postulated (Fig. 6.6). Similarly, low yields of 5a-alkoxy-α-tocopherols were obtained by oxidation of α-tocopherol with *tert*-butyl hydroperoxide or other peroxides in inert solvents containing various alcohols,[23,24] although the involvement of 5a-C-centered radicals in the formation mechanism was not evoked for explanation in these cases.

The second observation cited as evidence for a radical mechanism involving radical **5** is the frequent occurrence of ethano-dimer **12**, proposed to proceed by recombination of two 5a-C-centered radicals **10** (Fig. 6.7).[21,25,26]

The third fact that seemed to argue in favor of the occurrence of radicals **10** was the observation that reactions of α-tocopherol under typical radical conditions, that is, at the presence of radical initiators in inert solvents or under irradiation, provided also large amounts of two-electron oxidation products such as *o*-QM **3** and its spiro dimerization product **9** (Fig. 6.8).[16,25,26] This was taken as support of a disproportionation reaction involving α-tocopheroxyl radical **2** and its hypothetical tautomeric chromanol methide radical **10**, affording one molecule of *o*-QM **3** (oxidation) and regenerating one molecule of **1** (reduction). The term "disproportionation" was used here to describe a one-electron redox process with concomitant transfer of a proton, that is, basically a H-atom transfer from hypothetical **10** to radical **2**.

By a combination of synthetic approaches, isotopic labeling, using tocopherols with ^{13}C-labeling at C-5a and C-7a, EPR spectroscopy, and high-level DFT computations, it was shown that there is no radical formation at either C-5a or C-7a and that chromanol methide radical **10** does not occur in tocopherol.[11] EPR failed to detect

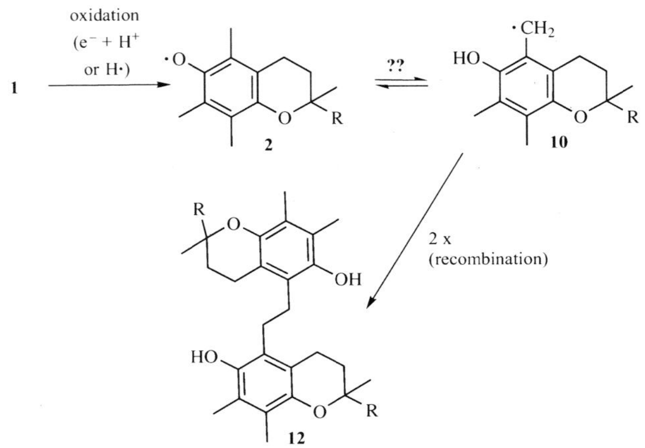

FIGURE 6.7 Formation of α-tocopherol ethano-dimer (**12**) as the result of a hypothetical radical recombination of two radicals **10.**

those species, and computations predicted only negligibly different formation energies for the hypothetical C-5a radicals and C-7a radicals, respectively. Thus, if radicals at 5a-C and 7a-C were involved in tocopherol chemistry, then products of both species would have to be expected. The fact that in reality, products of 5a-C are highly preferred over those of 7a-C—or are even formed exclusively—were

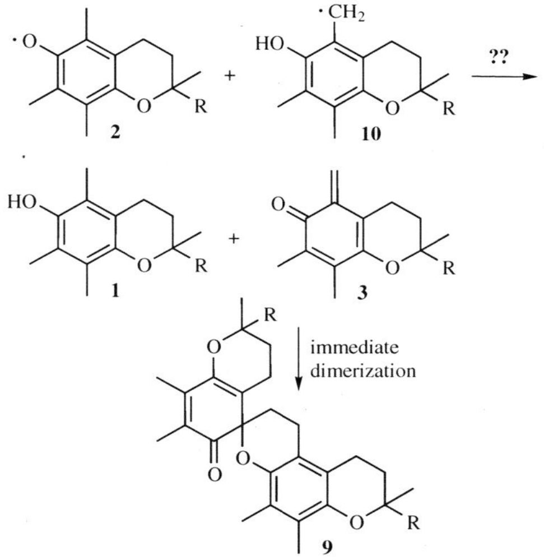

FIGURE 6.8 Hypothetical disproportionation of two α-tocopherol-derived radicals **2** and **10** in the absence of other coreactants to account for the formation of typical two-electron oxidation products (*o*-QM **3**, α-tocopherol spiro dimer **9**).

already seen as an indirect proof of the underlying chemistry not being radical by nature.

The formation of 5a-α-tocopheryl benzoate (**11**) upon reaction of α-tocopherol (**1**) with dibenzoyl peroxide, which has usually been taken as "solid proof" of the involvement of 5a-C-centered radicals in tocopherol chemistry (see Fig. 6.6), was shown to proceed according to a nonradical, heterolytic mechanism involving *o*-QM **3** (Fig. 6.9).

The initiator-derived radical products generate α-tocopheroxyl radicals (**2**) from α-tocopherol (**1**). The radicals **2** are further oxidized to *ortho*-quinone methide **3** in a formal H-atom abstraction, thereby converting benzoyloxy radicals to benzoic acid and phenyl radicals to benzene. The generated *o*-QM **3** adds benzoic acid in a [1,4]-addition process, whereas it cannot add benzene in such a fashion. This pathway accounts for the observed occurrence of benzoate **11** and simultaneous absence of a 5a-phenyl derivative and readily explains the observed products without having to involve the hypothetical C-centered radical **10**.

To conclusively disprove the involvement of the chromanol methide radical, the reaction of α-tocopherol with dibenzoyl peroxide was conducted in the presence of a large excess of ethyl vinyl ether used as a solvent component. If 5a-α-tocopheryl benzoate (**11**) was formed homolytically according to Fig. 6.6, the presence of ethyl vinyl ether should have no large influence on the product distribution. However, if (**11**) was formed heterolytically according to Fig. 6.9, the intermediate *o*-QM **3** would be readily trapped by ethyl vinyl ether in a hetero-Diels–Alder process with inverse electron demand,[27] thus drastically reducing the amount of **11** formed. Exactly the latter outcome was observed experimentally. In fact, using a 10-fold excess of ethyl vinyl ether relative to α-tocopherol and azobis(isobutyronitrile) (AIBN) as radical

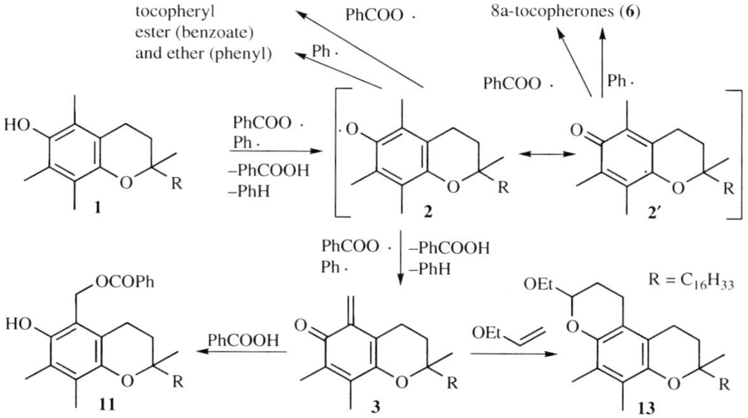

FIGURE 6.9 Confirmed heterolytic formation pathway for 5a-α-tocopheryl benzoate (**11**) without involvement of 5a-C-centered radicals and its proof by trapping of *ortho*-quinone methide **3** with ethyl vinyl ether to pyranochroman **13.** Shown are the major products of the reaction of α-tocopherol (**1**) with dibenzoyl peroxide.

starter, no 5a-α-tocopheryl benzoate (**11**) at all was formed but only the corresponding trapping product **13**.

Also for the reaction that was described as "dimerization" of the chromanol methide radicals **10** to the ethano-dimer of α-tocopherol **12**, the involvement of the C-centered radicals has been disproven and these intermediates lost their role as key intermediates in favor of the *o*-QM **3**. It was experimentally shown that ethano-dimer **12** in hydroperoxide reaction mixtures of α-tocopherol was formed according to a more complex pathway involving the reduction of the spiro dimer **9** by α-tocopheroxyl radicals **2**, which can also be replaced by other phenoxyl radicals (Fig. 6.10).[11] Neither the hydroperoxides themselves, nor radical initiators such as AIBN, nor tocopherol alone were able to perform this reaction, but combinations of tocopherol with radical initiators generating a high flux of tocopheroxyl radicals **2** afforded high yields of the ethano-dimer **12** from the spiro dimer **9**.

The last reaction commonly evoked to support the involvement of radical species **10** in tocopherol chemistry is the "disproportionation" of two molecules into the phenol α-tocopherol and the *ortho*-quinone methide **3** (Fig. 6.8), the latter immediately dimerizing into spiro dimer **9**. This dimerization is actually a hetero-Diels–Alder process with inverse electron demand. It is largely favored, which is also reflected by the fact that spiro dimer **9** is an almost ubiquitous product and by-product in vitamin E chemistry.[28,29] The disproportionation mechanism was proposed to account for the fact that in reactions of tocopheroxyl radical **2** generated without chemical coreactants, that is, by irradiation, the spiro dimer **9** was the only major product found.

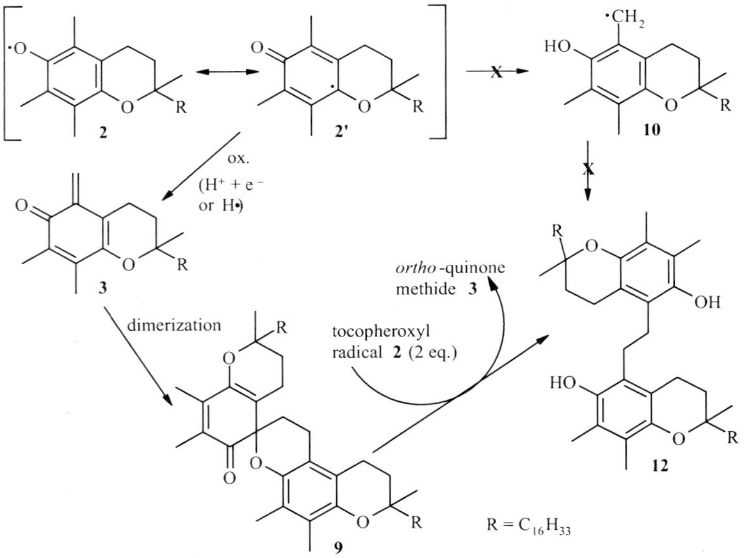

FIGURE 6.10 Confirmed formation pathway of ethano-dimer **12** by reduction of spiro dimer **9** in different reaction systems. 5a-C-centered radicals **10** are not involved in this process.

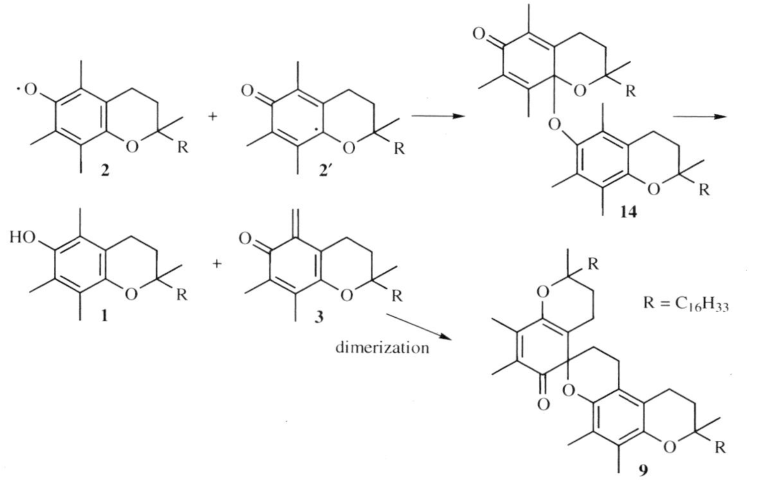

FIGURE 6.11 Confirmed pathway for the observed "disproportionation" of tocopheroxyl radical **2** into α-tocopherol (**1**) and *o*-QM **3**, the latter immediately dimerizing into α-tocopherol spiro dimer (**9**). 5a-C-centered radicals **10** are not involved in this process.

An alternative pathway (Fig. 6.11) was proved, which did not involve the dubious 5a-C-centered radical **10**, but instead only the well-documented and theoretically sound structure of tocopheroxyl radical **2** and its major canonic form **2′**. According to general tocopherol chemistry (*cf.* Fig. 6.3), **2** and **2′** will recombine in the absence of other coreactants to afford a labile 8a-α-tocopheryl-tocopherone (**14**), which undergoes [1,4]-elimination to afford α-tocopherol (**1**) and *ortho*-quinone methide **3**, by analogy to other 8a-tocopherones (**6**).[30–32] *o*-QM **3**, once formed, will immediately dimerize into spiro dimer **9** in inert media. This pathway explains the observed product readily on the basis of general tocopherol chemistry without the need to evoke the 5a-C-centered radical **10**. The important coupling intermediate **14** was isolated under special chromatographic conditions—elution from finely powdered potassium carbonate with *n*-hexane—and was shown to decompose into α-tocopherol and its spiro dimer extremely readily, this decomposition being exactly the outcome of the alleged "radical disproportionation."

The above-described experiments, calculations, and theroretical considerations showed that there is no theoretical or experimental evidence whatsoever for the 5a-C-centered radical **10**. All relevant reactions can be traced back to occurrence and reactions of *o*-QM **3** as the central intermediate. The three reactions commonly cited to support the occurrence of the chromanol methide radical **10** in vitamin E chemistry (Figs 6.6–6.8) are actually typical processes of the *o*-QM intermediate (Figs 6.9–6.11).

The questions whether 5a-C-centered radicals exist in oxidation chemistry of α-tocopherol and whether mechanisms proposed in early days of vitamin E research are correct might appear academic at a first glance, but as soon as one recalls the immense medical, physiological, and economic importance of α-tocopherol and its derivatives, the significance of an exact knowledge about their basic chemistry becomes obvious. By analogy to the Mills–Nixon theory in tocopherol chemistry having been replaced by the concept of strain-induced bond localization (SIBL) recently (see Section 6.3.2), the condoned involvement of 5a-C-centered radicals in oxidation reactions of α-tocopherol must be revised. The confirmed alternative heterolytic pathways involving the *o*-QM **3** will certainly soon find their general acceptance in tocopherol chemistry.

6.3.2 Regioselectivity in the Oxidation of α-Tocopherol: Up-*o*-QMs Versus Down-*o*-QMs

According to literature accounts, oxidation chemistry of α-tocopherol (**1**) and PMC (**1a**) regioselectively involves C-5a, where the *ortho*-quinone methide **3** is formed, the so-called up-*o*-QM (Fig. 6.12). The isomeric compound with the *exo*-methylene group at C-7a (down-*o*-QM) was reportedly not observed.[33,34] Usually, the so-called "Mills–Nixon effect" reported for the first time in 1930[35] is given as explanation for the regioselectivity observed,[36,37] whereas the formation of other regioisomers than the "up"-*o*-QM was reported only very rarely.[28] The original work by Mills and Nixon, after which the effect is named, is based on three theories, today known as erroneous:

1. aromatic systems consist of two bond-shift isomers that are in equilibrium,
2. the van't-Hoff model of carbon, which implies that all the angles around the carbon are tetrahedral. Together with the first assumption, annulation of differently sized rings will shift the equilibrium between the equilibrating mesomers of benzene to that isomer that possesses the least strained angle, that is, an angle as close as possible to 109.5°,
3. The mechanism for electrophilic aromatic substitution is addition–elimination. Using these working hypotheses, Mills and Nixon explained the regioselectivity of electrophilic substitution in 5-hydroxyindan versus 6-hydroxytetralin.

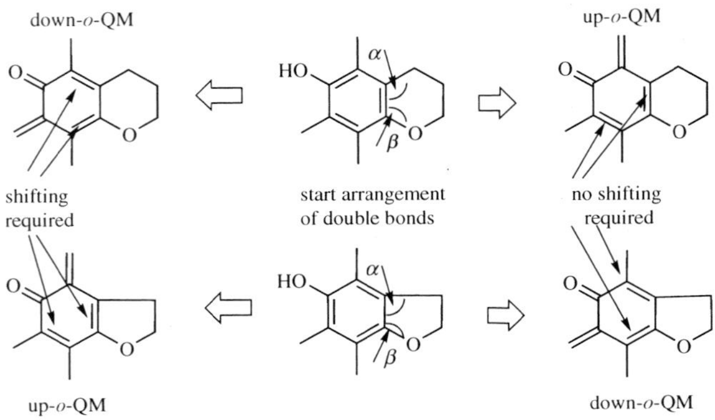

FIGURE 6.12 Traditional application of the "Mills–Nixon effect" theory to α-tocopherol-type benzopyranols and benzofuranols, having an anullation angle sum of $(\alpha + \beta)$.

Applying the Mills–Nixon explanation to vitamin E, it is usually argued that the annulation of a pyran or furan structure to the trimethyl-substituted phenol ring causes bond localization in the aromatic part of the corresponding α-tocopherol-type benzochromanol or benzofuranol. The term "α-tocopherol-type" refers to compounds derived from trimethylhydroquinone, that is, having three methyl substituents at the aromatic ring. Upon oxidation, only that one of the two possible *o*-QMs is formed, which requires as little rearrangement of the π-frame (double bonds) as possible (Fig. 6.12). In α-tocopherol-type benzopyranols (chromanols), the three double bonds in the aromatic ring are positioned so that one is placed at the annulation site: the *ortho*-quinone methide will thus be formed involving C-5a (the up-*o*-QM). In α-tocopherol-type benzofuranols, the three aromatic double bonds are positioned in a way that the annulation bond is a single bond: the favored *ortho*-quinone methide will be at C-6a (the "down"-*o*-QM). Thus, a widely accepted postulate was derived, which was frequently repeated throughout the literature: tocopherol-type chromanols are regioselectively oxidized at C-5a to form up-*o*-QMs, whereas tocopherol-type benzofuranols are always oxidized at C-6a to form down-*o*-QMs. The strict regioselectivity is due to the different ring size of the alicycle and due to the electronic effect of the alicyclic ring exerted on the aromat. Clearly, this explanation, which rested on theories today known as erroneous, was considered insufficient.

The issue of regioselectivity in oxidations of α-tocopherol-type antioxidants—being an open question for more than 70 years—was recently clarified by a combined experimental and theoretical study.[38] The approach was based on measuring the ratio between the down-*o*-QM and up-*o*-QM products obtained upon oxidation of 10 different α-tocopherol-type antioxidants (**1**, **15a–i**), which carried differently sized alicycles (Fig. 6.13). Thus, in these compounds the electronic effects were kept

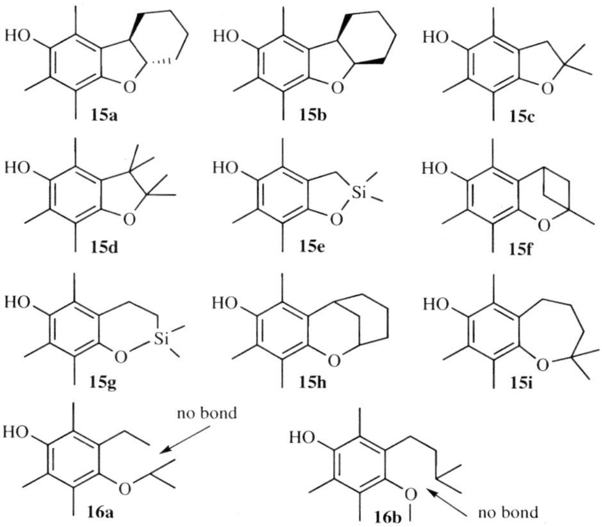

FIGURE 6.13 α-Tocopherol-type benzofurans and benzopyrans having different strains in the alicyclic ring (**15a-i**) and nonanullated α-tocopherol-type hydroquinones (**16a-b**).

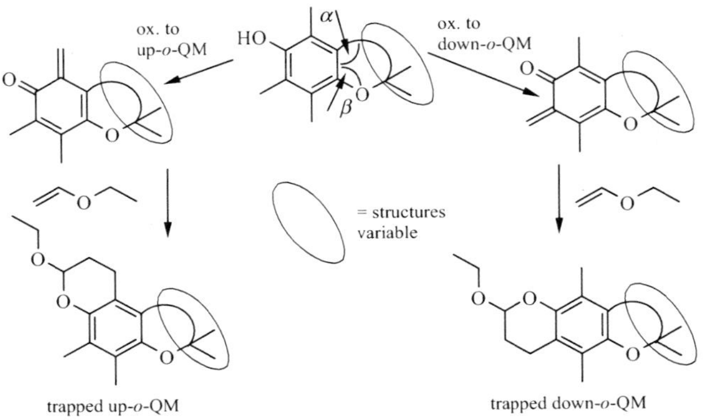

FIGURE 6.14 Oxidation of PMC-derivatives with different ring strains to mixtures of two possible *ortho*-quinone methides: the oxidation behavior and the ratio of the formed *o*-QMs agreed fully with the theory of strain-induced bond localization (SIBL).

constant and only the angular strain of the systems was changed, as seen by the ($\alpha + \beta$)-values, which cover an angle range between 219° (for **15a**) and 246° (for **15i**) (see Figs 6.12 and 6.14 and Table 6.1). In addition, tetrasubstituted hydroquinones (**16a–b**) carrying similar substituents as α-tocopherol—but no annulated ring—were used, which thus exhibited the same electronic substituent effects but no angular strain influence.

The model compounds were oxidized to the corresponding *o*-QMs, which were trapped by the fast reaction with ethyl vinyl ether (Fig. 6.14). Product analysis provided the ratio between the two *o*-QMs intermediates and by measuring the ratio between the trapped up-*o*-QMs and down-*o*-QMs at different temperatures, the activation energy difference for the formation of the two intermediates was obtained. The outcome proved unambiguously that the regioselectivity, that is, the ratio between up-*o*-QM and down-*o*-QM, was not a function of the ring size: it changed gradually,

TABLE 6.1 Ratio Between the Trapped Up-*o*-QMs and Down-*o*-QMs, and Kinetically Determined Activation Enthalpy Difference

Cpd	($\alpha + \beta$)	Up-*o*-QM (%, 373 K)	Down-*o*-QM (%, 373 K)	$\Delta\Delta H^{\ddagger}$ (kcal/mol)
15a	219	0.9	99.1	3.49 ± 0.12
15b	221	2.3	97.7	2.86 ± 0.08
15c	221	14.9	85.1	1.22 ± 0.04
15d	223	43.3	56.7	0.18 ± 0.01
15e	231	54.3	45.7	-0.109 ± 0.002
15f	233	66.9	30.1	-0.458 ± 0.009
15g	240	94.2	5.8	-1.96 ± 0.06
1	242	97.9	2.1	-2.768 ± 0.005
15h	244	99.3	0.7	-3.24 ± 0.12
15i	246	99.8	0.2	-4.77 ± 0.13

but not abruptly when going from a six-membered to a five-membered ring systems, in contrast to what has been assumed so far and what was derived from the Mills–Nixon postulate (Table 6.1). The up-*o*-QMs were increasingly favored when going from small ($\alpha + \beta$)-values to large ones, with down-*o*-QMs showing the opposite trend. The data clearly disproved the notion that vitamin E-related benzofuranols form only one *o*-QM (the down-form), whereas the chromanols give only the opposite one (up-*o*-QMs).

There was a clear linear correlation between ($\alpha + \beta$)-values and the differences in activation enthalpies. The absolute value of the activation enthalpy difference went through a minimum, meaning that at "medium angle sum" there was no distinct preference of one or the other *o*-QM type, whereas at "extreme angle sums"—either very large or very small ones—one of the *o*-QM types is largely preferred over the other one.

The regioselectivity in *o*-QM formation was also not a consequence of substitution, as hitherto assumed. The tetrasubstituted hydroquinones **16a** and **16b** represent the "open-ring version" of the truncated α-tocopherol model PMC (**1a**), having the same substituents and thus the same inductive electronic substituent effects as this chromanol, but no annulated ring (Fig. 6.13). Upon oxidation, both compounds afforded the up-*o*-QM and down-*o*-QMs in a nearly perfect 50/50 ratio. This proved that the regioselectivity in *o*-QM formation from PMC-type oxidants is also not a consequence of simple substitution. Regioselectivity in oxidative *o*-QM formation is observed only if trimethylhydroquinone (TMHQ) is annulated, that is, attached to another ring structure. It is not an intrinsic property of chain-substituted trimethylhydroquinone or caused by electronic substituent effects, rather it is caused by strain imposed through annulation.

The oxidation selectivity of α-tocopherol (**1**)—having an ($\alpha + \beta$) angle sum of 242°—is about 98.8/1.2 between up-*o*-QM (**3**) and the corresponding down-*o*-QM at room temperature; that of the corresponding benzofuranol (**15c**) is quite opposite at 11.8/88.2.

A comparison of the experimental data to the theoretical model of SIBL was made. To sort out the factors governing the regioselectivity, comprehensive calculations were carried out using models that underwent angle deformations to mimic the angular strain imposed by annulation of the ring, according to the SIBL principle[39] that was recently reviewed.[40–43] The agreement between the experimental results, such as the kinetics and the values for the activation energy differences derived from the up/down ratio of the two possible *ortho*-quinone methides, and the theoretical data according to the SIBL model was excellent, showing that the observed regioselectivity in oxidations of PMC-type antioxidants is simply a function of angular strain. This peculiar oxidation selectivity of α-tocopherol was thus fully explained by the SIBL theory, ending the decade-long Mills–Nixon controversy.

6.3.3 Detailed Formation Pathway and Stabilization of the Tocopherol-Derived *o*-QM 3 and Other *o*-QMs

Formation of *o*-QM **3** from α-tocopherol (**1**) means an overall loss of H_2, that is, two electrons and two protons are lost or transferred to the oxidant, respectively. The electrons and protons need not be transferred as individual species, but, for instance, also as two hydrogen atoms (one electron and one proton, two times) or as a hydride anion (two electrons and one proton) and a proton. For the detailed, stepwise mechanism, several sequences are thus conceivable. The respective mechanism is obviously not only dependent on the substrate tocopherol but also on the oxidant and the reaction conditions. So far, only one specific case has been studied in detail and clarified, the oxidation of α-tocopherol—and *ortho*-methylphenols in general—by silver oxide (Ag_2O). Computational treatment by DFT methods predicted the formation of the *o*-QM to proceed in three steps. First, a proton is released forming the corresponding phenolate anion. Second, a hydride ion is released from the *ortho*-methyl group and transferred to the oxidant. In this process, two equivalents of elemental silver and one molecule of water are generated. The hydride is transferred in a way that the two remaining hydrogens at the resulting methylene group are located at both sides of the aromatic plane, so that a perpendicular benzyl cation is formed, which at the same time is also a phenolate anion. The intermediate is thus a zwitterion carrying both a positive (benzylic position) and a negative (phenolic oxygen) charge. It should be noted that the latter structure is *not* a resonance structure of the *o*-QM. Such canonic structures differ *only* in the arrangement of multiple bonds. However, the zwitterionic intermediate and the *o*-QM are additionally distinguished by the different conformation of the exocyclic methylene group. Only the third step in the overall formation pathway, the rotation of the methylene group into the ring plane, eventually generates the *o*-QM and is coupled to the immediate aromatic–quinone conversion. While the intermediate is stabilized by through-space interaction of the two oppositely charged centers, the *o*-QM is stabilized by resonance and is by far the more stable species so that an experimental proof of the occurrence of the zwitterionic intermediate seemed to be rather unlikely (Fig. 6.15).

The first indication[27b] that a verification of its occurrence might be indeed possible was provided with the observation that oxidation of α-tocopherol by excess Ag_2O at −78 °C caused immediate formation of the spiro dimer via the *o*-QM **3** within less than

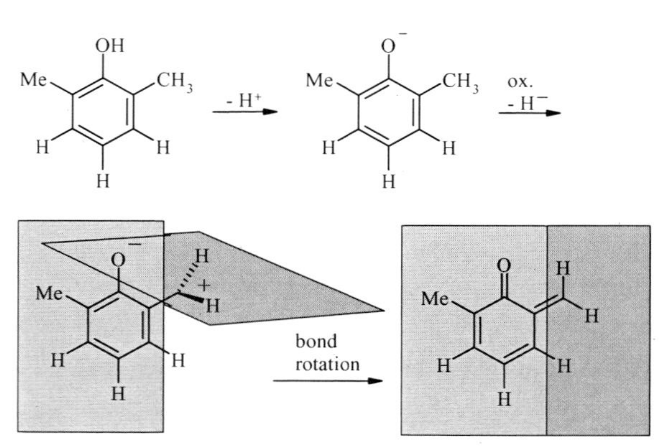

FIGURE 6.15 Detailed mechanism of the oxidation of *ortho*-methylphenols to *ortho*-quinone methides by Ag_2O according to DFT computations.[38]

10 s, whereas in the presence of an amine *N*-oxide, the tocopherol was consumed equally fast, but the generation of the spiro dimer was considerable retarded. It was likely that the zwitterionic amine *N*-oxide stabilized the zwitterionic oxidation intermediate by electrostatic interactions in a way that rotation of the exocyclic methylene group into the ring plane was prevented and thus *o*-QM formation (and its dimerization) was retarded. By using *N*-methylmorpholine-*N*-oxide (NMMO), the direct spectroscopic evidence for the zwitterionic intermediate in the formation of *o*-QM **3** was provided[27b] for the first time. At low temperatures, a complex (**17**) with NMMO was formed that slowly decomposed into *o*-QM **3** and unchanged NMMO. This degradation was immediate at temperatures above −30 °C.[27b]

The formation of spiro dimer **9** from complex **17** was significantly retarded as compared to noncomplexed *o*-QM **3**. Oxidation of α-tocopherol (**1**) in the presence of one equivalent of NMMO at −78 °C gave complete spiro dimer formation only after about 20 min, as compared to about 10 s in the absence of the amine *N*-oxide. NMR spectroscopy at low temperature confirmed an interaction between *o*-QM **3** and NMMO. The prominent signal of the proton spectrum was a singlet (2H) at 5.71 ppm, corresponding to the exocyclic methylene group. Its ^{13}C resonance at 182 ppm was indicative of a cationic species.[44] Furthermore, the proton resonances of the 7a-C and 8b-C methyl groups and the 4-C methylene group indicated the presence of an aromatic system, as did the ^{13}C NMR data for C-4a (118 ppm), C-5 (129 ppm), and C-6 (163 ppm), of which the latter strongly disagreed with a quinoid carbonyl carbon. Also the NMMO moiety was influenced by the interaction with *o*-QM **3** (Fig. 6.16).

Thus, in complex **17,** the stabilized *ortho*-quinone methide **3** was evidently not present in its "traditional" quinoid form, but in the form of a zwitterionic, aromatic

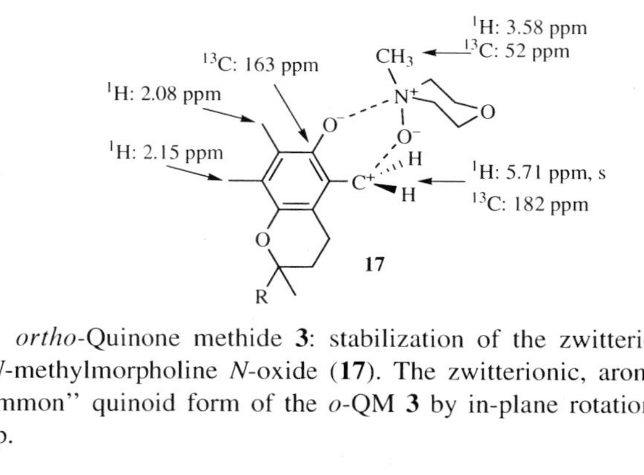

FIGURE 6.16 *ortho*-Quinone methide **3**: stabilization of the zwitterionic rotamer in a complex with *N*-methylmorpholine *N*-oxide (**17**). The zwitterionic, aromatic precursor **3a** affords the "common" quinoid form of the *o*-QM **3** by in-plane rotation of the exocyclic methylene group.

structure with an exocyclic methylene group rotated out of the aromatic plane. With the 5a-CH_2 group standing out-of-plane or even perpendicular to the plane, the positive charge is localized at C-5a and cannot dissipate into the aromatic ring by resonance. It was proposed that the primary stabilization effect was an increase in the rotational barrier of the exocyclic, cationic methylene group by electrostatic interactions with the negative charge of NMMO. This stabilized the aromatic structure and resulted in a restricted rotation into the in-plane form, which impeded the formation of the quinoid resonance form of *o*-QM **3**, so that the dimerization to spiro dimer **9** was retarded as observed. The activation parameters for the formation of free *o*-QM **3** from the complex **17** were estimated to a $\Delta H^{\neq}$ of 47 kJ/mol, a value which is comparable to the cleavage of a strong hydrogen bond.

The geometry of the zwitterions with its exocylic out-of-plane methylene group was quasi-preserved in the recently reported dibenzodioxocine derivative (**18**) that was formed in rather small amounts by rapidly degrading the NMMO complex at elevated temperatures.[45] Strictly speaking, dibenzodioxocine dimer **18** is actually not a dimer of *ortho*-quinone methide **3**, but of its zwitterionic precursor or rotamer **3a** (Fig. 6.17). As soon as the out-of-plane methylene group in this intermediate rotates into the ring plane, the *o*-QM **3** is formed irreversibly and the spiro dimer **9** results

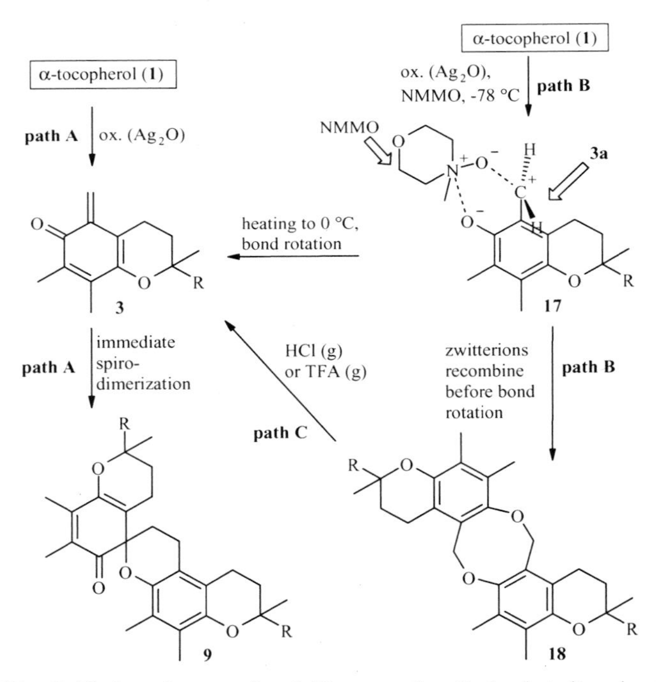

FIGURE 6.17 Oxidation of α-tocopherol (**1**) conventionally leads to its spiro dimer (**9**) via *ortho*-quinone methide **3** (path A). The zwitterionic *o*-QM precursor **3a** is stabilized by NMMO in complex **17**, which upon rapid heating produces small amounts of new dioxocine dimer **18** (path B). Acid treatment of **18** causes quantitative conversion into spiro dimer **9**, via *o*-QM **3** (path C).

inevitably. The formation of the dibenzodioxocine dimer from NMMO-complex **17**, which consists actually of two simultaneous etherification reactions driven by charge recombination, is thus competing with this bond rotation and must occur faster for the dioxocine dimer to form. Bond rotations have kinetic rate constants k_{rot} from 10^{-12}/s to 10^{-14}/s, the recombination rate of the two zwitterions ions in solution k_{rec} cannot be faster than diffusion controlled and is limited from about 10^{-7} L/mol/s to 10^{-9} L/mol/s. The faster rate of the bond rotation accounted for the fact that the yields in the dioxocine dimer were naturally limited, ranging below 5%.

Treatment of dioxocine dimer **18** with acid at 50 °C caused its decomposition and neat formation of spiro dimer **9** (Fig. 6.17, path C). Evidently, the two benzyl ether functions were cleaved, the resulting methylene groups immediately rotated into the plane forming *o*-QM **3**, and this intermediate dimerizes according to the "conventional" pathway into the spiro dimer **9**. A similar reaction, although accompanied by formation of several minor by-products, was affected by heating compound **18** in neat form above 155 °C. Interestingly, the conversion of dioxocine dimer → spiro dimer proceeded in neat substance, that is, also in solid state for the truncated model compounds, for example, by exposing the dioxocine dimers to an atmosphere of HCl or TFA. Within minutes, the dioxocine dimer **18** was converted neatly into spiro dimer without any side reactions, as was followed by IR spectroscopy, which was the first report of solid state processes involving *ortho*-quinone methides.

It was shown that complexes **19** of the zwitterionic precursors of *ortho*-quinone methides and a bis(sulfonium ylide) derived from 2,5-dihydroxy[1,4]benzoquinone[46] were even more stable than those with amine *N*-oxides. The bis(sulfonium ylide) complexes were formed in a strict 2:1 ratio (*o*-QM/ylide) and were unaltered at −78 °C for 10 h and stable at room temperature under inert conditions for as long as 15−30 min (Fig. 6.18).[47] The *o*-QM precursor was produced from α-tocopherol (**1**), its truncated model compound (**1a**), or a respective *ortho*-methylphenol in general by Ag_2O oxidation in a solution containing 0.50–0.55 equivalents of bis(sulfonium ylide) at −78 °C. Although the species interacting with the ylide was actually the zwitterionic oxidation intermediate **3a** and not the *o*-QM itself, the term "stabilized *o*-QM" was introduced for the complexes, since these reacted similar to the *o*-QMs themselves but in a well defined way without dimerization reactions.

In the 2:1 complexes formed, both *o*-QMs adopt a zwitterionic, aromatic structure with the exocyclic methylene group in perpendicular arrangement to the ring plane, stabilized by the negatively charged phenolic oxygen. Simultaneously, the negatively charged oxygens in the *o*-QM parts interact with the positively charged sulfur to provide additional stabilization.

The electrostatic interactions in the complexes **19** were obviously sufficient to "favor" the zwitterionic structure in a manner that formation of the usual *o*-QM was "suspended," so that all reactions typical of *o*-QMs in their quinoid form (such as [4 + 2]-cycloadditions) were suppressed or at least slowed down. Decomposition of the complex of α-tocopherol was immediate by fast heating to 40 °C or above. This caused disintegration of the complex **19**, immediate rotation of the methylene group into the ring plane, and thus formation of the *o*-QM, which then showed the "classical" chemistry of such compounds.

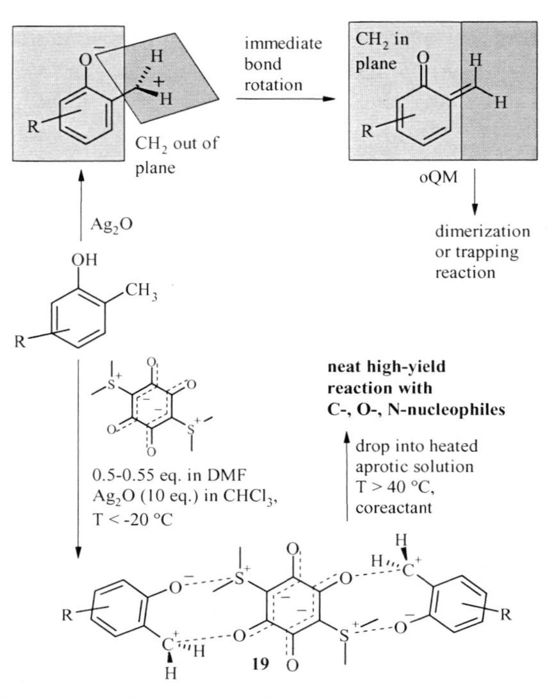

FIGURE 6.18 Oxidation of *ortho*-methylphenols to the corresponding *ortho*-quinone methide via transient zwitterionic intermediates that are stabilized by forming a complex **19** with the 2,5-dihydroxy[1,4]benzoquinone-derived bis(sulfonium ylide).

The complex **20** obtained from the truncated α-tocopherol model compound (**1a**) by Ag_2O oxidation in the presence of the sulfonium ylide was isolated at −30 °C as an amorphous addition product and was comprehensively characterized. It showed the exact ratio of 2:1, and a structural image was obtained by refining a quantum-chemical prediction (DFT) of the crystal structure according to X-ray powder diffraction data. Proton NMR spectroscopy of the complex showed a singlet (2H) at 5.85 ppm, corresponding to the exocyclic methylene group. This peak showed a heteronuclear correlation to a carbon at 191.8 ppm (C-5a), and HMBC cross-peaks at 129.9 (C-5, $^2J_{H\text{–}C}$), 117.2 (C-4a, $^3J_{H\text{–}C}$), and 154.1 ppm (C-6, $^3J_{H\text{–}C}$). The proton resonances of the 7a-CH_3, 8b-CH_3 methyl groups, and the 4-CH_2 methylene group at 11.8, 12.0, and 20.4 ppm indicated the presence of an aromatic system: due to the ring current effect, the resonances of the protons at C-7a, C-8b, and C-4 experience a down-field shift in tocopherol (**1**) and related derivatives, which evidently seems to be still operative in complex **20**. The high down-field shift of the carbon resonance at 191.8 ppm for the exocyclic methylene group is especially indicative of a cationic species (the ^{13}C resonances of carbocations can range between 100 and above 300 ppm),[44] and the peak at 154 ppm for C-5 agrees with a phenolic carbon, but not with a quinoid carbonyl carbon, as ^{13}C resonances of quinoid carbons are usually found between 180 and 195 ppm. Also the bis(sulfonium ylide) moiety was

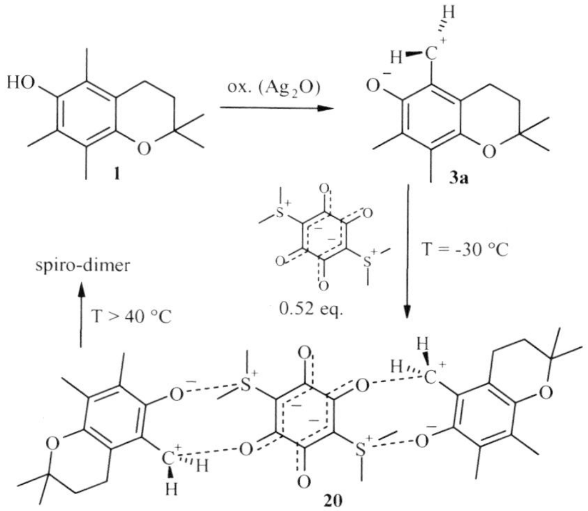

FIGURE 6.19 Formula and molecular structure of the 2:1 complex **20**, formed between the zwitterionic *o*-QM precursor derived from PMC (**1a**) and a bis(sulfonium ylide).

influenced, albeit rather weakly. The four magnetically equivalent methyl groups in bis(sulfonium ylide) resonating at 3.02 ppm in DMSO-d_6[46] appeared as singlet at 2.94 ppm in complex **20**. The carbon resonances changed from 94.0 (C–S) and 176.2 (C–O) ppm in the neat ylide to 88.4 (C–S), 172.2 (C–O), and 193.2 (C=O) ppm, respectively, in the complex. The stabilized zwitterions **3a** actually represent conformers of the *o*-QM **3**, with the conformational change—rotation of the exocyclic methylene group—being coupled to a fundamental change in the electronic structure, the transition from an aromatic into a quinoid system.

The stabilization of *o*-QMs as in Figs 6.18 and 6.19 had two implications. At first, the zwitterionic intermediate in *o*-QM formation mechanism—albeit strictly speaking only for the oxidation by Ag_2O—was confirmed. The stabilization approach might thus be useful also for other *o*-QM as those occurring in tocopherol chemistry and might allow to detect hitherto elusive *o*-QMs by trapping them and converting them into their more stable and better analyzable complexes. The second application of the stabilization approach lies in organic synthesis. The general advantage is that *o*-QMs in the form of their ylide complexes can be used and handled like stable, stoichiometrically usable, dosable reagents. They can be reacted in a controlled way without the danger of immediate self-dimerization or other uncontrolled side reactions.

The stabilization of the zwitterionic *o*-QM precursors is due to electrostatic interactions. It was reasonable to assume that also the other methods of stabilizing the zwitterions might be viable, and indeed it was confirmed that both steric and electronic effects are able to stabilize such intermediates. In 5-(4-octyl)-γ-tocopherol (5a-butyl-5a-propyl-α-tocopherol, **21**), the octyl group acts as a flywheel, which impedes the rotation of the C-5a moiety into the ring plane as compared to the parent zwitterions with the unsubstituted exocyclic methylene group. The situation is comparable with a figure skater performing a pirouette. With outstretched arms, the rotation will be much slower than with angled arms due to the torsional moment. Hence, an exocyclic methylene group will undergo rotation much faster than the propyl-butyl-substituted C-5a in the zwitterion derived from **21**. In Ag_2O oxidation at −78 °C, this "pirouette effect" caused the formation of the *o*-QM from **21** to be about 18 times slower than from **1a**. Also in the presence of the bisylide, the complexes derived from octyl derivative **21** degraded about 10 times slower than complex **20**.

A most illustrative example for the stabilization of the zwitterionic intermediate by electronic and steric effects in the formation of *o*-QMs is the 5a,5a-diphenyl-α-tocopherol derivative **22**. In the zwitterionic intermediate **23a** derived from this compound, both steric and electronic effects are active, which stabilize the out-of-plane zwitterions and destabilize the in-plane *o*-QM **23**. The C-5—C-5a bond in **23a** cannot freely rotate unless the two phenyl rings are concomitantly moved in a position perpendicular to the chroman plane that allows their passage over the chroman system. In this orientation, the phenyl ring cannot add to a conjugative stabilization of the *o*-QM. The *o*-QM **23** is thus disfavored by the steric effect and by the flywheel effect that was also active in the case of the octyl derivative **21** discussed above, and also by missing conjugative stabilization. However, the most decisive influence on the reactivity of this compound is the electronic effect of the two phenyl rings that is strongly stabilizing the positive charge. In principle, the compound is a triphenyl methane derivative in which the positive charge usually experiences strong stabilization by the phenyl substituents. The zwitterionic intermediate is thus favored by impeded rotation and by strong resonance stabilization of the positive charge. This favoring of the zwitterions **23a** and the disfavoring of the *o*-QM **23** are so strong that the usually wide energetic gap between the two forms is diminished and the zwitterion even becomes the energetically favored form. This was demonstrated by means of isotopic labeling. When oxidized at low temperature, the 5a-^{13}C-labeled compound showed one resonance at 205 ppm, corresponding to the zwitterionic intermediate **23a**. With increasing temperature, a second resonance- that of the *o*-QM form **23**- appeared at 128 ppm. From this form, the main reaction path leads to the xanthene derivative **24** (43 ppm), which is the final stable product. This product is not formed from the zwitterionic derivative **23a**, but only from the *o*-QM form **23**. The interrelation between the three compounds was nicely shown with the help of the ^{13}C resonances by the following sequence: low-temperature oxidation afforded the zwitterionic **23a** only. When this was heated to rt for a few seconds, an equilibrium between the **23a** and the *o*-QM **23** was established (two ^{13}C resonances) and at the same time generation of the xanthene **24** set in (the third signal appearing). Renewed cooling to −78 °C stopped the conversion of the zwitterion **23a** into *o*-QM **23**, but conversion of the *o*-QM **23** into xanthene **24** continued: the resonance at 205 ppm was unchanged, while the resonance at 128 ppm disappeared at the expense of that at 43 ppm. Upon heating to room temperature for several hours, only the resonance of the xanthene **24** remained, with all zwitterion **23a** being converted into this compound via the *o*-QM **23** (Fig. 6.20).

The diphenyl derivative **22**—due to its peculiar property to form equally stable zwitterion and *o*-QM species—was a nice probe to search for conditions stabilizing the

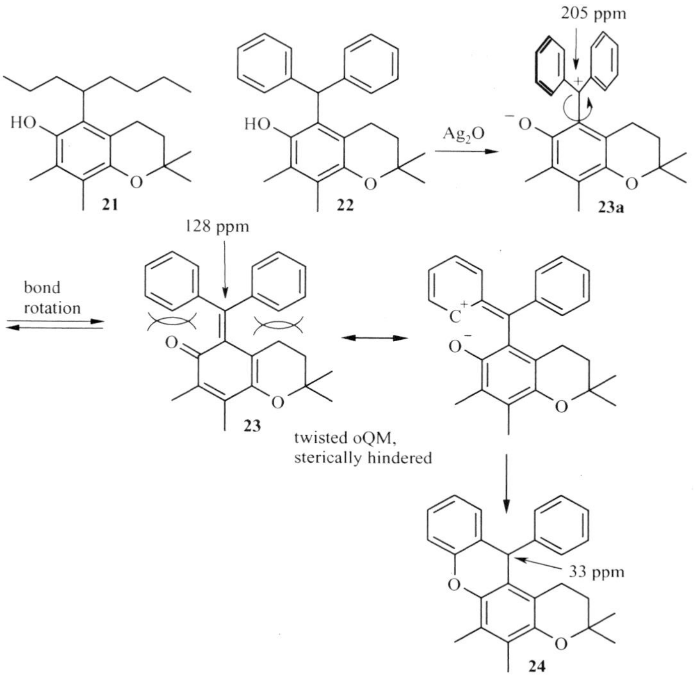

FIGURE 6.20 Octyl derivative **21**, diphenyl derivative **22** and its oxidation to zwitterion **23a**, and sterically hindered *o*-QM **23**.

zwitterionic form. Solvent effects, for instance, were readily detected. The rates of conversion from the zwitterions into the *o*-QM were roughly 13:8:1 when going from C_6D_6 to $CDCl_3$ to DMSO-d_6, evidently reflecting the stabilizing effect of polar solvents on the zwitterionic stage. Concluded from the stabilizing effect of amine *N*-oxide and the bis(sulfonium ylide), also a stabilizing effect of solvents containing salts and of ionic liquids was expected. However, this stabilizing effect was only moderate, with the zwitterion-to-*o*-QM conversion being about 22–34 times faster in C_6D_6 than in common BMIM-type ionic liquids with different anions. It was speculated that the two opposite charges in the stabilizing agent must be arranged in proximity and must be quite localized to exert a large stabilizing effect. This is the case in amine *N*-oxides and in the sulfonium ylide, but not in the case of the bulky ions and additionally delocalized charges, as in the case of ions with solvent shells and ionic liquids, respectively.

Stabilization of the zwitterionic intermediate in *o*-QM formation can also occur intramolecularly. In this case, the stabilizing moieties must be able to dissipate the positive charge at the benzylic group by a resonance effect and prevent rotation of the exocyclic methylene group by a steric blocking. One example for such a temporary stabilization is the nitration of α-tocopheryl acetate (**25**) by concentrated HNO_3, which produced 6-*O*-acetyl-5-nitro-α-tocopherol (**27**) in quite good yields,[48] the

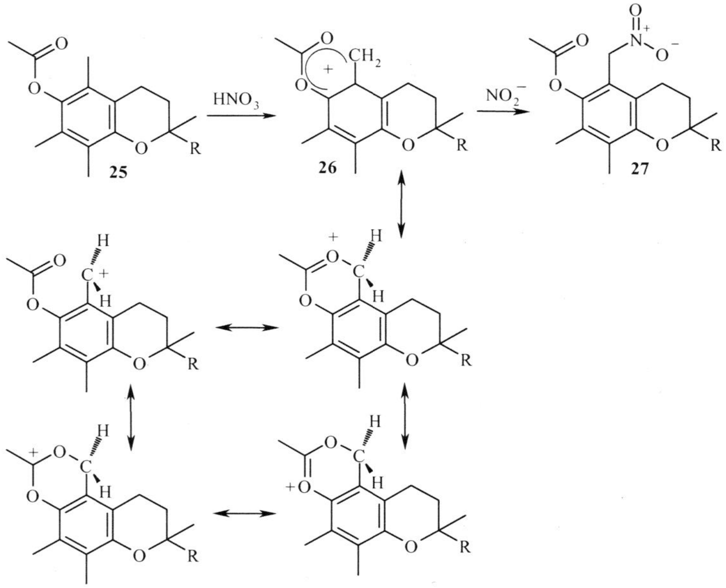

FIGURE 6.21 Synthesis of 6-*O*-acetyl-5-nitro-α-tocopherol (**27**) and four resonance forms of the cationic intermediate (**26**).

acetyl group not being cleaved during the reaction. The mechanism of the unusual formation reaction was studied in more detail,[49] and was shown to proceed via a 1,3,8-trioxa-phenanthrylium cation intermediate (**26**), which eventually added nitrite to afford **27** according to a nonradical, heterolytic course (Fig. 6.21). A deacetylation–oxidation–reacylation mechanism was ruled out by performing the reaction in propionic acid as the solvent and confirming the presence of an acetyl group—but not a propionyl moiety—in the product. In intermediate **26**, an effective charge delocalization over four atoms was effected through spatial interaction of the partially negative acyl oxygen with the positive benzylic position, resulting in strong resonance stabilization (Fig. 6.21). The 1,3,8-trioxa-phenanthrylium cation can be imagined as *O*-acylated zwitterionic precursor of the *o*-QM **3**. The benzylic methylene group is arranged perpendicular to the aromatic plane, so that the compound possesses four aromatic resonance hybrids involving the acetyl group, but no quinoid canonic forms. Generally, *O*-acyl substituents were shown to be crucial for the nitration reaction to proceed as they stabilize the cationic intermediate by resonance.

The reactions and compound presented in this chapter support the notion that the formation of *o*-QMs from the parent phenols is a quite complex process. In the case of the oxidation by Ag_2O but also likely in other oxidations, a zwitterionic intermediate is involved that can be stabilized intermolecularly, for example, by electrostatic interaction with other suitable zwitterions, or intramolecularly by neighboring groups or inductive/mesomeric effects. By stabilizing the zwitterionic intermediate and destabilizing the *o*-QM, the energetic gap between these two intermediates is lowered and both become observable at the same time. The stabilized zwitterionic precursors can be regarded as "stabilized *o*-QMs," as they are converted into *o*-QMs just by a bond rotation. This stabilization might have interesting applications in the identification of transient *o*-QMs and in organic synthesis.

6.4 REACTIONS OF THE "COMMON" TOCOPHEROL-DERIVED *ORTHO*-QUINONE METHIDE 3

6.4.1 Self-Reaction of the *o*-QM: Spiro Dimers and Spiro Trimers

The oxidation of α-tocopherol (**1**) to dimers[29,50] and trimers[15,51] has been reported already in the early days of vitamin E chemistry, including standard procedures for near-quantitative preparation of these compounds. The formation generally proceeds via *ortho*-quinone methide **3** as the key intermediate. The dimerization of **3** into spiro dimer **9** is one of the most frequently occurring reactions in tocopherol chemistry, being almost ubiquitous as side reaction as soon as the *o*-QM **3** occurs as reaction intermediate. Early accounts proposed numerous incorrect structures,[52] which found entry into review articles and thus survived in the literature until today.[22] Also several different proposals as to the formation mechanisms of these compounds existed. Only recently, a consistent model of their formation pathways and interconversions as well as a complete NMR assignment of the different diastereomers was achieved.[28]

The spiro dimer of α-tocopherol (**9**, see also Fig. 6.4) is formed as mixture of two diastereomers by dimerization of the *o*-QM **3** in a hetero-Diels–Alder reaction with inverse electron demand. Both isomers are linked by a fluxion process (Fig. 6.22), which was proven by NMR spectroscopy.[53] The detailed mechanism of the interconversion, which is catalyzed by acids, was proposed to be either stepwise or concerted.[53–55]

Formation of the ethano-dimer of α-tocopherol (**12**) by reduction of spiro dimer (**9**) proceeds readily almost independently of the reductant used. This reduction step can also be performed by tocopheroxyl radicals as occurring upon treatment of tocopherol with high concentrations of radical initiators (see Fig. 6.10). The ready reduction can be explained by the energy gain upon rearomatization of the cyclohexadienone system. Since the reverse process, oxidation from **12** to **9** by various oxidants, proceeds also quantitatively, spiro dimer **9** and ethano-dimer **12** can be regarded as a reversible redox system (Fig. 6.22).

The methano-dimer of α-tocopherol (**28**)[50] was formed by the reaction of *o*-QM **3** as an alkylating agent toward excess γ-tocopherol. It is also the reduction product of the furano-spiro dimer **29**, which by analogy to spiro dimer **9** occurred as two interconvertible diastereomers,[28] see Fig. 6.23. However, the interconversion rate was found to be slower than in the case of spiro dimer **9**. While the reduction of furano-spiro dimer **29** to methano-dimer **28** proceeded largely quantitatively and independently of the reductant, the products of the reverse reaction, oxidation of **28** to **29**, depended on oxidant and reaction conditions, so that those two compounds do not constitute a reversible redox pair in contrast to **9** and **12**.

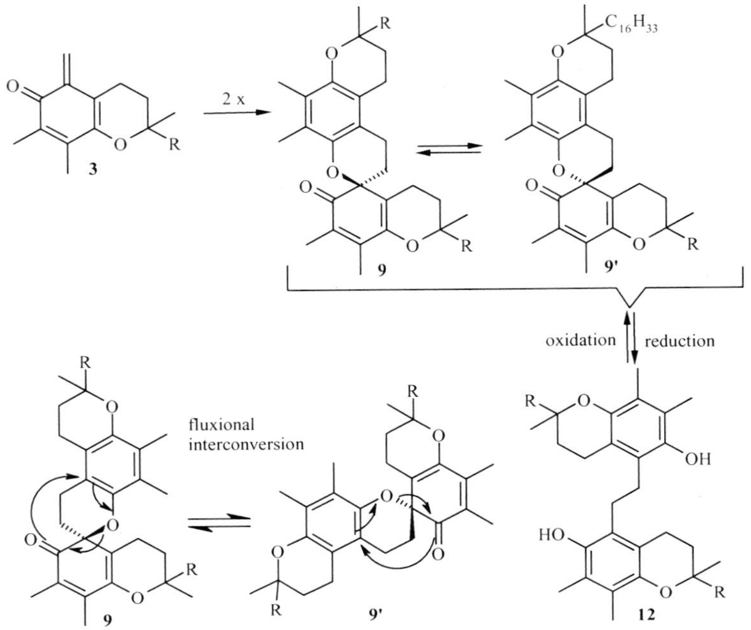

FIGURE 6.22 Spiro dimer of α-tocopherol (**9**): formation, redox reactions, and fluxional nature.

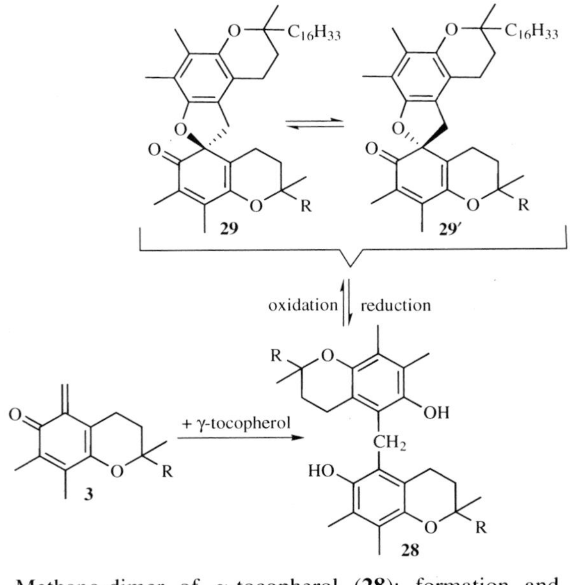

FIGURE 6.23 Methano-dimer of α-tocopherol (**28**): formation and redox reactions, including oxidation to the two fluxationally interconvertible diastereomers of furano-spiro dimer **29**.

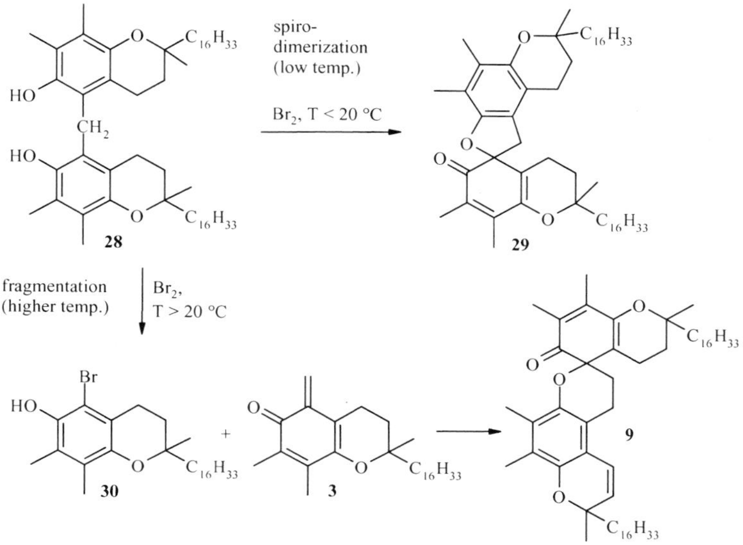

FIGURE 6.24 Redox behavior of the methano-dimer of α-tocopherol (bis(5-tocopheryl) methane, **28**): temperature dependence of the oxidation with bromine.

Treatment of methano-dimer **28** with elemental bromine revealed a remarkable reactivity: at low temperatures it proceeded quantitatively to the furano-spiro dimer **29**, by analogy with the ethano-dimer **12** giving spiro dimer **9** upon oxidation. With increasing temperatures, the reaction mechanism changed, however, now affording a mixture of 5-bromo-γ-tocopherol (**30**) and spiro dimer **9** (Fig. 6.24). Thus, the methano-dimer **28** fragmented into an "α-tocopherol part," in the form of *o*-QM **3** that dimerized into **9**, and a "γ-tocopherol part," which was present as the 5-bromo derivative **30** after the reaction. Thus, the overall reaction can be regarded as oxidative dealkylation.

A further hetero-Diels–Alder reaction with inverse electron demand between *o*-QM **3** as the dienophile and either of the two diastereomers of spiro dimer **9** as the diene provided the spiro trimers **31** and **32** (Fig. 6.25). The absolute configuration was derived from NMR experiments. It was moreover shown that only two of the four possible stereoisomeric trimers were formed in the hetero-Diels–Alder reaction: the attack of the *o*-QM **3** occurred only from the side *syn* to the spiro ring oxygen.[28]

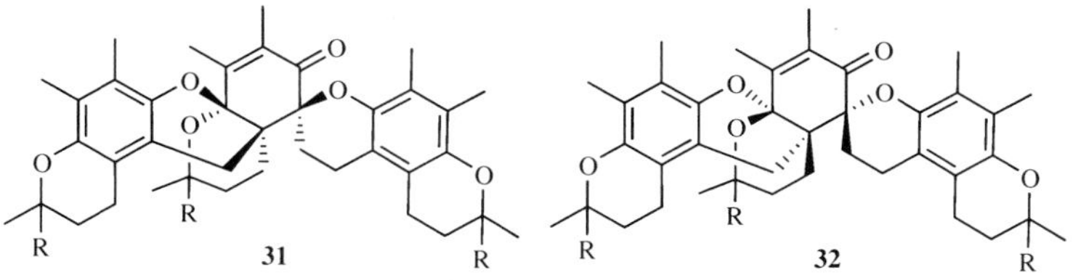

FIGURE 6.25 Spiro trimers of α-tocopherol (**31**, **32**) formed by reaction of *o*-QM **3** with the two diastereomers of spiro dimer **9**.

6.4.2 Spiro Oligomerization/Spiro Polymerization of Tocopherol Derivatives

If a compound possesses more than one *ortho*-alkylphenol unit capable of being oxidized to an *o*-QM, the spiro dimerization process occurs more than once at the molecule and becomes a spiro oligomerization or even spiro polymerization process eventually. Tetracycle **33**, which was obtained by condensation of trimethylhydroquinone with 1,1,3,3-tetramethoxypropane,[56] proved to be a very appropriate means to study such multiple spiro dimerization processes (Fig. 6.27). The compound consists of two chroman units of α-tocopherol-type that are connected with each other at the alicyclic pyran ring having C-2, C-3, and C-4 in common, which gave the compound its name "Siamese twin" tocopherol. Compound **33** is a vitamin E model, which "locks" the alicyclic chroman ring into a specific geometry, but without achieving this by means of sterically demanding, large substituents. Any conformational change in one of the two chromanol moieties in **33**, which would influence *o*-QM or radical stability,[57] is accompanied by the reverse change in the second "twin" moiety, causing the opposite effect there. The chromanoxyl radical derived from **33** gave EPR spectra that resembled those of α-tocopherol (**1**), but exhibit additional hyperfine structure due to the other "half" of the molecule.[56] The compound showed antioxidant properties that were superior to that of the truncated tocopherol model compound in several test systems.[58]

Since it contained *two* chromanol moieties, twin-tocopherol **33** gave all reactions characteristic of tocopherol twice, most notably *ortho*-quinone methide formation. Each one of the two "twin parts" in **33** was able to undergo a reaction similar to the spiro dimerization of tocopherol. Thus, this simple spiro dimerization eventually became a spiro oligomerization/spiro polymerization in the case of **33**: after both sides of the twin molecule had reacted in a spiro dimerization, each of the two newly attached twin molecules again possessed an end capable of undergoing spiro dimerization and so on, finally affording linear molecules consisting of twin molecules connected by spiro links that were formed in sequential hetero-Diels–Alder reactions. The lengths of the spiro polymers as well as the molecular weight distribution varied according to reaction time and reaction temperature. Different oxidants, solvents, reaction times, and reaction temperatures afforded polymers with 4–215 units depending on the conditions (Fig. 6.26).[56]

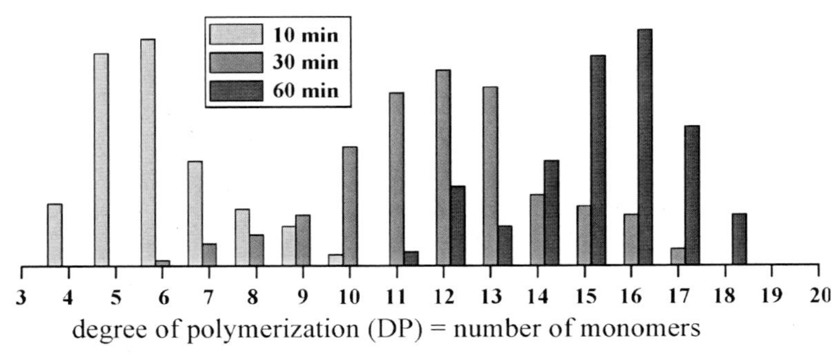

FIGURE 6.26 Oxidation of twin-tocopherol **33** by $AgNO_3$ in toluene at 25 °C: molecular weight distributions of the resulting oligomers depend on the reaction time.[73]

After reaction of the first "twin side" of **33** by spiro dimerization, for instance as dienophile, the second half can theoretically react either as dienophile or as diene (Fig. 6.27). Thus, a pyrano/spiro pair (reaction of the "left twin" as diene and the "right twin" as dienophile), a pyrano/pyrano pair, or a spiro/spiro couple (reaction of

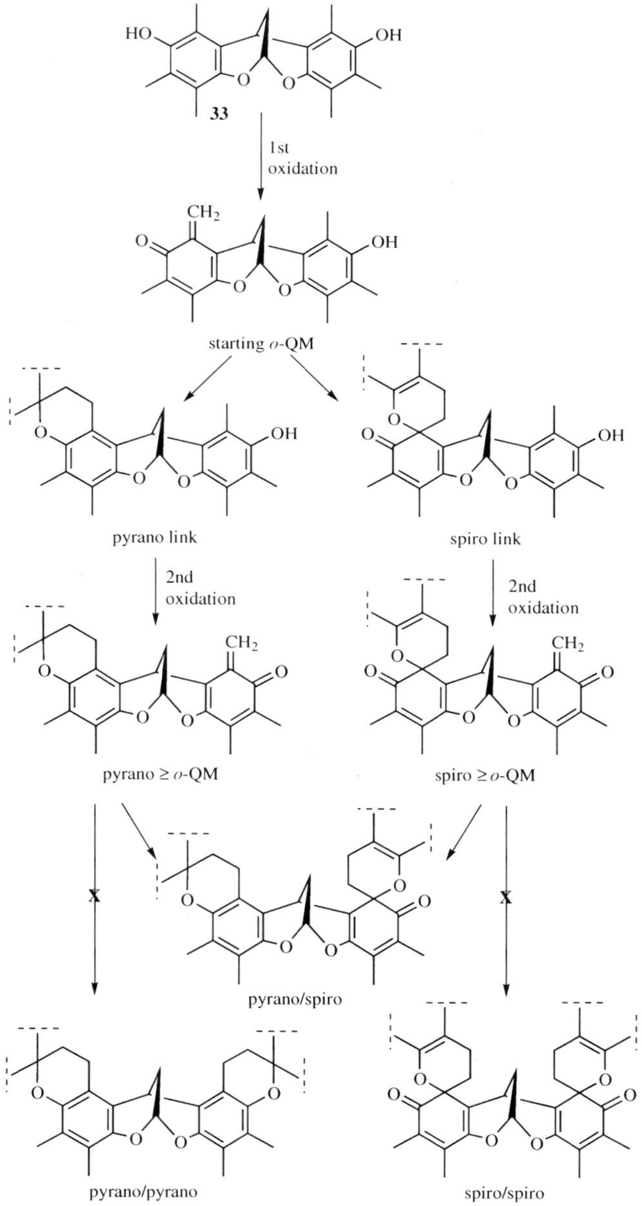

FIGURE 6.27 Regioselectivity of spiro–pyrano link formation upon dimerization of the *o*-QMs derived from twin-tocopherol **33**.

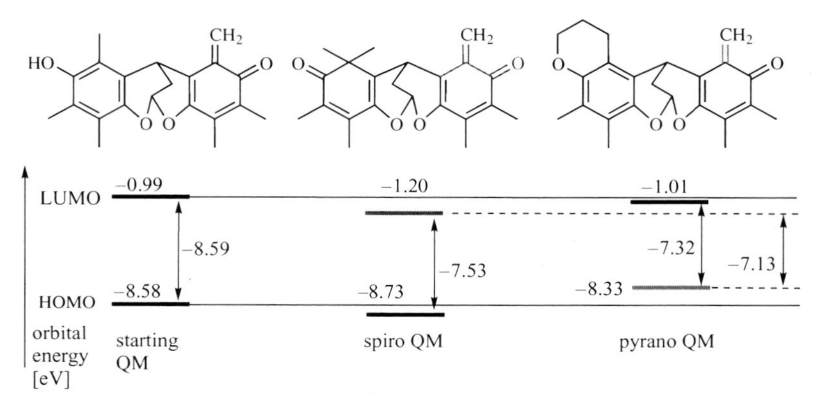

FIGURE 6.28 Influence of the reaction of the first *o*-QM (to a pyrano or spiro structure) on the reactivity of the second *o*-QM from **33** according to computations: only spiro/pyrano, but no pyrano/pyrano or spiro/spiro pairs are formed in the oligomeric products.[56]

"both twins" as dienes or dienophiles, respectively) is formed. However, only products were observed that contained exclusively asymmetric pairs (pyrano/spiro spiro/pyrano pairs) as the building blocks—but no symmetric (pyrano/pyrano or spiro/spiro) couples.[56]

The spiro–pyrano regioselectivity was rationalized in terms of frontier orbital theory. Reaction of the first *o*-QM as a diene (pyrano structure) resulted in an increase of the HOMO energy of the neighboring *o*-QM. Therefore, this *o*-QM will react as dienophile due to increased π-donor ability, and will thus form a spiro structure. By analogy, reaction of the first *o*-QM as a dienophile (spiro structure) decreased the LUMO energy of the neighboring *o*-QM leading to increased π-acceptor capability and subsequent reaction as a diene (pyrano structure). In both cases, spiro/pyrano couples resulted but no spiro/spiro or pyrano/pyrano neighbors, since there was a significantly decreased HOMO–LUMO energy difference for the asymmetric pairs as compared to the symmetric couples (Fig. 6.28).

When the oxidative spiro oligomerization starting from twin-tocopherol **33** was carried out at low temperatures, the cyclic tetramer **34** was obtained instead of linear products.[56] **34** contained only pyrano/spiro (spiro/pyrano) pairs as the building blocks too, but no pyrano/pyrano or spiro/spiro couples. Each spiro link in linear or cyclic spiro oligomers and spiro polymers was reduced in analogy to the spiro dimer of tocopherol (**9**). Consequently, reduction of **34** provided the macrocycle **35** (Fig. 6.29), which showed some similarities to calixarenes. In the presence of excess oxidant and reductant, respectively, **34** and **35** exchanged eight electron equivalents per molecule, by analogy to the reversible redox pair formed by the tocopherol-derived spiro dimer (**9**) and ethano-dimer (**12**) exchanging only two-electron equivalents.

Spiro dimerization of the tocopherol-derived *o*-QM seemed to be a quite favored process, which proceeds also in the case of moderately bulky substituents at C-5a.

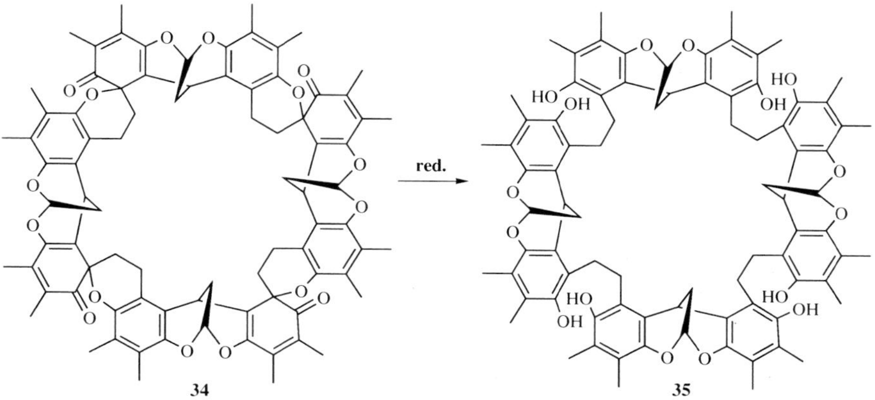

FIGURE 6.29 Cyclic spiro tetramer **34** and its reduction product, macrocycle **35**.

A particularly interesting case was the oxidative spiro dimerization of α,ω-bis (tocopheryl)alkanes (**36**), which basically present two α-tocopherol units linked at C-5a by an alkyl bridge.[59] The reaction of other α,ω-bis(hydroxyphenyl)-alkanes, such as **37–40**, proceeded similarly (Fig. 6.30).

Subject to typical conditions for spiro polymerization, for example, treatment with Ag_2O in inert solvents, each tocopherol unit of the bis(tocopheryl)alkanes **36** underwent oxidation to the respective *o*-QM. The subsequent dimerization process involves two different molecules as intramolecular dimerization is impossible due to steric reasons. Each "side" of the starting bis(tocopheryl)alkane forms a spiro dimer unit. The two newly attached molecules carry another tocopherol at the opposite end, which are again oxidized and react with two other molecules, and so on. This way, the spiro dimerization process became a spiro polymerization process, affording linear oligomers/polymers (**41**) consisting of spiro dimers linked by alkyl chain bridges of differing lengths (Fig. 6.31). Basically, these oligomers consist of spiro dimers of α-tocopherol (**9**) linked at the ethano-bridges by alkyl chains. The degree of

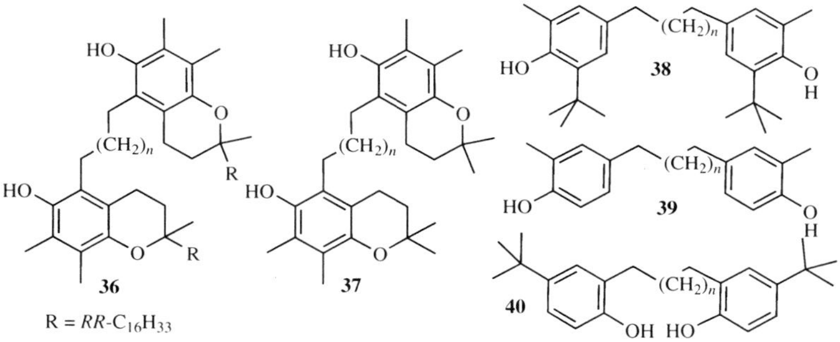

FIGURE 6.30 α,ω-Bis(tocopheryl)alkanes (**36**) and other α,ω-bis(hydroxyphenyl)-alkanes (**37–40**) as starting materials for the spiro oligomerization/spiro polymerization reaction.

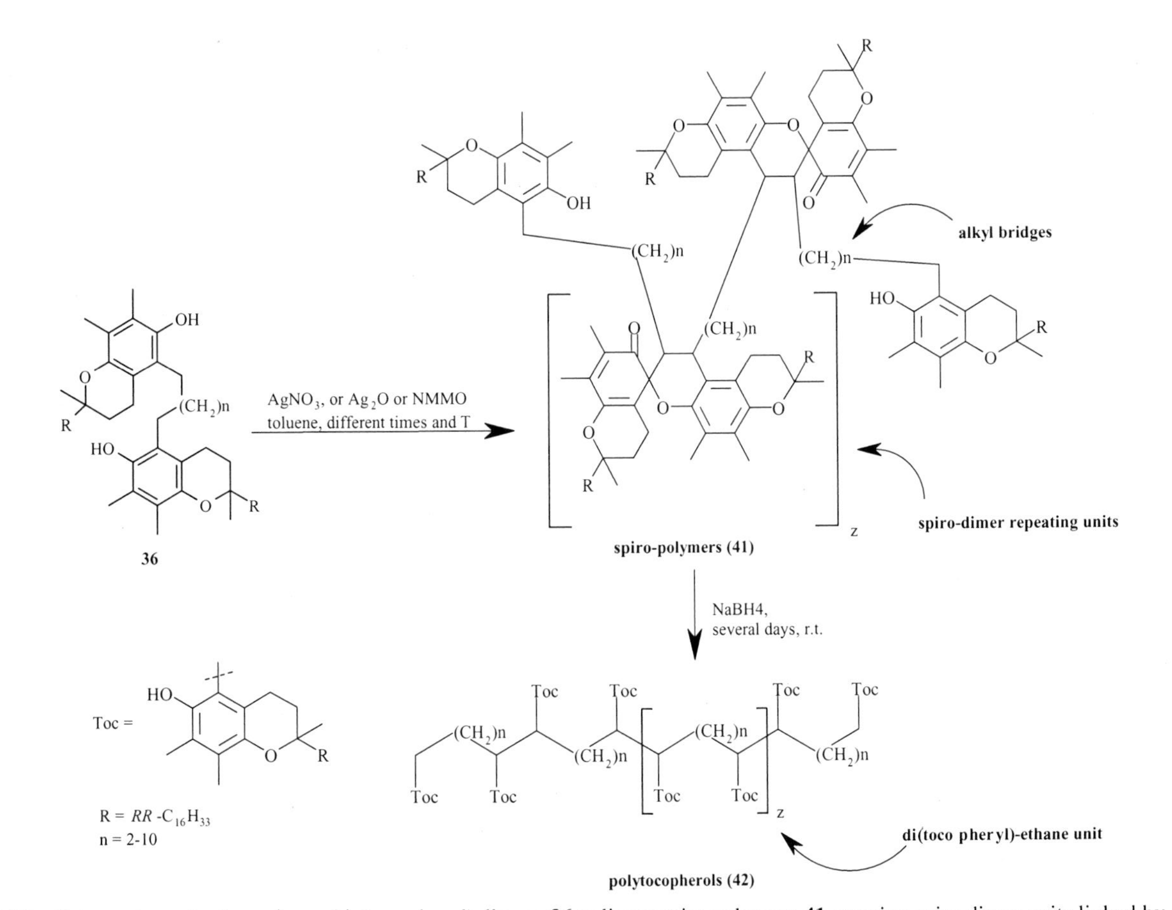

FIGURE 6.31 Spiro polymerization of α,ω-bis(tocopheryl)alkanes **36** to linear spiro polymers **41** carrying spiro dimer units linked by alkyl chains. Reduction of **41** converts the spiro dimer units into ethano-dimer units, resulting in polytocopherols **42**.

polymerization was adjustable by the choice of solvent, reaction temperature, reaction time, and oxidant (Fig. 6.26). As an example, 1,6-bis(5-tocopheryl)hexane (**36**, $n = 4$) afforded different spiro polymers, the maximum of the molecular weight distributions ranging between 16 and 534 spiro dimer units corresponding to molecular weights between about 14,500 and 483,000, the distributions themselves being rather narrow.[59]

Reduction of the spiro polymers proceeded by analogy to the reduction of the spiro dimer (**9**) giving the ethano-dimer of α-tocopherol (**12**). Each spiro dimer unit in the spiro polymer was thus converted into 1,2-bis(5-tocopheryl)-ethane elements, the ethane unit being a part of the alkyl chain bridges in **36**. The resulting products **42** are basically long alkane chains carrying vicinal 5-tocopheryl or other *ortho*-alkylphenol substituents in regular distances, which are set by the chain length of the alkyl bridges. The products thus represent polytocopherols (Fig. 6.31).

The spiro polymerization is a novel reaction type that uses the spiro dimerization of *o*-QMs to build up linear oligomers and polymers. The basic properties of the spiro dimer of α-tocopherol, that is, its fluxional structure and its ready reduction to the ethano-dimer, remain also active when such structural units are bound in the polymer. The products of the reaction, both in its poly(spiro dimeric) form (**41**) and in the form of the reduced polytocopherols (**42**), are interesting materials for application as high-capacity antioxidants, polyradical precursors, or organic metals, to name but a few.

A special case of the spiro oligomerization reaction is the oxidation of 1,3-bis(5-tocopheryl)propane (**43**) that did not cause spiro polymerization but formation of spiro tetramers (**44a-b**), distinguished by the arrangement of pyrano/spiro links. The regioselectivity of spiro–pyrano link formation as well as the influence of already formed spiro links on the ones to be formed was rationalized in terms of frontier orbital theory, similar to the case of the Siamese twin tocopherol (Fig. 6.28). According to the theory, spiro/pyrano neighbor couples would result but no spiro/spiro or pyrano/pyrano neighbors, predicting a clear preference of **44a** over **44b**. The validity of this hypothesis was confirmed experimentally since the two compounds were formed upon Ag_2O oxidation at room temperature in a ratio of 15:1. Both spiro tetramers were reduced to the same product, *cis*-1,2-*cis*-4,5-tetrakis(5-tocopheryl)cyclohexane (**45**). Also, the reduced tocopherol and the spiro tetramers establish a reversible redox pair, with the ratio of the two spiro tetramers **44a** and **44b** remaining constant during repeated redox cycles (Fig. 6.32).

6.4.3 Bromination of α-Tocopherol and Further Reactions of 5a-Bromo-α-Tocopherol and Other 5a-Substituted Tocopherols

Treatment of α-tocopherol (**1**) with elemental bromine provided quantitative yields of 5a-bromo-α-tocopherol (**46**). The reaction was assumed to proceed according to a radical mechanism, but later a nonradical oxidation–addition mechanism was proven (Fig. 6.33). Bromine oxidized α-tocopherol (**1**) to the intermediate *ortho*-quinone methide (**3**), which in turn added the HBr produced in the oxidation step.[60] If the HBr was removed by flushing with nitrogen, the spiro dimer (**9**) became the main product, and if it was purged by HCl gas, mainly 5a-chloro-α-tocopherol was produced.

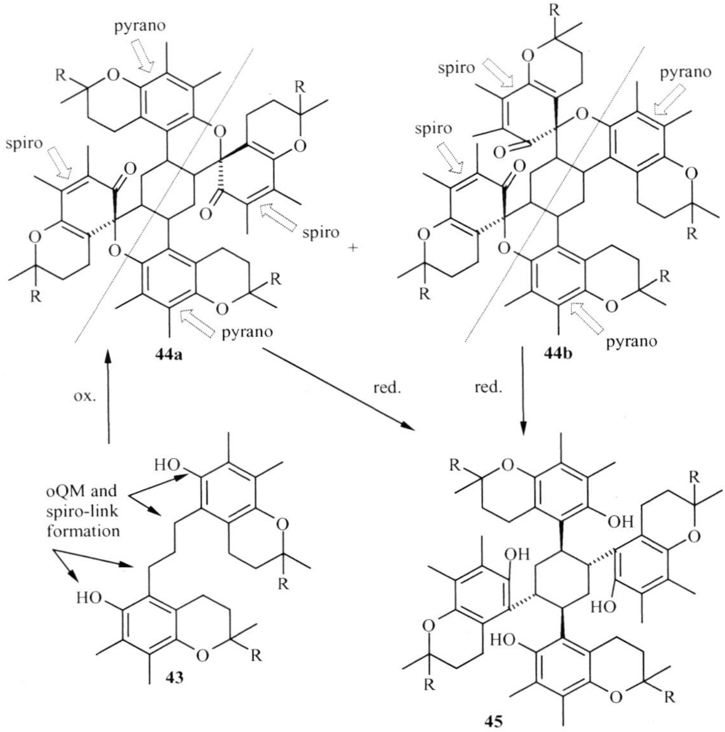

FIGURE 6.32 Configurationally different spiro tetramers **44a** and **44b** formed by spiro tetramerization of 1,3-bis(5-tocopheryl)propane (**43**) in a 15:1 ratio in agreement with theory. Reduction of each of the tetramers provides *cis*-1,2-*cis*-4,5-tetrakis(5-tocopheryl)cyclohexane (**45**). The dashed lines separate the two former units of **43**, indicating where bond formation during the spiro oligomerization process occurred.

5a-Bromo-α-tocopherol (**46**) has become a most frequently used starting material in the synthesis of 5a-substituted tocopherols. This is due to its ready availability and facile preparation in quantitative yield, which makes time-consuming and tedious purification procedures unnecessary. Because of its inherent benzyl bromide structure, 5a-bromo-α-tocopherol shows high reactivity and is amenable to facile nucleophilic

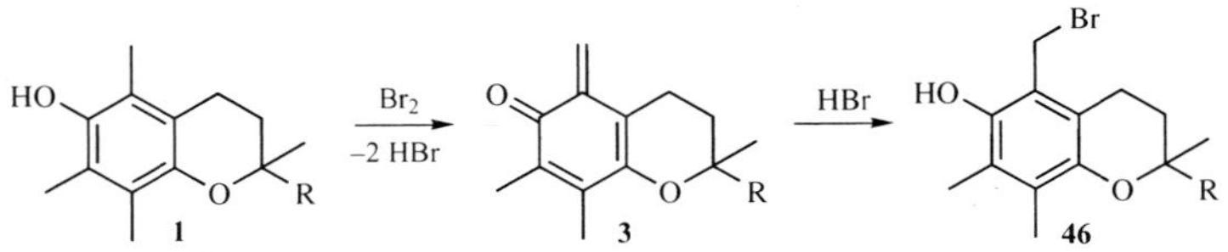

FIGURE 6.33 Synthesis of 5a-bromo-α-tocopherol (**46**) from α-tocopherol (**1**) according to an oxidation–addition mechanism involving the *o*-QM intermediate **3**.

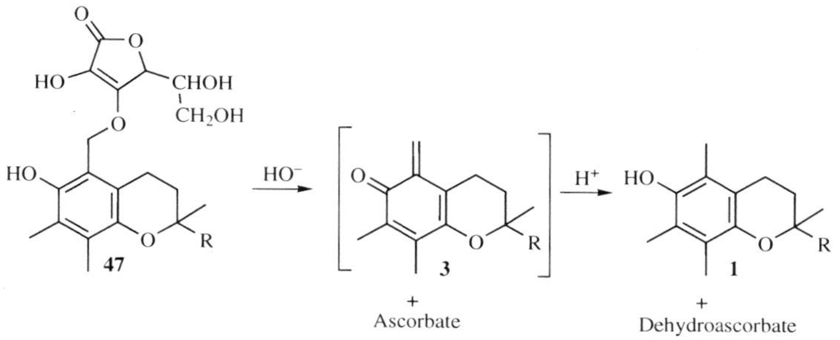

FIGURE 6.34 Regeneration of α-tocopherol by base-catalyzed fragmentation of 5a-tocopheryl ascorbate (**47**) followed by a redox process.

substitution,[61] although preparative difficulties arose because of the ready elimination of hydrogen bromide in basic media or at slightly elevated temperatures—approximately above 50 °C—causing re-formation of the *o*-QM **3**, and thus the α-tocopherol spiro dimer (**9**) as the most frequently observed by-product. Ready elimination occurred with all types of *O*-, *N*-, and *S*-substituents at C-5a, inducible by treatment with alkali or at elevated temperatures.

The stability of 5a-substituted tocopherols in acidic media and their lability in basic media was used as the basis for the development of prodrugs, which showed pH-dependent drug release: while the prodrugs are stable in the acidic medium of the stomach, they are readily cleaved in the basic media of the intestinal tract, where the drug—a 5a-substituent bound to the tocopherol moiety through an oxygen or amino functionality—is released.[62] The tocopherol moiety acts as a lipophilic drug carrier. Upon release of the drug and cleavage of the carrier, either spiro dimer **9**, *para*-tocopherylquinone (**7**), or, by reduction, α-tocopherol (**1**) is formed, all of them being physiologically fully compatible compounds. One illustrative example in this respect is tocopheryl ascorbate **47**. This compound is stable under neutral and acidic conditions, but eliminates ascorbate at a pH above 8 with concomitant formation of *o*-QM **3**. Both intermediates join in a redox reaction, and finally regenerated α-tocopherol and dehydroascorbate are produced in high yield (Fig. 6.34). Addition of sodium ascorbate rendered the tocopherol yields quantitative. Kinetic experiments showed the main reaction to proceed in the pH range of 8–11 under simulated physiological conditions, and the tocopherol to be generated in a finely dispersed and thus readily absorbable manner.

The reaction of 5a-bromo-α-tocopherol (**46**) with amines was further elaborated into a procedure to use this compound as a protecting group "Toc" for amines and amino acids (Fig. 6.35).[62] The protection effect was due to a steric blocking of the amino function by the bulky tocopheryl moiety rather than due to conversion into a non-nucleophilic amide derivative, and the Toc-protected amino acids were employed in the synthesis of dipeptides according to the dicyclohexylcarbodiimide (DCC) coupling method.[64] The overall yield of the reaction sequence was reported to be largely dependent on the coupling reaction, since both installation and removal of the

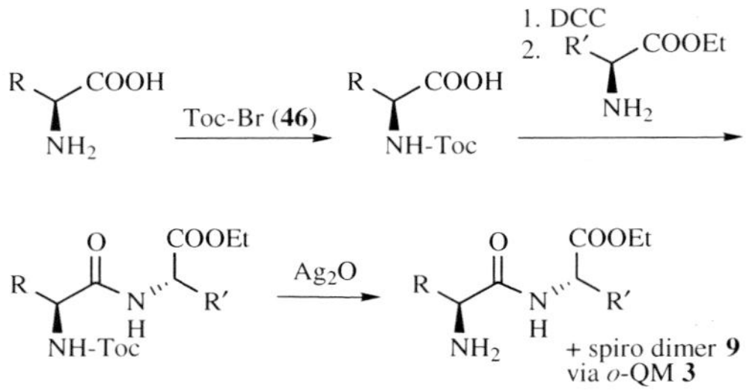

FIGURE 6.35 5a-Bromo-α-tocopherol (**46**) as auxiliary in the synthesis of dipeptides: the "Toc" protecting group.

protecting group were near-quantitative steps. Especially, the cleavage by treatment with silver oxide or mild bases could be performed under quite mild conditions. The Toc group is cleaved off as the *o*-QM **3** that immediately dimerizes into spiro dimer **9**, which is the only product derived from the protecting group that due to its high lipophilicity can conveniently be separated. The Toc-protected amines and amino acids represent 5a-substituted tocopherols, and can thus be cleaved under mildly oxidative, alkaline or thermal conditions.

6.4.4 Cyclization of para-Tocopherylquinone 7 into *o*-QM 3

As mentioned earlier in Section 6.2, the chromanoxylium cation **4** can be converted into *o*-QM **3** by elimination of a proton from the 5a-methyl group. When starting from α-tocopherol (**1**), the oxidative generation of the *o*-QM **3** can also be performed directly (under apolar, aprotic conditions) so that there is no need to take a "detour" via the chromanoxylium **4** from the synthetic point of view. However, this reaction is highly valuable when the chromanoxylium is not produced from α-tocopherol by oxidation (loss of two electrons and a proton), but from *para*-tocopherylquinone (**7**) by acid-induced cyclization. For this cyclization, two mechanisms are possible, the first one involving protonation of the tertiary hydroxyl group that is released as water, followed by cyclization and deprotonation at C-5a in a concerted process (Fig. 6.36, path A). The alternative path B starts with the attack of the hydroxyl group at the quinone carbonyl facilitated by protonation of the carbonyl oxygen. The intermediate 8a-hydroxytocopherone (Fig. 6.4, compound **6**, Nu = OH) undergoes [1,4]-elimination to afford the *o*-QM **3** (Fig. 6.36, path B). Whether this elimination is a simultaneous process or a stepwise one involving the chromanoxylium cation **4** as an intermediate cannot be answered at present. Although the first mechanism has not been disproven to occur, the second one is supported by the fact that 8a-substituted tocopherones can be isolated under special conditions and were indeed shown to produce the *o*-QM by [1,4]-elimination quite readily.[11] In addition, occurrence of a tertiary carbenium ion according to the former mechanism would involve side reaction such as elimination and competitive recyclization to furan derivatives, however, none of these products were found.

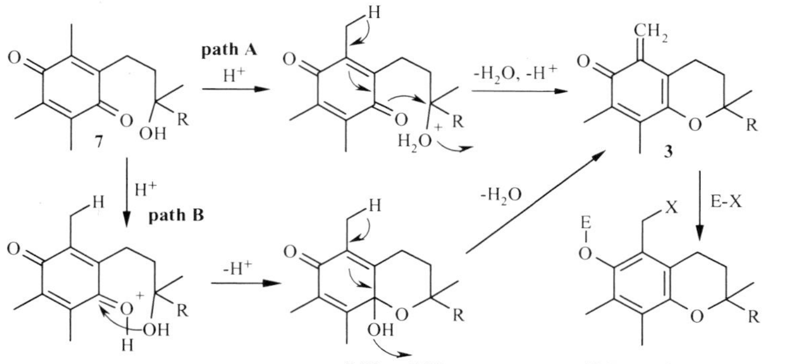

FIGURE 6.36 Two mechanistic alternatives for the cyclization of *para*-tocopherylquinone **7** into *o*-QM **3** and subsequent reaction with acyl or trimethylsilyl halides.

The cyclization of *para*-tocopherylquinone (**7**) and conversion into the *o*-QM **3** converts a *para*-quinoid system into an *ortho*-quinoid one. The preparative applicability of this reaction is high as the intermediate *o*-QM immediately reacts further with acyl halides and trimethylsilyl halides into the corresponding 5a-halo-tocopheryl esters (**48**)[65] and 5a-halo-*O*-trimethylsilyltocopherols (**49**),[66] respectively. The reagents are used in excess and are also responsible for generating traces of hydrogen halide to produce the acidic conditions necessary for triggering the cyclization. For example, by treating *para*-tocopherylquinone **7** with trimethylsilyl bromide, the corresponding *O*-trimethylsilyl-5a-bromo-α-tocopherol (**49**, $E = Me_3Si$, $X = Br$) was obtained.[66] Analogously, treatment with acetyl chloride provided 5a-chloro-α-tocopheryl acetate (**48**, E = Ac, X = Cl).[65]

6.4.5 Synthesis via *o*-QM 3 and Reaction Behavior of 3-(5-Tocopheryl) propionic Acid

3-(5-Tocopheryl)propionic acid (**50**) is one of the rare examples that the *o*-QM **3** is involved in a direct synthesis rather than as a nonintentionally used intermediate or by-product. $ZnCl_2$-catalyzed, inverse hetero-Diels–Alder reaction between *ortho*-quinone methide **3** and an excess of *O*-methyl-*C*,*O*-bis-(trimethylsilyl)ketene acetal provided the acid in fair yields (Fig. 6.37).[67] The *o*-QM **3** was prepared *in situ* by thermal degradation of 5a-bromo-α-tocopherol (**46**). The primary cyclization product, an *ortho*-ester derivative, was not isolated, but immediately hydrolyzed to methyl 3-(5-tocopheryl)-2-trimethylsilyl-propionate, subsequently desilylated, and finally hydrolyzed into **50**.

While tocopherylacetic aicd (**51**), the lower C_1-homologue of 3-(5-tocopheryl)-propionic acid (**50**) showed a changed redox behavior (see Section 6.5.1), compound **50** displayed the usual redox behavior of tocopherol derivatives, that is, formation of both *ortho*- and *para*-quinoid oxidation intermediates and products depending on the respective reaction conditions. Evidently, the electronic substituent effects that

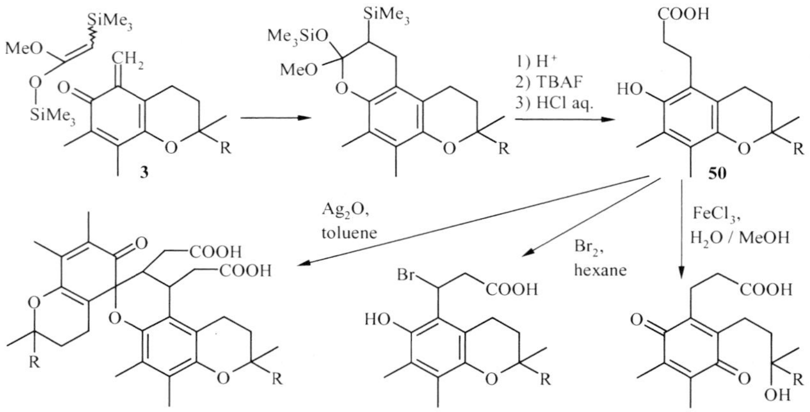

FIGURE 6.37 Synthesis of 3-(5-tocopheryl)-propionic acid (**50**) by trapping the intermediate *ortho*-QM **3** with a ketene acetal. Reaction products of **50** are formed in complete analogy to α-tocopherol (**1**).

changed the reactivity and oxidation behavior of tocopherylacetic acid (**51**) and its derivatives were neutralized by homologation, so that the system returned to its "normal" behavior. All three oxidation reactions typical of α-tocopherol (**1**)—bromination with elemental bromine to the 5a-bromo derivative, oxidation with Ag_2O to an intermediate *ortho*-quinone methide that dimerized into the spiro dimer, and oxidation in aqueous media to a *para*-quinone—proceeded with 3-(5-tocopheryl)-propionic acid (**50**) in complete similarity to α-tocopherol itself, demonstrating the analogous chemical behavior of the two compounds (Fig. 6.37).[67]

6.5 FORMATION OF TOCOPHEROL-DERIVED *o*-QMs INVOLVING OTHER POSITIONS THAN C-5A

6.5.1 5-(γ-Tocopheryl)acetic Acid

5-γ-Tocopherylacetic acid (**51**) was produced by hydrolysis of the corresponding nitrile precursor in aqueous dioxane with gaseous HCl, the precursor being obtained by reaction of 5a-bromo-α-tocopherol (**46**) with potassium cyanide in DMSO.[68] The nitrile was also the starting material for the preparation of different esters, amides, and the corresponding lactone.[68]

An interesting feature of 5-tocopherylacetic acid (**51**) and its derivatives was their appreciable thermal stability up to 200 °C. In contrast to 5a-substituted tocopherols carrying an electronegative substituent at C-5a, the homopolar C–C bond in the C_2-unit at the 5-position of the tocopherol skeleton was shown to be very stable. Thermal decomposition of **51** at temperatures above 250 °C caused a complete breakdown of the chroman structure, the C_3-unit consisting of C-2, C-2a, and C-3 being eliminated as propyne, the side chain as 4,8,12-trimethyltridec-1-ene (Fig. 6.38). Fragmentation

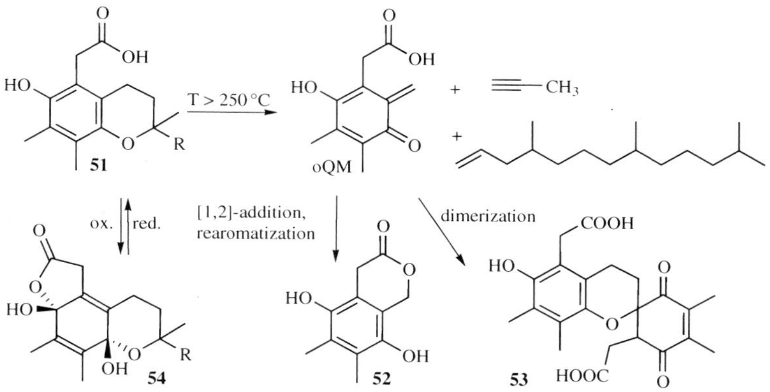

FIGURE 6.38 Redox behavior and thermal degradation of 5-tocopherylacetic acid (**51**), involving an *o*-QM formed by involvement of C-4 and O-1.

occurred with formation of an intermediate *ortho*-quinone methide involving C-4 and O-1, which was stabilized immediately in subsequent reactions, either by [1,2]-addition of the carboxylic OH-group followed by rearomatization into lactone **52**, or by dimerization to a spiro compound **53**, the latter reaction being a parallel of the common self-dimerization of *o*-QM **3** into spiro dimer **9**. From the historic perspective, this was the first report of an *o*-QM formed from α-tocopherol other than the "usual" *o*-QM **3**.[68]

Interestingly, tocopherylacetic acid underwent no reaction with bromine or Ag_2O—conditions that produce *o*-QM **3** from α-tocopherol readily. Apparently, the substituent effect prohibited formation of the typical 5a-*o*-QM structure, whereas the ability to form *para*-quinoid structures was not impaired: compound **51** was neatly oxidized into its corresponding *para*-quinone, present as lactono-semiketal **54** in aqueous media, both compounds forming a reversible redox pair (Fig. 6.38). Evidently, the electronic effects exerted by the carboxylic acid function in α-position to C-5a changed the oxidation chemistry of the tocopherol system in a way that *o*-QM formation at C-5a was largely, if not completely, disfavored and *para*-quinoid oxidation products were largely preferred. This notion is supported by the fact that homologation in the 5-substituent, that is, presence of a propionic acid rather than an acetic acid moiety, returns the reactivity of the system to "normal", that is, to that of α-tocopherol. This has been shown above (Section 6.4.5) for 3-(5-tocopheryl) propionic acid (**50**).

6.5.2 4-Oxo-α-Tocopherol

4-Oxo-α-tocopherol (**55**) proved to be a very interesting compound with regard to forming various intermediate tautomeric and quinoid structures. It undergoes an intriguing rearrangement of its skeleton under involvement of different *o*-QM structures. The 4-oxo-compound was prepared from 3,4-dehydro-α-tocopheryl acetate via its bromohydrin, which was treated with ZnO to afford 4-oxo-α-tocopherol (**55**).

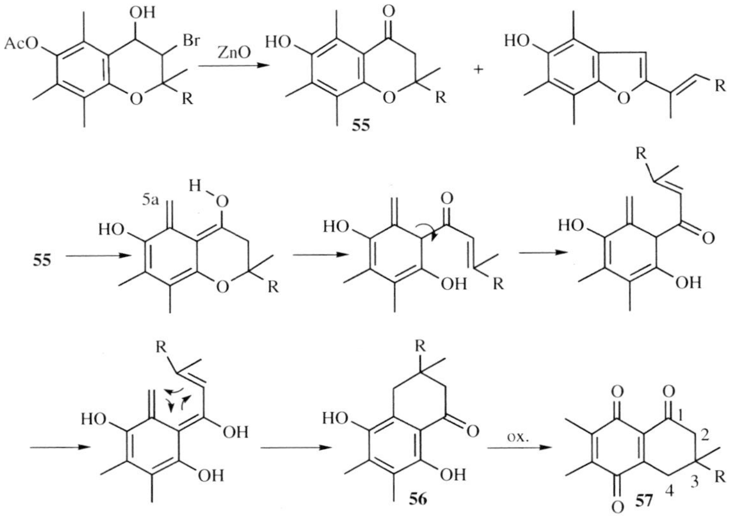

FIGURE 6.39 Synthesis of 4-oxo-α-tocopherol (**55**) and its oxidative rearrangement into naphthalenetrione **57**.

The ZnO was a very effective reagent as it caused dehydrobromination, simultaneous deacetylation, and tautomerization of the resulting enol intermediate.[69]

4-Oxo-α-tocopherol (**55**) was rearranged under simulated physiological, oxidative conditions into hydroquinone **56**, which was immediately oxidized into naphthalenetrione **57**, the final oxidation product, in yields of about 10%.[69] The rearrangement product possessed a carbon skeleton completely different from that of the starting tocopherol **55**. The rearrangement mechanism was shown to involve opening of the alicyclic ring, followed by formation of different tautomers, bond rotation, and electrocyclic ring closure as the key step (Fig. 6.39). The incorporation of C-5a of 4-oxo-α-tocopherol (**55**) into the alicyclic ring in **57** was demonstrated by means of isotopic labeling: 4-oxo-α-tocopherol trideuterated at C-5a produced compound **57** bisdeuterated at C-4, the "former" C-5a-position.

Apparently, introduction of the C-4 oxo-group changed the reactivity of the tocopherol system quite drastically. Enolization of the 4-carbonyl is coupled to formation of quinone dimethide structure involving C-5a and C-4. This reaction can be seen as a [1,5]-sigmatropic proton shift; it does not involve external oxidants. A similar quinone dimethide, after cleavage of the alicyclic ring and bond rotation, undergoes an intramolecular electrocyclic reaction with the ene structure in the former pyran unit. This reaction is somehow comparable to the trapping of *o*-QM **3** with ethyl vinyl ether, although this is of course an intermolecular process. The oxidation of the resulting annelated hydroquinone into the corresponding naphthoquinone **57** is the last step of the reaction, and probably also the driving force of the whole sequence that caused a far-reaching rearrangement of the carbon skeleton of **55**.

6.5.3 3-Oxa-Chromanols

3-Oxa-chromanols of the general formula **58**—termed 5,7,8-trimethyl-4*H*-benzo [1,3]dioxin-6-ols according to IUPAC rules—exhibit a structure quite close to tocopherols: only the C-3 methylene group is exchanged for an oxygen. However, their reactivity is quite different from that of α-tocopherol (**1**). Their remarkable feature is the dependence of the oxidation chemistry on the available amount of water as the coreactant and the involvement of a rich quinone methide chemistry. Therefore, although they, strictly speaking, do not represent conventional tocopherol derivatives, they were regarded as oxa-derivatives of the vitamin, and their chemistry shall be included in the discussion on tocopherol-derived *o*-QMs within this chapter.

3-Oxa-chromanols were obtained as diastereomeric mixtures by condensation of trimethylhydroquinone with the double equivalent of aldehydes in a straightforward one-pot reaction (Fig. 6.40).[70] 3-Oxa-chromanols have recently been tested for their antioxidative properties, as they represent an interesting novel class of phenolic antioxidants.[58] EPR measurements of the radicals derived from 3-oxa-chromanol derivatives revealed similar stabilities as compared to the α-tocopheroxyl radical (**2**), producing well-resolved multiline spectra, the hyperfine coupling constants for the methyl substituents at the aromatic ring being quite similar to those of **2**. Distinct effects of the configuration on the long-range couplings into the heterocyclic ring were observed.[27a,70]

The oxidation behavior of 3-oxa-chromanols was mainly studied by means of the 2,4-dimethyl-substituted compound 2,4,5,7,8-pentamethyl-4*H*-benzo[1,3]dioxin-6-ol (**59**) applied as mixture of isomers;[27a] it showed an extreme dependence on the amount of coreacting water present. In aqueous media, **59** was oxidized by one oxidation equivalent to 2,5-dihydroxy-3,4,6-trimethyl-acetophenone (**61**) via 2-(1-hydroxyethyl)-3,5,6-trimethylbenzo-1,4-quinone (**60**) that could be isolated at low temperatures (Fig. 6.41). This "detour" explained why the seemingly quite inert benzyl ether position was oxidized while the labile hydroquinone structure remained intact. Two oxidation equivalents gave directly the corresponding *para*-quinone **62**. Upon oxidation, C-2 of the 3-oxa-chroman system carrying the methyl substituent was always lost in the form of acetaldehyde.

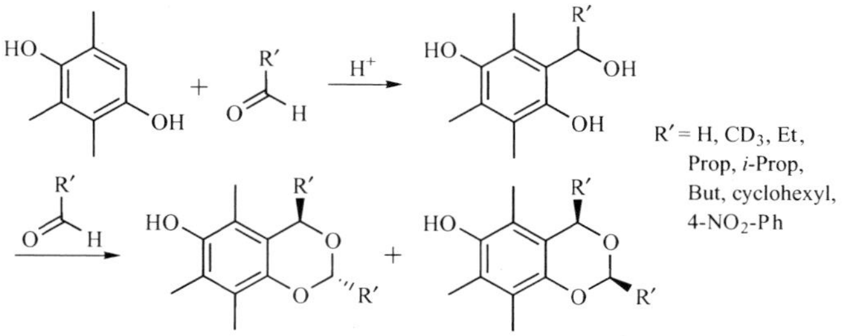

R′ = H, CD_3, Et, Prop, *i*-Prop, But, cyclohexyl, 4-NO_2-Ph

58 (*trans-isomer*) **58** (*cis-isomer*)

FIGURE 6.40 Synthesis of 3-oxa-chromanols (**58**) as mixture of *cis/trans*-isomers.

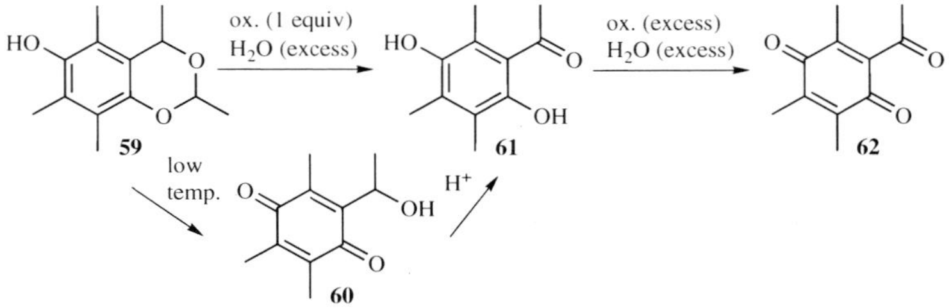

FIGURE 6.41 Oxidation of 3-oxa-chromanol **59** in aqueous media (excess water present), leading to acetophenone **61** with an equimolar amount of oxidant, and further to *para*-quinone **62** in the presence of excess oxidant.

Oxidation of 3-oxa-chromanol **59** in the presence of just one equivalent of water produced acetophenone **61** as well, but according to a different mechanism not involving *para*-quinone **60**. The process was elucidated by employing isotopically labeled starting material, selectively trideuterated at the 2- and 4-methyl groups (Fig. 6.42). The reaction involved an *ortho*-quinone dimethide intermediate **63**.[27a] Interestingly, such an intermediate was observed in the case of the chemistry of 4-oxo-α-tocopherol (**55**), which also possessed a strongly electronegative oxygen substituent at C-4, similar to the 3-oxa-chromanols. Intermediate **63** underwent a [1,5]-sigmatropic proton shift in a concerted way to give styrene derivative **64**, from which finally acetaldehyde was released by reaction with water to afford acetophenone **61**. The overall outcome of the reaction was thus the same as in the presence of excess water, but the formation mechanisms were completely different from each other. By means of deuterated starting material, the selective [1,5]-sigmatropic proton shift from the C-4a methyl group to the exocyclic methylene group was demonstrated, and the occurrence of both intermediates, *ortho*-quinone dimethide intermediate **63**

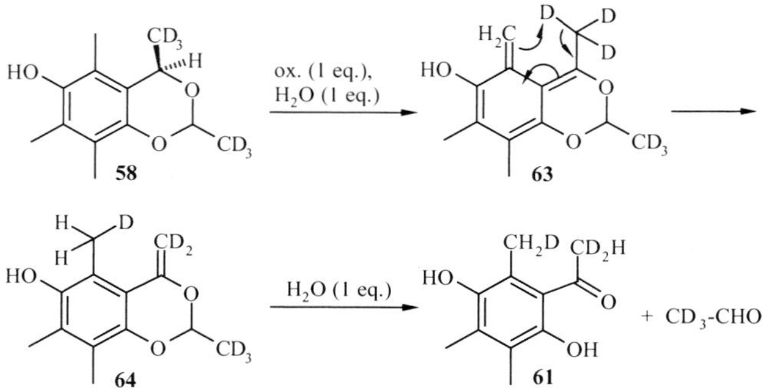

FIGURE 6.42 Oxidation of 3-oxa-chromanol **58** in the presence of 1 equivalent of water: mechanistic study by means of selectively deuterated starting material. The initially formed *ortho*-quinone dimethide **63** rearranges into styrene derivative **64**, which then reacts with water to provide acetophenone **61**.

and styrene derivative **64**, was additionally confirmed by trapping in hetero-Diels–Alder reactions.[27a]

In the absence of water, oxidation of **59** proceeded also via the *ortho*-quinone dimethide **63** and the styrene derivative **64**. However, as no water was present to react with the latter intermediate to release acetaldehyde, the C-2–C-3 bond was broken and a bond rotation occurred in the zwitterionic intermediate followed by C–C bond formation that established a chromanone system **65**. The zwitterionic intermediate in this reaction is remarkable as it somehow represents the "opposite" of the zwitterionic intermediate encountered in the formation of *o*-QMs from the parent phenols (see Fig. 6.15 and Section 6.3.3). In the former intermediate formed from the 3-oxa-chromanols, the negative charge is placed at the benzylic position, the positive one at the ring oxygen, both charges being stabilized by resonance. In the latter intermediate formed upon *o*-QM production from the phenols, the charge placement is opposite: a benzyl cation and a phenolate anion, which evidently rendered this intermediate much more stable than the former one. Chromanone **65** was immediately further oxidized to chromenone **66**, the probable driving force of the reaction. The overall process from 3-oxa-chromanol **59** to chromenone **66** required two equivalents of a two-electron oxidant. In the absence of water, evidently the chromanone–chromenone oxidation was favored over the oxidation of the starting material **59**. Chromanone **65** was consumed before the oxa-chromanol **59** was affected, so that upon addition of less than two equivalents of oxidant, chromenone **66** was present besides nonreacted starting material (Fig. 6.43).

If the formation of an exocyclic methylene group at C-4, and thus the formation of a styrene intermediate such as **64**, is impossible due to structural prerequisites, oxidation of the corresponding 3-oxa-chromanols will involve the *o*-QM formed

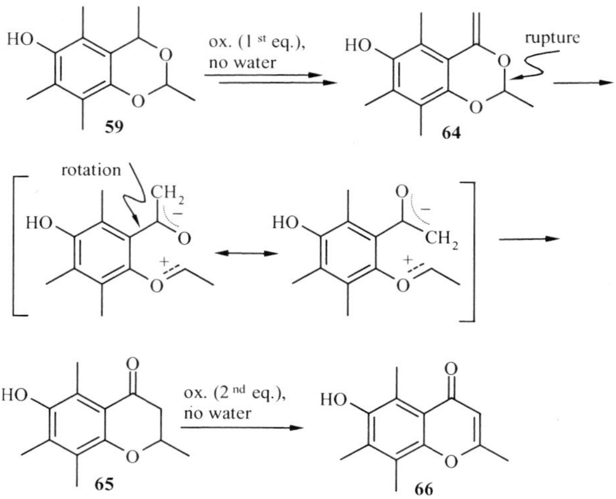

FIGURE 6.43 Oxidation of 3-oxa-chromanol **59** in the absence of water, providing chromenone **66** as the final product: mechanism and reaction intermediates.

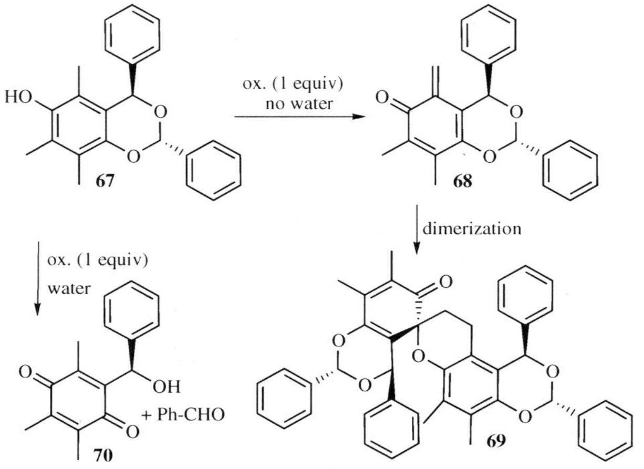

FIGURE 6.44 Oxidation of 3-oxa-chromanol **67**, having no protons at position C-4a able to undergo rearrangements by analogy to 3-oxa-chromanol **59** with its oxidation intermediates **63** and **64**. Due to this blocking at C-4/C-4a, the oxidation behavior of **67** resembles that of α-tocopherol (**1**).

involving C-5a, which will react to the corresponding spiro dimer (Fig. 6.44), by analogy to the reactivity of the α-tocopherol-derived *o*-QM **3**. For example, this chemical behavior was observed for 2,4-diphenyl-3-oxachromanol **67**, which upon oxidation in nonaqueous media with excess oxidant (Ag_2O), provided the sterically crowded tetraphenyl spiro dimer **69** via the intermediate C-5a-*o*-QM **68**.[27a] Due to the phenyl substituent at C-4, a hypothetical *ortho*-quinone dimethide intermediate (analogous to **63**) cannot rearrange to a styrene intermediate involving C-4a. This "blocking" of C-4a returned the system to a reactivity similar to α-tocopherol: an *o*-QM involving O-6 and C-5a as in α-tocopherol (**1**), but not the *ortho*-quinone dimethide involving C-5a and C-4 as in the 3-oxa-chromanols **58** and **59** was formed. Oxidation of 2,4-diphenyl-3-oxachromanol **67** in aqueous media also produced *para*-quinone **70** by analogy to α-tocopherol chemistry (Fig. 6.4).

The oxidation behavior of 3-oxa-chromanols showed both differences and similarities to that of α-tocopherol (**1**). Paralleling the chemistry of α-tocopherol, one-electron oxidation caused formation of the corresponding chromanoxyl radicals, which were relatively stable. Also, in the absence of a C-4a substituent with protons, the oxidation behavior entirely resembled that of α-tocopherol. In the presence of a C-4a substituent with protons, the oxidation behavior changed fundamentally. The primary *ortho*-quinone dimethide formed involving C-5a and C-4 (**63**) underwent different subsequent reactions depending on the water content present. The proton transfer from C-4a to C-5a in a [1,5]-sigmatropic rearrangement giving a styrene derivative **64** with the olefinic double bond between C-4 and C-4a was the preferred reaction. The further chemistry of this styrene intermediate is then highly dependent on the amount of coreacting water available (see Figs 6.42 and 6.43).

6.5.4 Selected Substituent-Stabilized Tocopherols and Conjugatively Stabilized *Ortho*-Quinone Methides

Similar to 3-oxa-chromanols, several other derivatives of α-tocopherol also exhibit a substantially changed oxidation behavior as compared to α-tocopherol itself. Especially 5-substituted derivatives with no 5a-hydrogen belong into this class, as they are unable to form an *o*-QM involving C-5a, which is most characteristic of α-tocopherol (**1**). In 5a,5a,5a-trimethyl-α-tocopherol (5-*tert*-butyl-γ-tocopherol, **71**) and other 5a,5a,5a-trialkyl tocopherols, the oxidation resistance in apolar media was greatly increased; in aqueous media the corresponding *para*-quinone was formed very readily.[71] Interestingly, blocking of C-5a with regard to *o*-QM formation did not lead to an increased involvement of C-7a in *o*-QM formation, but rather caused increased stability toward oxidation. This is somehow similar to the case of γ-tocopherol, where the missing 5a-methyl group (and thus the inability to form the α-tocopherol-type *o*-QM **3**) also causes increased stability toward two-electron oxidation in apolar media, but not involvement of the 7a-*o*-QM.

The reaction behavior of 5-phenyl-γ-tocopherol (**72**) was similar to that of the 5a,5a,5a-trialkyl tocopherols: the 5a-*o*-QM cannot be formed, and the oxidative stability was increased. However, if a *para*-OH group was introduced into the 5-phenyl substituent as in 5-(*p*-hydroxyphenyl)-γ-tocopherol (**73**), oxidation in apolar media proceeded quite readily—comparably fast to α-tocopherol—and produced the quinoid structure **74** that can be regarded as an *ortho*-quinone methide with regard to the basic tocopherol moiety, as *para*-quinone methide with regard to the phenyl substituent, and also as phenylogous 5,6-tocopherylquinone, the latter compound being quite well known in tocopherol chemistry also as α-tocored (**75**).[72] Quinoid compound **74** is a stable compound when stored at −20 °C in inert atmosphere that does not undergo cycloaddition reactions as *o*-QM **3** does, for example, spiro dimerization or reaction with ethyl vinyl ether. It is neatly reduced to the starting tocopherol **73** without side reactions. Under ambient conditions, in the presence of air, it undergoes autoxidation to a complex product mixture (Fig. 6.45).

Also 5-(4-methylphenyl)-γ-tocopherol (**76**), which can be imagined as "phenylogous α-tocopherol," was as readily oxidized as α-tocopherol (**1**) in aprotic media, providing the quinone structure **77**, which represents a phenylogous *o*-QM **3** by analogy.[73] The conjugative extension of *o*-QM **3** caused an appreciable stabilization, so that intermediate **77** was stable in inert solvents at rt without undergoing dimerization or conjugation reactions. In the presence of HBr, a [1,8]-addition process occurred that afforded 5-(4-bromomethylphenyl)-γ-tocopherol (**78**), other nucleophile such as water reacting in a similar way by [1,8]-addition. Also bromination of **76** with elemental bromine afforded 5-(4-bromomethylphenyl)-γ-tocopherol (**78**) quantitatively. The reaction proceeded according to an oxidation–addition mechanism via the phenylogous quinone methide **77** that added the HBr generated in the oxidation step (Fig. 6.46).[60] The reaction was thus the "phenylogous version" of the bromination of α-tocopherol (**1**) (see Fig. 6.33), with quinone **77** being the phenylogous *o*-QM **3** and bromide **78** acting as phenylogous product 5a-bromo-α-tocopherol (**46**)

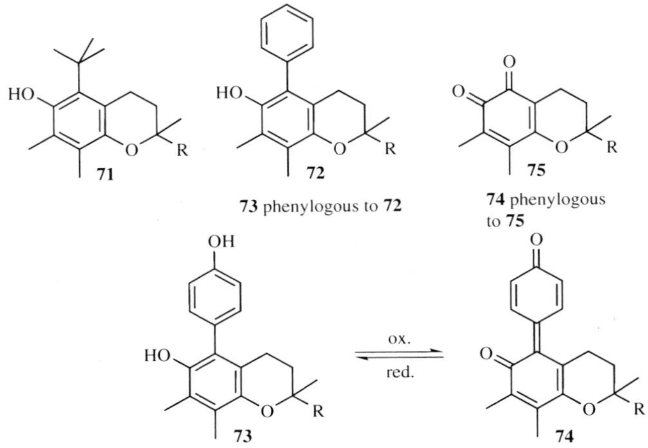

FIGURE 6.45 Inability of 5a-substituted derivatives to form structures analogous to *o*-QM **3** causes increased oxidative stability as in compounds **71** and **72**. 5-(*p*-Hydroxyphenyl)-γ-tocopherol (**73**) is oxidized to the conjugatively stabilized *o*-QM **74**, the phenylogous α-tocored (**75**).

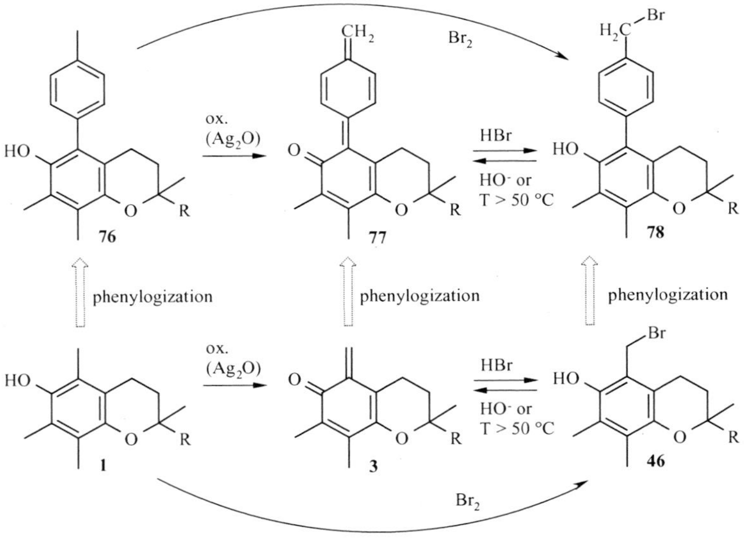

FIGURE 6.46 Oxidation chemistry of 5-(4-methylphenyl)-γ-tocopherol (**76**), establishing a reaction system "phenylogous" to α-tocopherol (**1**), with quinone methide **77** and benzyl bromide **78** being the conjugatively stabilized, phenylogous counterparts of *o*-QM **3** and 5a-bromo-α-tocopherol (**46**), respectively.

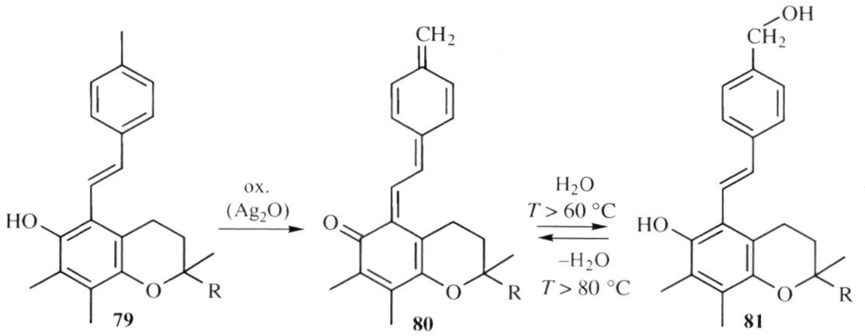

FIGURE 6.47 Oxidation of styryl-γ-tocopherol **79** to the stable *o*-QM **80**, a "styrylogous" *o*-QM **3**.

(see Fig. 6.46). Also, the ready elimination of HBr from **78** upon alkali treatment or thermal stress finds its parallels in the chemistry of 5a-bromo-α-tocopherol (**46**).

The stability of the phenylogous *o*-QM **77** led to the conclusion that the conjugative stabilization of *o*-QM **3** by introduction of the phenyl system was nearly as large as the rearomatization energy between *o*-QM **3** and α-tocopherol (**1**). This was confirmed by introducing an additional double bond to the conjugatively stabilized system: oxidation of the styryltocopherol **79** produced the *o*-QM **80**, which represents the "styrylogous" (vinylphenylogous) *o*-QM **3**.[73] This quinoid structure was completely stable at room temperature in the absence of oxygen and also in the presence of water. Neither did it undergo any dimerization nor cycloaddition reactions. Heating in the presence of water to 60 °C produced the corresponding hydroxymethyl derivative **81**, which at temperatures above 80 °C eliminated water again to regenerate the vinylphenylogous *o*-QM **80** (Fig. 6.47). In this compound, the conjugative stabilization is very similar to the rearomatization energy of *o*-QM **3** to α-tocopherol (**1**), which provided a nice means to approach this rearomatization energy experimentally.

In a similar fashion, 1,4-bis(5-γ-tocopheryl)benzene (**82**) was oxidized by Ag_2O in toluene into the quinoid compound **83**, a cross-conjugated dimer of *o*-QM **3**, and was reduced back to the parent phenol **82**. Both compounds formed a reversible redox system. Also quinone methide **83** was stable under ambient conditions and did not undergo the reactions typical of *o*-QM **3**. Interestingly, *o*-QM **83** was exclusively formed in the *cis*-form (both keto groups on the same side of the C-5–C-5 axis), but not as the corresponding *trans*-structure.[73] The reason for this was an interaction of the phenolic hydroxyl groups with each other and with the oxidant that placed the hydroxyls on the same side of the phenol, exerting a certain preorganizational effect. When oxidized by DDQ in dichloromethane at room temperature, 1,4-bis(5-γ-tocopheryl)benzene (**82**) formed a mixture of *o*-QM **83** and the corresponding *trans*-isomer **84** in an approximate 1:19 ratio. Stability and reduction of **84** to **82** were similar to the behavior of the *cis*-isomer **83** (Fig. 6.48).

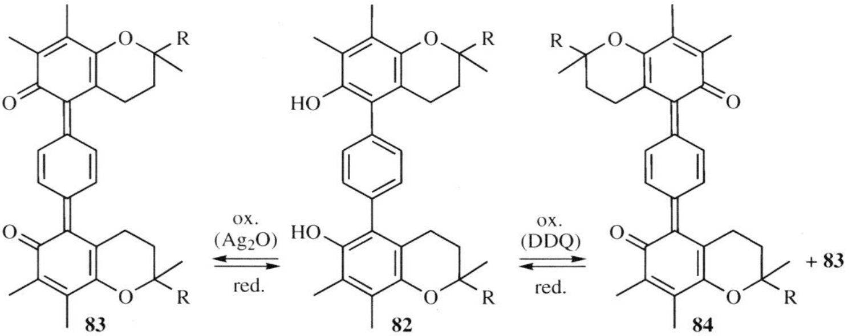

FIGURE 6.48 Oxidation of 1,4-bis(5-γ-tocopheryl)benzene (**82**) to the two stable *o*-QM isomers **83** (*cis*) and **84** (*trans*) depending on the oxidant used.

6.6 FUTURE PROSPECTS

Current research on tocopherol-derived *o*-QMs appears to be focusing on several areas as briefly outlined in the following. For theoretical considerations on *o*-QM stability, it will be interesting to estimate the conjugative stabilization of the *o*-QM that is necessary to balance or to outweigh the rearomatization energy gained upon spiro dimerization of the *o*-QM. Such a conjugatively stabilized or "overstabilized" *o*-QM would be stable, without tending to dimerize. A similar stabilization can be achieved not only by conjugation but also by external agents.[27b,47] Better ways to stabilize the *o*-QMs will open new ways to employ them in synthesis as uncomplicated reagents that can be synthesized, handled, and dosed easily and will help them to lose their "bad reputation" as reactive intermediate without much applicability in synthesis besides low-yield trapping reactions.

A research topic that slowly emerges is the formation of α-tocopherol-derived *ortho*-quinone imines, thus starting from tocopheramines that carry an amino function instead of the phenolic hydroxyl group. Formation, stability, and reactions of the corresponding tocopherol-derived *ortho*-quinone imines are of interest not only for pharmaceutical developments but also for the insights that they offer into the general chemistry of heteroanalogous quinoid structures. Although tocopheramines and *N*-alkyltocopheramines could well accommodate the negative charge of the zwitterionic mesomeric *o*-QM structure at the amino function, this is impossible for *N,N*-dialkyltocopheramines. Does this prevent *ortho*-quinone imine formation in such compounds from the beginning? Or are the corresponding *o*-QM-analogues just severely destabilized? Is there also an effect on formation and stability of the one-electron oxidation products, the corresponding aminyl radicals?

The oxidation chemistry of α-tocopherol and the regioselectivity of the *o*-QM formation are well investigated (Section 6.3.2). This is in complete contrast to β-tocopherol. At a first glance, the β-tocopherol case would appear much simpler as the free aromatic position at C-7 offers no alternative methyl group to be involved in *o*-QM formation. However, preliminary results show that the oxidation chemistry of β-tocopherol is rather complex, the free position undergoing alkylation by the *o*-QM and electrophilic substitution reactions. The *o*-QM chemistry of β-tocopherol thus offers an interesting field of activities as nearly no analogies from the α-tocopherol system can be applied.

In contrast to both α-tocopherol and β-tocopherol that offer a 5a-methyl group for *o*-QM formation, γ-tocopherol and δ-tocopherol lack this structural feature. So far it is still an enigma why those two tocopherols do not form analogous *o*-QMs involving C-7a, not even under quite drastic conditions. Ways to induce *o*-QM formation at these positions would open the whole range of *o*-QM chemistry also for these tocopherol congeners, be it by complexation, manipulation of the annulation angles to the heterocylic ring, or by introduction of electronic substitutent effects.

Finally, the spiro polymerization of molecules bearing two *ortho*-alkylphenol functions capable of forming *o*-QMs offers a wide field for reactivity and kinetic studies as well as multiple applications in polymer and material chemistry and science.

ACKNOWLEDGMENTS

For their valuable contributions, inspiring discussions, and encouraging enthusiasm, we would like to thank all past and present coworkers, especially Dr. Thomas Netscher, DSM Nutritional Products, Basel; Prof. Lars Gille, Research Institute for Biochemical Pharmacology and Toxicology, University of Veterinary Medicine Vienna; Dr. Francesco Mazzini, University of Pisa, Italy, and Prof. Kurt Mereiter, Vienna University of Technology. We are also thankful to all past and present students working in the field of tocopherols, tocopherol-derived *o*-QMs, and their oxidation chemistry, especially Dr. Christian Adelwöhrer, Dr. Wolfgang Gregor, Dr. Anjan Patel, and Dr. Elisabeth Kloser.

The financial support to this work by the Fonds zur Förderung der wissenschaftlichen Forschung (FWF, Austrian Science Foundation), Projects P-14687, P-17428, and P-19081 is gratefully acknowledged.

REFERENCES

1. (a) Preedy, V. R.; Watson, R. R. *Encyclopedia of Vitamin E*. CABI Publishing: Oxford, Cambridge. 2007. (b) Baldenius, K. U.; von dem Bussche-Hünnefeld, L.; Hilgemann, E.; Hoppe, P.; Stürmer, R. *Ullmann's Encyclopedia of Industrial Chemistry*, Vol. A27; VCH Verlagsgesellschaft: Weinheim 1996; pp. 478–488, 594–597. (c) Netscher, T. Stereoisomers of tocopherols: syntheses and analytics. *CHIMIA* 1996, 50, 563–567. (d) Packer, L.; Fuchs, J. *Vitamin E in Health and Disease;* Marcel Dekker Inc.: New York, 1993. (e) Isler, O.; Brubacher, G. *Vitamins I;* Georg Thieme Verlag: Stuttgart, 1982, p. 126.
2. IUPAC-IU. B. Nomenclature of tocopherols and related compounds recommendations 1981. *Eur. J. Biochem.* 1982, 123, 473–475.

3. IUPAC-IUB Commission on Biochemical Nomenclature (CBN). Nomenclature of quinones with isoprenoid side-chains recommendations (1973). *Arch. Biochim. Biophys.* 1974, 165, 1–8.
4. Matsuo, M.; Matsumoto, S. Electron spin resonance spectra of the chromanoxyl radicals derived from tocopherols (vitamin E) and their related compounds. *Lipids* 1983, 18, 81–86.
5. Burton, G. W.; Doba, T.; Gabe, E. J.; Hughes, L.; Lee, F. L.; Prasad, L.; Ingold, K. U. Autoxidation of biological molecules 4. Maximizing the antioxidant activity of phenols. *J. Am. Chem. Soc.* 1985, 107, 7053–7065.
6. Boguth, W.; Niemann, H. Electron pin resonance of chromanoxy free radicals from α-, ζ_2-, β-, γ- δ-tocopherol and tocol. *Biochim. Biophys. Acta* 1971, 248, 121–130.
7. Mukai, K.; Tsuzuki, N.; Ouchi, S.; Fukuzawa, K. Electron spin resonance studies of chromanoxyl radicals derived from tocopherols. *Chem. Phys. Lipids* 1982, 30, 337–345.
8. Tsuchija, J.; Niki, E.; Kamiya, Y. Oxidation of lipids IV. Formation and reaction of chromanoxyl radicals as studied by electron spin resonance. *Bull. Chem. Soc. Jpn.* 1983, 56, 229–232.
9. Mukai, K.; Tsuzuki, N.; Ishizu, K.; Ouchi, S.; Fukuzawa K Electron nuclear double resonance studies of radicals produced by the PbO_2 oxidation of α-tocopherol and its model compound in solution. *Chem. Phys. Lipids* 1981, 29, 129–135.
10. Lee, S. B.; Willis, A. C.; Webster, R. D. Synthesis of the phenoxonium cation of an α-tocopherol model compound crystallized with non-nucleophilic $[B(C_6F_5)_4]^-$ and $(CB_{11}H_6Br_6)^-$ anions. *J. Am. Chem. Soc.* 2006, 128, 9332–9333.
11. Rosenau, T.; Kloser, E.; Gille, L.; Mazzini, F.; Netscher, T. Vitamin E. Chemistry studies into initial oxidation intermediates of α-tocopherol: disproving the involvement of 5a-C-centered "Chromanol Methide" radicals. *J. Org. Chem.* 2007, 72(9), 3268–3281.
12. Inglett, G. E.; Mattill, H. A. Oxidation of hindered 6-hydroxychromans. *J. Am. Chem. Soc.* 1955, 77, 6552–6554.
13. Knapp, F. W.; Tappel, A. L. Some effects of *y*-radiation or linoleate peroxidation on α–tocopherol. *J. Am. Oil Chem. Soc.* 1961, 38, 151–156.
14. Skinner, W. A. Structure–activity relations in the vitamin E series. I. Effects of 5-methyl substitution on 6-hydroxy-2,2,5,7 8-pentamethylchroman. *J. Med. Chem.* 1967, 10, 657–661.
15. Skinner, W. A.; Alaupovic, P. Oxidation products of vitamin E and its model, 6-hydroxy-2,2,5,7 8-pentamethyl-chroman. V. Studies of the products of alkaline ferricyanide oxidation. *J. Org. Chem.* 1963, 28, 2854–2858.
16. Nilsson, J. L. G.; Daves, D. G.; Folkers, K. The oxidative dimerization of alpha-, beta-, gamma-, and delta-tocopherols. *Acta Chim. Scand.* 1968, 22, 207–218.
17. Fujimaki, M.; Kanamaru, K.; Kurata, T.; Igarashi, O. Studies on the oxidation mechanism of vitamin E. Part I. The oxidation of 2,2,5,7 8-pentamethyl-6-hydroxychromanol. *Agr. Biol. Chem.* 1970, 34, 1781–1786.
18. Goodhue, C. T.; Risley, H. A. Reactions of vitamin E with peroxides I. Reaction of benzoyl peroxide with D-α-tocopherol in hydrocarbons. *Biochem. Biophys. Res. Commun.* 1964, 17, 549–553.
19. Skinner, W. A. Vitamin E oxidation with free radical initiators Azobis(isobutyronitrile). *Biochem. Biophys. Res. Commun.* 1964, 15, 469–472.
20. Fukuzawa, K.; Gebicki, J. M. Oxidation of alpha-tocopherol in micelles and liposomes by the hydroxyl, perhydroxyl, and superoxide free radicals. *Arch. Biochem. Biophys.* 1983, 226, 242–251.

21. Matsuo, M.; Matsumoto, S.; Iitaka, Y.; Niki, E. Radical-scavenging reactions of vitamin E and its model compound 2 2,5,7,8-pentamethylchroman-6-ol, in a *tert*-butylperoxyl radical generating system. *J. Am. Chem. Soc.* 1989, 111, 7179–7185.

22. Kamal-Eldin, A.; Appelqvist, L. A. The chemistry and antioxidant properties of tocopherols and tocotrienols. *Lipids* 1996, 31, 671–701.

23. Suarna, C.; Southwell-Keely, P. T. Effects of alcohols on the oxidation of the vitamin E model compound 2,2,5,7 8-pentamethyl-6-chromanol. *Lipids* 1989, 24, 56–60.

24. Suarna, C.; Southwell-Keely, P. T. New oxidation products of α-tocopherol. *Lipids* 1988, 23, 137–139.

25. Rousseau-Richard, C.; Richard, C.; Martin, R. Kinetics of bimolecular decay of α-ocopheroxyl free radicals studied by ESR. *FEBS Lett.* 1988, 233, 307–310.

26. Nilsson, J. L. G.; Daves, D. G.; Folkers, K. New tocopherol dimers. *Acta Chim. Scand.* 1968, 22, 200–206.

27. (a) Rosenau, T.; Potthast, A.; Elder, T.; Lange, T.; Sixta, H.; Kosma, P. Synthesis and oxidation behavior of 2,4,5,7,8-pentamethyl-4*H*-1,3-benzodioxin-6-ol, a multifunctional oxatocopherol-type antioxidant. *J. Org. Chem.* 2002, 67, 3607–3614. (b) Rosenau, T.; Potthast, A.; Elder, T.; Kosma, P. Stabilization and first direct spectroscopic evidence of the *o*-quinone methide derived from vitamin E. *Org. Lett.* 2002, 4, 4285–4288.

28. Schröder, H.; Netscher, T. Determination of the absolute stereochemistry of vitamin E derived oxa-spiro compounds by NMR spectroscopy. *Magn. Reson. Chem.* 2001, 39, 701–708.

29. Schudel, P.; Mayer, H.; Metzger, J.; Rüegg, R.; Isler, O. Chemistry of vitamin E II. Structure of potassium ferricyanide oxidation product of tocopherol. *Helv. Chim. Acta* 1963, 46, 636–649.

30. Dürckheimer, W.; Cohen, L. A. Mechanisms of α-tocopherol oxidation; synthesis of the highly labile 9-hydroxy-α-tocopherone. *Biochem. Biophys. Res. Commun.* 1962, 9, 262–265.

31. Omura, K. Iodine oxidation of alpha -tocopherol and its model compound in alkaline methanol: unexpected isomerization of the product quinone monoketals. *J. Org. Chem.* 1989, 54, 1987–1990.

32. Kohar, I.; Southwell-Keely, P. T. Reduction of 8a-hydroxy-2,2,5,7 8-pentamethyl-6-chromanone. *Redox Report* 2002, 7, 251–255.

33. Machlin, L. J. *Vitamin E: A Comprehensive Treatise;* Marcel Dekker Inc: New York, 1980.

34. Parkhurst, R. M.; Skinner, W. A. Chromans and tocopherols. In *Chemistry of Heterocyclic Compounds;* Ellis, G. P.; Lockhardt, I. M. Eds.; Wiley: New York, 1981; Vol. 36.

35. Mills, W. H.; Nixon, I. G. Stereochemical influences on aromatic substitution. Substitution derivatives of 5-hydroxyhydrindene. *J. Chem. Soc.* 1930, 2510–2524.

36. Badger, G. M. Q. The aromatic bond. *Rev. Chem. Soc.* 1951, 5, 147.

37. Behan, J. M.; Dean, F. M.; Johnstone, R. A. W. Photoelectron spectra of cyclic aromatic ethers: the question of the Mills–Nixon effect. *Tetrahedron* 1976, 32, 167–171.

38. Rosenau, T.; Ebner, G.; Stanger, A.; Perl, S.; Nuri, L. From a theoretical concept to biochemical reactions: strain-induced bond localization (SIBL) in oxidation of vitamin E. *Chem. Eur. J.* 2005, 11(1), 280–287.

39. (a) Stanger, A.; Ashkenazi, N.; Boese, R.; Stellberg, P. J. Evidence for metal induced bond localization in cyclobutabenzenes: the crystal and molecular structures of η^6-$Cr(CO)_3$ and η^4-$Fe(CO)_3$ complexes of cyclobutabenzene. *Organomet. Chem.* 1997, 542, 19. (b) Stanger, A.; Ashkenazi, N.; Boese, R.; Stellberg, P. J. Erratum to "Evidence for metal

induced bond localization in cyclobutabenzenes: the crystal and molecular structures of (6-Cr(CO)3 and (4-Fe(CO)3 complexes of cyclobutabenzene" [*J. Organomet. Chem.* 542 (1997) 21]. *Organomet. Chem.* 1997, 548, 113. (c) Stanger, A.; Ashkenazi, N.; Boese, R.; Stellberg, P. J.; Erratum. *Organomet. Chem.* 1998, 556, 249–250.

40. Frank, N. L.; Siegel, J. S. *Advances in Theoretically Interesting Molecules;* AI Press Inc: Greenwich (CT), 1995; Vol. 3, 209–260.
41. Maksić, Z. B.; Eckert-Maksić, M.; Mó, O.; Yáñez, M. Pauling's Legacy: Modern Modelling of the Chemical Bond, Theoretical and Computational Chemistry; Elsevier: Amsterdam, 1999; Vol. 6, p. 47.
42. Stanger, A.; Vollhardt, K. P. C. The origin of the symmetrical structure of benzene. Is the sigma or the pi frame responsible? An *ab initio* study of the effect of HCC bond angle distortion. *J. Org. Chem.* 1988, 53, 4889–4890.
43. Rappoport, Z.; Kobayashi, S.; Stanger, A.; Boese, R. Crystal structure of 1,2-diphenyl-5,7-di-*tert*-butylspiro[2.5]octa-1,4,7-trien-6-one, a possible model for diphenylvinylidene-phenonium ions. *J. Org. Chem.* 1999, 64, 4370–4375.
44. Kalinowski, H. O.; Berger, S.; Braun, S. *^{13}C-NMR-Spektroskopie;* Georg Thieme Verlag: Stuttgart, 1984; p. 370.
45. Patel A.; Mazzini F.; Netscher T.; Rosenau T. On a novel dimer of α-tocopherol. *Research letters in Organic Chemistry* 2008, article ID 742590, doi: 10.1155/2008/742590.
46. Rosenau, T.; Mereiter, K.; Jäger, C.; Schmid, P.; Kosma, P. Sulfonium ylides derived from 2-hydroxy-benzoquinones: crystal and molecular structure and their one-step conversion into Mannich bases by amine *N*-oxides. *Tetrahedron* 2004, 60(27), 5719–5723.
47. Patel, A.; Netscher, T.; Rosenau, T. Stabilization of *ortho*-quinone methides by a bis (sulfonium ylide) derived from 2,5-dihydroxy-[1,4]benzoquinone. *Tetrahedron Lett.* 2008, 49, 2442–2445.
48. Witkowski, S.; Markowska, A. Nitration of α-tocopherol acetate. *Pol. J. Chem.* 1996, 70 (5), 656–657.
49. Adelwöhrer, C.; Rosenau, T.; Kosma, P. Novel tocopheryl compounds Part 16: nitration of α-tocopheryl acetate—a mechanistic study. *Tetrahedron* 2003, 59(41), 8177–8182.
50. Nelan, D. R.; Robeson, C. D. The oxidation product from α-tocopherol and potassium ferricyanide and its reaction with ascorbic and hydrochloric acids. *J. Am. Chem. Soc.* 1962, 84, 2963–2965.
51. Skinner, W. A.; Parkhurst, R. M. Oxidation products of vitamin E and its model, 6-hydroxy-2,2,5,7 8-pentamethylchroman. VII. Trimer formed by alkaline ferricyanide oxidation. *J. Org. Chem.* 1964, 29, 3601–3603.
52. Skinner, W. A.; Parkhurst, R. M. Oxidation products of vitamin E and its model, 6-hydroxy-2,2,5,7 8-pentamethylchroman. VIII. Oxidation with benzoyl peroxide. *J. Org. Chem.* 1966, 31, 1248–1251.
53. Fales H. M.; Lloyd H. A.; ; Ferretti J. A.; Silverton J. V.; Davis D. G.; Kon H. J. Optical resolution of the α-tocopherol spiro dimer and demonstration of its fluxional nature. *J. Chem. Soc. Perkin Trans.* 1990, 2, 1005–1010.
54. Dixie, C. J.; Sutherland, I. O. Effects of transition-state geometry on the rates of a [3.3] sigmatropic rearrangement. *J. Chem. Soc., Chem. Commun.* 1972, 646–647.
55. Chauhan, M.; Dean, F. M.; Hindley, K.; Robinson, M. Phenoxylium ion from α-tocopherol spirodimer and its significance for the mechanism of oxidative coupling in phenols. *Chem. Commun.* 1971, 19, 1141–1143.

56. (a) Rosenau, T.; Potthast, A.; Hofinger, A.; Kosma, P. Calixarene-type macrocycles by oxidation of phenols related to vitamin E. *Angew. Chem. Int. Ed.* 2002, 41(7), 1171–1173. (b) Rosenau T.; Potthast A.; Ebner G.; Hofinger A.; Kosma P. Calixarenartige makrocyclen durch oxidation phenolischer vitamin-E-derivate. *Angew. Chem.* 2002, 114(7), 1219–1221.
57. Nagaoka, S.; Mukai, K.; Itoh, T.; Katsumata, S. Mechanism of antioxidant reaction of vitamin E. 2. Photoelectron spectroscopy and *ab initio* calculation. *J. Phys. Chem.* 1992, 96, 8184–8187.
58. Gregor, W.; Grabner, G.; Adelwöhrer, C.; Rosenau, T.; Gille, L. Antioxidant properties of natural and synthetic chromanol derivatives: study by fast kinetics and electron spin resonance spectroscopy. *J. Org. Chem.* 2005, 70(9), 3472–3483.
59. Rosenau, T.; Netscher, T.; Adelwöhrer, C.; Ebner, G. Facile synthesis of α,ω-bis(5-γ-tocopheryl)alkanes. *Synlett* 2005, 2, 243–246.
60. Rosenau, T.; Habicher, W. D. Novel tocopherol compounds I. Bromination of α-tocopherol — reaction mechanism and synthetic applications. *Tetrahedron* 1995, 51, 7919–7926.
61. Rosenau, T.; Habicher, W. D. Novel tocopherol compounds III. Reaction of 5a-bromo-α-tocopherol with nucleophiles. *J. Prakt. Chem.* 1996, 338, 647–653.
62. Rosenau, T.; Kosma, P.; "The tocopherol-acetaminophen reaction": a new [1 4]-rearrangement discovered in vitamin E chemistry. *Eur. J. Org. Chem.* 2001, 5, 947–955.
63. Rosenau, T.; Chen, C. L.; Habicher, W. D. A vitamin E derivative as a novel extremely advantageous amino-protecting group. *J. Org. Chem.* 1995, 60, 8120–8121.
64. Bodanszky, M.; Bodanszky, A. *The Practice of Peptide Synthesis*, 2nd edition; Springer Verlag: Berlin, New York, 1994.
65. Rosenau, T.; Habicher, W. D. Novel tocopherol compounds-X. A facile synthesis of *O*-trimethylsilyl-5a-halo-α-tocopherols. *Tetrahedron Lett.* 1997, 38, 5959–5960.
66. Dallacker, F.; Eisbach, R.; Holschbach, M. Derivatives of vitamin E series 1. Preparation and reaction of all-rac-5-formyl-γ-tocopherol. *Chem. Ztg.* 1991, 115(4), 113–116.
67. Rosenau, T.; Potthast, A.; Kosma, P.; Habicher, W. D. Novel tocopherol compounds XI. Synthesis bromination and oxidation reactions of 3-(5-tocopheryl)propionic acid. *Synlett* 1999, 3, 291–294.
68. Rosenau, T.; Habicher, W. D.; Chen, C. L. Novel tocopherol compounds IV. 5-Tocopherylacetic acid and its derivatives. *Heterocycles* 1996, 43, 787–798.
69. Rosenau, T.; Potthast, A.; Ebner, G.; Hofinger, A.; Kosma, P. On a novel chromanone--naphthalenetrione rearrangement related to vitamin E. *Org. Lett.* 2002, 4(8), 1257–1258.
70. Adelwöhrer, C.; Rosenau, T.; Gille, L.; Kosma, P. Synthesis of novel 3-oxa-chromanol type antioxidants. *Tetrahedron* 2003, 59(15), 2687–2691.
71. Rosenau T. PhD Thesis, Novel tocopherol derivatives. Dresden University of Technology, Dresden. 1997.
72. Rosenau, T.; Gruner, M.; Habicher, W. D. Novel tocopherol compounds VIII. Reaction mechanism of the formation of α-tocored. *Tetrahedron* 1997, 53, 3571–3576.
73. Patel A.; Böhmdorfer S.; Rosenau T. Unpublished results.

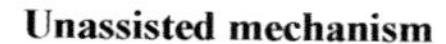

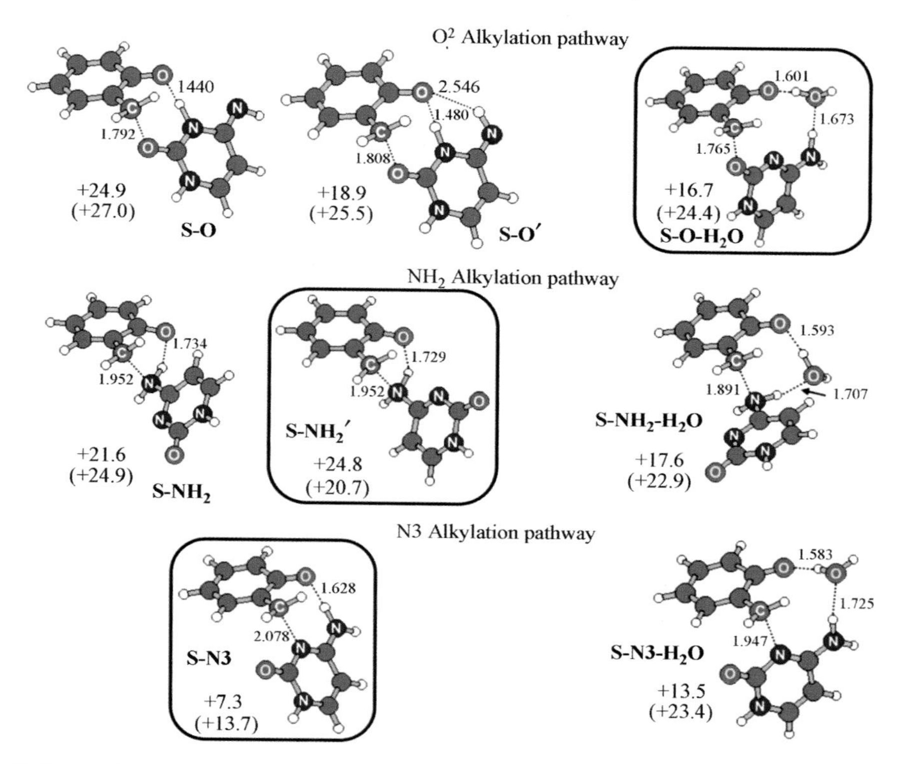

FIGURE 2.2 Optimized TS geometries of the cytosine alkylation reaction by *o*-QM, without and with water assistance. Bond lengths (in Å) and activation Gibbs free energies (in kcal/mol) in the gas phase and in aqueous solution (in parentheses) at the B3LYP/6-311 + G(d,p)// B3LYP/6-31G(d) level of theory with respect to the reactants have been taken from Ref. [14].

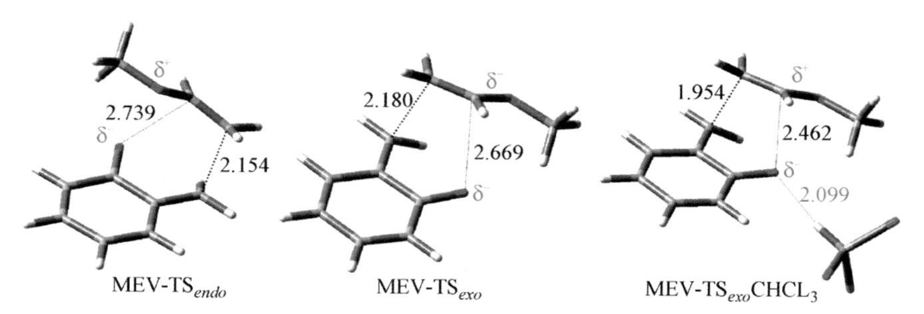

FIGURE 2.3 *Endo* and *exo* TS geometries of the Diels–Alder reaction between *o*-QM and **MVE**, also in the presence of an ancillary $CHCl_3$ molecule (**MEV-TS$_{exo}$CHCl$_3$**), at B3LYP/ 6-31G(d,p) level, in the gas phase. Forming bond lengths are in Å.

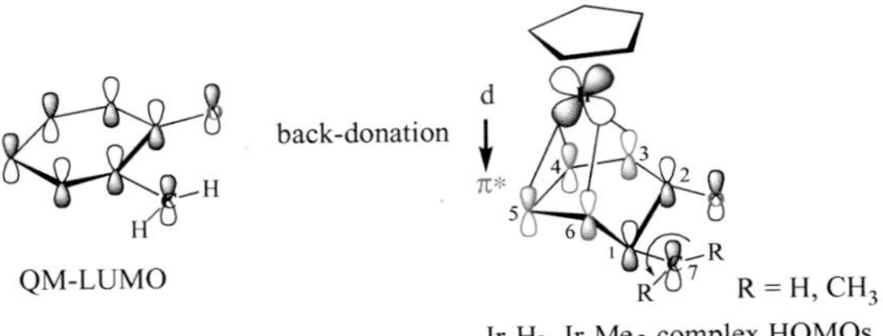

FIGURE 2.4 Schematic representation of the back donation from the metal to the QM-system, in the complexes **Ir-H$_2$** and **Ir-Me$_2$**.

SCHEME 2.14 Uncatalyzed, base-catalyzed, and reductive generation of **QMs** tethered to a naphthalene imide core, through the TSs **TS–NI**, **TS–NI$^-$**, and **TS–NI$^-$**, respectively [bond lengths are in Å; data in parentheses are for full R(U)B3LYP/6-31 + G(d,p) optimization in aqueous solution] (reproduced from Ref. [47] with permission from American Chemical Society).

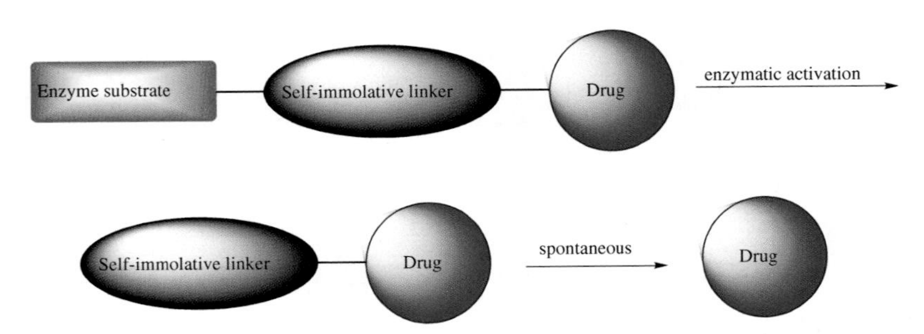

FIGURE 5.1 Graphical structure of a tripartite prodrug with self-immolative linker.

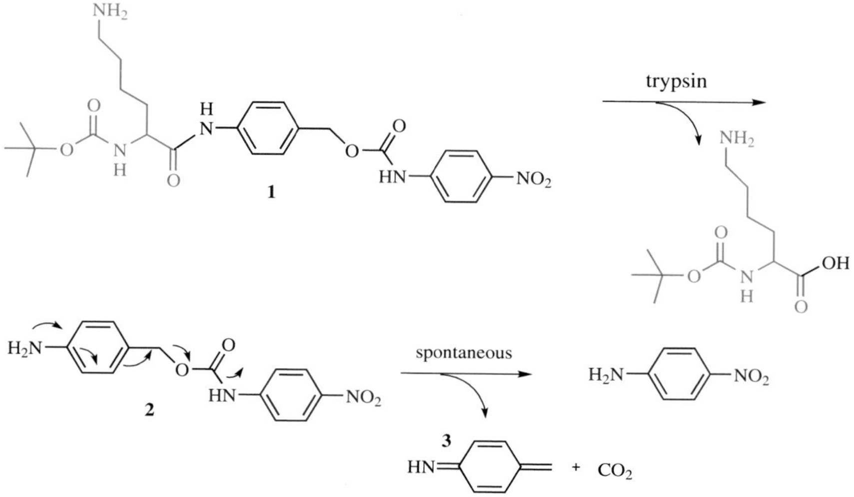

NH₂

H

O

O

N

O

N

H

O

O

O

HN

NO$_2$

1

trypsin

NH$_2$

O

O

N

H

OH

O

H$_2$N

O

O

HN

NO$_2$

2

spontaneous

H$_2$N

NO$_2$

3

HN

+ CO$_2$

FIGURE 5.2 Activation mechanism of a tripartite prodrug through enzymatic cleavage and elimination of an azaquinone methide.

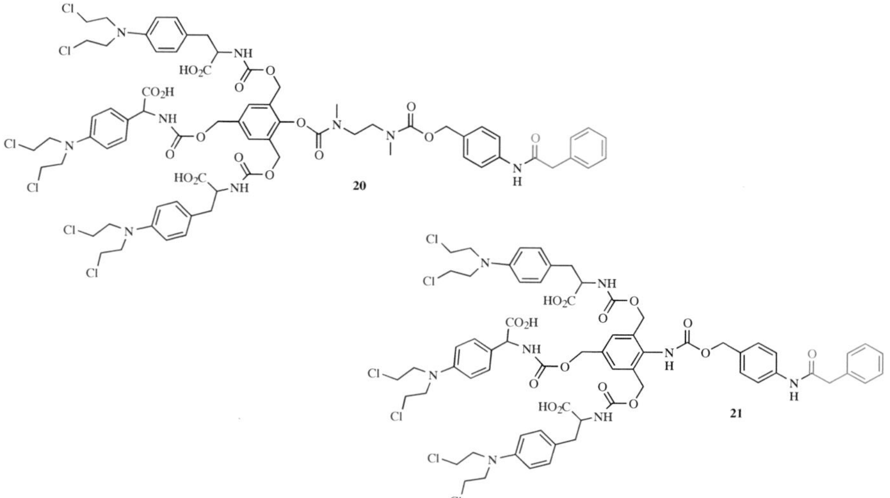

Cl

N

Cl

NH

HO$_2$C

O

O

CO$_2$H

NH

O

O

Cl

N

Cl

O

O

N

N

O

O

O

N

H

HO$_2$C

NH

O

O

20

Cl

N

Cl

Cl

N

Cl

NH

HO$_2$C

O

O

CO$_2$H

NH

O

O

O

N

H

O

O

N

H

O

Cl

N

Cl

HO$_2$C

NH

O

O

21

Cl

N

Cl

FIGURE 5.14 Chemical structures of AB_3 self-immolative dendritic prodrugs with melphalan tail units and a trigger that is activated by PGA.

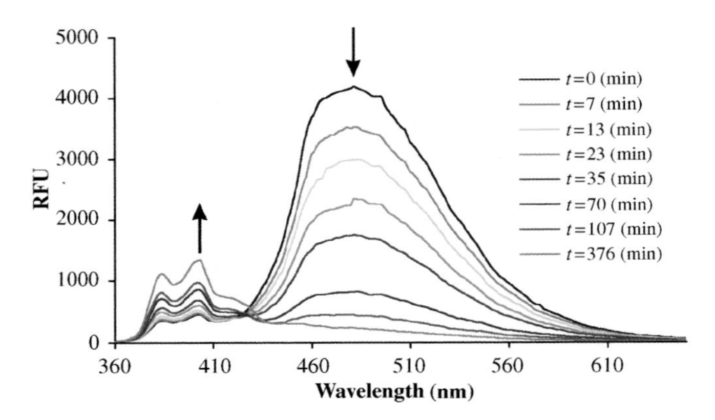

FIGURE 5.25 Emission fluorescence spectra ($l_{ex} = 340$ nm) of dendron **31** (50 mM) upon removal of the Boc trigger upon incubation in 1:1 MeOH/DMSO mixture with 2% aqueous solution of Bu_4NOH.

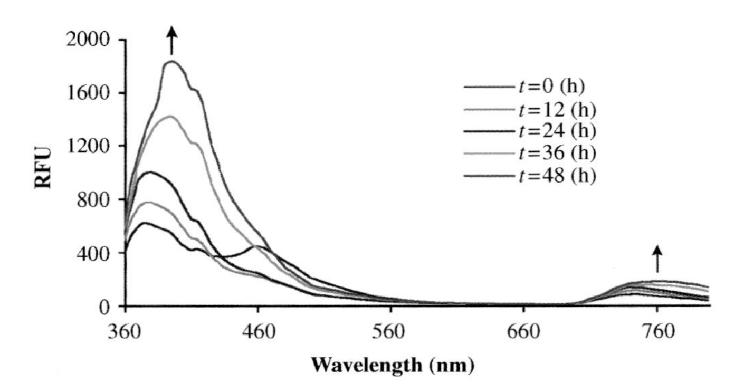

FIGURE 5.28 Emission fluorescence spectra ($l_{ex} = 260$ nm) of dendron **32** (12 mM) upon incubation in PBS, 37°C with PGA (0.1 mg/mL).

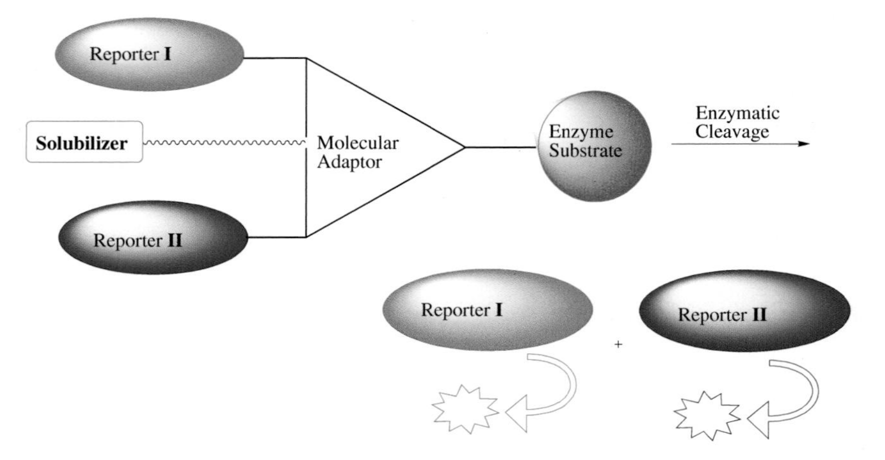

FIGURE 5.36 A graphical illustration of a molecular probe for detection of enzymatic activity with dual output.

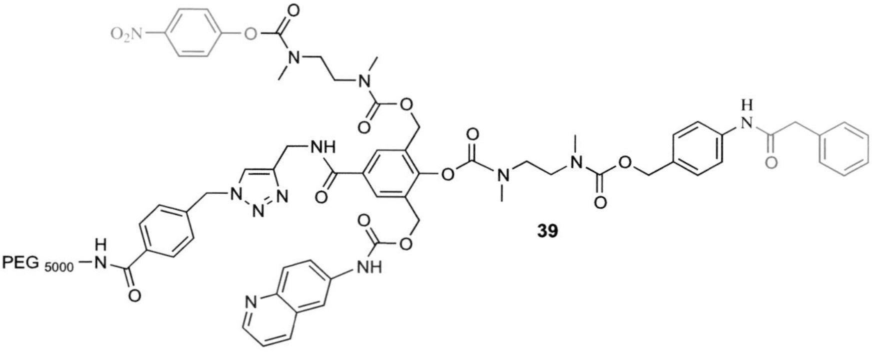

FIGURE 5.37 Chemical structure of a molecular probe with UV–Vis and fluorescence outputs for penicillin G amidase activity. The phenylacetamide group (red) is a substrate for PGA. The reporter units, 4-nitrophenol and 6-aminoquinoline, provide a visible signal and a fluorescence signal, respectively, upon release.

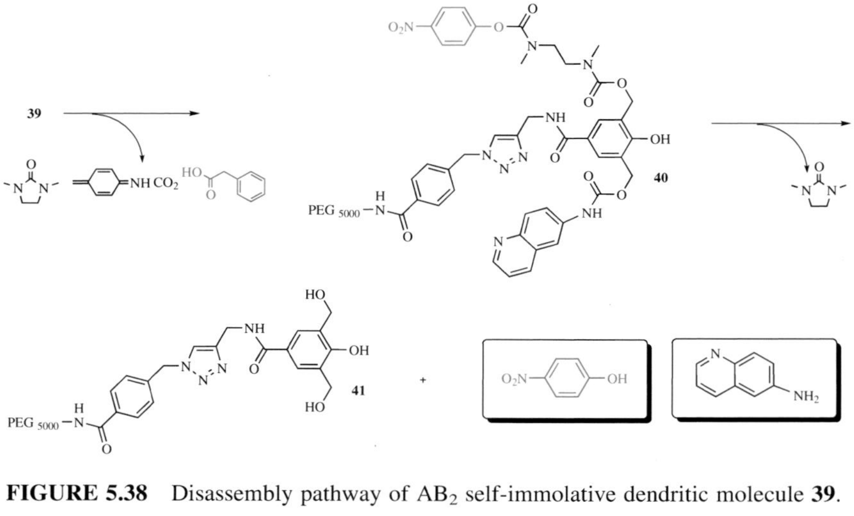

FIGURE 5.38 Disassembly pathway of AB_2 self-immolative dendritic molecule **39**.

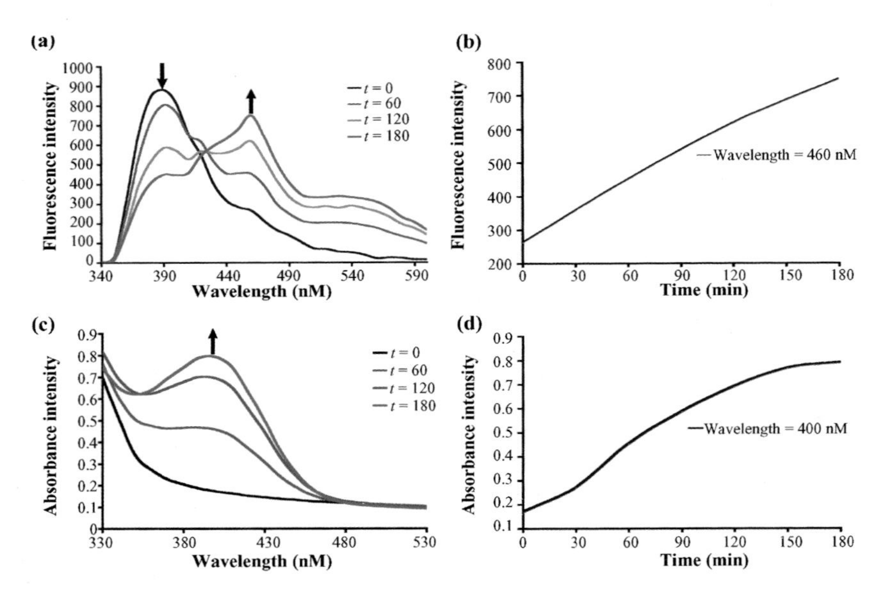

FIGURE 5.39 (a) Emission fluorescence ($l_{ex} = 250$ nm) of **39** (500 mM) upon addition of PGA (1.0 mg/mL). (b) Fluorescence of **39** at 460 nm in the presence of PGA (1.0 mg/mL) as a function of time. (c) UV–Vis absorbance spectra of **39** (500 mM) in the presence of PGA (1.0 mg/mL). (d) Absorbance of **39** at 400 nm after addition of PGA (1.0 mg/mL) as a function of time.

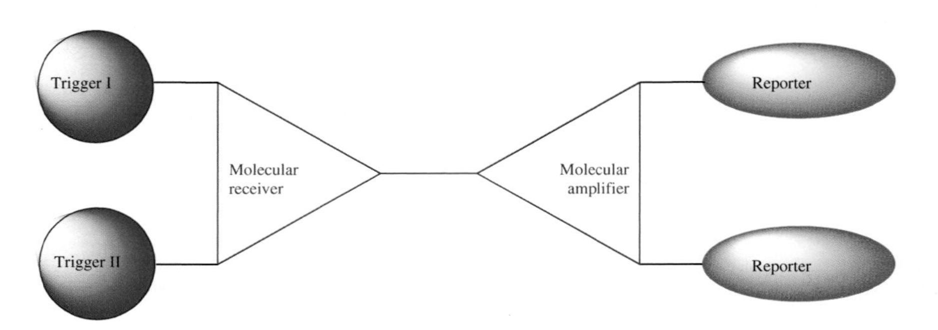

FIGURE 5.41 Graphical structure of a receiver–amplifier dendritic molecule.

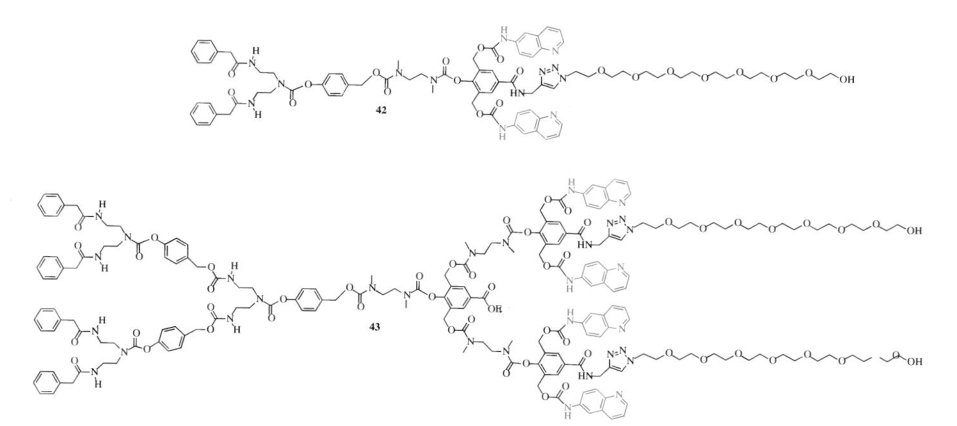

FIGURE 5.42 Chemical structures of first-generation (**42**) and second-generation (**43**) self-immolative, receiver–amplifier dendritic molecules with an enzymatic trigger (blue), cleaved by PGA and 6-aminoquinoline (red) reporter groups.

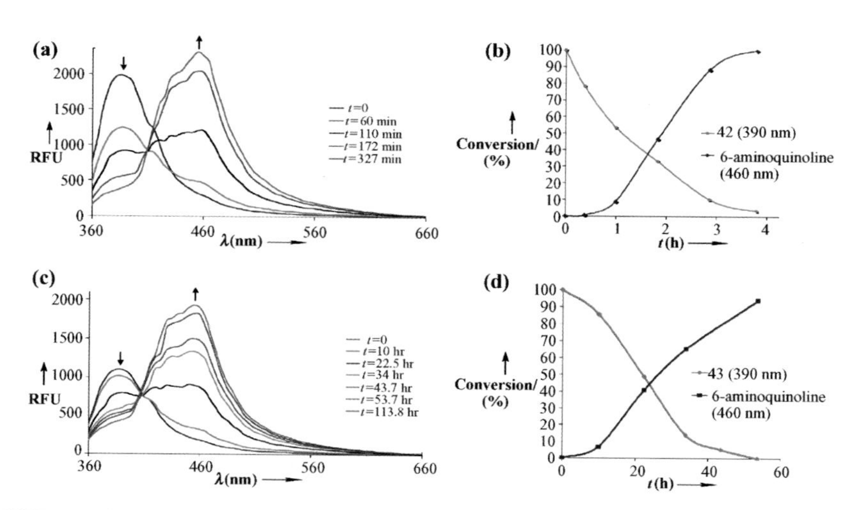

FIGURE 5.45 Emission fluorescence spectra ($\lambda_{ex} = 250$ nm) of **42** (a and b) and **43** (c and d) upon addition of PGA (0.1 mg/mL). The concentrations of dendritic molecules **42** and **43** were 25 and 12 mM, respectively.

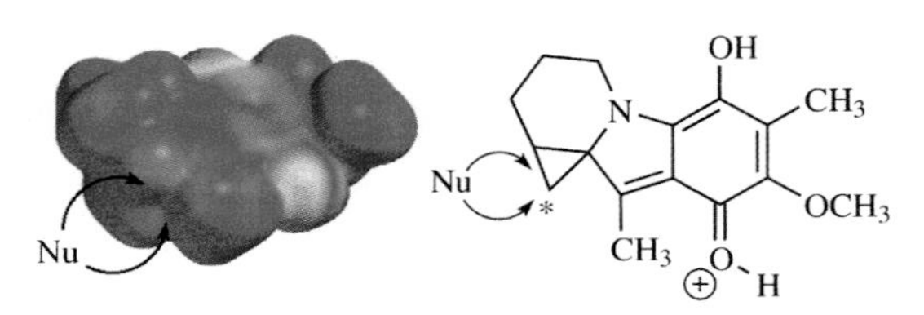

SCHEME 7.17 Electrostatic potential map of the protonated pyrido[1,2-*a*]indole-based cyclopropyl quinone methide. The two possible nucleophile-trapping paths with the respective products are shown.

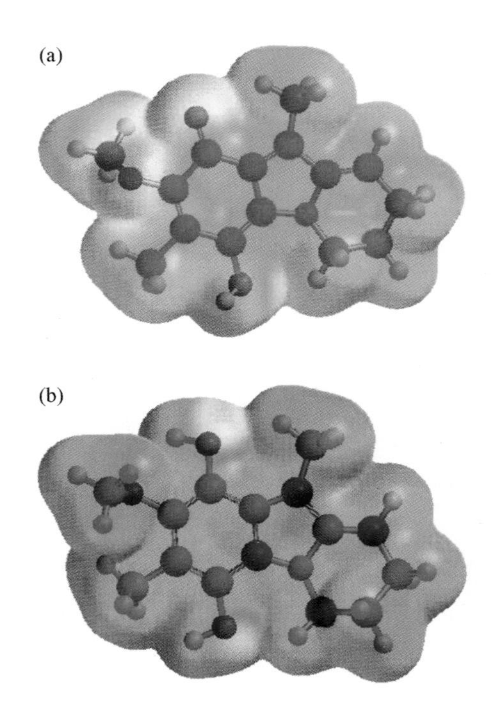

FIGURE 7.20 Potential–density maps for the *O*-protonated pyridoindole quinone methide and the corresponding neutral species. The charge density color codes are anionic (red), cationic (blue) and neutral (green).

7

CHARACTERIZING QUINONE METHIDES BY SPECTRAL GLOBAL FITTING AND ^{13}C LABELING

EDWARD B. SKIBO
Department of Chemistry and Biochemistry, Arizona State University, Tempe, AZ 85287-1604, USA

7.1 INTRODUCTION

Some naturally occurring quinones are functionalized to permit quinone methide formation upon two-electron reduction and leaving group elimination.[1,2] Since the reduction can occur via cellular enzymes and the quinone methide species can alkylate cellular nucleophiles, this process was termed "bioreductive alkylation" by Lin et al.[3] The clinically used antibiotic mitomycin C is the classic example of a bioreductive alkylating agent.[4–8] Scheme 7.1 shows that the two-electron reduction of mitomycin C causes aromatization to the indole (or mitosene) ring system, which then undergoes acid-catalyzed opening of the aziridine ring to afford the quinone methide species. The cytotoxic lesion of mitomycin C is the cross-linking of DNA at guanine bases resulting from quinone methide trapping of the 2-amino group followed by iminium ion trapping of another guanine 2-amino group.

The continuing interest in bioreductive alkylation is largely due to the clinical success of mitomycin C and the low reduction potentials observed in many tumors.[9] The low reduction potentials favor the quinone to hydroquinone conversion necessary for bioreductive alkylation. Hypoxia due to low blood flow[3] and/or the unusually high expression of the quinone two-electron reducing enzyme DT-diaphorase in some histological cancer types[10–14] contribute to the tumor's tendency to reduce quinones.

Quinone Methides, Edited by Steven E. Rokita

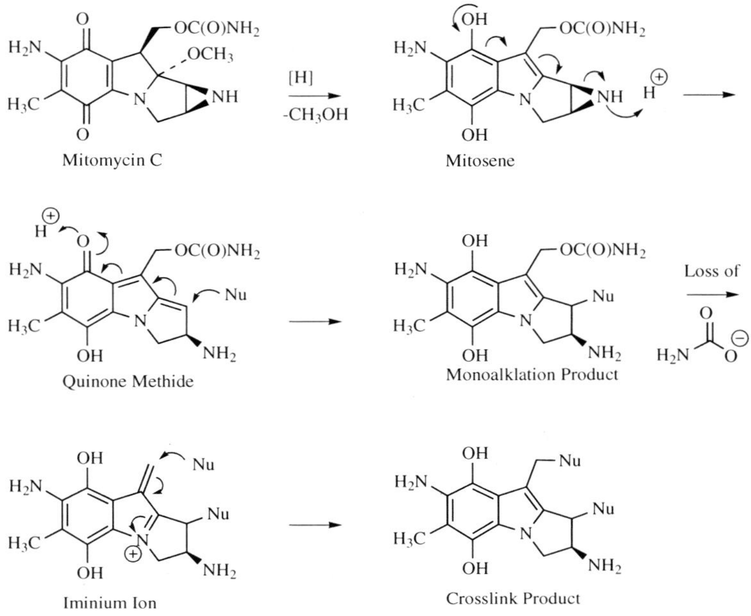

SCHEME 7.1 Mitomycin C reductive activation mechanism.

Selective bioreductive alkylation in high DT-diaphorase cancer types (melanoma renal and nonsmall-cell lung cancers)[15] would exhibit maximal antitumor activity with minimal side effects.

The discovery of the bioreductive alkylation pharmacophore prompted the design of relatively simple quinone systems that could afford a quinone methide intermediate upon reduction and leaving group elimination. In the 1970s, Lin et al. designed bioreductive alkylating agents based on benzoquinone, naphthoquinone, and other ring systems.[3,16–23]Scheme 7.2 shows examples of agents that could monoalkylate DNA or other cellular nucleophiles specifically in low potential tumor cells. Similarly, Lin et al. designed bioreductive cross-linkers by placement of two leaving groups on the quinone system. In the 1980s, this research group designed heterocyclic bioreductive alkylating agents that we considered purine ring mimics (Scheme 7.3). We postulated that these agents could affect purine metabolism specifically in low reduction potential tumor cells.[24–28]

For over 35 years, the quinone methide species has been invoked as a reactive intermediate in bioreductive alkylation and in other biological processes.[8,29] Generally, there is only circumstantial evidence that the quinone methide species forms in solution. Conceivably, the *O*-protonated quinone methide (i.e., the hydroquinone carbocation) could be the electrophilic species. If so, bioreductive alkylation may simply be an S_N1 reaction. Also, there are questions concerning the mechanism of quinone methide

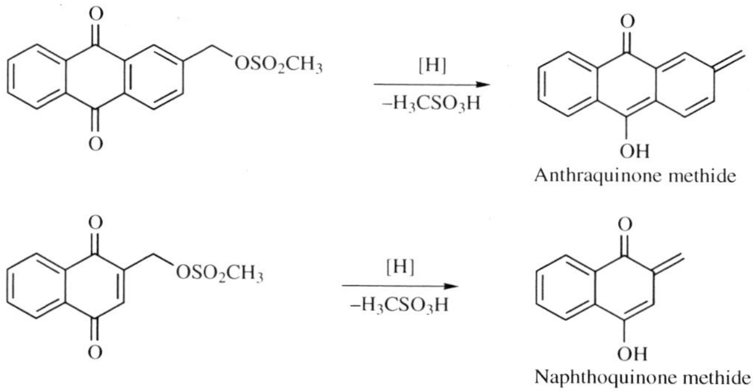

[H]
$-H_3CSO_3H$

Anthraquinone methide

[H]
$-H_3CSO_3H$

Naphthoquinone methide

SCHEME 7.2 Examples of potential bioreductive alkylating agents reported by Lin et al.[17].

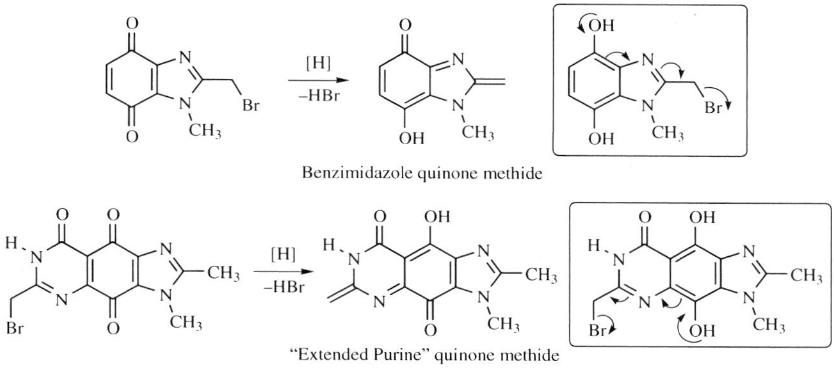

[H]
–HBr

Benzimidazole quinone methide

[H]
–HBr

"Extended Purine" quinone methide

SCHEME 7.3 Examples potential bioreductive alkylating purines reported by Skibo et al.[24,27,30,31].

reactions. What are the pH–rate laws for quinone methide reactions? Are there acid dissociations (pK_a values) involved? Finally, what are the products of quinone methide reactions? With a myriad of biological nucleophiles available, these are not easily answered question. This chapter outlines our efforts to answer these questions.

7.2 STUDYING THE TRANSIENT QUINONE METHIDE INTERMEDIATE

The bioreductive alkylating agents developed in this laboratory did not afford observable quinone methide species upon quinone reduction and leaving group elimination. Even repetitive scanning UV–Vis spectroscopy failed to detect a transient quinone methide species. We had to resort to kinetic and product studies to provide indirect evidence of the transient quinone methide species. These studies provided compelling evidence of quinone methide formation from benzimidazoles,[30,31] quinazoline,[25] imidazoquinazoline,[24,27] and mitomycin[32] based bioreductive alkylating agents. Lin and Sartorelli obtained evidence of quinone methide intermediates in bioreductive alkylating agents by trapping of these intermediates employing both Diels–Alder and nucleophile addition reactions.[23]

The quinone methide is not necessarily a transient nonobservable species. For example, the anthracycline antitumor agent daunomycin readily affords a spectrally observable quinone methide intermediate (Scheme 7.4) upon two-electron reduction.[33–35] In fact, stable terpene-based quinone methides have been isolated from a variety of natural sources, see Scheme 7.4 for examples.[36] These reports of observable quinone methide species suggested that even the so-called transient quinone methide species can be observed and studied. To do so, we would have to generate the quinone methide species rapidly such that spectral detection is possible before decomposition. An example of this approach is rapid generation of the quinone methide species by flash photolysis.[37,38] To study the transient quinone methide, we developed the spectral global fitting technique where the hydroquinone precursor of the quinone methide is rapidly formed via catalytic reduction while the UV–Vis spectrum of the reaction is repetitively scanned. In tandem with this technique, we employed ^{13}C labeling at the methide center and ^{13}C-NMR to verify the existence of the quinone methide and to identify its reaction products. We describe below both of these techniques for studying the transient quinone methide species.

Daunorubicin —[H]→ ... → Quinone methide

Examples of Quinone Methide Natural Products

Pristimerine Puupehenone

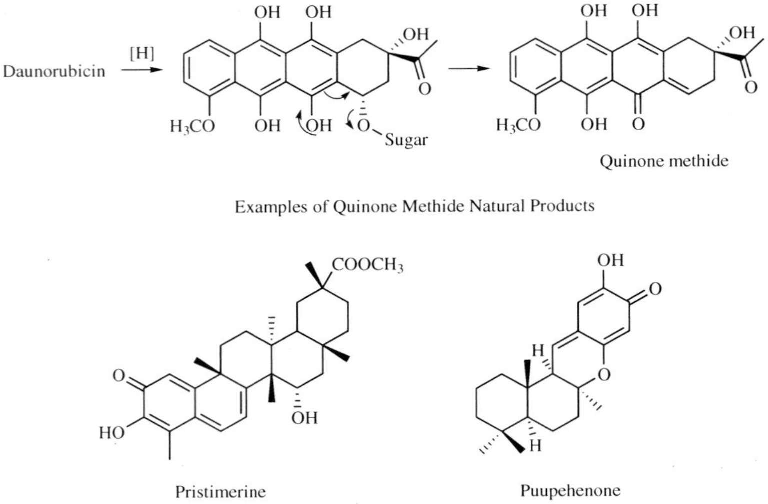

SCHEME 7.4 Stable quinone methide species.

7.2.1 Using Spectral Global Fitting to Study Transient Quinone Methides

Spectral global fitting is a technique that has been used to visualize and study transient photochemical intermediates.[39,40] The fitting process involves repetitive UV–Vis scanning of a reaction mixture over time resulting in a surface showing spectral changes associated with the reaction. This surface has the coordinates of absorbance versus time versus wavelength. Fitting the entire range of wavelengths (global fitting) to a rate law provides accurate rate constants and intermediate spectra associated with the reaction under study. We recently reported the use of spectral global fitting to study a transient intermediate involved in CC-1065 A-ring opening[41] and to study mitosene-like quinone methides.[42]

In order to generate the hydroquinone form of a bioreductive alkylating agent rapidly, we utilized a Thunberg cuvette with a DMSO stock of the agent in the top port and a buffer solution containing suspended 5% Pd on carbon in the bottom port (Fig. 7.1). The amount of catalyst added to the bottom port was typically 0.5 μg, which was pipetted from a well-sonicated aqueous stock solution. Air was then removed from the cuvette by purging with hydrogen. After sealing the cuvette, the reduction reaction is then initiated by mixing the ports. Repetitive UV–Vis scanning of the reaction with time affords an absorbance versus time versus wavelength surface.

We illustrate the spectral global fitting process with the catalytic reductive activation of prekinamycin (Scheme 7.5). The reduction reaction is fast and is complete in a single scan (~2 min) followed by the formation and disappearance of the red-colored quinone methide intermediate (Fig. 7.2). Global fitting of the entire surface to a three-exponential rate law ($\text{Abs} = A\,e^{-kt} + B\,e^{-k't} + C\,e^{-k''t} + D$) afforded the three rate constants associated with quinone methide formation from the hydroquinone (k_{obsd}), quinone methide disappearance (k'_{obsd}), and the slow precipitation of the finally divided catalyst (k''_{obsd}). In this rate law, Abs refers to the total absorbance at a given time and wavelength and constants A–D refers to the absorbance

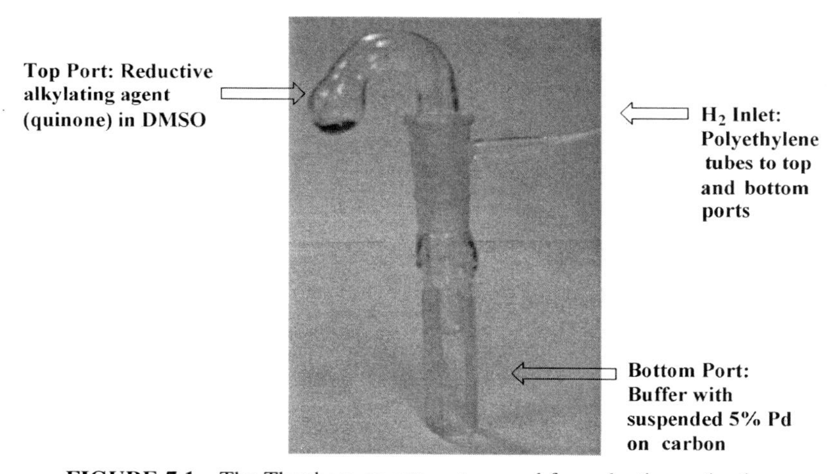

FIGURE 7.1 The Thunberg cuvette setup used for reductive activation.

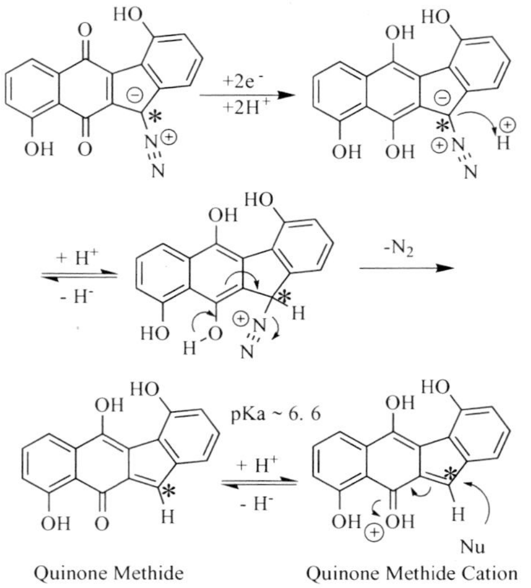

SCHEME 7.5 Generation of the prekinamycin quinone methide. The ^{13}C label is designated with an asterisk (*).

at this wavelength each reacting species. In all reactions, the relative magnitude of rate constants was quinone methide formation > quinone methide disappearance ≫ catalyst precipitation. We should point out that fits to three-exponential rate laws could provide solutions that are not unique. However, fitting each wavelength from 240 to 600 nm to the same rate law will provide a unique solution. These rate constants, obtained over a range of pH values, provide the pH–rate laws for quinone methide formation and fate as well as any pK_a values. Furthermore, the values of constants A–D as a function of pH will provide the UV–Vis spectrum of the transient quinone methide. In Section 7.3, we illustrate the utility of spectral global fitting in studying the prekinamycins and other bioreductive alkylating agents.

7.2.2 Enriched ^{13}C NMR Spectroscopy

The UV–Vis spectral detection of an intermediate in the catalytic reductive alkylation reaction provides only circumstantial evidence of the quinone methide species. If the bioreductive alkylating agent has a ^{13}C label at the methide center, then a ^{13}C-NMR could provide chemical shift evidence of the methide intermediate. Although this concept is simple, the synthesis of such ^{13}C-labeled materials may not be trivial. We carried out the synthesis of the ^{13}C-labeled prekinamycin shown in Scheme 7.5 and prepared its quinone methide by catalytic reduction in an N_2 glove box. An enriched ^{13}C-NMR spectrum of this reaction mixture was obtained within 100 min of the catalytic reduction (the time of the peak intermediate concentration in Fig. 7.2). This spectrum clearly shows the chemical shift associated with the quinone methide along with those of decomposition products (Fig. 7.3).

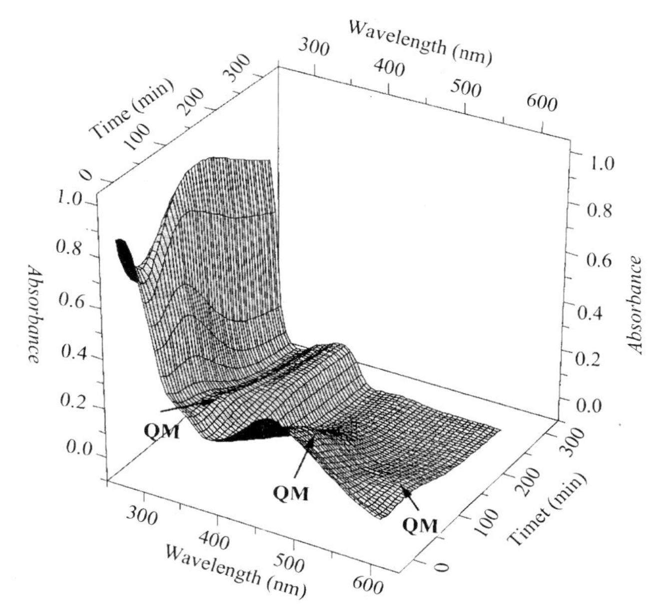

FIGURE 7.2 Absorbance versus time versus wavelength surface for the reaction sequence in Scheme 7.5 carried out in anaerobic pH 7.5 phosphate buffer in the presence of H_2 and 5% Pd on charcoal.

The chemistry of reactive species is often complex because various decomposition paths afford a multitude of products. However, substitution of reactive carbon centers with a single ^{13}C will permit a rapid assessment of the number and type of products formed in such reactions. Although ^{13}C labeling has been used to investigate biological and chemical processes in this way,[43] studies of DNA alkylation were limited only to the ^{13}C-labeled methylating agent *N*-nitrosourea.[44] In this laboratory, it has been possible to prepare ^{13}C-labeled reactive species and gain insights into the chemistry of mitosenes,[45] reactive cyclopropane species,[46,47] and aziridinyl

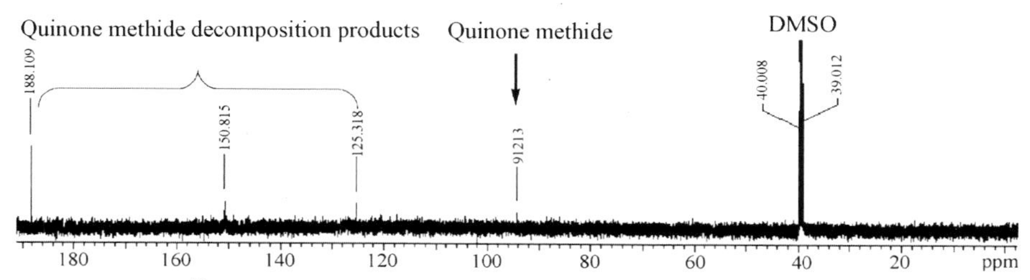

FIGURE 7.3 ^{13}C-NMR of the product mixture from the reaction shown in Scheme 7.5 carried out in anaerobic pH 7.5 phosphate buffer in the presence of H_2 and 5% Pd on charcoal.

indoloquinones.[48] The results of our ^{13}C and spectral global fitting studies are summarized in the following section.

7.3 NEW INSIGHTS INTO METHIDE CHEMISTRY

Our studies of quinone methides and related species using ^{13}C labeling and spectral global fitting started in the late 1990s and have continued to the present with two papers on prekinamycins slated for publication in 2009. In this section, we summarize what we learned from both our published and unpublished studies.

7.3.1 Novel Methide Polymerization Reactions

The fate of a quinone methide intermediate often involves the formation of polymeric products along with the expected nucleophile and proton-trapping reactions. A case in point is the reductive alkylation reaction of the mitosene-like antitumor agent WV-15 with guanosine to afford the 2-amino adduct shown in Scheme 7.6. Maliepaard et al. carried out this reaction in conjunction with an investigation of WV-15 cross-linking adducts with DNA.[49] We carried out the same reaction employing ^{13}C-labeled WV-15

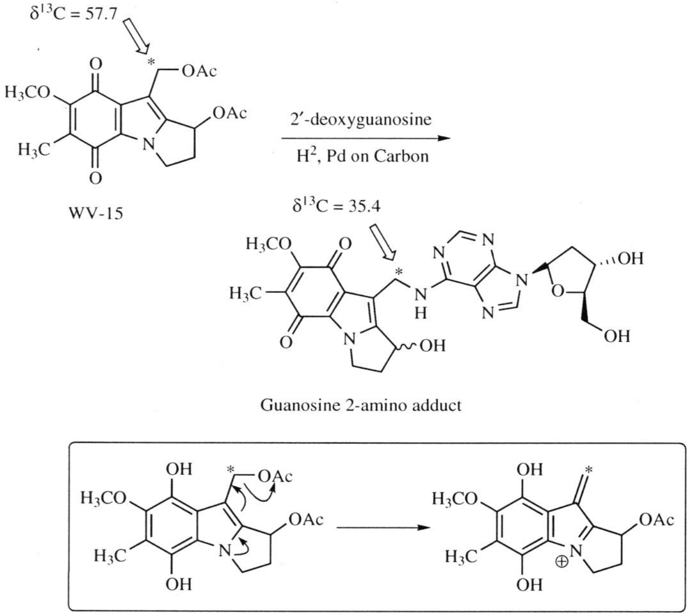

$\delta^{13}C = 57.7$

WV-15

2′-deoxyguanosine

H^2, Pd on Carbon

$\delta^{13}C = 35.4$

Guanosine 2-amino adduct

SCHEME 7.6 Catalytic reductive alkylation of guanosine with ^{13}C at the C-10 position of WV-15. The inset shows generation of the iminium methide upon reduction.

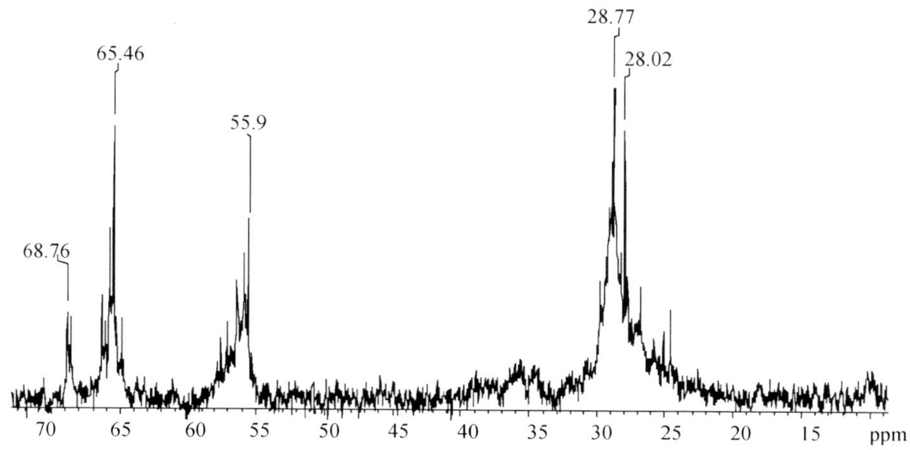

FIGURE 7.4 ^{13}C-NMR spectrum of the crude product mixture from the reaction shown in Scheme 7.6. Each chemical shift represents the C-10 center of a reaction product.

to assess the other reactions of the iminium methide intermediate shown in the inset of Scheme 7.6.[45]

We succeeded in isolating the same adduct reported by Maliepaard et al. in trace quantities, but a ^{13}C-NMR spectrum of the crude reaction mixture revealed the presence of hundreds of additional products (Fig. 7.4). The chemical shift of the guanosine 2-amino adduct (35.4 ppm) is not obvious in this spectrum. The bands of chemical shifts centered at 65 and 55 ppm correspond to structures with oxygen attachment to the C-10 center, but the band of products centered at 28 ppm represented unknown structures.

To simplify product studies, we subjected an analogue of WV-15 with a single acetate-leaving group located at the 10-position to reductive activation (Scheme 7.7).[45] There were still polymeric products formed in this reaction mixture, from which a dimer and pentamer were isolated and identified. In addition, other multimeric species structurally related to these products were present in the reaction mixture. The ^{13}C-NMR spectrum of the pentamer clearly shows the four ^{13}C-labeled methylene linkages, with chemical shifts in the 27–29 ppm range, and a ^{13}C-labeled methyl terminus (Fig. 7.5). We invoked the head-to-tail-coupling mechanism shown in Scheme 7.8 to explain the formation of these species. Likewise, we attributed the formation of products centered at 28 ppm upon WV-15 reductive activation (Fig. 7.4) to head-to-tail coupling.

Illustrated in Scheme 7.8 are the mechanisms that give rise to the products shown in Scheme 7.7. These mechanisms involve either electrophilic attack or an internal redox reaction. The internal redox reaction shown in Scheme 7.8 involves proton trapping from the solvent or from the hydroquinone hydroxyl group as shown. This process has been documented for the mitomycin system[50] and also occurs in many quinone methide systems.[25,30,31]

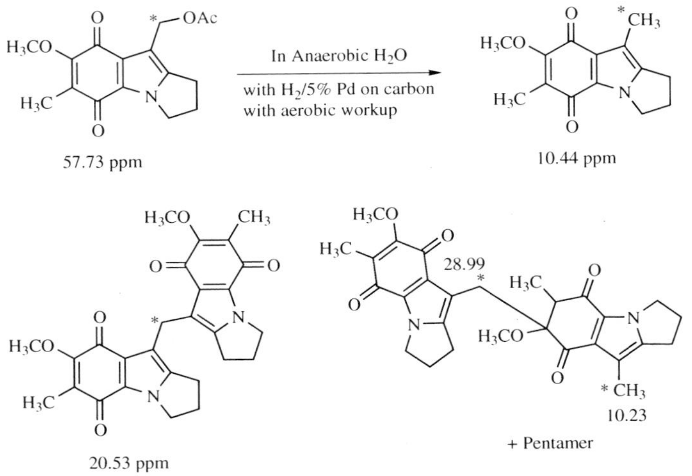

In Anaerobic H_2O
with H_2/5% Pd on carbon
with aerobic workup

57.73 ppm
10.44 ppm
20.53 ppm
28.99
10.23
+ Pentamer

SCHEME 7.7 Products of reductive activation of a simplified WV-15 analogue. The ^{13}C labels are designated with asterisks (*).

Alkylation reactions by the iminium methide species are well known in the mitomycin and mitosene literature [4,49,51–53] and are largely responsible for the cytotoxicity/antitumor activity of these compounds. As illustrated in Scheme 7.8, the electron-rich hydroquinone intermediate can also be attacked by the iminium ion resulting in either head-to-head or head-to-tail coupling. The head-to-head coupling illustrated in Scheme 7.8 is followed by a loss of formaldehyde to afford the coupled hydroquinone species that oxidizes to the head-to-head dimer upon aerobic workup. Analogous dimerization processes have been documented in the indole literature, [54–56] while the head-to-tail mechanism is unreported. In order to

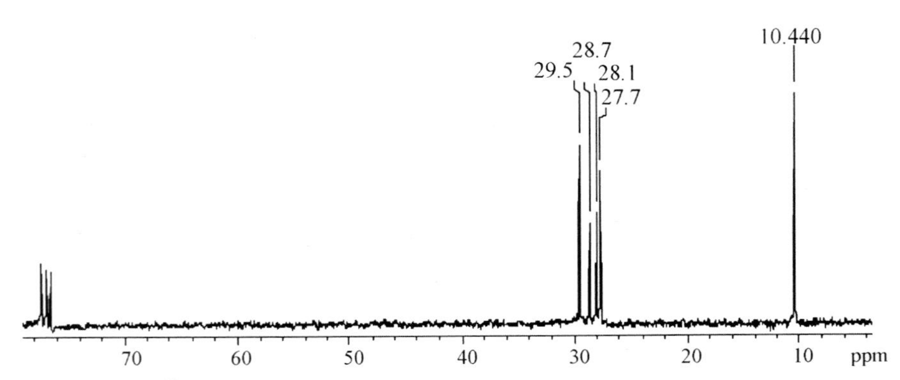

FIGURE 7.5 ^{13}C-NMR spectrum of the pentamer showing four methylene linkages and a methyl terminus.

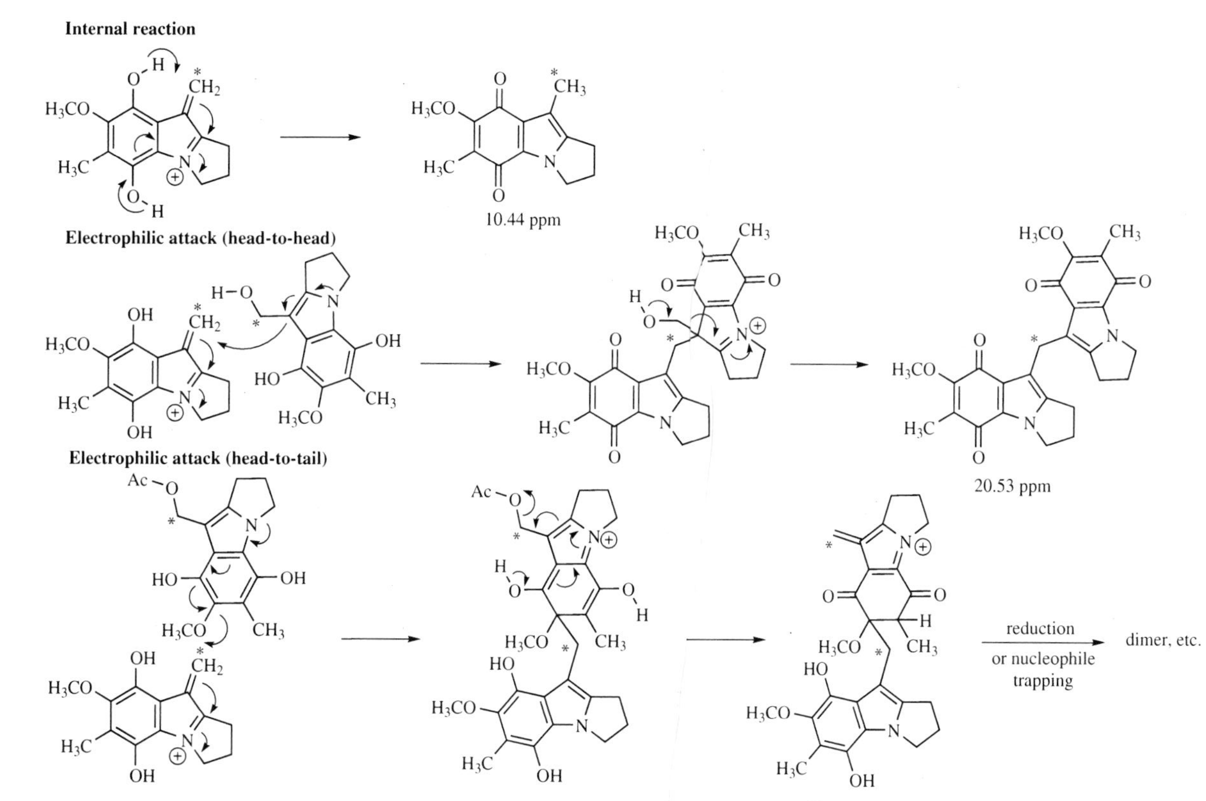

SCHEME 7.8 Mechanisms of formation for products shown in Scheme 7.7. The ^{13}C labels are designated with asterisks (*).

explain the observed head-to-tail dimer, electrophilic attack must occur at the 7-position followed by tautomerization steps and finally reduction of the iminium methide with reducing equivalents originating from the hydroquinone ring. This iminium methide could also trap water to afford a 10-hydroxyl substituent. The mass balance of the reaction shown in Scheme 7.7 was only ~20% indicating the presence of other reaction products. Indeed, a ^{13}C-NMR of the crude reaction mixture showed a multitude of products; an envelope of resonances was present between 25 and 30 ppm indicating a range of head-to-tail polymeric species.

The results presented above indicate that the previously unknown head-to-tail polymerization is the major reaction product of the iminium methide species. To investigate the generality of this reaction, we next studied a neutral "ene-imine" species shown in Scheme 7.9.[48] As illustrated in this scheme, the generation of this reactive species requires quinone reduction followed by elimination of acetic acid. The ene-imine is structurally related to the methyleneindolenine reactive species that is a metabolic oxidation product of 3-methylindole (Scheme 7.9).[57–59]

The toxicity of 3-methylindole has been attributed to methyleneindolenine trapping of nitrogen and sulfur nucleophiles.[57,60–62] Likewise, the ene-imine shown in Scheme 7.9 readily reacted with hydroquinone nucleophiles, resulting in head-to-tail products. Shown in Fig. 7.6 is the ^{13}C-NMR spectrum of a ^{13}C-labeled ene-imine generated by reductive activation. The presence of the methylene center of the ene-imine is apparent at 98 ppm, along with starting material at 58 ppm and an internal redox reaction product at 18 ppm. Thus, the reactive ene-imine actually builds up in solution and can be used as a synthetic reagent.

Reductive activation of the quinone shown in Scheme 7.9 and incubation in methanol afforded a complex mixture of products consisting mainly of head-to-tail coupling at C-5 or C-7 (Scheme 7.10). Minor reactions involve transfer of H_2 from the hydroquinone to the ene-imine (internal redox reaction) and methanol trapping. The structures of the dimers and trimers in Scheme 7.10 were derived from ^{1}H-NMR,

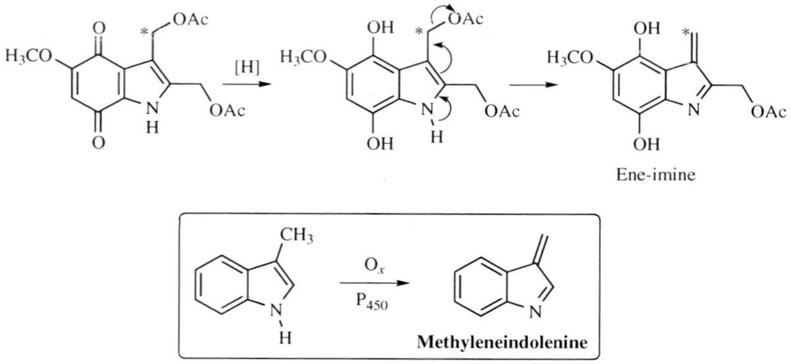

SCHEME 7.9 Ene-imine formation by reductive activation and by metabolic activation. The ^{13}C labels are designated with asterisks (*).

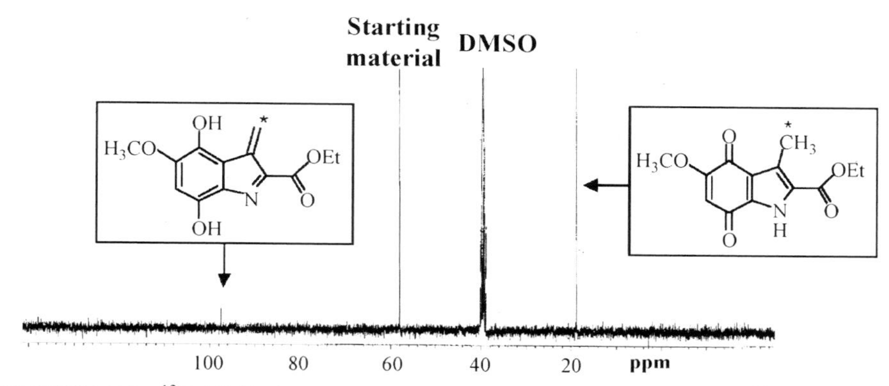

FIGURE 7.6 ^{13}C-NMR documentation of ene-imine intermediate. The ^{13}C labels are designated with asterisks (*).

^{13}C-NMR, COSY, HMQC (heteronuclear multiple quantum coherence), and HMBC (heteronuclear multiple bond correlation).[48] Furthermore, the structure of trimer was confirmed by X-ray crystallography.[48] The incorporation of ^{13}C into the indole 3α position proved valuable in these structural determinations and in documenting the ene-imine intermediate. For example, the presence of a trimer was readily determined from its ^{13}C-NMR spectrum (Fig. 7.7).

The head-to-tail-coupling reactions described above are potentially useful in the design of dynamic combinatorial libraries. Features of these reactions include the rapid and reversible formation of carbon–carbon bonds, multifunctional ene-imine building blocks, and formation of stereo centers upon ene-imine linkage. Support for template-directed synthesis utilizing ene-imine building blocks is the formation of a poly ene-imine species that could recognize 3′-GGA-5′ sequences of DNA.[48] It is noteworthy that some polyene-imines are helical and could form a "triple helix" with DNA.

7.3.2 Products of Dithionite Reductive Activation

Dithionite-mediated reductive activation of mitomycin C has been employed in the study of its DNA alkylation chemistry.[6,63] However, dithionite activated mitomycin C possesses different DNA alkylation properties than that activated by catalytic hydrogenation and enzymatic reduction. We postulated that a new alkylating species is produced by dithionite reductive activation resulting in different reactivity than the iminium methide species. To investigate dithionite-mediated reductive activation further, we treated ^{13}C-labeled analogues of WV-15 with dithionite and carried out spectral and product studies.

Typically, the dithionite species disproportionates in aqueous media to afford the hydrogen sulfite and thiosulfonate nucleophiles.[64] This finding suggests that sulfite esters ($-OSO_2^-$) and Bunte salts ($-SSO_3^-$)[65] could be formed upon iminium methide

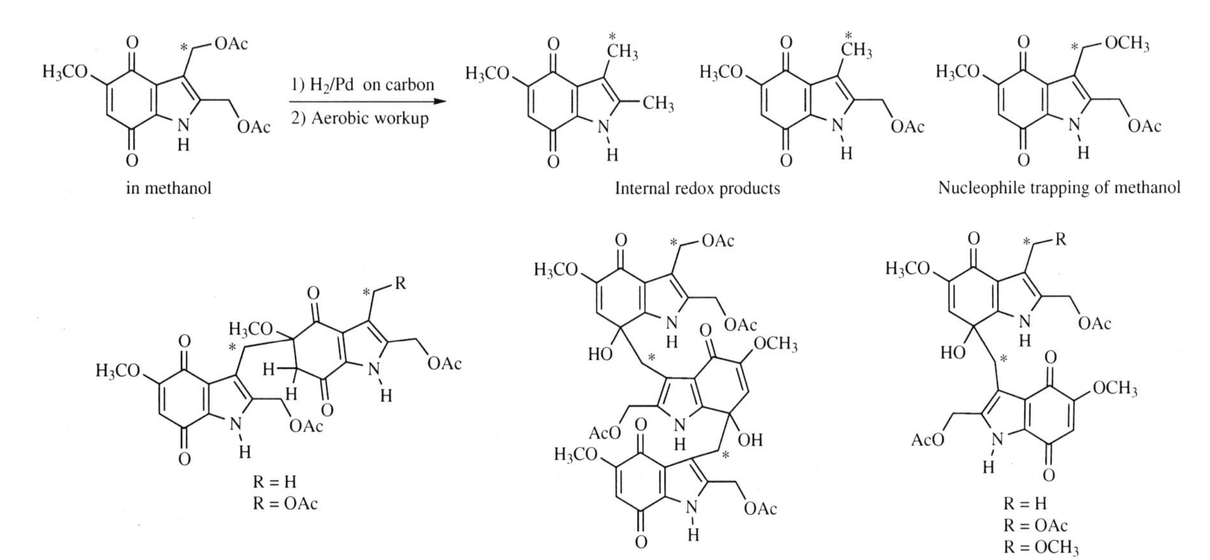

SCHEME 7.10 Fate of the ene-imine reactive species. The ^{13}C labels are designated with asterisks (*).

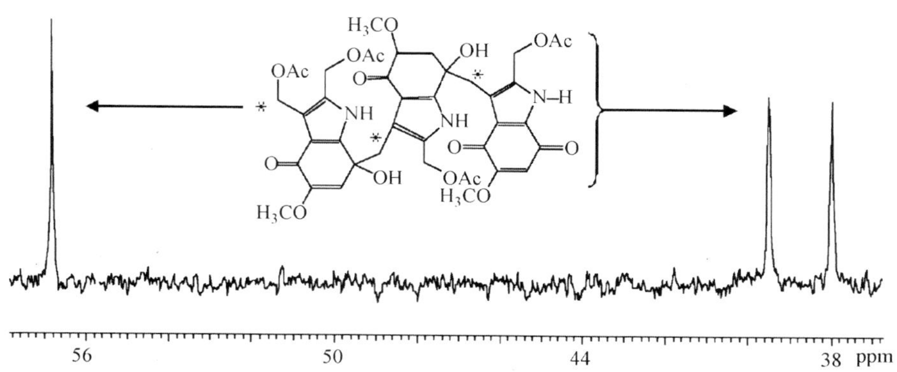

FIGURE 7.7 ^{13}C-NMR spectrum of an ene-imine trimer. The ^{13}C labels are designated with asterisks (*).

trapping of these nucleophiles. Under high-concentration preparative conditions, even sulfite diesters (–OSO_2–) could be present in these reactions. Finally, the anionic sulfite monoester (–OSO_2^-) can rearrange to the sulfonate (–SO_3^-) residue. Since many of these products may only be observed in solution, approximate ^{13}C shifts were obtained from the literature. A sulfite ester possesses a ^{13}C chemical shift of 58 ppm for the carbon attached to oxygen while alkylsulfonates possess ^{13}C chemical shifts from 41 ppm to as high as 49 ppm.[66] In fact, mitosene sulfonates possess a C-10α-^{13}C chemical shift of ~47 ppm.[67,68] A reference for the ^{13}C chemical shift of the Bunte salt is not available, but a calculation indicates that the ^{13}C chemical shift should be ~30 ppm.

Scheme 7.11 shows the product structures resulting from the dithionite reduction of a simplified version of WV-15. The symmetric sulfite diester was extracted from the reaction mixture with methylene chloride. The isolation and characterization of the sulfite diester confirmed that this species can form in dithionite reductive activation reactions and provided the chemical shift for the 10α-^{13}C center of a mitosene sulfite ester (49.37 ppm). The aqueous fraction of the reaction contained the mitosene sulfonate and trace amounts of Bunte salt, based on their ^{13}C chemical shifts.

Dithionite reduction of ^{13}C-labeled WV-15 (structure in Scheme 7.6) was carried out in anaerobic D_2O and the resulting products evaluated by *quantitative* ^{13}C-NMR spectroscopy immediately after the reduction (Fig. 7.8).[45] The spectrum indicates that both the sulfite ester and the sulfonate are formed in a 60:40 ratio. Also, no long-lived methide species are observed in this reaction.

The above studies show that dithionite reduction of mitosenes results in the formation of 10α-sulfite esters as well as sulfonates. The presence of the excellent sulfite-leaving group at the mitosene 10α-position suggests that alkylation reactions at this position could still occur. The subsequent sulfite to sulfonate rearrangement results in loss of alkylation capability by this position.

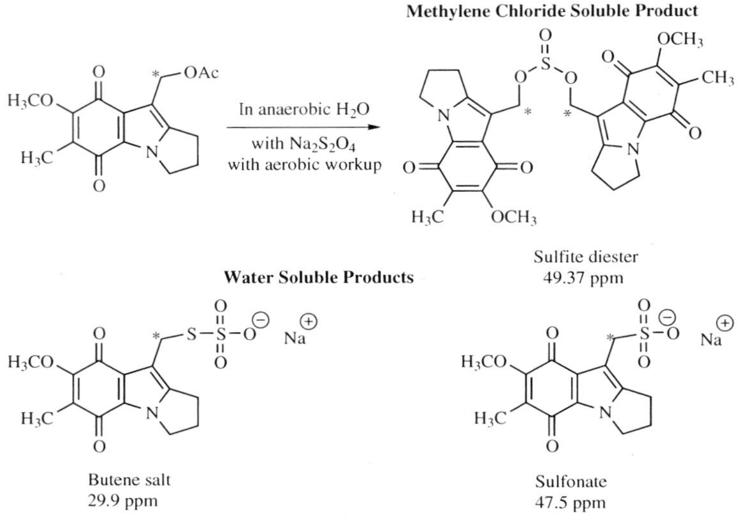

Methylene Chloride Soluble Product

In anaerobic H_2O

with $Na_2S_2O_4$
with aerobic workup

Sulfite diester
49.37 ppm

Water Soluble Products

Butene salt
29.9 ppm

Sulfonate
47.5 ppm

SCHEME 7.11 Dithionite reduction of a mitosene bearing an acetate-leaving group at the 10α position (a simplified WV-15 analogue). The ^{13}C labels are designated with asterisks (*).

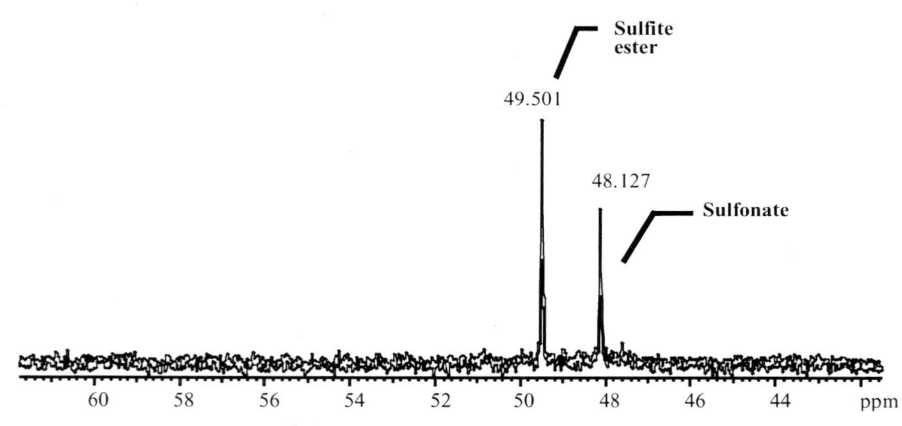

FIGURE 7.8 Quantitative ^{13}C-NMR (peak areas proportional to concentration) of dithionite-reduced 10α-^{13}C-labeled WV-15 (structure in Scheme 7.6).

7.3.3 Probing DNA Adduct Structures with ^{13}C-Labeled Methides

Many of the contributors to this volume have addressed the reactions of quinone methides with DNA nucleophiles. The ^{13}C-labeled methide center has the potential of identifying the type and number of such adducts using ^{13}C-NMR. An obvious drawback to this approach is the possible dilution of the ^{13}C label to natural abundance levels. We are currently developing a new technique utilizing the unnatural ^{13}C double label to study labeled adducts at high dilution. Ongoing work in this area is summarized below.

The mitosene WV-15 has been shown to form DNA adducts that are linked to the 2-amino group of a guanine base at the 10α center, see Scheme 7.6.[49] We employed DNA hexamers d(AAATTT) and d(GGGCCC) and 10α-^{13}C-labeled WV-15 to inventory the DNA adducts.[45] These self-complementary hexamers permitted an assessment of WV-15 reductive alkylation of AT versus GC base pairs and also kept the number of unique adducts low. The reductive alkylation reactions were carried out in anaerobic aqueous buffer at 25 °C, which is nearly equal to the hexamer melting point. Thus, there was duplex hexamer present in solution during alkylation reactions ensuring the interaction of reduced WV-15 with either DNA groove. Reduced WV-15 does not react with nonduplex DNAs such as 2-mers. The alkylated DNAs were separated from the WV-15 hydrolysis and polymerization products by two precipitations and then dried to afford an orange pellet. The quinone chromophore ($\lambda_{max} = 450$ nm, $\varepsilon = 1100\ M^{-1}\ cm^{-1}$) permitted a determination of the percent alkylation by weight: 2% for d(AAATTT) and 2.4% for d(GGGCCC).

Figure 7.9 shows the ^{13}C spectrum of the WV-15 reductively alkylated d(GGGCCC). The reductively alkylated d(AAATTT) spectrum is similar to the spectrum in Fig. 7.9 and is not shown. Both spectra show the iminium methide trapping of nitrogen and oxygen, along with a mystery adduct at δ 85 ppm. Possible nitrogen centers include the 2-amino group of guanine, the 6-amino of adenine, and the N(7) centers. Significantly, studies have confirmed that the 6-amino of adenine is reductively alkylated by mitomycin C.[69] The ^{13}C-alkylated N(7) center has a chemical shift similar to those of alkylated amines[43] and an assessment cannot be made as to which type of nitrogen is alkylated. The reductive alkylation of oxygen by WV-15 has not been documented; phosphate and guanine *O*(6) alkylations are likely. These are reasonable choices since

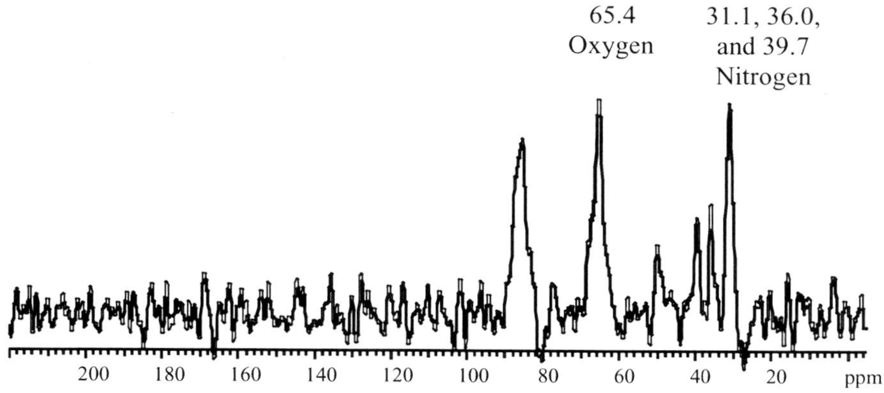

FIGURE 7.9 ^{13}C-NMR spectrum of 10α-^{13}C-labeled WV-15 reductively alkylated d (GGGCCC).

quinone methides can alkylate phosphate oxygens,[70] while cationic alkylating agents are known to alkylate phosphate oxygens and guanine *O*(6) centers.[44] The resonances at δ85 can be explained by both nitrogen and oxygen substitution at the electrophilic center.

The results of this study indicate that the iminium methide derived from reduced WV-15 is not a selective alkylating agent of DNA as originally proposed.[49] Significantly, the presence of oxygen adducts could not have been determined with the wet methods used in this original study because hydrolysis of the oxygen adducts would have prevented their isolation. Thus, ^{13}C labeling and NMR studies provide a reliable means of inventorying and identifying DNA adducts, even if they are not stable.

The ^{13}C labeling of methide centers has also proven useful in characterizing a new DNA-cleaving agent developed in this laboratory.[48,71] This agent affords an intrastrand cross-link by alkylating the guanine N(7) position and the phosphate backbone (Scheme 7.12). PAGE studies revealed the presence of a Maxam and Gilbert G ladder and a hydrolytic ladder due to presence of these respective alkylation sites. Although the PAGE results were convincing, we wished to inventory all the reactions of the ene-imine electrophile with DNA. In order to determine the DNA alkylation site, we treated hexamer DNA with the sequence d(ATGCAT) with two-electron reduced cleaving agent. The alkylated DNA was

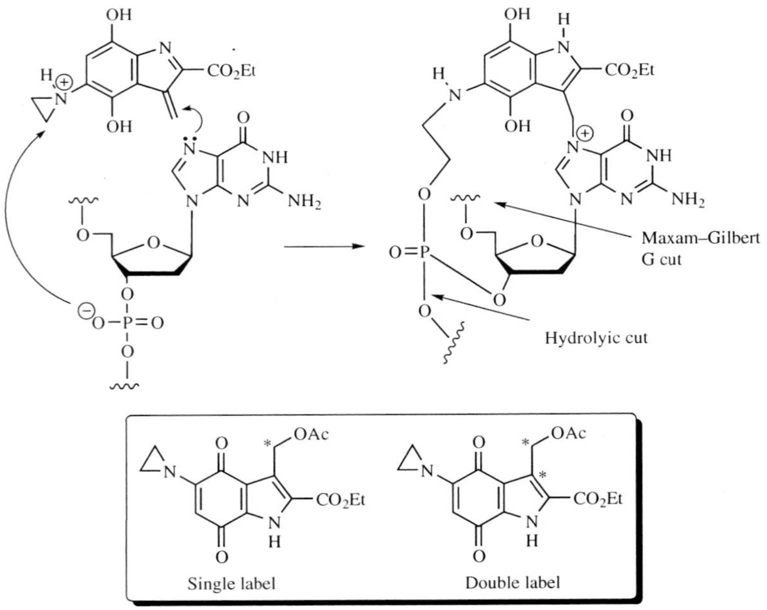

SCHEME 7.12 DNA intrastrand cleaving agent shown alkylating the guanine N(7) and phosphate positions. Single- and double-labeled analogues of the cleaving agent are shown in the inset.

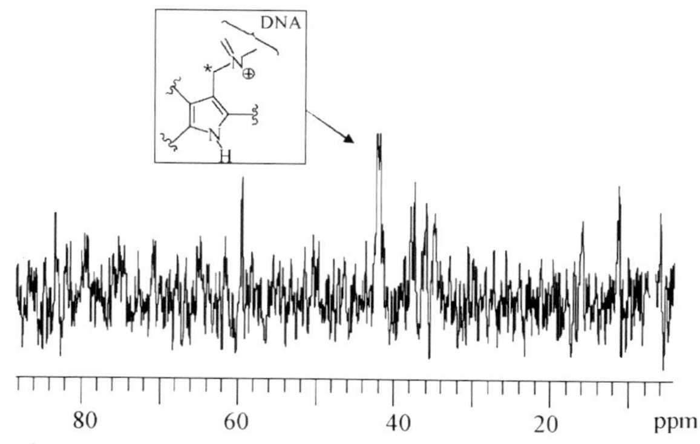

FIGURE 7.10 ^{13}C-NMR spectrum of d(ATGCAT) reductively alkylated with the 3α-^{13}C-labeled cleaving agent shown in the inset of Scheme 7.12.

precipitated twice to remove hydrolysis products. The alkylated DNA spectrum shown in Fig. 7.10 has a strong peak at 42 ppm indicating that the ^{13}C-methylene is bound to the guanine N(7) position. However, the enriched ^{13}C label was diluted to the extent that this resonance is hardly resolved from natural abundance ^{13}C labels.

To remedy the problem inherent in the dilution of a ^{13}C label, we prepared the "unnatural" double-labeled ^{13}C analogue shown in the inset of Scheme 7.12. The incorporation of the double label was carried out using a Japp–Klingmann/Fischer indole reaction[72] and a Vilsmeier formylation, respectively. The chances of two adjacent ^{13}C labels occurring naturally is 1×10^{-4}%; therefore, an enriched double-labeled analogue can be readily distinguished from natural abundance of^{13}C labels. The adjacent ^{13}C labels result in intense enriched ^{13}C-NMR resonances due to more efficient relaxation. This feature is apparent in Fig. 7.11, which shows the ^{13}C-NMR of the dodecamer d(CGCAAATTTGCG)$_2$ that was reductively alkylated with the double-labeled ^{13}C analogue. The ^{13}C chemical shift of the 3α center supports guanine N(7) alkylation by the ene-imine as the sole DNA reaction product.

To distinguish adjacent ^{13}C labels from natural abundance isotopes, proton-detected ^{13}C-NMR spectra (HMBC) will show cross peaks associated with the double label that are split into doublets. In contrast, natural abundance ^{13}C will show single cross peaks.

Another example of the utility of ^{13}C labels in studying methide reactions with DNA is our study of the antitumor agent shown in Scheme 7.13.[48,71] As illustrated in this scheme, this agent is capable of alkylating DNA via an ene-imine and a quinone methide reactive intermediate. To verify the presence of ene-imine and quinone methide alkylation products, the hexamer d(ATGCAT) was treated with reduced 3α-^{13}C and 2α-^{13}C-labeled analogues of the antitumor agent. ^{13}C-NMR spectra of the alkylation products were obtained and compared with the ^{13}C natural abundance spectrum of the hexamer (Fig. 7.12). The series of spectra in this figure illustrate the limitation of using enriched ^{13}C-NMR to study DNA alkylation reactions. The formation of multiple alkylation products as well as low yields could dilute

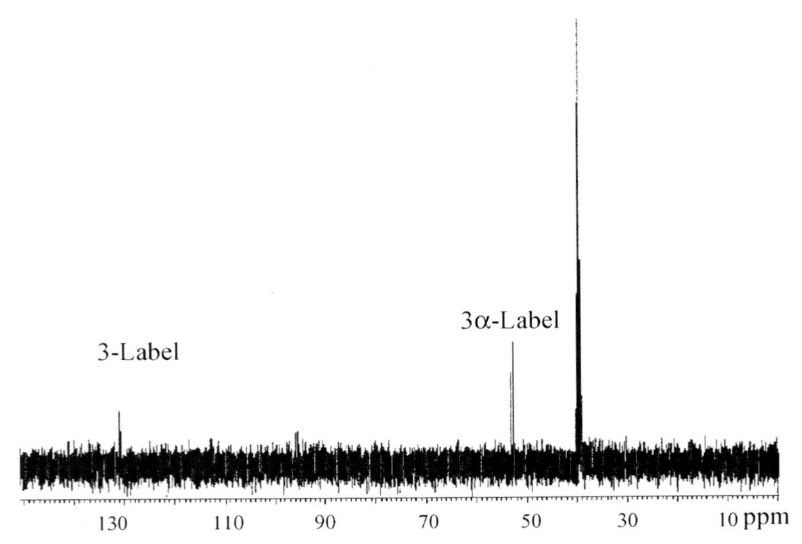

FIGURE 7.11 ^{13}C-NMR spectrum of d(CGCAAATTTGCG)$_2$ reductively alkylated with the 3α,3-^{13}C double-labeled cleaving agent shown in the inset of Scheme 7.12.

the incorporated ^{13}C labels to the point that they are as intense as the natural abundance ^{13}C spectrum. The solution is to compare the natural abundance and enriched spectra and note differences. Comparison of the natural abundance ^{13}C-NMR spectrum (a) with the ^{13}C spectrum of 2α-labeled alkylating agent (a) reveals that the oxygen of acetate was still attached at the 2α-^{13}C center. In contrast, the 3α-labeled alkylating agent afforded a ^{13}C spectrum showing that nitrogen alkylation had occurred at this position. These results are consistent with elimination of acetate

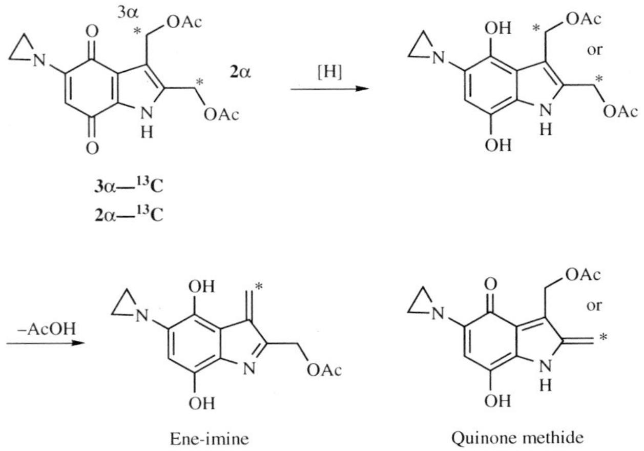

3α—^{13}C
2α—^{13}C

[H]

–AcOH

Ene-imine

Quinone methide

SCHEME 7.13 Probing the DNA reaction products of the ene-imine and quinone methide.

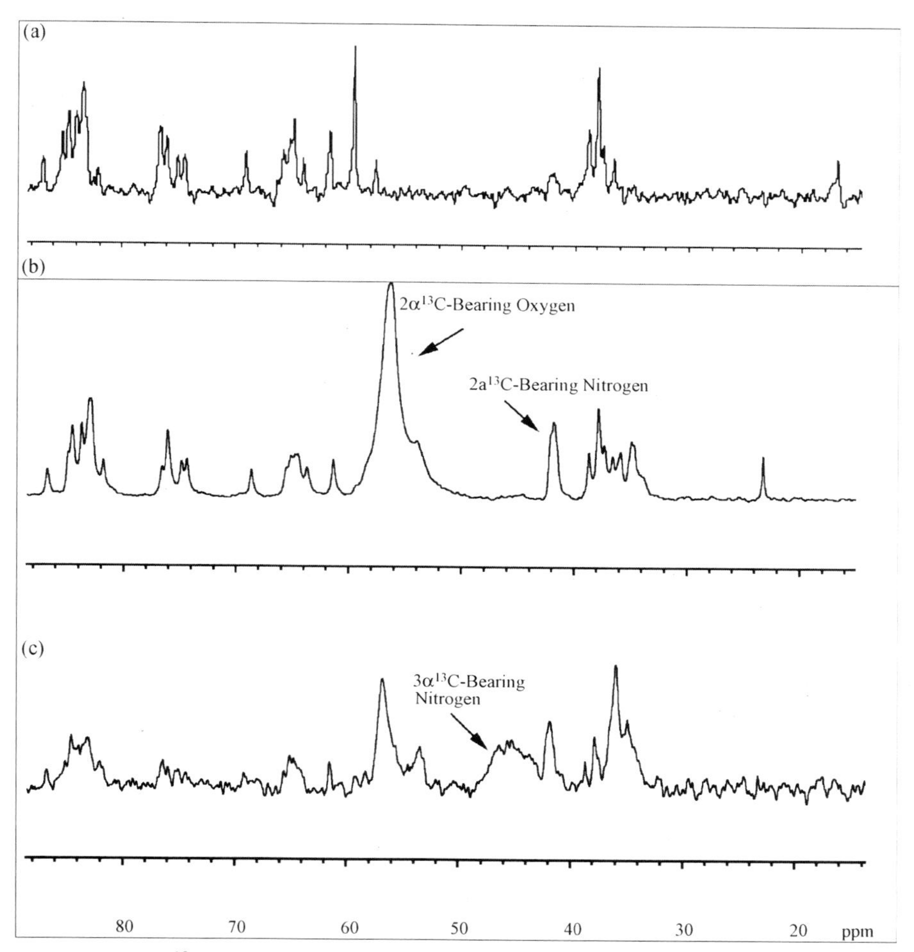

FIGURE 7.12 ^{13}C-NMR spectra of native hexamer d(ATGCAT) (a) and alkylation products obtained by treatment of the hexamer with two-electron reduced 2α-^{13}C- (b) and 3α-^{13}C-labeled reductive alkylating agents (c).

from the 3α-position rather than from the 2α-position as we have observed in the hydrolysis study.[48]

7.3.4 Design of a "Cyclopropyl Quinone Methide"

The inspiration for the cyclopropyl quinone methide design came from the mitosenes[4,32,73] and the A-ring of CC-1065.[74] While the mitosene hydroquinone affords the quinone methide by elimination of a leaving group, the present hydroquinone systems can eliminate the leaving group only by formation of a fused cyclopropane ring (Scheme 7.14). We used both pyrrolo[1,2-*a*]indole [46] and

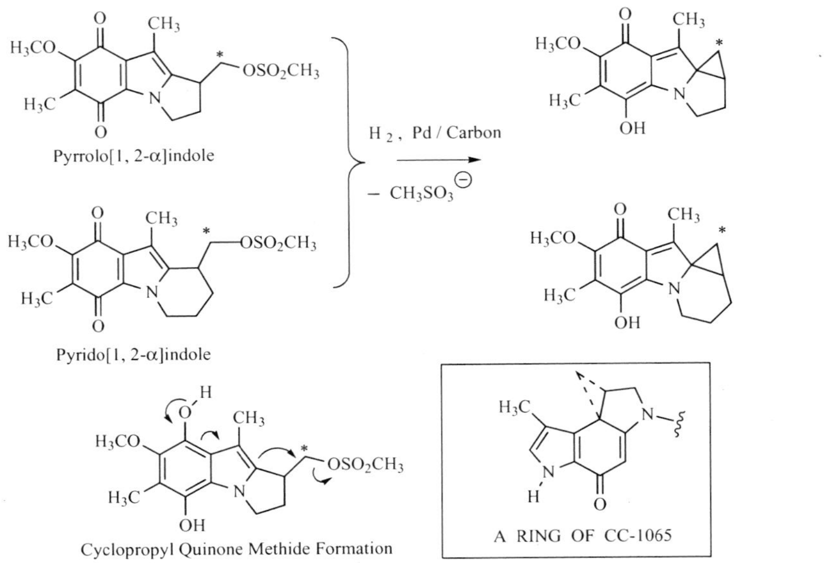

Pyrrolo[1, 2-α]indole

H_2, Pd / Carbon

$- CH_3SO_3^{\ominus}$

Pyrido[1, 2-α]indole

Cyclopropyl Quinone Methide Formation

A RING OF CC-1065

SCHEME 7.14 Cyclopropyl quinone methide formation upon reductive activation. The CC-1065 A-ring is shown in the inset.

pyrido[1,2-*a*]indole[47] systems to generate the unprecedented cyclopropyl quinone methide structure. The study of both ring systems was carried out to investigate the role of the stereoelectronic effect on alkylation reactions. The ^{13}C labels were crucial in following the complex reactions of the cyclopropyl quinone methides.

The stereoelectronic effect predicts that the course of a reaction will be determined by favorable orbital overlap. For example, nucleophilic attack of the pyrrolo[1,2-*a*]indole-based cyclopropyl quinone methide at the more substituted carbon, indicated with an arrow in Fig. 7.13, results in significant π-orbital overlap since the breaking bond is 45° out of the π system plane, inset of Fig. 7.13. In contrast, nucleophilic attack at the methylene center of the cyclopropane ring, where the ^{13}C label is located, results in no overlap because the developing π orbital is orthogonal to the neighboring π system. Similarly, nucleophilic attack of the CC-1065 A-ring occurs at the methylene center, because subsequent bond breaking results in good overlap of the developing π-orbital with the π system.[75] As a result of stereoelectronic control, the pyrrolo[1,2-*a*]indole-based cyclopropyl quinone methide species should undergo a ring expansion reaction upon nucleophile trapping. If so, the ^{13}C-labeled center will not bear the nucleophile.

The reductive activation reaction of the ^{13}C-labeled pyrrolo[1,2-*a*]indole shown in Scheme 7.14 was carried out in methanol and a ^{13}C-NMR spectrum was obtained for the crude organic extract. This ^{13}C-NMR spectrum, shown in Fig. 7.14, reveals the presence of starting material as well as products with ^{13}C-labeled alkene and alkane centers. We confirmed the ^{13}C assignments shown

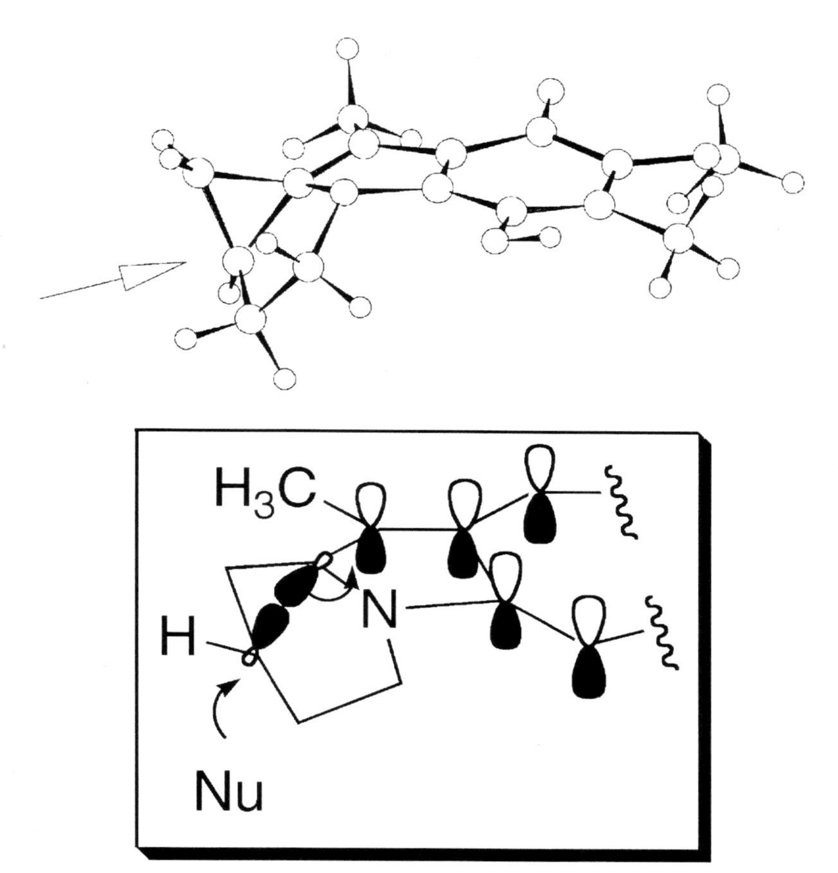

FIGURE 7.13 Molecular model of the pyrrolo[1,2-*a*]indole showing the site of nucleophile attack that provides a favorable stereoelectronic effect. The inset shows expected orbital interactions.

in this scheme by product isolation from the crude mixture and spectral identification. A close inspection of the ^{13}C-NMR spectrum reveals the presence of a trace product with a ^{13}C resonance at 74 ppm, which may be the result of methanol trapping without ring expansion. Its formation could arise by nucleophilic attack of water at the least hindered position of the cyclopropyl quinone methide or perhaps hydrolysis of the starting material. Likewise, treatment of the reduced ^{13}C-labeled pyrrolo[1,2-*a*]indole with pH 7.4 tris chloride buffer provided only the ring-expanded nucleophile trapping (water and chloride) products. Once again, we observed only a trace amount of nucleophile-trapping product without ring expansion in the ^{13}C-NMR.

The product structures shown in Fig. 7.14 indicate that the cyclopropyl quinone methide traps protons and nucleophiles like the classic quinone methides, see mechanisms in Scheme 7.15. However, the cyclopropyl quinone methides can also

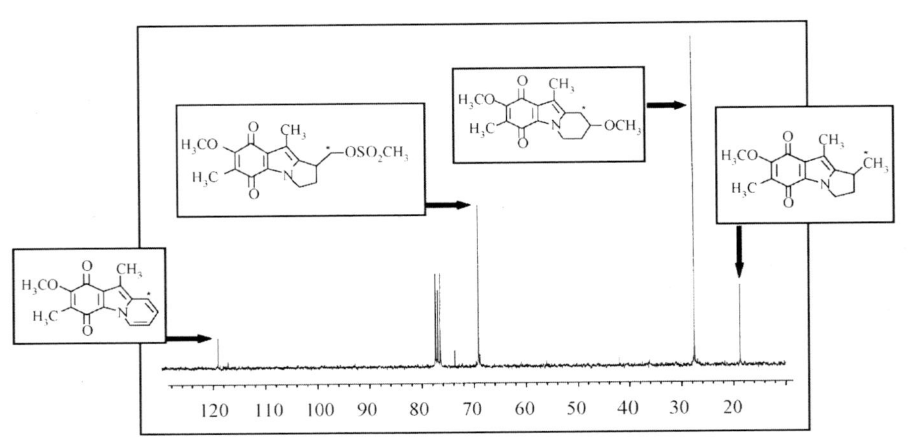

FIGURE 7.14 Product ^{13}C-NMR of the mixture resulting from the reductive activation of the pyrrolo[1,2-*a*]indole shown in Scheme 7.14 in methanol followed by aerobic workup.

undergo a prototropic shift to afford an alkene. Further oxidation of this alkene results in aromatization of the pyrido ring.

Screening on the pyrrolo[1,2-*a*]indole-based quinone shown in Scheme 7.14 against the National Cancer Institute's human cancer 60-cell line panel revealed a lack of significant cytostatic and cytotoxic activity. We postulated that the presence of ring expansion upon nucleophilic attack on the pyrrolo[1,2-*a*]indole-based cyclopropyl quinone methide is responsible for the lack of activity. The resulting six-membered ring can readily eliminate the nucleophile to afford the alkene that eventually aromatizes. The experiment described in Scheme 7.16 showed that reductive alkylation of the guanine N(7) position can occur to afford "orange DNA." Over a 2-day period, an elimination reaction affords native DNA and the alkene. The driving force for elimination of the quaternized nitrogen is the formation of a stable six-membered ring alkene, which is conjugated to an aromatic system.

In a recent study, we showed that the more flexible pyrido[1,2-*a*]indole-based cyclopropyl quinone methide is not subject to the stereoelectronic effect.[47] Scheme 7.17 shows an electrostatic potential map of the protonated cyclopropyl quinone methide with arrows indicating the two possible nucleophilic attack sites on the electron-deficient (blue-colored) cyclopropyl ring. The ^{13}C label allows both nucleophile attack products, the pyrido[1,2-*a*]indole and azepino [1,2-*a*]indole, to be distinguished without isolation. The site of nucleophilic is under steric control with pyrido[1,2-*a*]indole ring formation favored by large nucleophiles.

The results of the methanolic solvolysis study shown in Fig. 7.15 reveals that nucleophilic attack on the cyclopropyl quinone methide by methanol affords the pyrido[1,2-*a*]indole (73 ppm) and azepino[1,2-*a*]indole (29 ppm) trapping products. Initially, nucleophilic attack on the cyclopropane ring affords the hydroquinone derivatives (see Scheme 7.17) that oxidizes to the quinones upon aerobic workup.

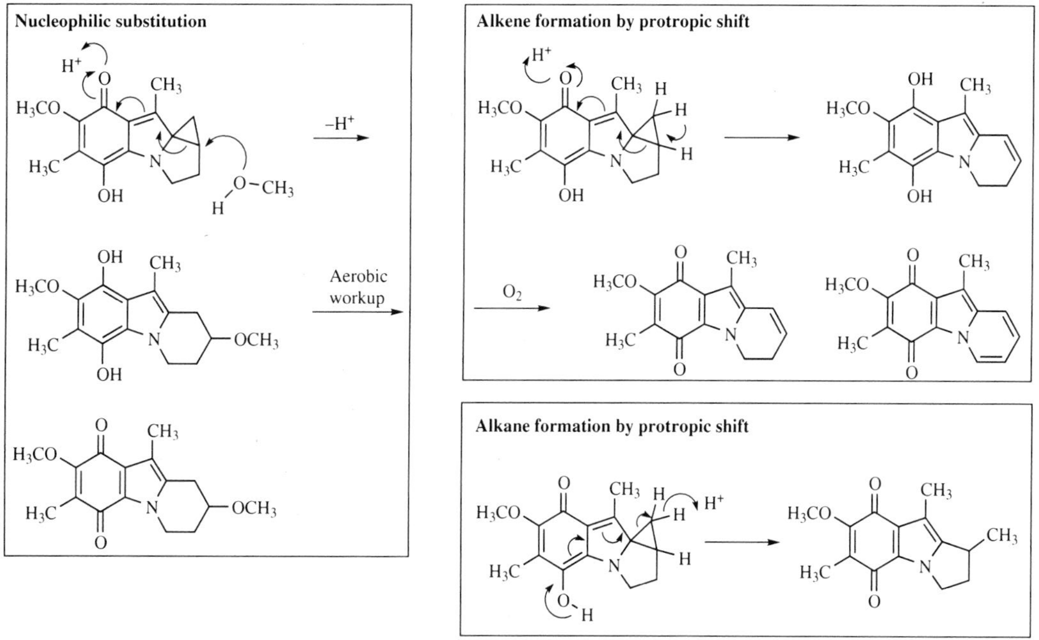

Nucleophilic substitution

$-H^+$

Aerobic workup

Alkene formation by protropic shift

O_2

Alkane formation by protropic shift

SCHEME 7.15 Mechanisms of cyclopropyl quinone methide fate.

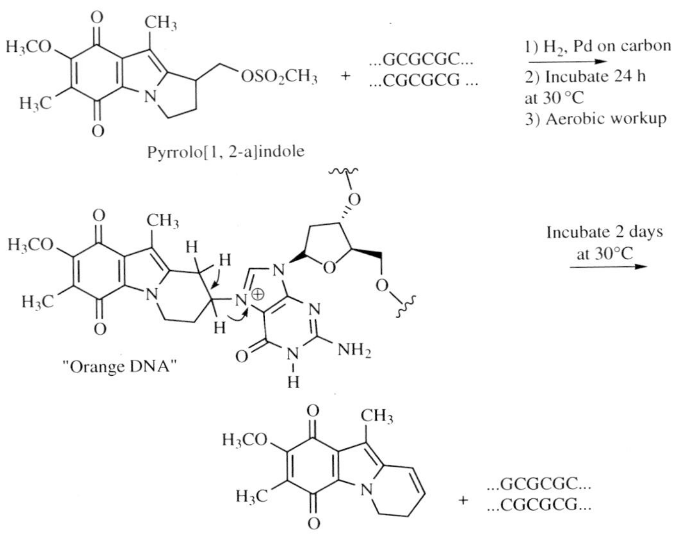

Pyrrolo[1, 2-a]indole

...GCGCGC...
...CGCGCG ...

1) H_2, Pd on carbon
2) Incubate 24 h at 30 °C
3) Aerobic workup

"Orange DNA"

Incubate 2 days at 30°C

...GCGCGC...
...CGCGCG...

SCHEME 7.16 Reductive alkylation of the alternating DNA polymer, poly d(GC), at the guanine N(7) position followed by elimination to afford the alkene.

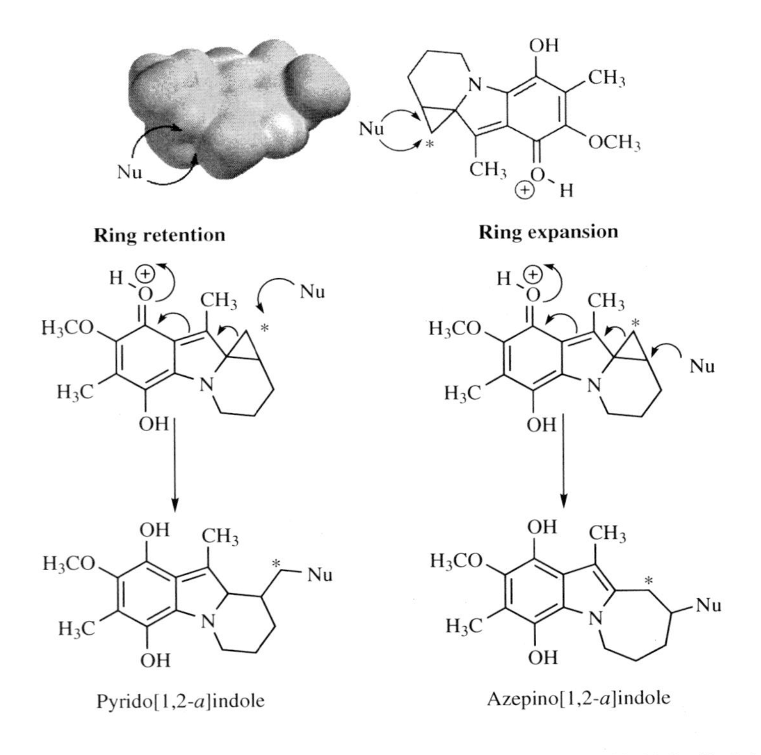

SCHEME 7.17 Electrostatic potential map of the protonated pyrido[1,2-*a*]indole-based cyclopropyl quinone methide. The two possible nucleophile-trapping paths with the respective products are shown. (See the color version of this scheme in Color Plates section.)

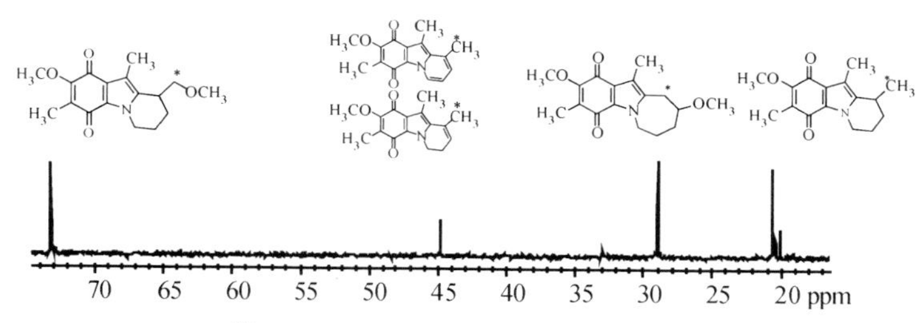

FIGURE 7.15 Enriched ^{13}C-NMR of the methanolic solvolysis pyrido[1,2-*a*]indole-based cyclopropyl quinone methide.

The process that affords the alkenes with $^{13}C = 45$ ppm is essentially an internal redox reaction wherein electrons flow from the fused tetrahydropyrido ring to the quinone ring by means of a series of tautomerizations.

To assess the trapping of biological nucleophiles, the pyrido[1,2-*a*]indole cyclopropyl quinone methide was generated in the presence of 5′-dGMP. The reaction afforded a mixture of phosphate adducts that could not be separated by reverse-phase chromatography (Fig. 7.16). The ^{13}C-NMR spectrum of the purified mixture shown in Fig. 7.16 reveals that the pyrido[1,2-*a*]indole was the major product with trace amounts of azepino[1,2-*a*]indole present. Since the stereoelectronic effect favors either product, steric effects must dictate nucleophilic attack at the least hindered cyclopropane carbon to afford the pyrido[1,2-*a*]indole product. Both adducts were stable with elimination and aromatization not observed. In fact, the pyrido[1,2-*a*]indole precursor (structure shown in Scheme 7.14) to the pyrido [1,2-*a*]indole cyclopropyl quinone methide possesses cytotoxic and cytostatic properties not observed with the pyrrolo[1,2-*a*]indole precursor.[47]

7.3.5 Kinetic Studies of the Mitosene Quinone Methide

In a recent study,[42] we compared the reductive activation of pyrido[1,2-*a*]indole ($n = 1$) and pyrrolo[1,2-*a*]indole ($n = 0$) based quinones (Scheme 7.18). Our goals were to determine the influence of the six-membered pyrido and the five-membered pyrrolo fused rings on quinone methide formation as well as cytostatic/cytotoxic activity. The postulate made at the outset of this study was that reactive species derived from the six-membered pyrido ring analogues would be less reactive than those derived from the five-membered pyrrolo ring analogues. The presence of axial hydrogens in the fused pyrido ring, in contrast to the relatively flat pyrrolo ring, was the basis for this postulate. The results of these studies would have a bearing on the documented lack of antitumor activity exhibited by pyrido[1,2-*a*] indole analogues[51] as well as the activity of the pyrrolo[1,2-*a*]indole-based antitumor agents.[76]

We studied the reactions shown in Scheme 7.18 using the global fitting methodology described in Section 7.2.1. The quinone methide species rapidly built up in solution upon quinone reduction and trapped by either water or a proton to afford the final products shown in Scheme 7.18. global fitting provided the rates of quinone methide

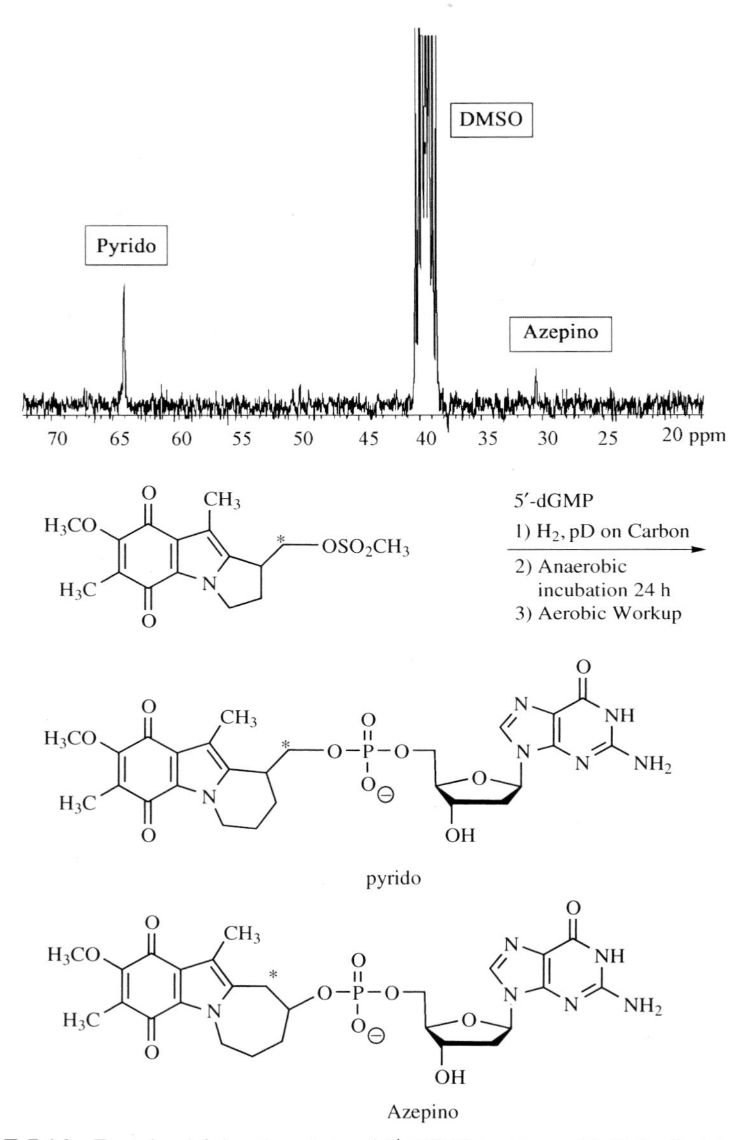

FIGURE 7.16 Trapping of the phosphate of 5′-dGMP by the pyrido[1,2-*a*]indole quinone methide. The ^{13}C-NMR shows most trapping with ring retention, labeled "pyrido," with trace amounts of ring expansion, labeled "azepino."

formation from the hydroquinone as well as the rates of quinone methide disappearance for both ring systems. The general mechanism of quinone methide formation and fate is outlined in Scheme 7.19. The rate laws for quinone methide formation and fate are shown in Eqs. 7.1 and 7.2, respectively.

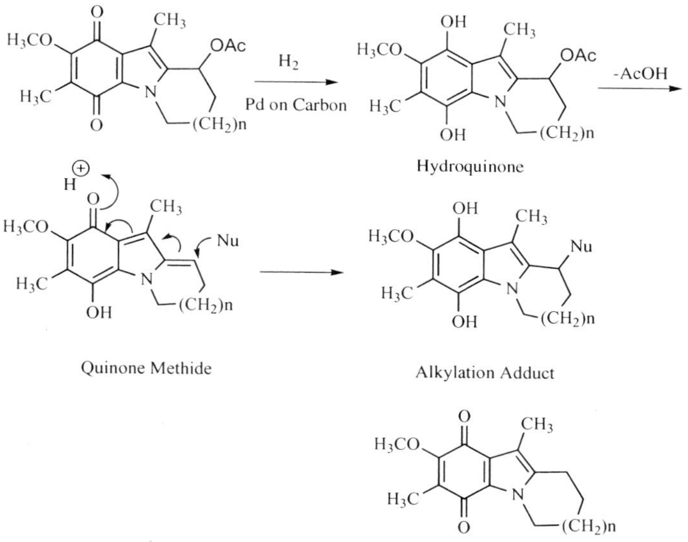

Hydroquinone

Quinone Methide

Alkylation Adduct

Proton Trapping Product

SCHEME 7.18 Formation and fate of the pyrrolo[1,2-*a*]indole ($n = 0$) and the pyrido[1,2-*a*] indole-based ($n = 1$) quinone methides.

Equation 7.1 is the rate law for quinone methide formation:

$$k_{\text{obsd}} = a_{\text{H}}k_1 + k_0. \tag{7.1}$$

Equation 7.2 is the rate law for quinone methide disappearance:

$$k'_{\text{obsd}} = \frac{a_{\text{H}}k_2 + K_{\text{w}}[k_3 + k_4]}{a_{\text{H}} + K_a}. \tag{7.2}$$

The rate constants associated with the conversion of the pyrrolo[1,2-*a*]indole hydroquinone to its quinone methide were fit to the rate law equation (7.1), see Fig. 7.17 for rate data and the fit. The solid line in Fig. 7.17 was generated with Eq. 7.1 where $k_0 = 0.09\ \text{min}^{-1}$ and $k_1 = 1.5 \times 10^5\ \text{M}^{-1}\ \text{min}^{-1}$. The mechanism consistent with the pH–rate profile is the spontaneous elimination of acetate (k_0 process) and the proton assisted elimination of acetate (k_1 process) from the electron-rich hydroquinone. The k_0 process is independent of pH and exhibits a zero slope while the k_1 process exhibits a -1 slope consistent with acid catalysis.

The rate constants associated with the acid-catalyzed conversion of the pyrido[1,2-*a*]indole hydroquinone to its quinone methide were too large to measure. We did manage to measure two rate constants at pH 7 and 8, both with a value of $0.36\ \text{min}^{-1}$. Based on the pH–rate profile obtained the pyrrolo[1,2-*a*]indole hydroquinone, these

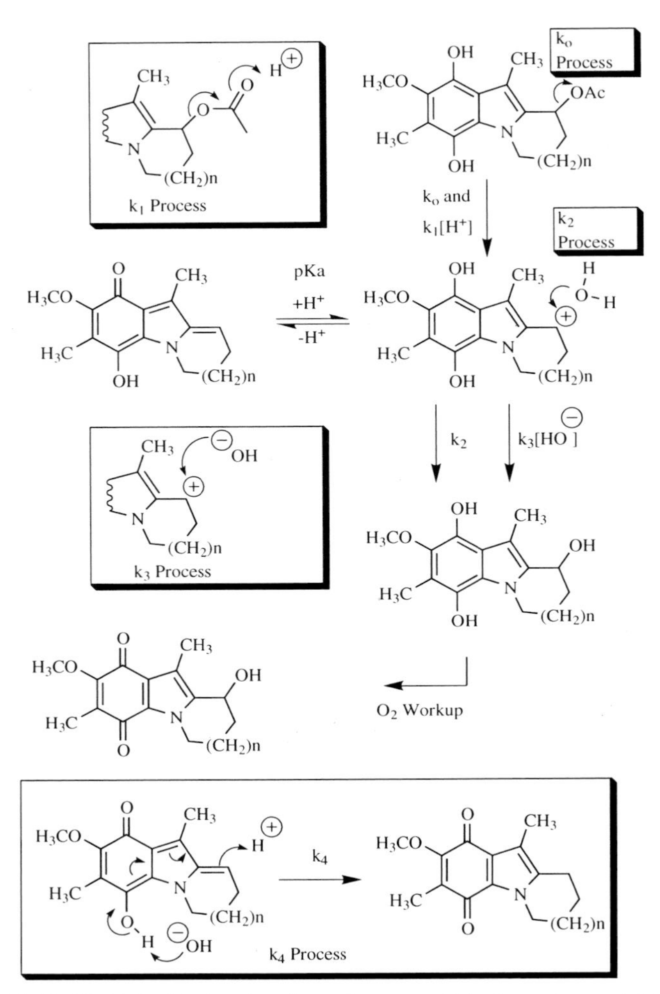

SCHEME 7.19 Mechanism for quinone methide formation and fate.

rate constants very likely represent the k_0. Even with two data points, we can safely conclude that acetate is eliminated ~4-fold faster from the pyrido[1,2-*a*]indole hydroquinone than from the pyrrolo[1,2-*a*]indole hydroquinone. The reason for the difference was obvious from molecular models; the steric effect of axial hydrogen in the pyrido ring promotes leaving group elimination.

Equation 7.2 represents the rate law for quinone methide disappearance. This equation was derived using material balance where reactions occur from and equilibrating mixture of neutral and protonated quinone methide. Both the protonated (k_2 process) and neutral equivalent (k_3 and k_4 processes) react to afford the observed major products shown in Scheme 7.18. Alternatively, the quinone methide can be protonated

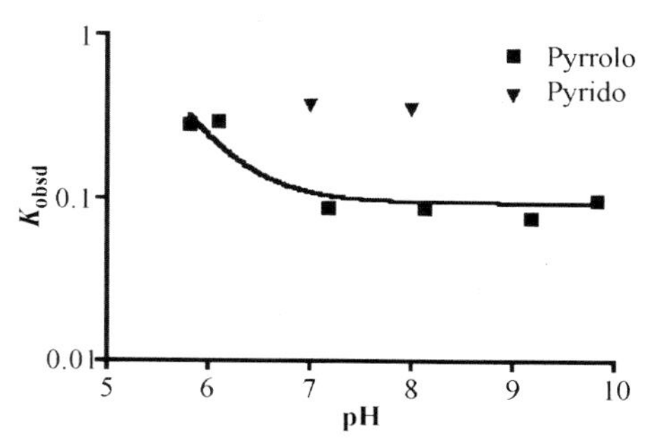

FIGURE 7.17 pH–rate profile for quinone methide formation.

at the methide carbon center to afford a proton-trapping product (k_4 process). The k_4 process in expected to parallel the k_3 process on the pH profile because both involve quinone methide protonation along with the participation of hydroxide. Essentially, the transition states for both k_3 and k_4 processes consist of neutral quinone methide and water. In fact, the pyrido[1,2-*a*]indole quinone methide traps both water and protons by these respective processes only at pH values $<pK_a$, where protonated quinone methide is the predominant species. The pyrroloindole quinone methide on the other hand does not exhibit a significant k_4 process.

The pH–rate profiles for the disappearance of the pyrrolo[1,2-*a*]indole and pyrido[1,2-*a*]indole quinone methides are shown in Figs. 7.18 and 7.19, respectively. The data in both profiles were fit to Eq. 7.2 generating the following constants for the pyrrolo[1,2-*a*]indole quinone methide: $k_2 = 0.032\ min^{-1}$, $K_a = 2.84 \times 10^{-7}$ ($pK_a = 6.55$), $k_3 = 8.98 \times 10^4\ M^{-1}\ min^{-1}$ and for the pyrido[1,2-*a*]indole quinone methide: $k_2 = 0.006\ min^{-1}$, $K_a = 3.5 \times 10^{-8}$ ($pK_a = 7.51$), $k_3 = 1.9 \times 10^4\ M^{-1}\ min^{-1}$.[42]

The mechanistic interpretation of the pH–rate profiles for quinone methide disappearance relied on Hartree–Fock calculations with 6–31G basis sets as well as on product studies. In Fig. 7.20, we show the potential–density maps of the protonated and neutral pyridoindole quinone methide with negative charge density colored red and positive charge density colored blue. Inspection of the methide

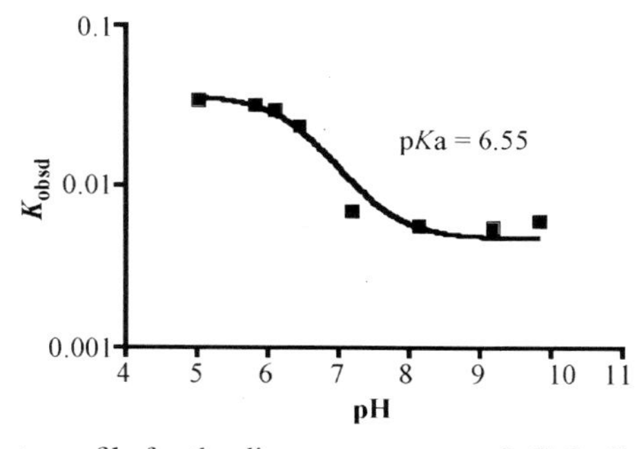

FIGURE 7.18 pH–rate profile for the disappearance pyrrolo[1,2-*a*]indole quinone methide.

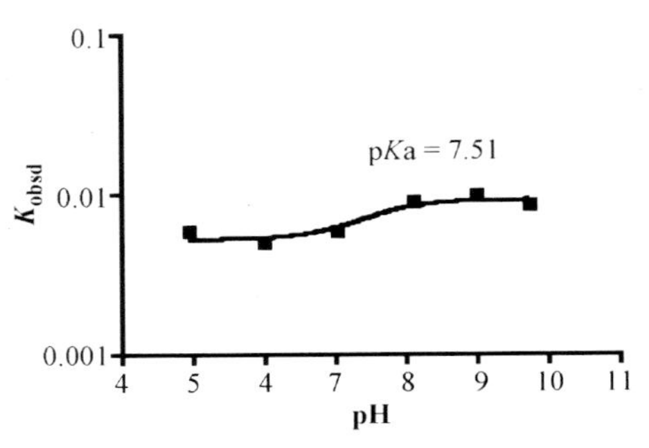

FIGURE 7.19 pH–rate profile for the disappearance pyrido[1,2-*a*]indole quinone methide.

reacting center of the neutral species reveals a slight negative charge density (yellow–green color). We interpret this observation as the combination of electron release to and electron withdrawal from the methide reacting center; see resonance structures in Fig. 7.20. This result suggests that water does not trap the neutral quinone methide but rather its kinetic equivalent occurs: the hydroxide trapping of the protonated quinone

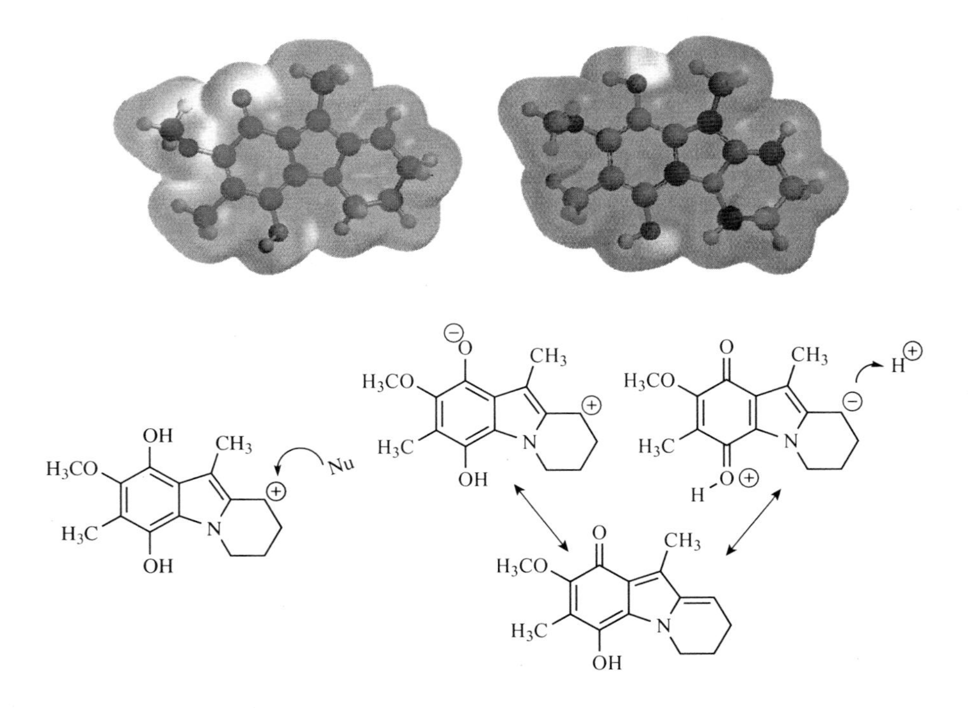

FIGURE 7.20 Potential–density maps for the *O*-protonated pyridoindole quinone methide and the corresponding neutral species. The charge density color codes are anionic (red), cationic (blue), and neutral (green). (See the color version of this figure in Color Plates section.)

methide. Thus, the rate law in Eq. 7.2 contains the autoprotolysis constant for water $K_w = 1.5 \times 10^{-14}$ at 30 °C. The numerator expression for hydroxide attack on the protonated quinone methide is $a_H k_3(K_w/a_H)$ which simplifies to $K_w k_3$ in Eq. 7.2, where k_3 is the second-order rate constant for hydroxide attack, a_H is the proton activity as measured with a pH meter, and K_w/a_H is the hydroxide concentration at a particular a_H value. Although the proton concentration decreased, the hydroxide concentration has increased, resulting in the plateau above the quinone methide pK_a. Consistent with this mechanism, water trapping is the only product observed at pH values above the quinone methide pK_a for both the pyrrolo[1,2-*a*]indole and pyrido [1,2-*a*]indole quinone methide.

The pH–rate studies and calculated potential–density maps indicate that the quinone methide species must be *O*-protonated to afford a carbocation before it can trap a nucleophile. Therefore, carbocation pK_a values at or near neutrality would be required for nucleophile trapping in a biological system (pH values ~7.4). Previously, we reported the pK_a for the mitomycin C carbocation–quinone methide equilibrium to be 7.1,[32] a value similar to those obtained from the pH–rate data shown in Figs. 7.18 and 7.19. In contrast, the pK_a values for the protonated carbonyl oxygen of a ketone or ester usually have negative values. We consider the relative high pK_a value of 6–8 to be typical value for a carbocation–quinone methide equilibrium. The formation of a resonance-stabilized aromatic carbocation is one reason for these high pK_a values. Another reason is the high energy of the quinone methide, which is close to that of the carbocation. The thermodynamic cycle shown in Section 7.3.7 illustrates the role of quinone methide energy on the carbocation pK_a. Investigators often refer to the quinone methide species derived from bioreductive activation as an electrophilic species. Actually, the neutral quinone methide is only somewhat nucleophilic. The *O*-protonated quinone methide (i.e., a carbocation) is the actual electrophilic species.

The pH–rate data also indicate that the pyrrolo[1,2-*a*]indole quinone methide traps water and hydroxide with larger rate constants than those observed for the pyrido[1,2-*a*]indole quinone methide. The explanation is that steric congestion by axial hydrogen of the pyrido ring slows nucleophile addition. Conversely, this steric congestion promotes leaving group elimination from the pyrido ring, see pH–rate profile in Fig. 7.17. Proton trapping of the pyrido[1,2-*a*]indole quinone methide center can then compete with the slowed water trapping affording significant amounts of proton trapping product (Scheme 7.19).[42] We expected that steric congestion would also destabilize biological adducts of the pyrido[1,2-*a*]indole quinone methide. Previously, we reported the preparation of an N(7)-pyrido[1,2-*a*] indole adduct of poly d(GC)[46] (Scheme 7.16). Under mild conditions, the N(7) center was eliminated from the pyrido[1,2-*a*]indole adduct to afford the alkene and native DNA. From the above findings, we postulated that the pyrido[1,2-*a*]indole quinone methide should exert less cytostatic activity than the pyrrolo[1,2--*a*]indole quinone methide. We did not anticipate any cytotoxic activity because neither system functioned as a cross-linker. Screening against five human cancer cell lines validated the above postulate with the precursor to the pyrido[1,2-*a*]indole quinone methide not even possessing cytostatic activity (Table 7.1).[42]

TABLE 7.1 Results of Screening of the Pyrido[1,2-*a*]indole and the Pyrrolo[1,2-*a*] indole Quinones (Structures Shown in Scheme 7.18)

Parameter log GI_{50}	BxPC-3 Pancreas	MCF-7 Breast	SF-268 CNS	H460 NSC Lung	KM20L2 Colorectal	DU-145 Prostrate
Pyrido	>−4	>−4	>−4	>−4	>−4	>−4
Pyrrolo	−4.8	−4.95	>−4	−4.95	>−4	−4.96

Log GI_{50} values of >−4 indicate concentrations too high to measure.

7.3.6 Cyclopent[*b*]indole-Based Quinone Methides

The cyclopent[*b*]indole analogues shown in Scheme 7.20 were designed *de novo* in this laboratory.[71,77] The design features of the cyclopent[*b*]indole quinone methide are outlined below. Calculations revealed that the cyclopent[*b*]indole quinone methide would be more stable than the mitosene-based quinone methide and therefore more likely to build up in solution. We believe the relief of strain at the indole nitrogen in the cyclopent[*b*]indole ring compared to the pyrrolo[1,2-*a*] indole ring is largely responsible for the increased stability of the latter. The use of the electron-rich indole ring ensures that the *O*-protonated cyclopent[*b*]indole quinone methide has a pK_a near neutrality so that the carbocation exists at physiological pH. The use of a fused pentane ring affords minimal steric congestion for nucleophile alkylation. Finally, the presence of an *N*-acetyl group in some analogues would serve to stabilize the quinone methide by internal acyl transfer (Scheme 7.20).

The rationale for the cyclopent[*b*]indole design discussed above was that the quinone methide would build up in solution and intercalate/alkylate DNA. Enriched ^{13}C-NMR studies indicate that the quinone methide builds up in solution and persists for hours, even under aerobic conditions (Fig. 7.21). In contrast, the quinone methide species formed by known antitumor agents (mitomycin C) are short lived and highly reactive. The spectrum shown in Fig. 7.21 also shows the N to O acyl transfer product that we isolated and identified. However, we could not determine if the quinone methide structure actually has the acetyl group on the N or O centers.

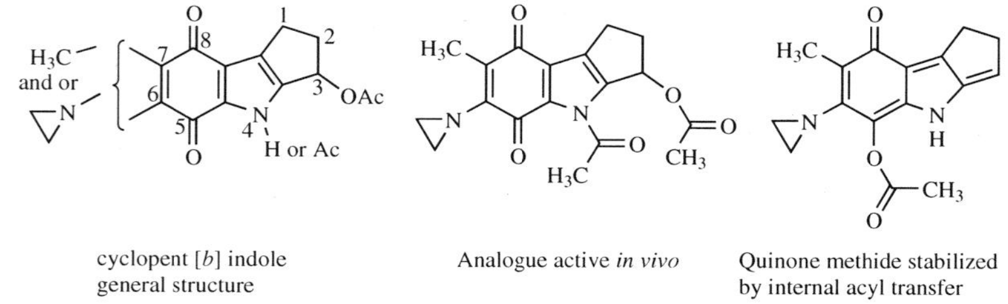

cyclopent [*b*] indole general structure

Analogue active *in vivo*

Quinone methide stabilized by internal acyl transfer

SCHEME 7.20 Cyclopent[*b*]indole analogues. Quinone methide structure shown with internal hydrogen bonding.

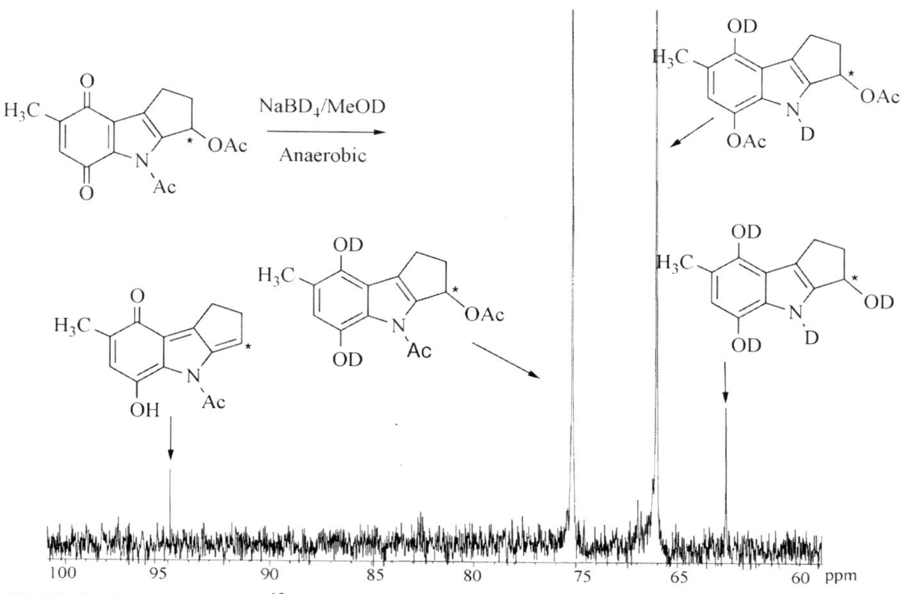

FIGURE 7.21 Enriched ^{13}C-NMR spectrum of cyclopent[*b*]indole reductive activation products.

DNA cleavage studies show that the cyclopent[*b*]indole quinone methides alkylate DNA predominately at 3′-GT-5′ steps with some alkylation at 3′-GG-5′ steps. The PAGE gel shown in Table 7.2 was run on a DNA restriction fragment that had been treated with reductively activated cyclopent[*b*]indoles. This laboratory has used hydrolytic cleavage to detect phosphate backbone alkylations.[78] Inspection of the hydrolytic cleavage lanes in Table 7.2 reveals that no phosphate backbone alkylations are present. Maxam and Gilbert G cleavage with piperdine[79] shows cleavage at G and T bases. This result suggests that the quinone methide can alkylate both at the guanine N

TABLE 7.2 Autoradiogram of the Reductive Cleavage of a 5′-^{32}P End-Labeled 514 bp Restriction Fragment from pBR322 DNA (EcorI/RsaI) by Cyclopent[*b*]indoles

Lane 1	$NaBH_4$ reductive activation, piperidine cleavage
Lane 2	$NaBH_4$ reductive activation, hydrolytic cleavage
Lane 3	$NaBH_4$ reductive activation, piperidine cleavage
Lane 4	$NaBH_4$ reductive activation, hydrolytic cleavage

Lane 7 is a Maxam and Gilbert G + A ladder. Cleavage occurs predominately at 3′-GT-5′ with some cleavage at 3′-GG-5′.

(7) and perhaps at the thymine O(6). Both alkylation reactions will make these respective bases susceptible to piperdine-mediated base removal. Modeling of the postulated quinone methide species at these steps revealed that access to either nucleophilic site is possible upon intercalation. The computer model also correctly predicted that an aziridinyl ring was not required for DNA alkylation. Note that the PAGE results shown in Table 7.2 indicate that the aziridinyl ring has no influence on DNA alkylation and cleavage. The absence of hydrolytic cleavage is consistent with the absence of aziridinyl alkylation of the phosphate backbone. The origin of GT step selectivity was not obvious, but HOMO–LUMO interactions play a role in the intercalators selectivity.[80]

In vivo studies were carried on the aziridinated cyclopent[*b*]indole quinone out before it was discovered that the aziridinyl ring did not participate in DNA alkylation. The results in Fig. 7.22 for the B16 melanoma syngraft model reveal that there was substantial reduction of tumor mass at 3 mg/kg. However, toxicity (animal deaths) became apparent at 5 mg/kg. On the other hand, human lung cancer xenografts in SCID (severe combined immunodeficient) mice were reduced to 50% mass with 3×1 mg/kg doses without any animal deaths.

The design of a bioreductive alkylating agent possessing antitumor activity described above validates our approach to rational drug design. Namely, the putative

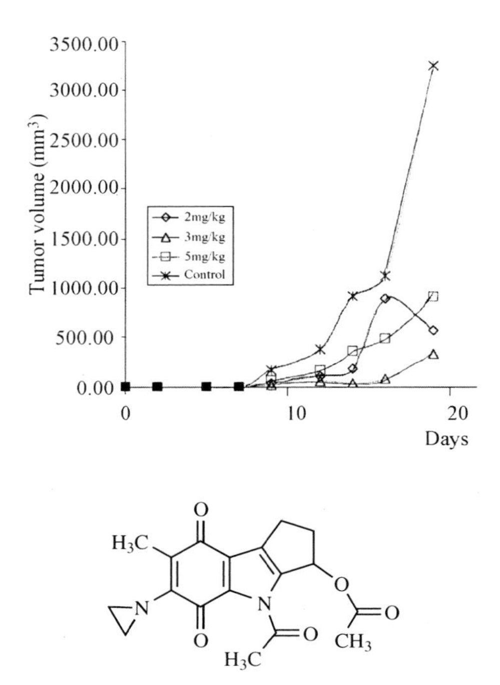

FIGURE 7.22 Results of the cyclopent[*b*]indole below in the B-16 melanoma model in C57/bl.

quinone methide species must be sterically unencumbered at the reacting center, *O*-protonation of the quinone methide must be possible at physiological pH so that alkylation can occur, and the quinone methide must be long lived in aqueous solution.

7.3.7 Prekinamycin-Based Quinone Methides

The chemistry and biology of the diazobenzo[*b*]fluorene-based natural products (Fig. 7.23) have been the subject of intense study due to the presence of the unusual diazo functional group.[81] Much of these efforts have centered on the mechanistic role of the diazo group in exerting antitumor[82] and antibacterial activity.[83] In recent articles,[84] we addressed the chemistry and biological activity of prekinamycin analogues using the techniques of spectral global fitting, ^{13}C labeling of the reactive center, Hartree–Fock calculations,[85] and *in vitro* assays against human cancer cell lines. From these studies, a mechanism for prekinamycin biological activity emerged that predicts the presence (or absence) cytostatic activity against human cancer cell lines.

Kinamycin A

Prekinamycin

R = H
R = CH_3

R = H
R = CH_3

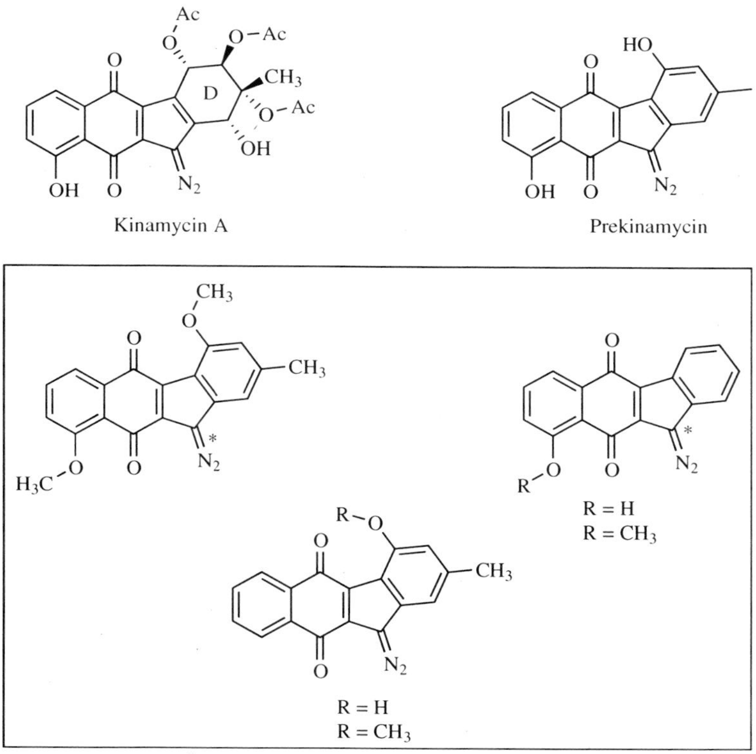

FIGURE 7.23 Prototype diazobenzo[*b*]fluorene-based natural products kinamycin A and prekinamycin. Compounds prepared for this study are shown in the inset.

Figure 7.23 shows the prototype diazobenzo[*b*]fluorene-based natural products kinamycin A and prekinamycin. The kinamycin A–D family were first isolated from *Streptomyces murayamaensis*, but the structures were incorrectly characterized as having a cyanobenzo[*b*]carbazole ring.[83,86,87] Since the initial discovery of the kinamycins, many new analogues have been discovered from natural sources.[88–92]

The global fitting provided rate data as a function of pH as well as intermediate spectra. The buildup of the quinone methide species observed with global fitting (see Fig. 7.2) correctly suggested that ^{13}C-NMR could detect the ^{13}C label of the 5-position of the quinone methide. The buildup of a species with a chemical shift of 94 ppm at the 5-position (see Fig. 7.3) confirmed that this transient intermediate is actually the quinone methide. Figure 7.24 shows the spectrum of the intermediate quinone methide species at pH 7.5 obtained from the global fit. Consistent with our structural assignment as the quinone methide, this intermediate spectrum has nearly identical λ_{max} and extinction coefficient values to those of a stable quinone methide analogue prepared as outlined in Scheme 7.21.

The global fitting studies revealed that the hydroquinone species shown in Scheme 7.5 affords the quinone methide at rates of $1–2 \times 10^{-3}$ min^{-1} that are independent of both pH (from 7 to 9.5) and the concentration of buffers used to hold pH. We interpreted this observation as protonation of the C-5 center followed by the slow loss on the nitrogen-leaving group. The anionic C-5 center of the electron-rich hydroquinone ring would be very basic resulting in complete protonation near neutrality.

The ^{13}C-NMR spectrum of the products arising from prekinamycin reductive activation is shown in Fig. 7.3. The identity of the compounds giving rise to the

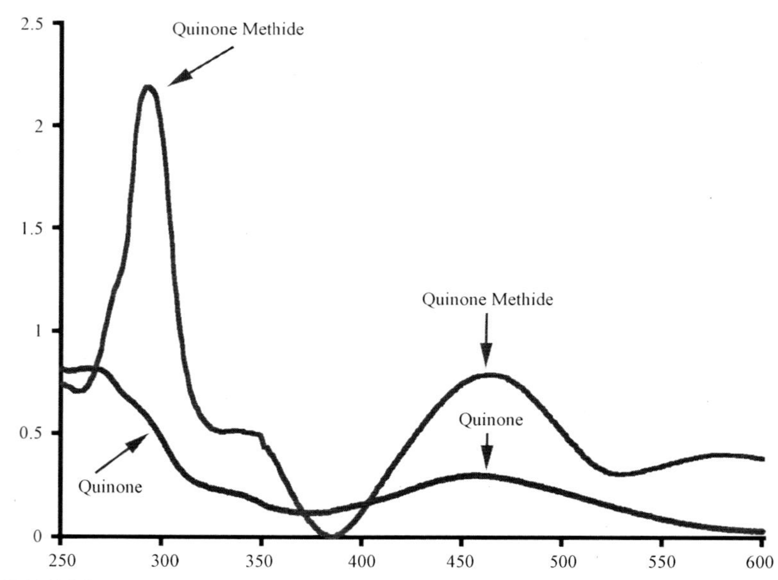

FIGURE 7.24 Spectra obtained from spectral global fitting to the surface shown in Fig. 7.2: spectrum of quinone at zero time; and the spectrum of the quinone methide intermediate.

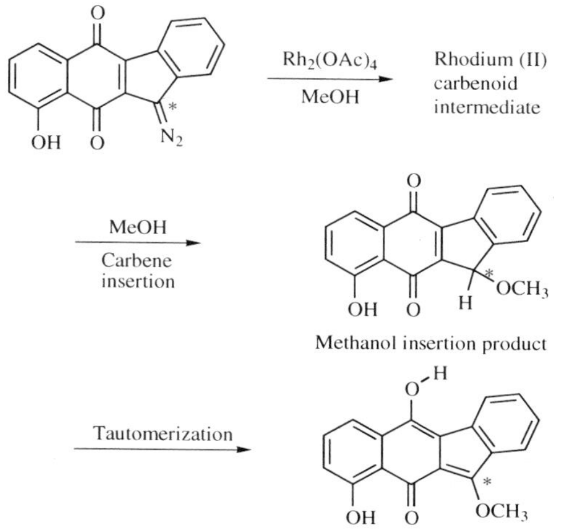

$Rh_2(OAc)_4$
MeOH
Rhodium (II) carbenoid intermediate
MeOH
Carbene insertion
Methanol insertion product
Tautomerization

SCHEME 7.21 Synthesis of a stable quinone methide using rhodium (II) acetate in methanol.[84,93] The ^{13}C label is designated with an asterisk (*).

resonances in this figure was determined based on their ^{13}C chemical shift and high-resolution MS of the isolates. Shown in Scheme 7.22 are the structures determined from these studies. The quinone methide species is apparent in Fig. 7.3 with its C-5 chemical shift of 94.2 ppm. The high-resolution MS of this species supports this assignment, although tautomerization to its keto form may have occurred during workup. The formation of the δ 125 ppm product results from the trapping of the quinone methide by water followed by oxidation and tautomerization, see Scheme 7.23 for the mechanism. Tautomerization of the δ 125 ppm product affords

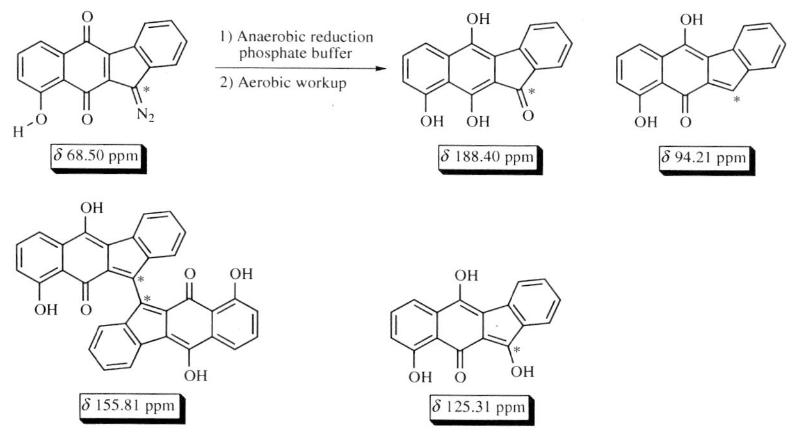

SCHEME 7.22 Products arising from prekinamycin reductive activation at pH 7.5. The ^{13}C labels are designated with asterisks (*).

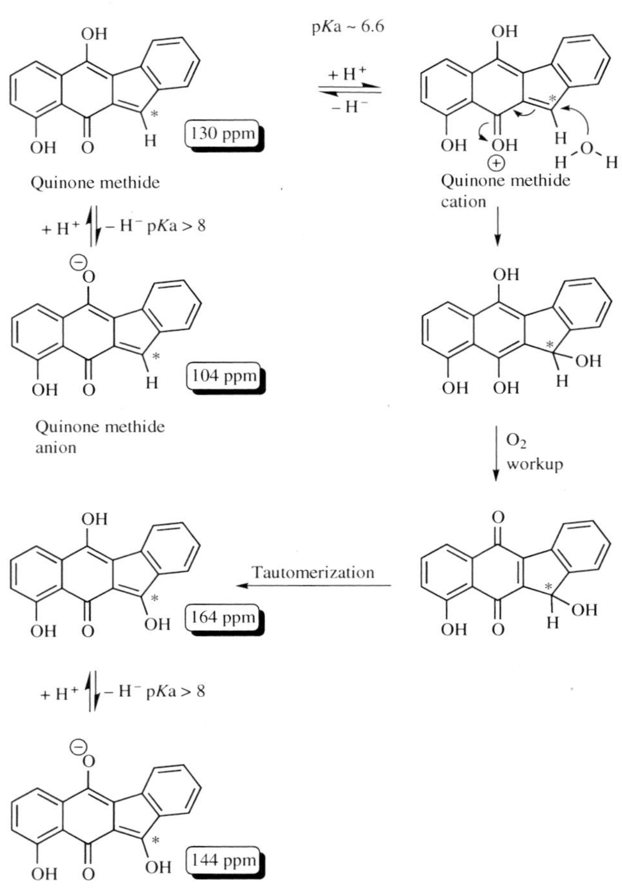

SCHEME 7.23 Mechanism of quinone methide trapping of water with calculated ^{13}C chemical shifts of the five-carbon center of methide enolates.

the keto form at δ 188 ppm that was readily isolated and identified without decomposition. The dimer at δ 155 ppm was isolated as a green solid that is NMR-silent due to radical formation. In summary, we observed reactions that are typical of quinone methides: nucleophile trapping, proton trapping, and dimerization.[25,30,31,33,94,95]

Shown in Scheme 7.23 are the mechanism of quinone methide trapping by water and subsequent oxidative and tautomerization reactions. Also shown in Scheme 7.23 are the calculated ^{13}C chemical shifts of the five-carbon center. Some of the chemical shifts observed in solution, see Fig. 7.3 and Scheme 7.22, are upfield shifted from the calculated values. For example, the ^{13}C chemical shifts of the five-carbon center of the quinone methide should possess a chemical shift of 130 ppm not 94 ppm. Similarly, the 5-hydroxy quinone methide should possess a chemical shift of 164 ppm not 125 ppm. The explanation for the upfield shifts is the equilibrium formation of quinone methide enolates by ionization of the 11-hydroxyl group. The formation of a quinone methide enolate was previously observed in the benzimidazole system (Scheme 7.3)[30] and is feasible at the reaction pH of 7.5.

We carried out a hydrolytic study of the stable prekinamycin quinone methide (prepared as shown in Scheme 7.21) to provide insights into its reactivity. Spectral global fitting studies indicated that this quinone methide was converted to its ketone by a single first-order process in aqueous buffers. Reactions held at pH values greater than 7 exhibited general acid buffer catalysis and buffer dilutions were carried out at constant pH values to determine lyate-only (water, hydroxide and hydronium) rate constants. The hydrolysis mechanism is shown in Scheme 7.24 and the inset shows the role of general acid catalysis. Plotting the log of lyate rate constants versus pH provided the pH–rate profile shown in this figure. These data were computer fit to the rate law shown in Fig. 7.25 and the solution was used to generalize the solid line passing through the data points.

The results enumerated above indicate that the quinone methide species must be protonated, by either a specific or general acid, to afford a cation before it can trap a nucleophile. The pK_a determined from pH–rate profile (p$K_a = 6.66$) is consistent with *O*-protonated quinone methide pK_a values of 6–7 discussed in Section 7.3.5.

We consider the relatively high pK_a values of 6–8 to be typical value for a cation–quinone methide equilibrium. The formation of a resonance-stabilized aromatic carbocation is one reason for these high pK_a values. Another reason is the high energy of the quinone methide. The thermodynamic cycle shown in

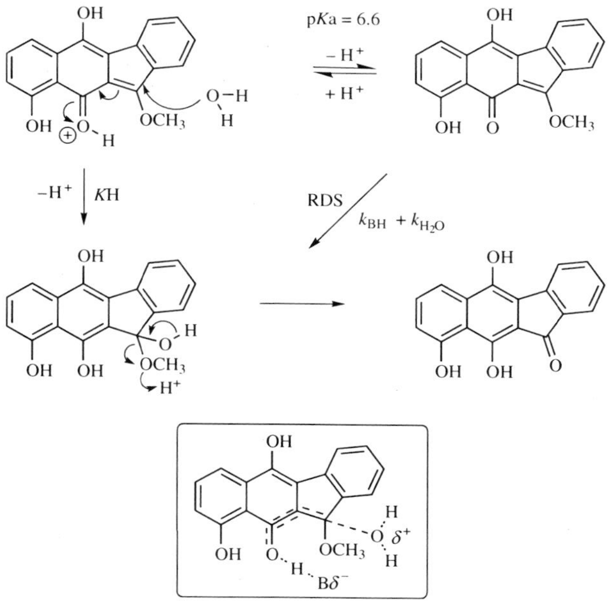

SCHEME 7.24 Hydrolysis mechanism of the prekinamycin quinone methide.

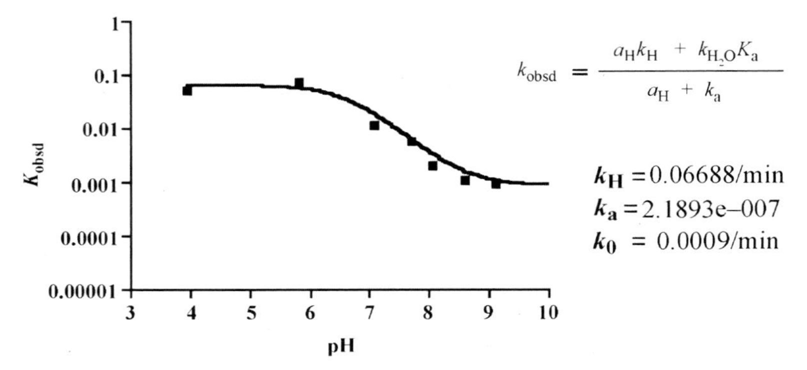

FIGURE 7.25 pH–rate profile for prekinamycin quinone methide hydrolysis.

Scheme 7.25 shows the role of quinone methide energy on the cation–quinone methide equilibrium. A high pK_a value for this equilibrium is expected if the energy of the quinone methide approaches that of the carbocation. To construct this cycle, we used the K_a values that we determined for the protonated ketone (p$K_a = -0.9$) and quinone methide (p$K_a = 6.6$). This pK_a difference requires that the keto form be more stable than the quinone methide by -10.2 kcal/mol. We obtained the calculated energy difference of 10.1 kcal/mol from Hartree–Fock calculations using 6–31G and STO-3G basis sets, inset of Scheme 7.25.

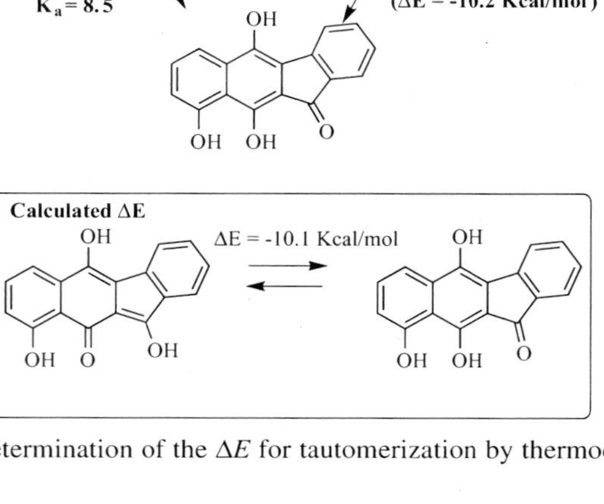

K_a = 2.19e^{-07}
-H$^+$
+H$^+$
K_a = 8.5
K_{ketone} = 3.4e^{07}
(ΔE = -10.2 Kcal/mol)

Calculated ΔE
ΔE = -10.1 Kcal/mol

SCHEME 7.25 Determination of the ΔE for tautomerization by thermodynamic cycle and calculation (inset).

The results cited above indicate that protonation of the prekinamycin quinone methide will afford a cation capable of trapping nucleophiles. The trapping of cellular nucleophiles by the prekinamycin quinone methide could result in cytostatic and perhaps cytotoxic activity. If the prekinamycin quinone methide readily tautomerizes to its keto form instead, nucleophile trapping will not be possible and there should be no activity against cells. To test this hypothesis, we correlated biological activity against five human cancer cell lines with the calculated ΔE (kcal/mol) values for quinone methide tautomerization. The results described below indicated a relationship where compounds with relatively stable keto forms are less active than those with a stable quinone methide (enol) form.

The results of ΔE calculations shown in Scheme 7.26 show that internal hydrogen bonds can influence the thermodynamics of quinone methide tautomerization in some instances. For the prekinamycin quinone methide without internal hydrogen bonds

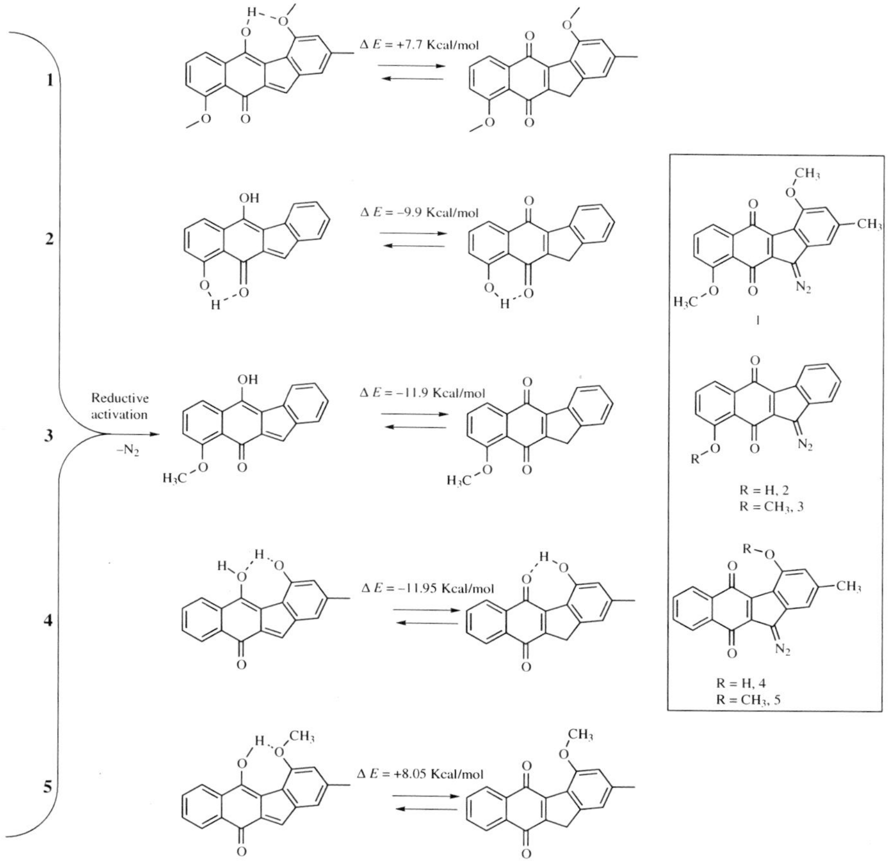

SCHEME 7.26 Calculated ΔE (kcal/mol) values for quinone methide (enol)/keto equilibrium.

TABLE 7.3 Concentrations in μg/mL that Cause 50% Growth Inhibition (GI_{50})[a]

Parameter GI_{50}	BxPC-3 Pancreas	MCF-7 Breast	SF-268 CNS	H460 NSC Lung	KM20L2 Colorectal	DU-145 Prostrate
1	0.52	0.43	0.53	1.2	0.46	4.4
2	1.3	1.1	2.8	0.6	3.1	0.6
3	>10	>10	>10	>10	>10	1.6
4	4.2	7.3	>10	>10	>10	>10
5	0.5	0.41	0.89	0.59	1.3	0.66

[a] GI_{50} values marked as >10 μg/mL were too high to measure. Compounds **3** and **4** are considered inactive. Key to compound numbers in Scheme 7.26.

(from prekinamycin **3**), the keto tautomer is favored by −11.9 kcal/mol. The negative ΔE values for tautomerization of quinone methides formed from prekinamycins **2** and **4** likewise indicate that the corresponding keto forms are favored. Internal hydrogen bonding is present in both of these quinone methides as well as in their corresponding keto tautomers, balancing out any stabilizing interactions. In contrast, the quinone methides derived from **1** and **5** possess an internal hydrogen bonding interaction involving the 1-methoxy group, but none in the corresponding keto forms. Furthermore, the angle of the internal hydrogen bond in both quinone methides (162°) is close to the colinearity (180°) of the ideal hydrogen bond. As a consequence, these quinone methides are more thermodynamically stable than the corresponding keto forms.

Table 7.3 shows the concentrations of **1–5** that result in 50% growth inhibition (GI_{50}) of five human cancer cell lines. Inspection of these data reveals that cytostatic activity of **1** and **3–5** depends on the thermodynamic favorability of the quinone methide species compared to the corresponding keto form. The most cytostatic prekinamycins **1** and **5** are associated with the thermodynamically stable quinone methides. In contrast, the inactive prekinamycins **3** and **4** are associated with thermodynamically stable keto tautomers. The exception is prekinamycin **2,** which is cytostatic and possesses a relatively stable keto tautomer **3** compared to its quinone methide. Although the ΔE value for quinone methide tautomerization can predict cytostatic properties, prekinamycin **2** shows that there must be other factors determining biological activity.

7.4 CONCLUSIONS AND FUTURE PROSPECTS

7.4.1 Quinone Methide O-Protonation

The most important conclusion of this research program is that quinone methide *O*-protonation is required for alkylation to occur. The quinone methide species is often referred to in the literature as an electrophilic species. Actually, the quinone methides obtained from reductive activation possess a slightly electron-rich methide center. There is electron release to the methide center by the hydroxyl, which is balanced by electron withdrawal to the carbonyl center, see Fig. 7.20. Thus, protonation of the quinone methide carbonyl oxygen affords a cationic species that is the actual electrophilic species.

The quinone methide reactive species is close in energy to the cationic species formed upon *O*-protonation. Consequently, the pK_a values of the *O*-protonated quinone methides have values that are close to neutrality. For example, we reported the pK_a for the mitomycin C carbocation–quinone methide equilibrium to be 7.1 [32] Recently, we reported pK_a values of 6.55 and 7.51 for protonated mitosene and pyrido[1,2-*a*]indole quinone methides, respectively.[42] If a bioreductive alkylating agent is to possess antitumor activity, the pK_a value for the cation–quinone methide equilibrium must be at or near neutrality so that the cationic alkylating agent is present in significant quantities.

7.4.2 Antitumor Agent Design

Our spectral global fitting and ^{13}C-labeling studies provided insights into how to design and active bioreductive alkylating agent. The quinone methide species must be sterically unencumbered at the reacting center, *O*-protonation of the quinone methide must be possible at physiological pH so that alkylation can occur, and the quinone methide must be long lived in aqueous solution.

Quinone methide formation in a five-membered ring will be sterically unencumbered and accessible to large biological nucleophiles. Examples include mitosene and the cyclopent[*b*]indole-based bioreductive alkylating agents, see Sections 7.3.5 and 7.3.6. The pK_a for the *O*-protonated of quinone methide should be at or near neutrality. The indole-based bioreductive alkylating agents will meet this requirement, but the electron-deficient benzimidazole (*O*-protonated quinone methide $pK_a = 5.5$) will not.[31] These pK_a values can be determined directly using spectral global fitting. Finally, the longevity of he quinone methide in solution can be determined using ^{13}C-NMR provided the methide centers are ^{13}C-labeled. Stable quinone methides can be rationally designed employing internal hydrogen bonds to stabilize the quinone methide tautomer (Scheme 7.25) or internal acyl transfer to stabilize the quinone methide by covalent modification (Scheme 7.20). Quinone methide stability, and hence activity against human cancer cell lines, can be assessed from the calculated ΔE (kcal/mol) values for the quinone methide (enol)/keto equilibrium (Scheme 7.26).

7.4.3 Enriched ^{13}C NMR Monitoring of Methide Reactions

With a ^{13}C label at the methide center, the presence of reactive methide intermediate can be verified and complex reaction products can be inventoried and eventually identified. The only limitations are the synthesis and cost involved in incorporation of the ^{13}C label. As a rule we, only use ^{13}C-labeled dimethylformamide and NaCN as starting materials because of their low cost and availability. Another limitation of enriched ^{13}C-NMR monitoring is dilution of the enriched label to natural abundance levels. Currently, we are developing isotope-editing techniques that utilize unnatural ^{13}C double labels to solve this problem.

The discoveries made possible by the use of ^{13}C labels are outlined below. A new ene-imine carbon–carbon bond coupling reaction was discovered that afforded novel polymers (Schemes 7.7 and 7.8). The presence of ^{13}C-labeled monomers permitted elucidation of the number of repeating units and type of linkages of the polymeric products. Dithionite reductive activation of mitosenes afforded products with sulfide leaving groups that could function as alkylating centers (Scheme 7.11 and Fig. 7.8). DNA adducts with methides were detected without laborious DNA degradation and adduct isolation (Figs. 9–12). Complex cyclopropyl quinone methide reaction products, which in some cases could not be separated, were readily identified using ^{13}C chemical shifts (Figs. 14–16). Finally, the prekinamycin quinone methide reaction products, which were converted to radicals upon contact with air, were identified using ^{13}C chemical shifts (Scheme 7.22). Finally, enriched ^{13}C-NMR aided in the identification of transient species that were only postulated to exist in solution (Figs. 3, 6 and 7.21).

7.4.4 Future Prospects

The spectral global fitting and enriched ^{13}C-NMR techniques are applicable to studying other processes besides methide forming reactions. Currently, this laboratory is investigating the antibiotic leinamycin and the DNA alkylating agent and food toxin ptaquiloside using these techniques.

We postulate that the double ^{13}C-labeling technique presented in this chapter could be used to study adducts on large pieces of DNA and even follow the chemical details cellular metabolic processes in real time. The double ^{13}C-labeling technique is currently being developed to solve problems in metabolism and toxicology.

This ene-imine coupling reaction presented in this chapter has been shown to afford polymers possessing diverse shapes, hydrogen bond donors/acceptor residues, and hydrophilic/hydrophobic residues. The reversibility of the coupling reaction suggests that ene-imine coupling can be employed in template-directed synthesis. Template-directed synthesis of receptors for DNA sequences and other targets is currently being developed.

REFERENCES

1. Moore, H. W.; Czerniak, R. Naturally occurring quinones as potential bioreductive alkylating agents. *Med. Res. Rev.* 1981, 1, 249–280.
2. Moore, H. W. Bioactivation as a model for drug design bioreductive alkylation. *Science* 1977, 197, 527–532.
3. Lin, A. J.; Cosby, L. A.; Shansky, C. W.; Sartorelli, A. C. Potential bioreductive alkylating agents. 1. Benzoquinone derivatives. *J. Med. Chem.* 1972, 15, 1247–1252.
4. Franck, R. W.; Tomasz, M. The chemistry of mitomycins. In *The Chemistry of Antitumor Agents;* Wilman, D. E. Ed.; Blackie & Sons, Ltd.: Glasgow, Scotland, 1990; pp. 379–394.
5. Siegel, D.; Beall, H.; Senekowitsch, C.; Kasai, M.; Arai, H.; Gibson, N. W.; Ross, D. Bioreductive activation of mitomycin C by DT-diaphorase. *Biochemistry* 1992, 31, 7879–7889.

6. Schiltz P.; Kohn H. Studies on the reactivity of reductively activity mitomycin C. *J. Am. Chem. Soc.* 1993, 115, 10510–10518.
7. Cummings, J. The role of reductive enzymes in cancer cell resistance to mitomycin C. *Drug Resist. Update* 2000, 3, 143–148.
8. Ping, W.; Yang, S.; Lixia, Z.; Hanping, H.; Xiang, Z. Quinone methide derivatives: important intermediates to DNA alkylating and DNA cross-linking actions. *Curr. Med. Chem.* 2005, 12, 2893–2913.
9. Belcourt, M. F.; Hodnick, W. F.; Rockwell, S.; Sartorelli, A. C. Bioactivation of mitomycin antibiotics by aerobic and hypoxic Chinese hamster ovary cells overexpressing DT-diaphorase. *Biochem. Pharm.* 1996, 51, 1669–1678.
10. Gutierrez, P. L. The role of NAD(P)H oxidoreductase (DT-diaphorase) in the bioactivation of quinone-containing antitumor agents: a review. *Free Radic. Biol. Med.* 2000, 29, 263–275.
11. Rauth, A. M.; Goldberg, Z.; Misra, V. DT-diaphorase: possible roles in cancer chemotherapy and carcinogenesis. *Oncol. Res.* 1997, 9, 339–349.
12. Garrett, M. D.; Workman, P. Discovering novel chemotherapeutic drugs for the third millennium. *Eur. J. Cancer* 1999, 35, 2010–2030.
13. Stratford, I. J.; Workman, P. Bioreductive drugs into the next millennium. *Anti-Cancer Drug Des.* 1998, 13, 519–528.
14. Workman, P. Enzyme-directed bioreductive drug development revisted: a commentary on recent progress and future prospects with emphasis on quinone anticancer agents and quinone metabolizing enzymes, particularly DT-diaphorase. *Oncol. Res.* 1994, 6, 461–475.
15. Ross, D. T.; Scherf, U.; Eisen, M. B.; Perou, C. M.; Rees, C.; Spellman, P.; Iyer, V.; Jeffrey, S. S.; Van De Rijn, M.; Waltham, M.; Pergamenschikov, A.; Lee, J. C. E.; Lashkari, D.; Shalon, D.; Myers, T. G.; Weinstein, J. N.; Botstein, D.; Brown, P. O. Systematic variation in gene expression patterns in human cancer cell lines. *Nat. Genet.* 2000, 24, 227–235.
16. Lin, T. S.; Antonini, I.; Cosby, L. A.; Sartorelli, A. C. 2,3-Dimethyl-1,4-naphthoquinone derivatives as bioreductive alkylating agents with crosslinking potential. *J. Med. Chem.* 1984, 27, 813–815.
17. Lin, T.-S.; Tiecher, B. A.; Sartorelli, A. C. 2-Methylanthraquinone derivatives as potential bioreductive alkylating agents. *J. Med. Chem.* 1980, 23, 1237–1242.
18. Antonini, I.; Lin, T. S.; Cosby, L. A.; Dai, Y. R.; Sartorelli, A. C. 2- and 6-Methyl-1,4-naphthoquinone derivatives as potential bioreductive alkylating agents. *J. Med. Chem.* 1982, 25, 730–735.
19. Lin, A. J.; Pardini, R. S.; Cosby, L. A.; Lillis, B. J.; Shansky, C. W.; Sartorelli, A. C. Potential bioreductive alkylating agents. 2. Antitumor effect and biochemical studies of naphthoquinone derivatives. *J. Med. Chem.* 1973, 16, 1268–1271.
20. Lin, A. J.; Shansky, C. W.; Sartorelli, A. C. Potential bioreductive alkylating agents. 3. Synthesis and antineoplastic activity of acetoxymethyl and corresponding ethyl carbamate derivatives of benzoquinones. *J. Med. Chem.* 1974, 17, 558–561.
21. Lin, A. J.; Pardini, R. S.; Lillis, B. J.; Sartorelli, A. C. Potential bioreductive alkylating agents. 4. Inhibition of coenzyme Q enzyme systems by lipoidal benzoquinone and naphthoquinone derivatives. *J. Med. Chem.* 1974, 17, 668–672.

22. Lin, A. J.; Lillis, B. J.; Sartorelli, A. C. Potential bioreductive alkylating agents. 5. Antineoplastic activity of quinoline-5,8-diones, naphthazarins, and naphthoquinones. *J. Med. Chem.* 1975, 18, 917–921.
23. Lin, A. J.; Sartorelli, A. C. 2,3-Dimethyl-5,6-bis(methylene)-1,4-benzoquinone. Active intermediate of bioreductive alkylating agents. *J. Org. Chem.* 1973, 38, 813–815.
24. Lee, C.-H.; Skibo, E. B. Active site directed reductive alkylation of xanthine oxidase by imidazo[4,5-*g*]quinazoline-4,9-diones functionalized with a leaving group. *Biochemistry* 1987, 26, 7355–7362.
25. Lemus, R. L.; Skibo, E. B. Studies of extended quinone methides. Design of reductive alkylating agents based on the quinazoline ring system. *J. Org. Chem.* 1988, 53, 6099–6105.
26. Skibo, E. B.; Gilchrist, J. H. Synthesis and electrochemistry of pyrimidoquinazoline-5,10-diones. design of hydrolytically stable high potential quinones and new reductive alkylation systems. *J. Org. Chem.* 1988, 53, 4209–4218.
27. Lemus, R. L.; Lee, C. H.; Skibo, E. B. Studies of extended quinone methides. Synthesis and physical studies of purine-like monofunctional and bifunctional imidazo[4,5-*g*]quinazoline reductive alkylating agents. *J. Org. Chem.* 1989, 54, 3611–3618.
28. Lemus, R. H.; Skibo, E. B. Design of pyrimido[4,5-*g*]quinazoline-based anthraquinone mimics. structure–activity relationship for quinone methide formation and the influence of internal hydrogen bonds on quinone methide fate. *J. Org. Chem.* 1992, 57, 5649–5660.
29. Thompson, D. C.; Thompson, J. A.; Sugumaran, M.; Moldeus, P. Biological and toxicological consequences of quinone methide formation. *Chem.-Biol. Interact.* 1993, 86, 129–162.
30. Skibo E. B. Studies of extended quinone methides, the hydrolysis mechanism of 1-methyl-2-(bromomethyl)-4,7-dihydroxybenzimidazole. *J. Org. Chem.* 1986, 51, 522–527.
31. Skibo, E. B. Formation and fate of benzimidazole-based quinone methides. Influence of pH on quinone methide fate. *J. Org. Chem.* 1992, 57, 5874–5878.
32. Boruah, R. C.; Skibo, E. B. Determination of the pK_a values for the mitomycin C redox couple by tritration, pH rate profile, and Nernst–Clark fits. Studies of methanol elimination, carbocation formation, and the carbocation/quinone methide equilibrium. *J. Org. Chem.* 1995, 60, 2232–2243.
33. Fisher, J.; Ramakrishnan, K.; Becvar, J. E. Direct enzyme-catalyzed reduction of anthracyclines by reduced nicotinamide adenine dinucleotide. *Biochemistry* 1983, 22, 1347–1355.
34. Brand, D. J.; Fisher, J. Tautomeric instability of 10-deoxydaunomycinone. *J. Am. Chem. Soc.* 1986, 108, 3088–3096.
35. Bird, D. M.; Gaudiano, G.; Koch, T. H. Leucodaunomycins: new intermediates in the redox chemistry of daunomycin. *J. Am. Chem. Soc.* 1991, 113, 308–315.
36. Ravelo, A. G.; Estevez-Braun, A.; Chavez-Orellana, H.; Perez-Sacau, E.; Mesa-Siverio, D. Recent studies on natural products as anticancer agents. *Curr. Top. Med. Chem.* 2004, 4, 241–265.
37. Chiang, Y.; Kresge, A. J.; Hellrung, B.; Schunemann, P.; Wirz J. Flash photolysis of 5-methyl-1,4-naphthoquinone in aqueous solution: kinetics and mechanism of photoenolization and of enol trapping. *Helv. Chim. Acta* 1997, 80, 1106–1121.
38. Chiang, Y.; Kresge, A. J.; Zhu, Y. Kinetics and mechanisms of hydration of *o*-quinone methides in aqueous solution. *J. Am. Chem. Soc.* 2000, 122, 9854–9855.

39. Beecham, J. M.; Brand, L. Global analysis of fluorescence decay: applications to some unusual experimental and theoretical studies. *Photochem. Photobiol.* 1986, 44, 323–329.

40. Matheson, I. B. C. The method of successive integration: a general technique for recasting kinetic equations in a readily soluble form which is linear in the coefficients and sufficiently rapid for real time instrumental use. *Anal. Instrum.* 1987, 16, 345–373.

41. LaBarbera, D. V.; Skibo, E. B. Solution kinetics of CC-1065 A-ring opening: substituent effects and general acid/base catalysis. *J. Am. Chem. Soc.* 2006, 128, 3722–3727.

42. Khdour O.; Skibo E. B. Chemistry of pyrrolo[1,2-*a*]indole- and pyrido[1,2-*a*]indole-based quinone methides. Mechanistic explanations for differences in cytostatic/cytotoxic properties. *J. Org. Chem.* 2007, 72, 8636–8647.

43. Beckmann, N. *Carbon-13 NMR Spectroscopy of Biological Systems;* Academic Press: New York, 1995.

44. Golding, B. T.; Bleasdale, C.; McGinnis, J.; Muller, S.; Rees, H. T.; Rees, N. H.; Farmer, P. B.; Watson, W. P. The mechanism of decomposition of *N*-methyl-*N*-nitrosourea (MNU) in water and a study of its reactions with 2′-deoxyguanosine, 2′-deoxyguanosine 5′-monophosphate and d(GTGCAC). *Tetrahedron* 1997, 53, 4063–4082.

45. Ouyang, A.; Skibo, E. B. The iminium ion chemistry of mitosene DNA alkylating agents. enriched ^{13}C-NMR studies. *Biochemistry* 2000, 39, 5817–5830.

46. Ouyang, A.; Skibo, E. B. Design of a cyclopropyl quinone methide reductive alkylating agent. *J. Org. Chem.* 1998, 63, 1893–1900.

47. Khdour, O.; Ouyang, O.; Skibo, E. B. Design of a 'cyclopropyl quinone methide' reductive alkylating agent II. *J. Org. Chem.* 2006, 71, 5855–5863.

48. Skibo, E. B.; Xing, C.; Groy, T. Recognition and cleavage at the DNA major groove. *Bioorg. Med. Chem.* 2001, 9, 2445–2459.

49. Maliepaard, M.; deMol, N. J.; Tomasz, M.; Gargiulo, D.; Janssen, L. H. M.; vanDuynhoven, J. P. M.; vanVelzen, E. J. J.; Verboom, W.; Reinhoudt, D. N. Mitosene–DNA adducts. Characterization of two major DNA monoadducts formed by 1,10-bis(acetoxy)-7-methoxymitosene upon reductive activation. *Biochemistry* 1997, 36, 9211–9220.

50. Kohn, H.; Zein, N.; Lin, X. Q.; Ding, J.-Q.; Kadish, K. M. Mechanistic studies on the mode of reaction of mitomycin c under catalytic and electrochemical reduction conditions. *J. Am. Chem. Soc.* 1987, 109, 1833–1840.

51. Orlemans, E. O. M.; Verboom, M. W.; Scheltinga, M. W.; Reinhoudt, D. N.; Lelieveld, P.; Fiebig, H. H.; Winterhalter, B. R.; Double, J. A.; Bibby, M. C. Synthesis, mechanism of action, and biological evaluation of mitosenes. *J. Med. Chem.* 1989, 32, 1612–1620.

52. Kohn, K. W. Beyond DNA cross-linking: history and prospects of DNA-targeted cancer treatment—Fifteenth Bruce F. Cain Memorial Award Lecture. *Cancer Res.* 1996, 56, 5533–5546.

53. Maliepaard, M.; Sitters, C. A. M. C.; de Mol, N. J.; Janssen, L. H.; Stratford, I. J.; Stephens, M.; Verboon, W.; Reinhoudt, D. N. Potential antitumor mitosenes: relationship between *in vitro* DNA interstrand cross-link and DNA damage in *Escherichia coli* K-12 strains. *Biochem. Pharmacol.* 1994, 48, 1371–1377.

54. Leete, E. 3-Hydroxymethylindoles. *J. Am. Chem. Soc.* 1959, 81, 6023–6026.

55. Remers, W. A.; Roth, R. H.; Weiss, M. J. The mitomycin antibiotics. Synthetic studies. IV. Introduction of the 9-hydroxymethyl group into the 1-ketopyrrolo[1,2-*a*]indole system. *J. Am. Chem. Soc.* 1964, 86, 4612–4617.
56. Allen, G. R.; Poletto, J. F.; Weiss, M. J. The mitomycin antibiotics. Synthetic studies. II. The synthesis of 7-mehtoxymitosene an antibacterial agent. *J. Am. Chem. Soc.* 1964, 86, 3877–3879.
57. Thornton-Manning, J.; Appleton, M. L.; Gonzalez, F. J.; Yost, G. S. Metabolism of 3-methylindole by vaccinia-expressed P450 enzymes: correlation of 3-methyleneindolenine formation and protein binding. *J. Pharmacol. Exp. Ther.* 1996, 276, 21–29.
58. Lanza, D. L.; Code, E.; Crespi, C. L.; Gonzalez, F. J.; Yost, G. S. Specific dehydrogenation of 3-methylindole and epoxidation of naphthalene by recombinant human CYP2F1 expressed in lymphoblastoid cells. *Drug Metab. Dispos.* 1999, 27, 798–803.
59. Lanza, D. L.; Yost, G. S. Selective dehydrogenation/oxygenation of 3-methylindole by cytochrome P450 enzymes. *Drug Metab. Dispos.* 2001, 29, 950–953.
60. Ruangyuttikarn, W.; Skiles, G. L.; Yost, G. S. Identification of a cysteinyl adduct of oxidized 3-methylindole from goat lung and human liver microsomal proteins. *Chem. Res. Toxicol.* 1992, 5, 713–719.
61. Loneragan, G. H.; Gould, D. H.; Mason, G. L.; Garry, F. B.; Yost, G. S.; Lanza, D. L.; Miles, D. G.; Hoffman, B. W.; Mills, L. J. Association of 3-methyleneindolenine, a toxic metabolite of 3-methylindole, with acute interstitial pneumonia in feedlot cattle. *Am. J. Vet. Res.* 2001, 62, 1525–1530.
62. Regal, K. A.; Laws, G. M.; Yuan, C.; Yost, G. S.; Skiles, G. L. Detection and characterization of DNA adducts of 3-methylindole. *Chem. Res. Toxicol.* 2001, 14, 1014–1024.
63. Tomasz, M.; Mercado, C. M.; Olson, J.; Chatterjie, N. The mode of interaction of mitomycin C with deoxyribonucleic acid and other polynucleotides *in vitro*. *Biochemistry* 1974, 13, 4878–4887.
64. Lyons, D.; Nickless, G. The lower oxy-acids of sulphur. In *Inorganic Sulphur Chemistry;* Nickless, G.Ed.; Elsevier: New York, 1968; Chapter 14, pp. 519–522.
65. Distler, H. The chemistry of Bunte salts. *Angew. Chem. Int. Ed.* 1967, 6, 544–553.
66. Kalinowski, H.-O.; Berger, S.; Braun, S. ^{13}C-*NMR-Spektroskopie;* Georg Thieme Verlag: Stuttgart 1984.
67. Schiltz, P.; Kohn, H. Sodium dithionite-mediate mitomycin C reductive activation processes. *Tetrahedron Lett.* 1992, 33, 4709–4712.
68. McGuinness, B. F.; Lipman, R.; Nakanishi, K.; Tomasz, M. J. Reaction of sodium dithionite activated mitomycin C with guanine at non-cross-linkable sequences of oligonucleotides. *J. Org. Chem.* 1991, 56.
69. Palom, Y.; Lipman, R.; Musser, S. M.; Tomasz, M. A mitomycin-*N*-6-deoxyadenosine adduct isolated from DNA. *Chem. Res. Toxicol.* 1998, 11, 203–210.
70. Zhou, Q. B.; Turnbull, K. D. Phosphodiester alkylation with a quinone methide. *J. Org. Chem.* 1999, 64, 2847–2851.
71. Xing, C.; Skibo, E. B.; Dorr, R. T. Aziridinyl quinone antitumor agents based on indoles and cyclopent[*b*]indoles: structure–activity relationships for cytotoxicity and antitumor activity. *J. Med. Chem.* 2001, 44, 3545–3562.
72. Liu, R.; Zhang, P.; Gan, T.; Cook, J. M. Regiospecific bromination of 3-methylindoles with NBS and its application to the concise synthesis of optically active unusual tryptophans present in marine cyclic peptides. *J. Org. Chem.* 1997, 62, 7447–7456.

73. Remers, W. A. Mitomycins and porfiromycin. In *The Chemistry of Antitumor Antibiotics;* John Wiley & Sons Inc.: New York, 1979; Vol. 1, pp. 221–276.

74. Boger, D. L.; Johnson, D. S. CC-1065 and the duocarmycins: unraveling the keys to a new class of naturally derived DNA alkylating agents. *Proc. Nat. Acad. Sci. USA* 1995, 92, 3642–3649.

75. Boger, D. L.; Johnson, D. S. CC-1065 and the duocarmycins: understanding their biological function through mechanistic studies. *Angew. Chem. Int. Ed.* 1996, 35, 1438–1474.

76. Tomasz, M.; Palom, Y. The mitomycin bioreductive antitumor agents: cross-linking and alkylation of DNA as the molecular basis of their activity. *Pharmacol. Ther.* 1997, 76, 73–87.

77. Xing, C. G.; Wu, P.; Skibo, E. B.; Dorr, R. T. Design of cancer-specific antitumor agents based on aziridlinylcyclopent[*b*]indoloquinones. *J. Med. Chem.* 2000, 43, 457–466.

78. Schulz, W. G.; Nieman, R. A.; Skibo, E. B. Evidence for DNA phosphate backbone alkylation and cleavage by pyrrolo[1,2-*a*] benzimidazoles, small molecules capable of causing sequence specific phosphodiester bond hydrolysis. *Proc. Natl. Acad. Sci. USA* 1995, 92, 11854–11858.

79. Maxam, A. M.; Gilbert, W. Sequencing end-labeled DNA with base-specific chemical cleavages. *Methods Enzymol.* 1980, 65, 499–560.

80. Nakatani, K.; Matsuno, T.; Adachi, K.; Hagihara, S.; Saito, I. Selective intercalation of charge neutral intercalators into GG and CG steps: implication of HOMO–LUMO interaction for sequence-selective drug intercalation into DNA. *J. Am. Chem. Soc.* 2001, 123, 5695–5702.

81. Marco-Contelles, J.; Molina, M. T. Naturally occurring diazo compounds: the kinamycins. *Curr. Org. Chem.* 2003, 7, 1433–1442.

82. Hasinoff, B. B.; Wu, X.; Yalowich, J. C.; Goodfellow, V.; Laufer, R. S.; Adedayo, O.; Dmitrienko, G. I. Kinamycins A and C, bacterial metabolites that contain an unusual diazo group, as potential new anticancer agents: antiproliferative and cell cycle effects. *Anti-Cancer Drugs* 2006, 17, 825–837.

83. Omura, S.; Nakagawa, A.; Yamada, H.; Hata, T.; Furusaki, A. Structures and biological properties of kinamycin A, B, C, and D. *Chem. Pharm. Bull.* 1973, 21, 931–940.

84. (a) Khdour, O.; Skibo, E. B., Quinone methide and cation chemistry of prekinamycins: In Vitro Correlations. *Org. Biomol Chem* 2009, 7. (b) Khdour, O.; Skibo, E. B., Evidence of the Prekinamycin Quinone Methide. Synthesis of 13C-Labelled Prekinamycins and Spectral Global Fitting. *Org. Biomol. Chem.* 2009, 7.

85. Durig, J. R.; Berry, R. J.; Durig, D. T.; Sullivan, J. F.; Little, T. S. The use of ab initio calculations in molecular spectroscopy. *Teubner-Texte Phys.* 1988, 20, 54–68.

86. Ito, S.; Matsuya, T.; Omura, S.; Otani, M.; Nakagawa, A. A new antibiotic, kinamycin. *J. Antibiot.* 1970, 23, 315–317.

87. Hata, T.; Omura, S.; Iwai, Y.; Nakagawa, A.; Otani, M.; Ito, S.; Matsuya, T. New antibiotic, kinamycin. Fermentation, isolation, purification, and properties. *J. Antibiot.* 1971, 24, 353–359.

88. Lin, H. C.; Chang, S. C.; Wang, N. L.; Chang, L. R. FL-120A.apprx.D′, new products related to kinamycin from *Streptomyces chattanoogensis* subsp. *taitungensis* subsp. nov. I. Taxonomy, fermentation and biological properties. *J. Antibiot.* 1994, 47, 675–680.

89. Young, J. J.; Ho, S. N.; Ju, W. M.; Chang, L. R. FL-120A-D′, new products related to kinamycin from *Streptomyces chattanoogensis* subsp. *taitungensis* subsp. nov. II. Isolation and structure determination. *J. Antibiot.* 1994, 47, 681–687.
90. Cone, M. C.; Seaton, P. J.; Halley, K. A.; Gould, S. J. New products related to kinamycin from *Streptomyces murayamaensis*. I. Taxonomy, production, isolation and biological properties. *J. Antibiot.* 1989, 42, 179–188.
91. Seaton, P. J.; Gould, S. J. New products related to kinamycin from *Streptomyces murayamaensis*. II. Structures of pre-kinamycin, keto-anhydrokinamycin, and kinamycins E and F. *J. Antibiot.* 1989, 42, 189–197.
92. Isshiki, K.; Sawa, T.; Naganawa, H.; Matsuda, N.; Hattori, S.; Hamada, M.; Takeuchi, T.; Oosono, M.; Ishizuka, M. 3-*O*-Isobutyrylkinamycin C and 4-deacetyl-4-*O*-isobutyrylkinamycin C, new antibiotics produced by a *Saccharothrix* species. *J. Antibiot.* 1989, 42, 467–469.
93. Paulissen, R.; Reimlinger, H.; Hayez, E.; Hubert, A. J.; Teyssie, P. Transition metal catalysed reactions of diazocompounds. II. Insertion in the hydroxylic bond. *Tetrahedron Lett.* 1973, 14, 2233–2236.
94. Angle, S. R.; Rainier, J. D.; Woytowicz, C. Synthesis and chemistry of quinone methide models for the anthracycline antitumor antibiotics. *J. Org. Chem.* 1997, 62, 5884–5892.
95. Rokita, S. E.; Yang, J. H.; Pande, P.; Greenberg, W. A. Quinone methide alkylation of deoxycytidine. *J. Org. Chem.* 1997, 62, 3010–3012.

8

NATURAL DITERPENE AND TRITERPENE QUINONE METHIDES: STRUCTURES, SYNTHESIS, AND BIOLOGICAL POTENTIALS

QIBING ZHOU

Department of Chemistry, Virginia Commonwealth University, 1001 West Main Street, Richmond, VA 23284-2006, USA

8.1 INTRODUCTION

Quinone methides (QMs) are reactive intermediates that have been involved in the covalent modification of biomolecules such as DNA and proteins to impact related biological events and subsequent processes.[1–6] The occurrence of the QM structure in natural products has been found with diterpenoids and triterpenoids as exemplified by diterpene QM taxodione **1** and triterpene QM pristimerin **2** (Scheme 8.1). As compared to a "simple" *p*-QM **3**, the QM structures of the terpenoids are highly conserved with some common features including a hydroxyl group *ortho* to the carbonyl group and extended conjugation at the *exo*-cyclic methylene of the QM. Structurally, the *o*-hydroxyl group enhances the reactivity of the QM through an intramolecular hydrogen bond while the extended conjugation counteracts such activation. In addition, the stability of the QM is further controlled by the steric and conformation of terpenoids. As a result, the reactivity of the QM is "fine-tuned" toward certain nucleophiles selectively, which is well consistent with the observed biological activities. In the discussion of this chapter, various natural diterpene QMs and their biological activities are

Quinone Methides, Edited by Steven E. Rokita

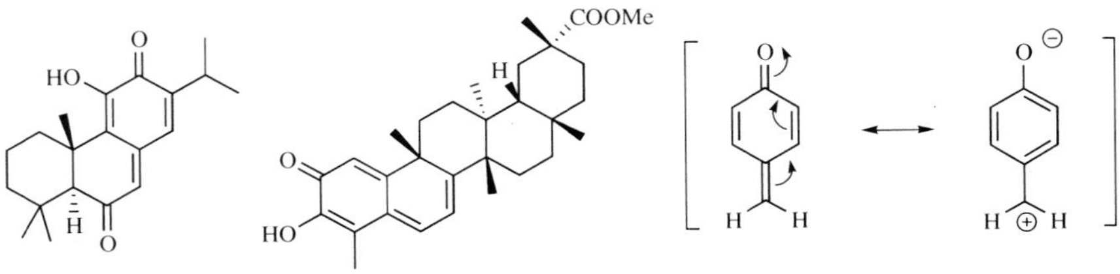

1 2 3 4

SCHEME 8.1 Examples of natural diterpene and triterpene QMs.

presented first, followed by three unique approaches in the total synthesis of diterpene QMs. The discussion of triterpene QMs mainly focuses on the recent progress of the biomolecular targets and pathways, and the potential in the treatment of neurodegenerative diseases. Finally, the *in situ* generation of reactive oxygen species is reviewed since this may also be involved in the biological activity of natural terpene QMs.

8.2 NATURAL DITERPENE QMs

Since its discovery in 1968, taxodione **1** has been found in many plants,[7–8] especially in the roots of the *Salvia* family including *S. aspera*,[9] *S. broussonetii*,[10] *S. lanigera*,[11] *S. montbretii*,[12] *S. moorcrafiana*,[13] *S. nipponica*,[14] *S. pachystachys*,[15] *S. prionitis*,[16] and *S. verbenaca*.[11] In addition, a variety of diterpenoids with similar QM structures have been identified in plants such as *Salvia*, *Coleus*, and *Plectranthus*, and most of them showed effective antimicrobial and antifungal activities.[17–45] In the light of these biological activities, it may be reasonably argued that diterpene QMs are involved in chemowarfare of these plants.

8.2.1 Chemical Structures and Biological Activity of Natural Diterpene QMs

Taxodone **5**, an unconjugated diterpene QM, was isolated along with taxodione **1** from the extractions of *Taxodium distichum* seeds[7–8] and *Salvia pachyphylla* and *S. munzii* roots[17–18] (Scheme 8.2). Structurally, taxodone **5** differs from taxodione **1** at the B ring with a hydroxyl group and is converted to taxodione **1** upon a mild acid treatment and subsequent oxidation with silica gel and O_2. Taxodone **5** exhibited a stronger tumor-suppressive effect than taxodione **1** in rats at doses between 14 and 25 mg/kg. However, toxicity was observed with taxodone **5** at a higher dose of 50 mg/kg.[7–8]

Related to taxodione **1**, diterpene QMs **6** and **7** were identified as the most active components for the observed antifungal activity from the root bark of *Bobgunnia madagascariensis*, a tree commonly found in tropical Africa and used in folklore medicine (Scheme 8.2).[19–20] Diterpene QM **6** exhibited a 50% inhibition of fungi *Candida albicans*, *Candida globrata*, and *Candida krusei* at a concentration of 0.2 μg/mL, which was lower than that of the known antifungal agents, amphotericin B and fluconazole. On the contrary, a concentration of 7.0 μg/mL was required for the isomeric mixture of **7** to show the same effect, possibly due to the presence of the hemiacetal group in the structure.

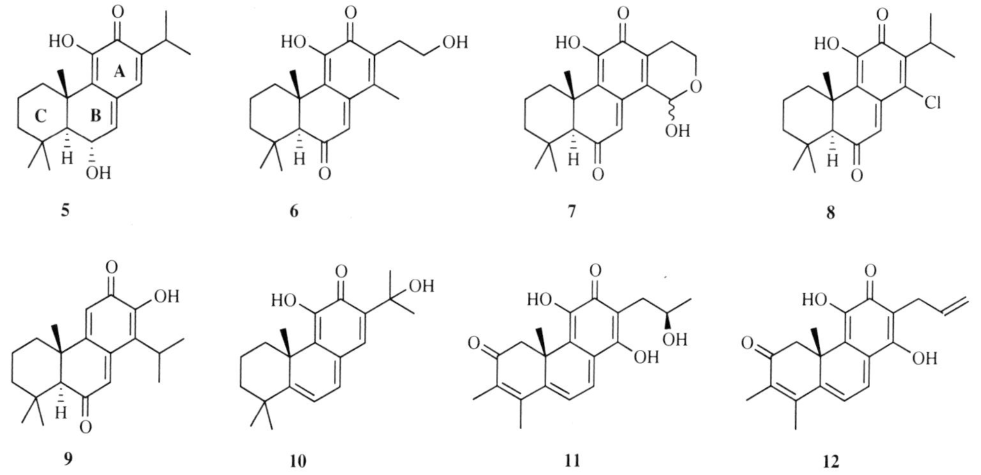

5 6 7 8

9 10 11 12

SCHEME 8.2 Chemical structures of various natural diterpene QMs.

Chlorinated taxodione **8** was also found along with taxodione **1** from the stem of *Rosemarinus officinalis*, although its biological activity is yet to be studied (Scheme 8.2).[21] Maytenoquinone **9**, a structural isomer of taxodione **1**, has been isolated from the roots of several medicinal plants such as *Maytenus dispermus*,[22] *Salvia melissodora*,[23] and *Harpagophytum procumbems* (devil's claw)[24] used in folklore medicine.

Diterpenoids containing the QM moiety conjugated with alkenes have been isolated as early as 1966.[25] Fuerstion **10** was first isolated from *Fuerstia solchen*, and similar derivatives were reported from the roots of *Salvia moorcrafiana* and *Hoslundia opposita* (Scheme 8.2).[25–28] In contrast to taxodione and its derivatives, there were few reported biological activities of fuerstion **10** and related derivatives; although antimalaria activity was found with fuerstion benzoic ester conjugates.[28] The lack of biological activity has been attributed to the conversion of fuerstion **10** under acidic conditions to an inactive metabolite **13** (Scheme 8.3),[25–27] which was formed through a rearrangement of the methyl group of the protonated QM **14**, followed by the opening of the C ring and oxidation of the resulting catechol **15**. Further extension of the conjugated QM with alkenes (Scheme 8.2) led to natural products coleons E **11** and F **12**, which were isolated from the gland of *Coleus barbatus* leaves.[29–31]

Coleons U **17** and V **18** are the structural tautomers of hydroxylated taxodione **19** in the enol–ketal forms and isolated from *Plectranthus hereroensis* and *Coleus xanthanthus* (Scheme 8.4).[32–35] Coleon U **17** exhibited antiproliferation effects against human breast, lung, glioblastoma, renal, and melanoma cancer cell lines at 5 μM range[33-34] and some antibacterial effect at 1 μg/mL,[35] although the role of hydroxylated taxodione **19** was not clear. One interesting point to note is that both coleons **17** and **18** were products of a diterpene QM **23**, possibly via the oxidation of the hydroxyl adduct **24** (Scheme 8.4).[36] Diterpene QM **23** was obtained from the oxidation of natural diterpene 6β-hydroxyroyleanone **20**[37] and found in equilibrium with its tautomer *para*-quinone **25**.[36] Closely related to this category of diterpene QM are coleon C **26**[38] with a hydroxylated isopropyl group on the A ring and coleon S **27**[39] containing a

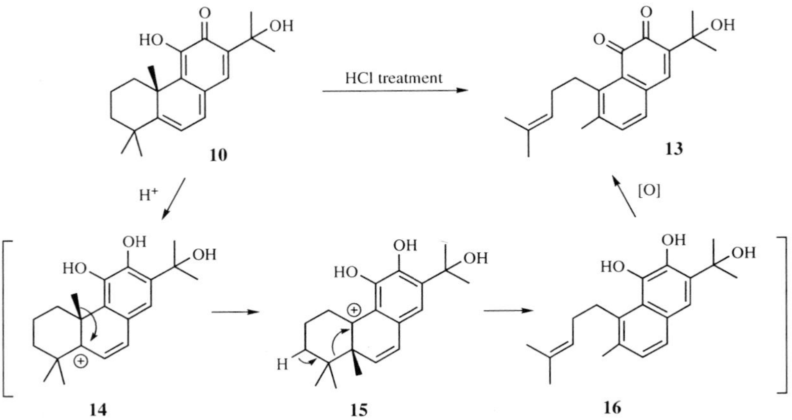

SCHEME 8.3 Acid-catalyzed rearrangement of fuerstion **10**.

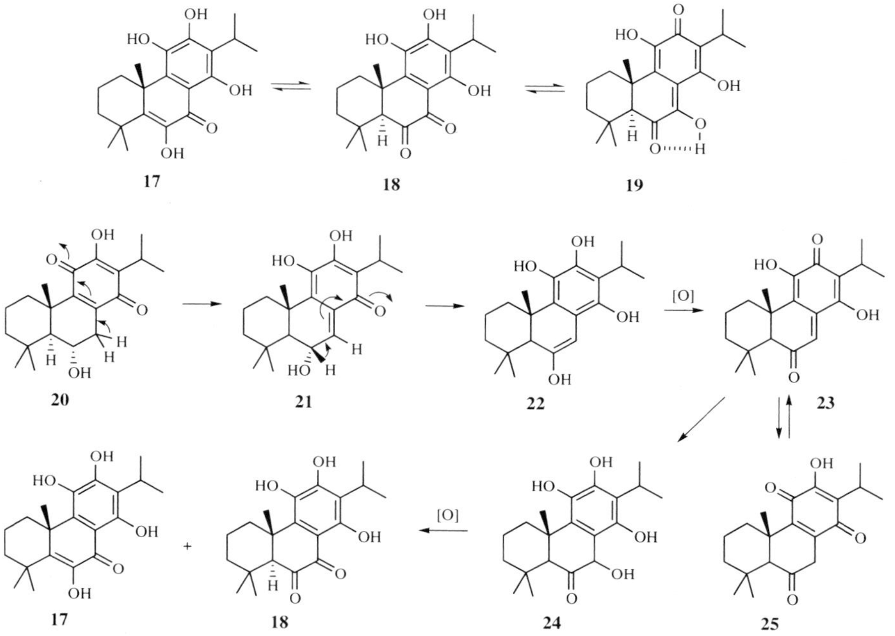

SCHEME 8.4 Tautomerization of coleons **17** and **18** and formation through QMs.

hydroxyl group on the C ring (Scheme 8.5). Similar derivatives are also reported with celaphanol A **29**[40] and dehydroxylated coleon U **28** (Scheme 8.5),[41] which is a selective insecticidal agent.[10]

Other variations of natural diterpene QM included unconjugated diterpene QMs **30–32** from the roots of *Salvia texana*[42] and carboxylated diterpenes **33–34** from *Salvia officialis*[43] and *Salvia pachyphylla*[17] (Scheme 8.6). It is obvious that all of these diterpenes **30–34** are secondary metabolites related to taxodione and taxodone. Diterpene **34** is an intramolecular carboxylate adduct of the corresponding QM and exhibited selective inhibition of human ovarian and breast cancer cell lines at 5 μM range.[17] Terpene QM **35** (Scheme 8.6) has a spiral structure and was isolated from *Chamaecyparis obtusa* and *Crytomeria japonica*.[44–45]

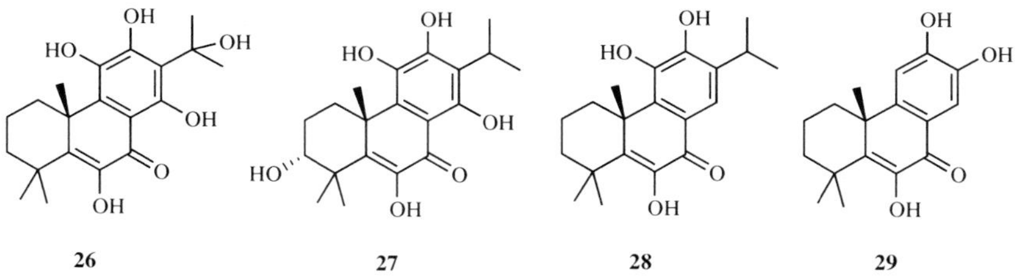

SCHEME 8.5 Diterpenoids related to coleon U.

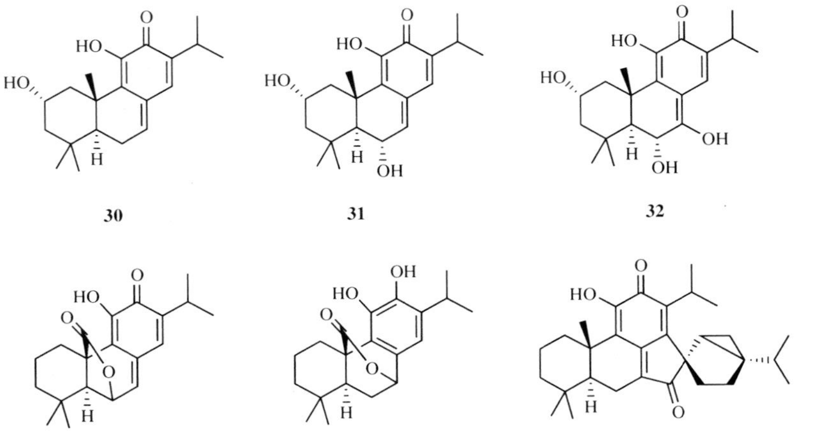

SCHEME 8.6 Other variations of natural diterpene QMs.

8.2.2 Dimers of Natural Diterpene QMs

Dimerization of natural diterpene QM has been observed with taxodione, dehydroxylated coleon V, and coleons E and F.[46–48] Dimers **36–38** are most likely formed as an artifact during the isolation process, and the mechanism of their formation is quite unique due to the presence of the *ortho*-hydroxyl group (Scheme 8.7). The formation of the dimers via the connection of the B rings is possibly achieved through a conjugated adol condensation followed by tautomerization and oxidation. Interestingly, some of these dimers are biological active against human cancer cell lines.[12]

8.3 TOTAL SYNTHESIS OF DITERPENE QMs

The first synthesis of (±)-taxodione **1** was achieved in 1970 by Mori and coworker[49] from podocarpic acid, which has the desired tricyclic carbon skeleton and yet is an expensive starting material even in today's market. The first total synthesis of (±)-taxodione **1** was reported by Motsumoto and coworkers[50] in 1971 and many significant improvements have been made in the following decades.[51–59] A key synthetic design was to construct the A, B, and C rings of taxodione efficiently with the correct *trans* stereochemistry. Three major strategies have been explored successfully using (1) a sequential construction of the rings as in Motsumoto's syntheses; [50–51] (2) 4 + 2 Diels–Alder cycloadditions in Engler's[57] and Konopelski's syntheses;[58] and (3) a polyene cyclization as a mimic of the biosynthesis in Livinghouse's synthesis.[59]

8.3.1 Stepwise Synthesis of Diterpene QMs

The total synthesis by Motsumoto et al. is an example of classic organic syntheses using the ring closure methodology to stepwise construct the tricyclic carbon

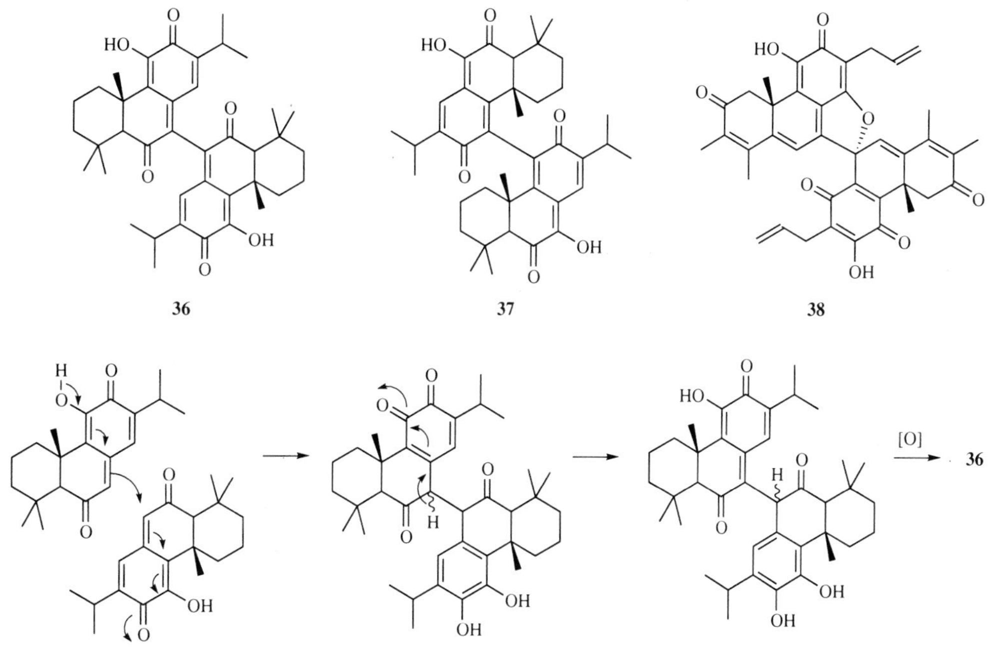

SCHEME 8.7 Dimers of natural diterpene QMs and mechanism of their formation.

skeleton.[50] The key intermediates of Motsumoto's synthesis are summarized in Scheme 8.8 and the reagents and detailed conditions can be found in the original paper.[50] Briefly, the formation of B ring was achieved by two sequential steps of acylation of 1,2-dimethoxy- 3-isopropylbenzene with succinic anhydride. It was found that bromination of the intermediate after the first step of acylation was necessary to block the *ortho*-position of the isopropyl group prior to the second step of acylation. The bromo group of the intermediate **42** was removed with hydrogenation, followed by methyl addition of the carbonyl group, elimination of the resulting hydroxyl group, and oxidation with lead tetraacetate to intermediate **43**. The C ring of taxodione was furnished through the Robinson annulation of intermediate **43** to **44**. The germinal methyl groups were incorporated through the methylation of enolates formed with *t*-butoxide, and the carbonyl group of the C ring of **45** was then reduced via a thiolketal intermediate to alkene **46**. The anti-Markovnikov hydration using hydroborane produced intermediate **47** with the desired *trans* stereochemistry at the B–C ring junctions. Finally, the oxidation of the hydroxyl group of the B ring, followed by the removal of methyl groups of the A ring and spontaneous oxidation with silica gel, completed the total synthesis of (±)-taxodione **1**.

In 1977, Motsumoto and coworkers reported a more succinct and improved total synthesis of taxodione **1** (Scheme 8.9),[51] in which the A and C rings were connected directly using β-cyclocitral **48** and a benzyl chloride **49** followed by oxidation with pyridine dichromate (PDC). The cyclization of the resulting intermediate **50** was achieved in the presence of pyrophosphoric acid at 100 °C to produce the *cis*-ketone **51** and the *trans*-ketone **52** at a 2:1 ratio in a total yield of 60%. The undesired *cis*-isomer was converted to the *trans*-ketone **52** in a series of steps of oxidation and reduction via an enol-acetate intermediate in an overall yield about 30%. The incorporation of the second phenol group on the A ring was achieved with perbenzoic acid after the reduction of the carbonyl group and the deprotection of the phenol group of the *trans*-ketone **53**. Reduction of intermediate **53** with $LiAlH_4$, followed by oxidation with the Jones reagent (chromic acid in acetone), led to the formation of (±)-taxodione **1** on silica gel during the chromatographic separation.

8.3.2 Diel–Alder Approach in the Diterpene QM Synthesis

The 4 + 2 Diels–Alder approach of the total synthesis of (±)-taxodione and taxodone was successfully achieved by two groups, Engler and coworkers in 1989[57] and Konopelski and coworkers in 1994[58]. Engler's synthesis utilized the 4 + 2 cycloaddition to join the A and C rings while Konopelski's synthesis constructed the A ring with a Diels–Alder reaction. In Engler's synthesis[57] (Scheme 8.10), 1,3-dimethoxybenzene **54** was converted to a *para*-quinone **56** by aromatic addition to acetone, followed by hydrogenation of the benzylic alcohol, demethylation of the phenol ether with BBr_3, and oxidation with Fremy's salt. The Diels–Alder reaction of the resulting quinone **56** with diene **57** was carried out under a high pressure at 12 kbar in the presence of 0.86 equiv $ZnBr_2$ for 5 days, resulting in the cycloadduct **58** and its isomer at 4:1 ratio in 74% yield. The stereoisomer mixture was not separated because two of the chiral centers were removed in the subsequent aromatization step with PPh_3 after the reduction of **58** with 1

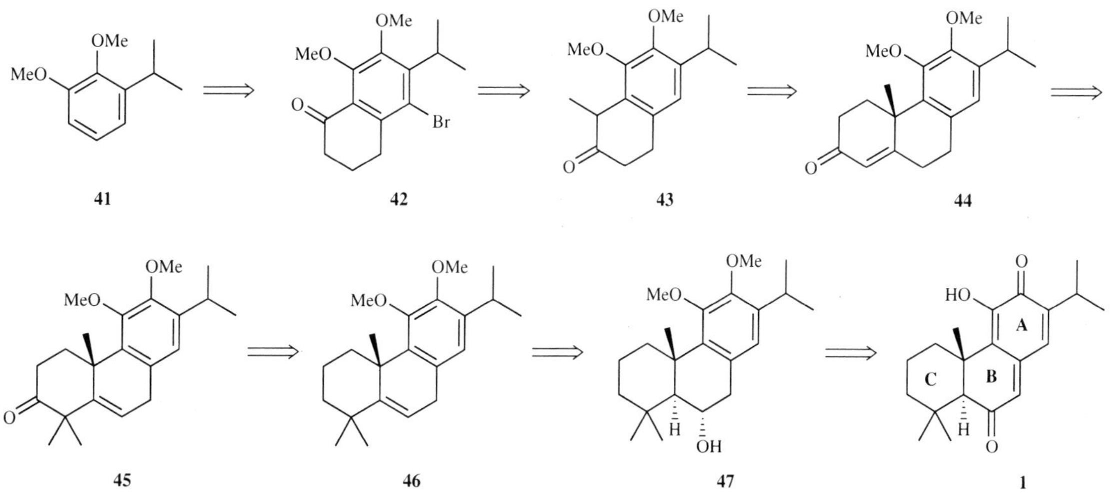

41 42 43 44

45 46 47 1

SCHEME 8.8 Key steps in the total synthesis of taxodione **1** by Motsumoto and coworkers.

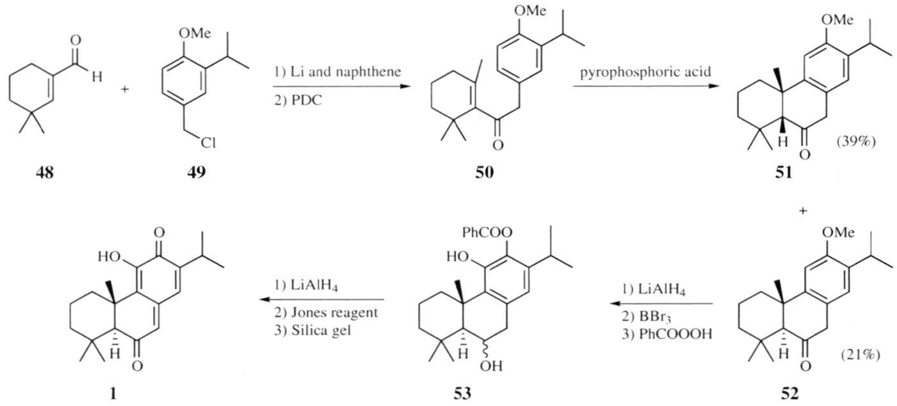

SCHEME 8.9 Improved synthesis of taxodione **1** by Motsumoto and coworkers.

equivalent of L-selectide. Further methylation of the phenol groups afforded the key intermediate **46**, which can be converted to taxodione by a series of hydroboration, oxidation, demethylation, and oxidation as reported in the Motsumoto's first synthesis.[50]

Konopelski and coworkers[58] first constructed B and C rings using a Lewis acid-catalyzed ring closure to intermediate **61** with the desire *trans* stereochemistry (Scheme 8.11). A Wittig reaction with the ketone **61** followed by epoxidation with mCPBA produced **62**, which was converted to diene **64** via lactonization and TBDMS silylation of the enolate. The Diels–Alder reaction of diene **64** and methyl acrylate was carried out in nitromethane for 18 h to **65**, followed by *in situ* aromatization of the A ring with HCl treatment, and the resulting phenol group was methylated. Intermediate **66** was then converted to phenol **67** with Baeyer–Villiger oxidation, and the isopropyl group was incorporated via a thiol intermediate. The introduction of the functional group on the B ring was achieved with oxidation of catechol with CrO_3, followed by reduction with $NaBH_4$ and

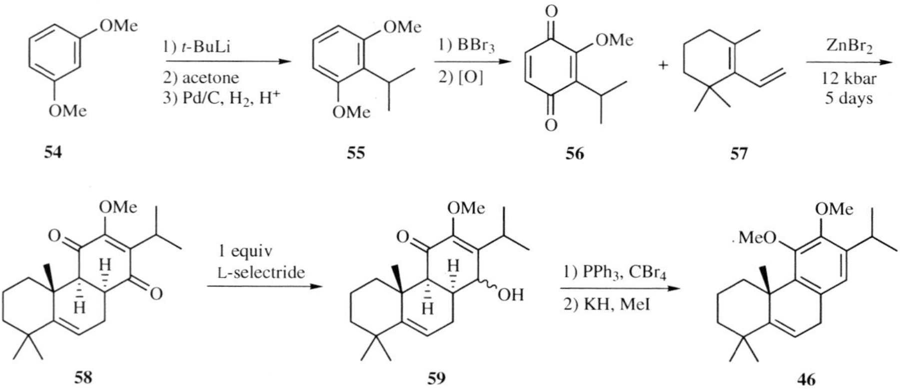

SCHEME 8.10 Engler's synthesis of taxodione **1** via Diels–Alder cycloaddition.

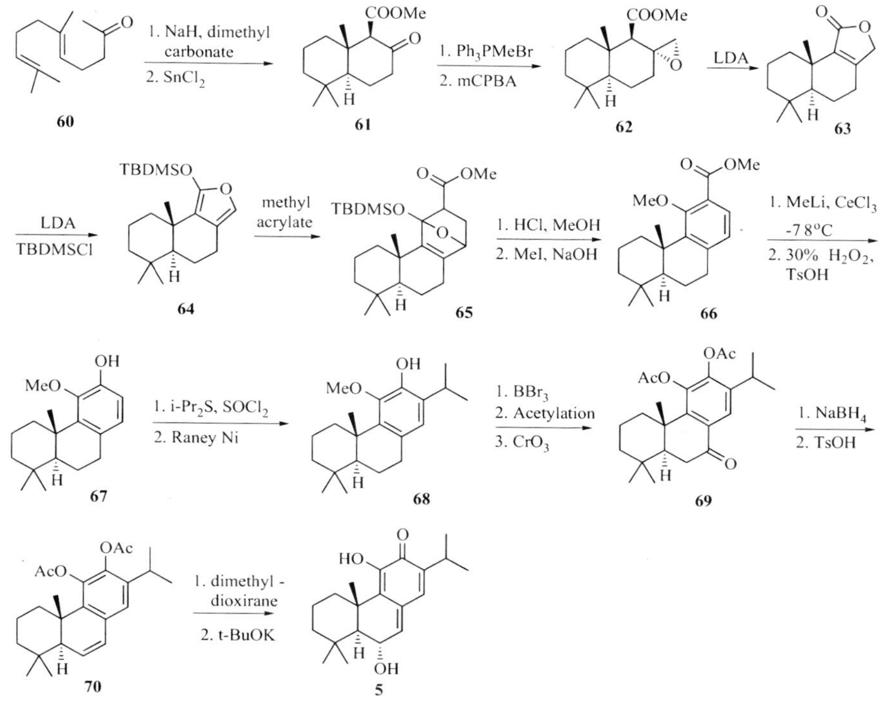

SCHEME 8.11 Konopelski's synthesis of taxodone **5** via Diels–Alder cycloaddition.

elimination to alkene **70**, which was converted to (±)-toxodone **5** upon epoxidation with dimethyldioxirane and subsequent base treatment.

8.3.3 Polyene Cyclization as a Mimic of Biosynthesis in Plants

The mimic of biosynthesis of (±)-taxodione was first investigated by Johnson and coworkers[54] and later accomplished by Livinghouse and coworker in 1994[59] using a Lewis acid-catalyzed polyene cyclization to form the tricylic ring with the desired *trans* stereochemistry (Scheme 8.12). In the Livinghouse's total synthesis,[59] the benzyl chloride **72** was obtained by the reduction of carboxylic acid **71** with borane to the corresponding alcohol, followed by chlorination. Coupling of the benzyl chloride **72** and geranyl cyanide with *n*-butyl lithium afforded the desired polyene **73** in 88% yield. The cyclization of **73** in the presence of Lewis acid BF_3–nitromethane complex produced the desired *trans*-isomer in an impressive 83% yield. The conversion of the cyano group to an alcohol was achieved using the nucleophilic addition of a carbanion to molecular oxygen followed by Lewis acid hydrolysis and subsequent reduction in an overall 54% yield. Oxidation of the alcohol of intermediate **75** with PDC followed by the removal of the methyl groups produced a catechol precursor, which was converted to (±)-taxodione **1** on silica gel via oxidation.

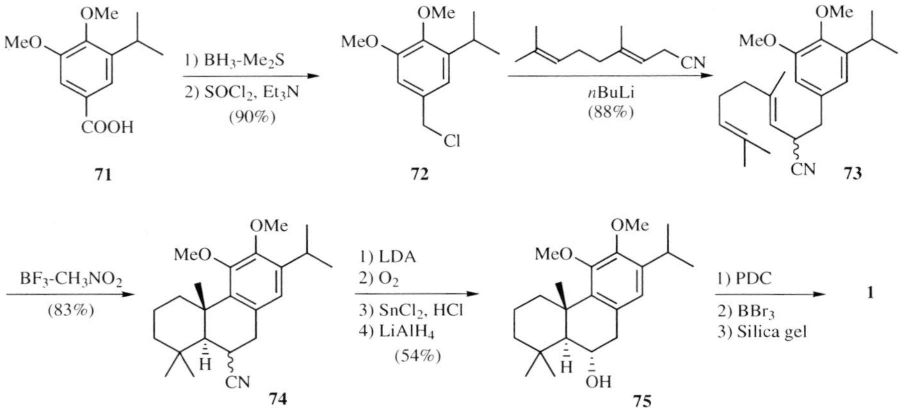

SCHEME 8.12 Synthesis of taxodione **1** via polyene cyclization.

Besides taxodione and taxodone, these synthetic approaches have been conveniently modified to synthesize other related derivatives such as ferruginols, sugiols, and royleanones.[51–52,56–57,60–62] In addition, the enantiomeric-selective synthesis of (+)-taxodione **1** has been reported using (−)-abietic acid as the starting material[63] or with a chiral ligand in the polyene cyclization process.[64]

8.4 NATURAL TRITERPENE QMs

The most studied natural triterpene QMs are pristimerin **2**, celastrol **76**, and tingenone **77**, which were identified as the major active ingredients isolated from plants used in folklore medicine (Scheme 8.13).[65] For example, celastrol **76** is the key active compound from the root of *Tripterygium wilfordii* Hook. f., which is used to treat anti-inflammatory symptoms in Chinese traditional herbal medicine and as insecticides in the crop field.[65] Recently, the prominent anti-inflammatory activity of these triterpene QMs has prompted them as potential therapeutic candidates for neurodegenerative diseases such as Alzheimer's and Huntington's diseases.[66–67] The structures of these natural triterpene QMs were established as conjugated alkenes in early 1970s and since then, many similar triterpene QMs have been reported.[65] Because a

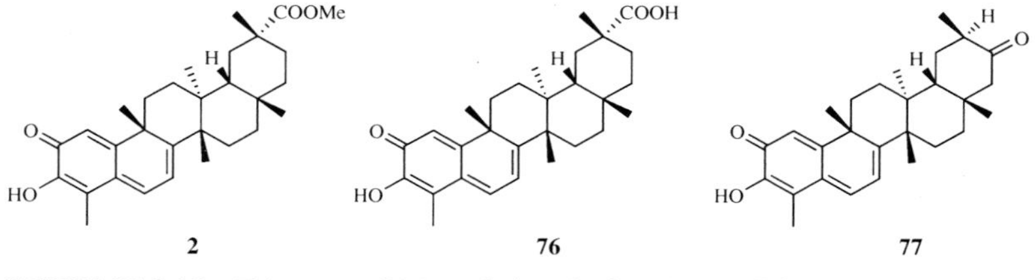

SCHEME 8.13 Triterpene QMs: pristimerin **2**, celastrol **76**, and tingenone **77**.

comprehensive review of natural triterpene QMs was made by Gaunatilaka[65] on the structures, analyses, natural occurrences, partial chemical reactions, and some biological activities in 1996, the focus of this discussion is mainly on the recent progress within last 10 years, especially the molecular targets and underlying mechanisms with triterpene QMs.

8.4.1 Cytotoxicity of Natural Triterpene QMs Against Cancer Cell Lines

Pristimerin **2**, celastrol **76**, and tingenone **77** were initially investigated as potent antitumor agents due to the high cytotoxicity observed in a panel of cancer cell lines including mouse lymphoma P-388, human lung carcinoma A-549, colon carcinoma HT-28, and melanoma MEL-28 with the IC_{50} values in the range of 0.2–0.3 μM.[68] The potent cytotoxicity led further investigations on the structure–activity relationship of triterpene QMs, which was summarized in the review by Ravelo and coworkers in 2004.[68] Briefly, several important conclusions have been drawn related to the triterpene QM structure **78** (Scheme 8.14): (1) low activity was observed if the A ring was replaced with a cycloalkane structure; (2) no significant impact on the biological activity was observed if the *ortho*-hydroxyl group of the A ring was acetated (R_1) and/or if there were multiple hydrogen-bond donors or receptors (R_2) on the E ring; (3) reduced biological activity was observed if the QM is conjugated with a carbonyl group in **79**, which is in contrast with that of natural diterpene QMs; (4) similar biological activity was observed if there was a leaving group (X = thiol or acetate) present at the benzylic position of the corresponding catechol **80** (R_1 = H) or acetated catechol (R_1 = acetate), possibly due to the regeneration of the QM through the elimination of the leaving group; (5) comparable biological activity was observed if the QM structure was converted to a diketo structure **81**; (6) an electron-withdrawing group must be present in the keto-catechol **82**; and (7) further conjugation of the QM in the C or D rings decreased the biological activity. Thus, the biological cytotoxicity is highly correlated to the presence of the QM moiety in the triterpene structure.

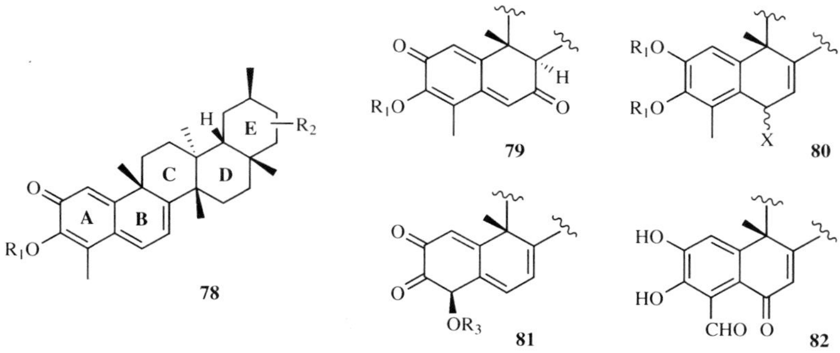

SCHEME 8.14 Important chemophores of triterpene QMs.

The cytotoxic mechanism has been confirmed through the triterpene QM-induced apoptotic pathways.[69–71] For example, apoptotic nuclear DNA fragmentation was observed in human leukemia HL-60 cells as early as 3 h after the treatment of 2.2 μM celastrol **76**, and activation of early apoptosis was detected time dependently with annexin V staining by flow cytometry.[69] Interestingly, celastrol **76** also acted as a topoisomerase II inhibitor *in vitro* with an IC_{50} of 7.4 μM. Similarly, apoptosis was evidently the key cytotoxic mechanism in human breast cancer MDA-MB-231 cells with pristimerin **2**.[70] Treatment of pristimerin **2** induced the activation of caspase-3, a key proapoptotic protein in the concentration- and time-dependent manner, along with the degradation of PARP, a DNA repair protein. On the contrary, the addition of z-VAD-fmk, a pan-caspase inhibitor as a blocker of the apoptotic pathway, significantly reduced the cytotoxicity induced by pristimerin **2** in MDA-MB-231 cells. In addition, the release of cytochrome *c* from mitochondria, a marked event of the mitochondrial controlled apoptosis pathway, was observed even in the presence of z-VAD-fmK or the mitochondrial pore inhibitor cyclosporine A, suggesting that mitochondria might be a direct target by pristimerin. In melanoma, both celastrol **76** and pristimerin **2** induced apoptosis and significant activation of JNK1/2 in the mitogen-activated protein kinase (MAPK) pathway, which may be the major reason for the selective killing of melanoma cells over the normal melanocytes.[71]

8.4.2 Anti-Inflammatory Effects of Natural Triterpene QMs

The anti-inflammatory effect of pristimerin **2**, celastrol **76**, and tingenone **77** was observed at much lower concentrations than that of the cytotoxicity. For example, all these three QMs inhibited the production of IL-1β, a proinflammatory cytokine with an IC_{50} value of 50 nM in the LPS (lipopolysaccharide) stimulated monocytes.[72–73] Until recently, the in-depth understanding of the anti-inflammatory mechanism was realized with celastrol **76** via the inhibition of the nuclear factor (NF)-κB signaling pathway (Fig. 8.1), although the final protein target needs to be confirmed.[74–75] In the absence of celastrol **76**, proinflammatory cytokine TNF-α or LPS binds to the cell membrane receptors and activates the phosphorylation of IκBα protein of the inactive NF-κB/IκBα complex by IKK complex via a cascade of stimulating pathway in the cytoplasm. The phosphorylated IκBα protein is then ubiquitinated and degraded by proteasome, resulting in the release of active NF-κB proteins, which are translocated in the nucleus to activate the transcription of a variety of genes related to inflammation, immunoresponses, and cell proliferation. With the treatment of celastrol **76**, the TNF- or LPS-induced expression of the NF-κB-dependent proteins such as IL-1, IL-6, TNF-α, and induced NO synthase (iNOS) was inhibited, and a decreased level of NF-κB/DNA complex was observed in the nuclear extract (Fig. 8.1).[74–75] It was found that celastrol **76** did not interfere with the binding of NF-κB to DNA, and thus the decreased level of NF-κB in nucleus was due to the decreased translocation of active NF-κB from cytoplasm. The low level of active NF-κB was then linked to the inhibition of both the phosphorylation and the degradation of the IκBα protein by celastrol in cytoplasm, and therefore, the blockage of the upper level kinase activity (IKKs) was indicated. Lee and coworkers

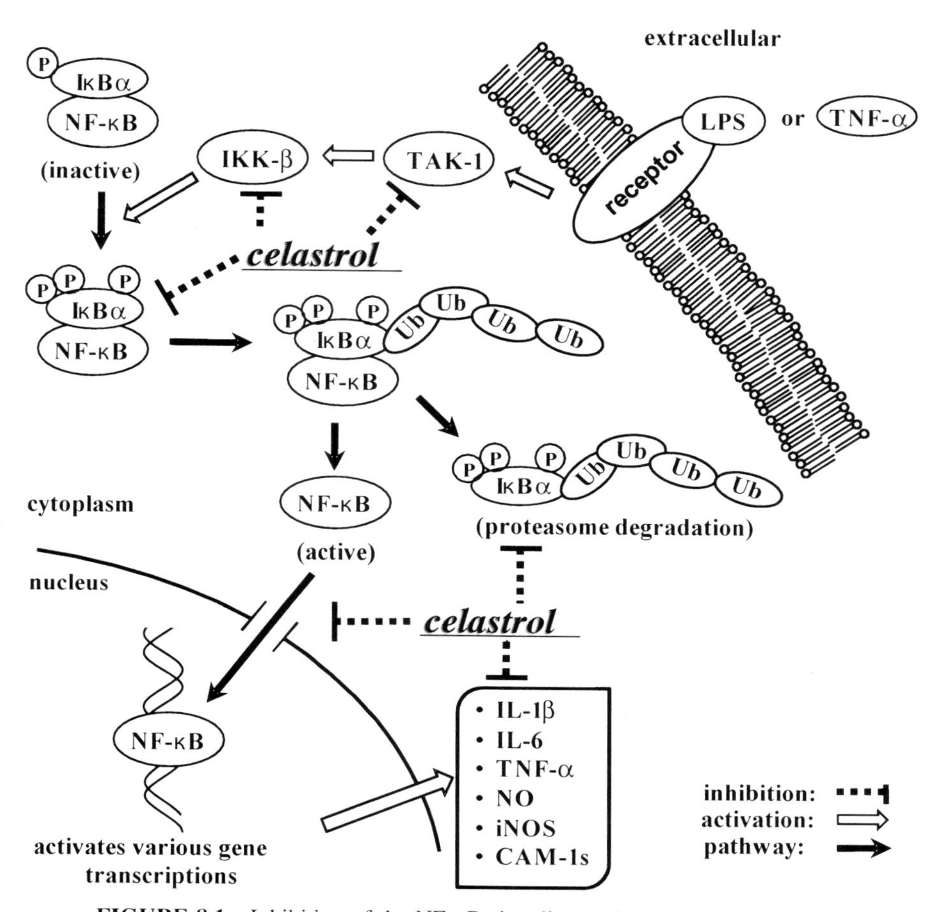

FIGURE 8.1 Inhibition of the NF-κB signaling pathway with celastrol **76**.

showed that celastrol **76** covalently modified the cysteine residue 179 in the active site of IKK-β and thus deactivated the kinase activity[74] while Aggarwal and coworker suggested that a higher level kinase TAK-1 was the actual inhibition target with celastrol **76**.[75] Even though the final molecular target in the NF-κB pathway needs to be confirmed, it is clear that natural triterpene QM celastrol **76** exhibits the anti-inflammatory effect via the inhibition of the phosphorylation of IκBα protein of the NF-κB complex in cytoplasm to block the downstream activation of NF-κB-dependent genes in the responses of proinflammatory stimuli.

The inhibition of the phosphorylation of IκBα by celastrol in the NF-κB signaling pathway is not cell specific[72–75] and thus significantly broadened the potential medicinal applications of natural triterpene QMs. Celastrol **76** has been evaluated in the treatment of Crohn's disease,[76] which is a chronic relapsing inflammatory bowel disease. In the peripheral blood mononuclear cells, celastrol **76** reduced the levels of the LPS-induced cytokines including TNF-α, IL-1β, IL-6, and IL-8 by 50% in the range of 3–30 nM and additionally inhibited the phosphorylation of p38 protein of the MAPK signaling pathway. In human endothelial cells, treatment with celastrol **76** resulted in a significant reduction in the surface expression of vascular and intercellular cell adhesion molecules (CAM-1s),[77] which are stimulated by inflammatory cytokines such as TNF-α, IL-1β, or IFN-γ through the NF-κB signaling pathway and are critical to the adhesion of leukocytes to endothelial cells. In both studies, reduced levels of NF-κB were consistently observed in nucleus with celastrol **76** in the presence of the inflammatory stimuli. In the mice study of lupus nephritis, celastrol **76** treatment considerably suppressed the levels of NO, TGF-β1 mRNA, IL-10, and inflammatory double-stranded DNA antibody and reduced the protein excretion in urine.[78–79]

The anti-inflammatory effect of celastrol **76** has recently been investigated in the treatment of neurodegenerative diseases such as Parkinson's and Alzheimer's diseases,[66–67,80] where an elevated inflammatory environment around the neuron cells with excessive activated microglia was observed. The addition of celastrol **76** (100 nM) in the LPS-stimulated microglia inhibited the inflammatory responses from the NF-κB pathway including the production of NO, proinflammatory cytokines IL-1β and TNF-α, iNOS and levels of activated NF-κB in the nucleus.[81] In addition, the phosphorylation of the ERK1/2, one of the MAPK signaling pathways was inhibited. In the mice model, celastrol **76** treatment effectively reduced the levels of NF-κB and TNF-α against the effects induced by neurotoxins such as 1-methyl-4-phenyl-1,2,3,6-tetrahydropyridine and 3-nitropropionic acid as models for Parkinson's and Huntington's diseases, respectively.[82] Besides the anti-inflammatory effect, celastrol **76** was found to be able to reduce the accumulation of aggregates in neurodegenerative diseases.[83–89] In Huntington's disease model, celastrol **76** was identified as one of the most effective drug candidates to reverse the polyglutamine aggregates in striatal cells from a library of 1040 compounds.[83–86] Similarly in the Alzheimer's disease model, amyloid-β peptide accumulation in the CHO cells was substantially reduced upon the addition of celastrol **76**. [87–88] The antiaggregate effect with celastrol **76** in neurons was revealed as the result of enhanced expression level of the heat shock proteins (HSP) such as HSP-27, HSP-32, and HSP-70, which are protein folding chaperons and are responsible for cell repair under stress, via the activation of heat shock protein transcript factor 1 (HSF-1).[83–89] Notably, the induction of HSPs was much more significant with celastrol **76** than that with pristimerin **2** (~35-fold versus ~10-fold), and it was found that the free carboxylic acid group on the E ring of the triterpene QMs was critical for the induction of HSPs through the investigation of a series of analogues.[83]

In the NF-κB pathway with celastrol **76**, the inhibition of the IκBα degradation is due to the upstream blockage of the kinase activity and not by the direct inhibition of proteasome activity. On the contrary, direct inhibition of proteasome activity was observed with celastrol **76** and pristimerin **2** in prostate cancer cells.[90–92] Both triterpene QMs directly inhibited the activity of the 20S subunits of proteasome at 2.5 μM and induced the accumulation of ubiquitinated proteins over time in cells, which consequently blocked androgen receptors and led to apoptosis in prostate cancer cells.

The structure–activity relation of triterpene QMs on the anti-inflammatory effect has been investigated with a series of analogues by comparing inhibition of the IL-1β production in the LPS-induced monocytes.[93–94] Clearly, the conjugated QM structure with alkenes is essential for the observed inhibition, and similar inhibitory effect was found with derivatives that are capable to form the conjugated QM through hydrolysis and/or oxidation.[93–94] Also, the presence of the E ring in the triterpene QM structure is an additional contributor.

8.4.3 Other Biological Activities of Natural Triterpene QMs

Besides anti-inflammatory and cytotoxic effect, natural triterpene QMs in general showed selective antibacterial activity against Gram-positive bacteria.[95] For example, suctione **83** (Scheme 8.15) was three times more potent than tingenone **77** against *Bacillus subtilis* at 0.1–0.16 μg/mL.[96] However, the loss of antimicrobial activity was observed with the derivatives **84** that have degraded E-ring structure of the triterpene or compound **85** formed from the rearrangement under acidic conditions (Scheme 8.15).[97–98] Antiparasital activity was observed with some triterpene QMs including anti-Giardia lambio,[99] antimalarial,[100] and antileishmania,[101] especially QM **86** with IC_{50} values below 50 ng/mL.[100–101] Effective antioxidant effect was reported with QM **87** that has extended conjugations.[102] Also, celastrol **76** was able to decrease the ionic current in the ion channel of membrane proteins.[130–104] Other additional biological activities included inhibition of lens aldose-reductase activity,[105] antiviral activity against human cytogalovirus (herpesvirus),[106] and antifeedant–insecticidal activity.[107]

8.4.4 Biosynthesis of Natural Triterpene QMs

Despite the broad medical potentials reported so far, the total synthesis of triterpene QMs is yet to be reported. On the contrary, the biosynthesis of triterpene QMs has recently been validated as from the oxidosqualene **88** (Scheme 8.16) in the plants including *Maytenus aquifolium* and *Salacia campestris*.[108] With the assistance of HPLC analysis and isotopic labeling, it was found that triterpene QMs **90** were formed only in the root of these plants from friedelin **89** and similar cyclized intermediates, which were synthesized in the leaves from oxidosqualene by cyclase.

8.5 TERPENE QMs AND REACTIVE OXYGEN SPECIES

In addition to the QM structure of the natural terpene QMs, the reactive oxygen species (ROS) may also play a significant role in the observed biological activities. In the synthesis of taxodione and taxodone, QMs were formed from the catechol precursors through the spontaneous oxidation in the presence of silica gel.[7, 8,49–51]

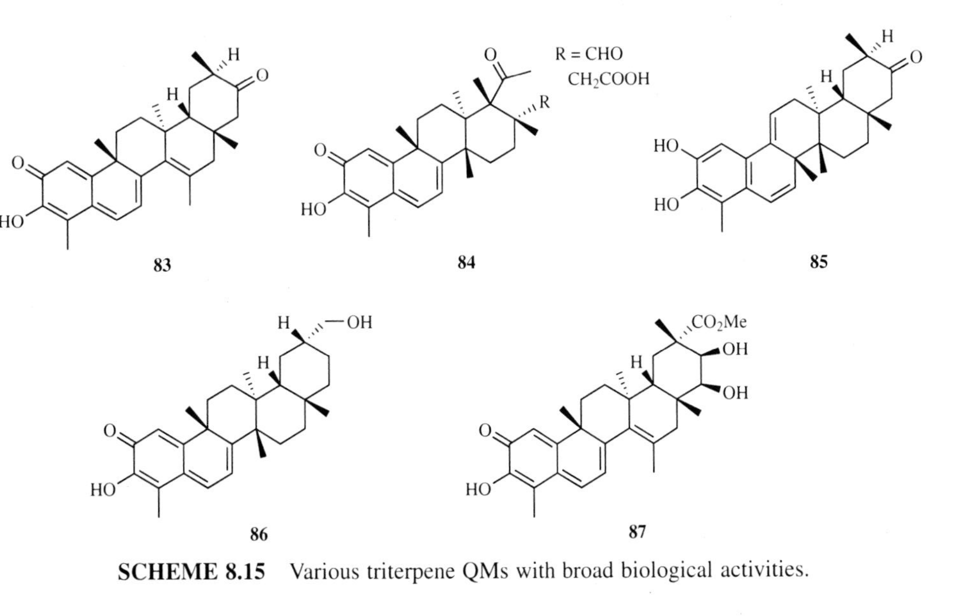

SCHEME 8.15 Various triterpene QMs with broad biological activities.

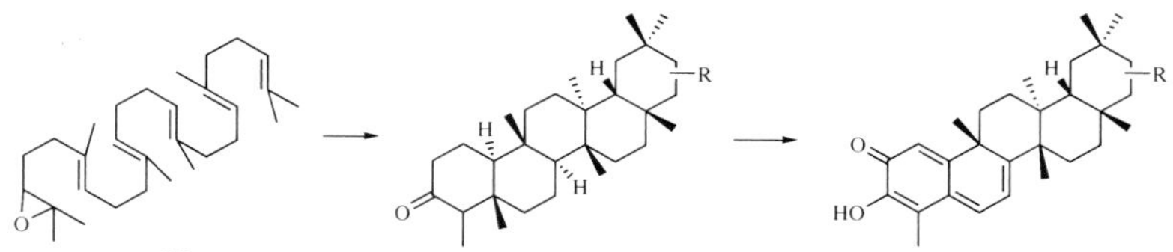

SCHEME 8.16 Biosynthesis of triterpene QMs from oxidosqualene **88**.

Under aqueous conditions, oxidation of similar catechol derivatives has been shown to have detrimental effects on cells due to *in situ* generation of ROS.[109–111] Zhou and coworker reported that extensive DNA damage on a short oligonucleotide target was observed in the Cu^{2+}-induced oxidation under buffered conditions with a simplified analogue of the natural terpene QM precursor **91** (Scheme 8.17).[112] The formation of the terpene QM **92** in the presence of Cu^{2+} was confirmed with a thiol-trapping study using HPLC–MS analysis, and the generation of a second QM **94** of the initial thiol adduct **93** was observed (Scheme 8.17), which was consistent with that of taxodione.[8]

The observed DNA damage was attributed to the production of ROS in the Cu(II)-induced oxidation through the investigation of a series of analogues with different stereochemistry and functional groups.[113] The DNA fragment patterns were observed similarly with all of these analogues despite the difference in the structures and oxidation products such as QM **92** versus *o*-quinone. The DNA damage also resembled that of the oxidative damage by H_2O_2/Fe(III)-EDTA and was confirmed due to the ROS generated in the disproportion of Cu(II)/(I)-O_2 redox cycle via a Cu superoxide complex with the study using a series of radical quenchers. Importantly, the addition of NADH in the Cu^{2+}-induced oxidation significantly enhanced the extent of DNA oxidative damage due to the reducing potential of NADH. This result suggested that coupling of the redox cycle of catechol/QM and the disproportion of Cu(II)/(I) with O_2 may continuously produce ROS until all of NADH was

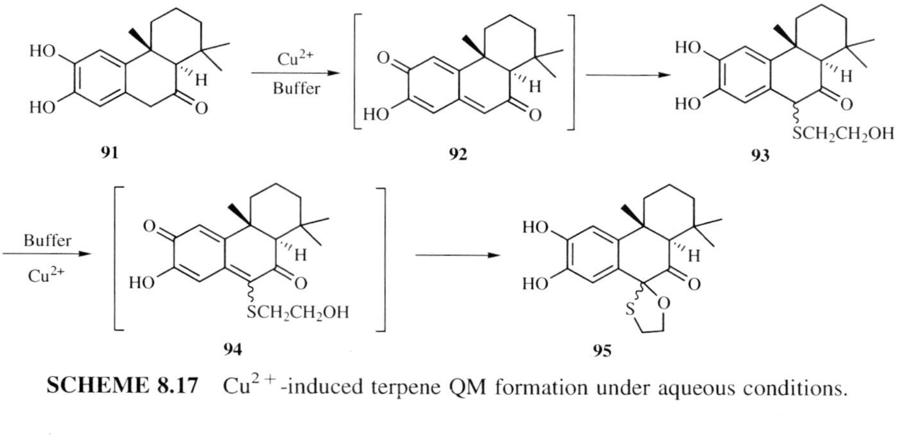

SCHEME 8.17 Cu^{2+}-induced terpene QM formation under aqueous conditions.

fully consumed, and thus significantly increased the oxidative stress level in the cells.

8.6 CONCLUSION AND FUTURE PROSPECTS

Natural diterpene and triterpene QMs belong to a small category of natural products with a highly conserved core structure, yet it is impressive to see the broad biological activities with such a unique QM structure in the terpenoids. The recent applications of celastrol in the treatment of neurodegenerative diseases such as Alzheimer's, Huntington's, and Parkinson's diseases excitingly expanded the QM chemistry into several key cell signaling pathways related to proliferation, apoptosis, and inflammatory responses. However, further mechanistic studies are needed to provide an in-depth understanding of the medicinal potentials of terpene QMs. For example, although the difference between the cytotoxicity of celastral and its anti-inflammatory effect can be explained simply as the result of increased concentrations, it was not clear why specific molecular targets such as HSPs and the kinase in the NF-κB pathway were selected by celastrol at low concentrations. In addition, the molecular target of the NF-κB pathway with celastrol needs to be confirmed and so does its interaction with proteins in the MAPK pathways. Finally, the cytotoxic potential of ROS generated in the reduction/oxidation cycle of terpene QMs needs to be clarified. Most recently, celastrol was found to induce the transcription of oxidative-response genes along with the heat shock proteins; however, these induced effects were significantly blocked with the addition of a free thiol such as DTT.[114] Overall, the advancing understanding of natural terpene QMs from the structure–activity relationship to the involvement in specific signaling pathways is an excellent example of the expansion of the QM chemistry in the twenty-first century.

REFERENCES

1. Gates, K. S. Covalent modification of DNA by natural products. In *Comprehensive Natural Products Chemistry;* Barton, D.; Nakanishi, K.; Meth-Cohn, O.; Kool, E. T., Eds.; Elsevier: New York, 1999; Vol. 7, pp 491–552.
2. Thompson, D. C.; Thompson, J. A.; Sugumaran, M.; Moldeus, P. Biological toxical consequences of quinone methide formation. *Chem.-Biol. Interact.* 1993, 86, 129–162.
3. Moore, H. W.; Czerniak, R. Naturally occurring quinones as potential bioreductive alkylating agents. *Med. Res. Rev.* 1981, 1, 249–280.
4. Grünanger, P.; Chinomethide. In *Methoden der Organisch Chemie;* Muller, E.; Bayer, O.,Eds.; Houben-Weyl: Stuttgart, 1979; Vol. VII/3b, pp 395–521.
5. Wagner, H. U.; Gompper, R. Quinone methides. In *The Chemistry of Quinonoid Compound;* Patai, S.,Ed.; John Wiley & Sons: New York, 1974; Vol. 1, pp 1145–1178.

6. Turner, A. B. Quinone methides. *Quart. Rev.* 1964, 18, 347–360.
7. Kupchan, S. M.; Karim, A.; Marcks, C. Taxodione and taxodone, two novel diterpenoid quinone methide tumor inhibitors from *Toxodium distichum*. *J. Am. Chem. Soc.* 1968, 90, 5923–5924.
8. Kupchan, S. M.; Karim, A.; Marcks, C. Tumor inhibitors. XLVIII. Taxodione and taxodone, two novel diterpenoid quinone methide tumor inhibitors from *Toxodium distichum*. *J. Org. Chem.* 1969, 34, 3912–3918.
9. Esquivel, B.; Flores, M.; Hernández-Ortega, S.; Toscano, R. A.; Ramamoorthy, T. P. Abietane and icetexane diterpenoids from the roots of *Salvia aspera*. *Phytochemistry* 1995, 39, 139–143.
10. Fraga, B. M.; Diaz, C. E.; Guadaño, A.; González-Coloma, A. Diterpenes from *Salvia broussonetii* transformed roots and their insecticidal activity. *J. Agric. Food Chem.* 2005, 53, 5200–5214.
11. Sabri, N. N.; Abou-Donia, A. A.; Assad, A. M.; Ghazy, N. M.; El-Lakany, A. M.; Tempesta, M. S.; Sanson, D. R. Abietane diterpene quinones from roots of *Salvia verbenaca* and *S. lanigera*. *Planta Med.* 1989, 55, 582.
12. Topcu, G.; Ulubelen, A. Abietane and rearranged abietane diterpenes from *Salvia montbretii*. *J. Nat. Prod.* 1996, 59, 734–737.
13. Simões, F.; Michavila, A.; Rodríguez, B.; García-Alvarez, M. C.; Hasan, M. A quinone methide diterpenoid from the root of *Salvia moorcrafiana*. *Phytochemistry* 1986, 25, 755–756.
14. Ikeshiro, Y.; Mase, I.; Tomita, Y. Abietane-type diterpene quinones from *Salvia nipponica*. *Planta Med.* 1991, 57, 588.
15. Ulubelen, A. New diterpenes from *Salvia pachystachys*. *J. Nat. Prod.* 1990, 53, 1597–1599.
16. Li, M.; Zhang, J.-S.; Ye, Y.-M.; Fang, J.-N. Constituents of the roots of *Salvia prionitis*. *J. Nat. Prod.* 2000, 63, 139–141.
17. Guerrero, I. C.; Andrés, L. S.; León, L. G.; Machin, R. P.; Padrón, J. M.; Luis, J. G.; Delgadillo, J. Abietane diterpenoids from *Salvia pachyphylla* and *S. clevelandii* with cytotoxic activity against human cancer cell lines. *J. Nat. Prod.* 2006, 69, 1803–1805.
18. Luis, J. G.; Grillo, T. A. New diterpenes from *Salvia munzii*: chemical and biogenetic aspects. *Tetrahedron* 1993, 49, 6277–6284.
19. Schaller, F.; Rahalison, L.; Islam, N.; Potterat, O.; Hostettmann, K.; Stoeckli-Evans, H.; Mavi, S. A new potent antifungal 'quinone methide' diterpene with a cassane skeleton from *Bobgunnia madagascariensis*. *Helv. Chim. Acta* 2000, 83, 407–413.
20. Schaller, F.; Wolfender, J.-L.; Hostettmann, K.; Mavi, S. New antifungal 'quinone methide' diterpenes from *Bobgunnia madagascariensis* and study of their interconversion by LC/NMR. *Helv. Chim. Acta* 2001, 84, 222–229.
21. El-Lakany, A. M. Chlorosmaridione: a novel chlorinated diterpene quinone methide from *Rosematinus officinalis* L. *Nat. Prod. Sci.* 2004, 10, 59–62.
22. Martín, J. D. New diterpenoids extractives of *Maytenus dispermus*. *Tetrahedron* 1973, 29, 2553–2559.
23. Esquivel, B.; Sánchez, A. A.; Vergara, F.; Matus, W.; Hemandez-Ortega, S.; Ramírez-Apan, M. T. Abietane diterpenoids from the roots of some Mexican *Salvia* species

(Labiatae): chemical diversity, phytogeographical significance, and cytotoxic activity. *Chem. Divers.* 2005, 2, 738–747.

24. Clarkson, C.; StæK, D.; Hansen, S. H.; Smith, P. J.; Jaroszewski, J. W. Identification of major and minor constitutes of *Harpagophytum procumbens* (devil's claw) using HPLC–SPE-NMR and HPLC–ESIMS/APCIMS. *J. Nat. Prod.* 2006, 69, 1280–1288.
25. Karanatsios, D.; Scarpa, J. S.; Eugster, C. H. Struktur von Fuerstion. *Helv. Chim. Acta* 1966, 49, 1151–1171.
26. Ulubelen, A.; Öksüz, S.; Kolak, U.; Bozok-Johansson, C.; Çelik, C.; Voelter, W. Antibacterial diterpenes from the roots of *Salvia viridis*. *Planta Med.* 2000, 66, 458–462.
27. Ulubelen, A.; Öksüz, S.; Kolak, U.; Birman, H.; Voelter, W. Cardioactive terpenoids and a new rearranged diterpene from *Salvia syriaca*. *Planta Med.* 2000, 66, 627–629.
28. Achebach, H.; Waibel, R.; Nkunya, M. H. H.; Weenen, H. Antimalarial compounds from *Hoslundia opposita*. *Phytochemistry* 1992, 31, 3781–3784.
29. Rüedi, P.; Eugster, C. H. Struktur von Coleon E, einem neuen ditepenoiden Methylenchinon aus der *Coleus barbatus*-Gruppe (*Labiatae*). *Helv. Chim. Acta* 1972, 55, 1994–2014.
30. Rüedi, P. Neue Diterpene aus Blattdrüsen von *Plectranthus barbatus* (Labiatae) Die absolute Konfiguration der 2-Hydropropyl-Seitenkette in Coleon E. *Helv. Chim. Acta* 1986, 69, 972–984.
31. Rüedi, P.; Eugster, C. H. Struktur von Coleon F. *Helv. Chim. Acta* 1973, 56, 1129–1132.
32. Mei, S.-X.; Jiang, B.; Niu, X.-M.; Li, M.-L.; Yang, H.; Na, Z.; Lin, Z.-W.; Li, C.-M.; Sun, H.-D. Abietane diterpenoids from *Coleus xanthanthus*. *J. Nat. Prod.* 2002, 65, 633–637.
33. Marques, C. G.; Pedro, M.; Simões, M. F. A.; Nascimento, M. S. J.; Pinto, M. M. M.; Rodríguez, B. Effect of abietane diterpenes from *Plectranthus grandidentatus* on the growth of human cancer cell line. *Planta Med.* 2002, 68, 839–840.
34. Cerqueira, F.; Corderio-Da-Silva, A.; Gaspar-Marques, C.; Simões, F.; Pinto, M. M. M.; Nascimento, M. S. J. Effect of abietane diterpenes from *Plectranthus grandidentatus* on T- and B-lymphocyte proliferation. *Bioorg. Med. Chem.* 2004, 12, 217–223.
35. Gaspar-Marques, C.; Rijo, P.; Simões, M. F.; Duarte, M. A.; Rodriguez, B. Abietanes from *Plectranthus grandidentatus* and *P. hereroensis* against methicillin- and vancomycin-resistant bacteria. *Phytomedicine* 2006, 13, 267–271.
36. Rüedi, P.; Eugster, C. H. 14-Hydroxytaxodion: Partialsynthesse und Reaktionen. *Helv. Chim. Acta* 1981, 64, 2219–2226.
37. Teixeira, A. P.; Batista, O. B.; Simões, M. F.; Nascimento, J.; Duarte, A.; de la Torre, M. C.; Rodríguez, B. Abietane diterpenoids from *Plectranthus grandidentatus*. *Phytochemistry* 1997, 44, 325–327.
38. Rüedi, P.; Eugster, C. H. Coleone C, D, I, I′ aus einer madeğassischen *Plectranthus sp. nov*. Interkonversion von *cis*- und *trans*-A/B-6,7-Diketoditerpenen. *Helv. Chim. Acta* 1975, 58, 1899–1912.
39. Arihara, S.; Rüedi, P.; Eugster, C. H. Diterpenoide Drüsenfarbstoffe: Coleone S und T aus *Plectranthus caninus* Roth (*Labiatae*), ein neues Diosphenol/*trans*-A/B-6,7-Diketon-Paar aus der Abietanreihe. *Helv. Chim. Acta* 1977, 60, 1443–1447.
40. Chen, B.; Duan, H.; Takaishi, Y. Triterpene caffeoyl esters and diterpenes from *Celastrus stephanotifolius*. *Phytochemistry* 1999, 51, 683–687.

41. Hueso-Rodríguez, J. A.; Jimeno, M. L.; Rodríguez, B.; Savona, G.; Bruno, M. Abietane diterpenoids from roots of *Salvia phlomoides*. *Phytochemistry* 1983, 22, 2005–2009.
42. González, A. G.; Aguiar, Z. E.; Luis, J. G.; Ravelo, A. G.; Domínguez, X. Quinone methide diterpenoids from the roots of *Salvia texana*. *Phytochemistry* 1988, 27, 1777–1781.
43. Tada, M.; Hara, T.; Hara, C.; Chiba, K. A quinone methide from *Salvia officinalis*. *Phytochemistry* 1997, 45, 1475–1477.
44. Hirose, Y.; Hasegawa, S.; Ozaki, N. Three new terpenoid quinone methides from the seed of *Chamaecyparis obtusa*. *Tetrahedron Lett.* 1983, 24, 1535–1538.
45. Su, W.-C.; Fang, J.-M.; Cheng, Y.-S. Hexacarbocyclic triterpenes from leaves of *Cryptomeria japonica*. *Phytochemistry* 1993, 34, 779–782.
46. Rüedi, P.; Uchida, M.; Eugster, C. H. Partialsynthese der Grandidone A, 7-Epi-A, B, 7-Epi-B, C, D und 7-Epi-D aus 14-Hydroxytaxodion. *Helv. Chim. Acta* 1981, 64, 2251–2256.
47. Uchida, M.; Miyase, T.; Yoshizaki, F.; Bieri, J. H.; Rüedi, P.; Eugster, C. H. 14-Hydroxytaxodion als Hauptditerpen in *Plectranthus grandidentatus* Gürke; Isolierung von sieben neuen dimeren Diteroenen aus P. *grandidentatus*, *P. myrianthus* Briq. und *Coleus carnosus* Hassk.; Strukuren der Grandidone A, 7-Epi-A, B, 7-Epi-B, C, D und 7-Epi-D. *Helv. Chim. Acta* 1981, 64, 2227–2250.
48. Rüedi P. Struktur und Partialsynthese von dimerem Coleon F. *Helv. Chim. Acta* 1988, 71, 1638–1641.
49. Mori, K.; Matsui, M. Diterpenoid total synthesis. XIII. Taxodione, a quinone methide tumor inhibitor. *Tetrahedron* 1970, 26, 3467–3473.
50. Mastumoto, T.; Tachibana, Y.; Uchida, J.; Fukui, K. The total synthesis of (±)-taxodione, a tumor inhibitor. *Bull. Chem. Soc. Jpn.* 1971, 44, 2766–2770.
51. Mastumoto, T.; Usui, S.; Morimoto, T. A. Convenient synthesis of (±)-taxodione, (±)-ferruginol, and (±)-sugiol. *Bull. Chem. Soc. Jpn.* 1977, 50, 1575–1579.
52. Snitman, D. L.; Himmelsbach, R. J.; Haltiwanger, R. C.; Watt, D. S. A synthesis of (±)-cryptojapanol and (±)-taxodione. *Tetrahedron Lett.* 1979, 2477–2480.
53. Johnson, W. S.; Shenvi, A. B.; Boots, S. G. An approach to taxodione involving biomimetic polyene cyclization methodology. *Tetrahedron* 1982, 38, 1397–1404.
54. Stevens, R. V.; Bisacchi, G. S. Benzocyclobutenones as synthons for the synthesis of C-11 oxygenated diterpenoids. Application to the total synthesis of (±)-taxodione. *J. Org. Chem.* 1982, 47, 2396–2399.
55. Burnell, R. H.; Jean, M.; Poirier, D. Synthesis of (±)-taxodione. *Can. J. Chem.* 1987, 65, 775–781.
56. Harring, S. R.; Livinghouse, T. Sulfenium ion promoted polyene cyclizations in natural product synthesis. An efficient biomimetic-like synthesis of (±)-nimbidiol. *Tetrahedron Lett.* 1989, 30, 1499–1502.
57. Engler, T. A.; Sampath, U.; Naganathan, S.; Velde, D. V.; Takusagawa, F.; Yohames, D. A new general synthetic approach to diterpenes: application to synthesis of (±)-taxodione and (±)-royleanone. *J. Org. Chem.* 1989, 54, 5712–5727.
58. Sánchez, A. J.; Konopelski, J. P. Phenol benzylic epoxide to quinone methide electron reorganization: synthesis of (±)-taxodone. *J. Org. Chem.* 1994, 59, 5445–5452.
59. Harring, S. R.; Livinghouse, T. Polyene cascade cyclizations mediated by $BF_3{\cdot}CH_3NO_2$. An unusual efficient method for the direct, stereospecific synthesis of polycyclic

intermediates via cationic initiation at non-functionalized 3° alkenes. An application to the total synthesis of (±)-taxodione. *Tetrahedron* 1994, 50, 9229–9254.

60. Burnell, R. H.; Jean, M.; Marceau, S. Synthesis of maytenoquinone. *Can. J. Chem.* 1988, 66, 227–230.
61. Yang, Z.; Kitano, Y.; Chiba, K.; Shibata, N.; Kurokawa, H.; Doi, Y.; Arakawa, Y.; Tada, M. Synthesis of variously oxidized abietane diterpenes and their antibacterial activities against MRSA and VRE. *Bioorg. Med. Chem.* 2001, 9, 347–356.
62. Li, A.; She, X.; Zhang, J.; Wu, T.; Pan, X. Synthesis of C-7 oxidized abietane diterpenes from racemic ferruginyl methyl ether. *Tetrahedron* 2003, 59, 5737–5741.
63. Haslinger, E.; Michl, G. Synthesis of (+)-taxodione from (−)-abietic acid. *Tetrahedron Lett.* 1988, 29, 5751–5754.
64. Tada, M.; Nishiiri, S.; Yang, Z.; Imai, Y.; Tajima, S.; Okazaki, N.; Kitano, Y.; Chiba, K. Synthesis of (+)- and (−)-ferruginol via asymmetric cyclization of a polyene. *J. Chem. Soc., Perkin Trans. 1* 2000, 2657–2664.
65. Gunatilaka, A. A. L. Triterpenoid quinonemethide and related compound (celastroloids). In *Progress in the Chemistry of Organic Natural Products;* Hertz, W.; Kirby, G. W.; Moore, R. E.; Steglich, W.; Tamm, C., Eds.; Springer-Verlag: New York, 1996; Vol. 67, 1–123.
66. Brown, I. R. Heat shock proteins and protection of the nervous system. *Ann. N.Y. Acad. Sci.* 2007, 1113, 147–158.
67. Corson, T. W.; Crews, C. M. Molecular understanding and modern application of traditional medicines: triumphs and trial. *Cell* 2007, 130, 769–774.
68. Ravelo, A. G.; Estévez-Braun, A.; Chávez-Orellana, H.; Pérez-Sacau, E.; Mesa-Siverio, D. Recent studies on natural products as anticancer agent. *Curr. Top. Med. Chem.* 2004, 4, 241–265.
69. Nagase, M.; Oto, J.; Sugiyama, S.; Yube, K.; Takaishi, Y.; Sakato, N. Apoptosis induction in HL-60 cells and inhibition of topoisomerase II by triterpene celastrol. *Biosci. Biotechnol. Biochem.* 2003, 67, 1883–1887.
70. Wu, C.-C.; Chan, M.-L.; Chen, W.-Y.; Tsai, C.-Y.; Chang, F.-R.; Wu, Y.-C. Pristimerin induced caspase-dependent apoptosis in MDA-MB-231 cells via direct effects on mitochondria. *Mol. Cancer Ther.* 2005, 4, 1277–1285.
71. Abbas, S.; Bhoumik, A.; Dahl, R.; Vasile, S.; Krajewski, S.; Cosford, N. D. P.; Ronai, Z. A. Preclinical studies of celastrol and acetyl isogambogic acid in melanoma. *Clin. Cancer Res.* 2007, 13, 6769–6778.
72. Dirsch, V. M.; Kiemer, A. K.; Wagner, H.; Vollmar, A. M. The triterpenoid quinonemethide pristimerin inhibits induction of inducible nitric oxide synthase in murine macrophages. *Eur. J. Pharm.* 1997, 336, 211–217.
73. Huang, F.-C.; Chan, W.-K; Moriarty, K. J.; Zhang, D.-C.; Chang, M. N.; He, W.; Yu, K.-T.; Zilberstein, A. Novel cytokine release inhibitors. Part I: triterpenes. *Bioorg. Med. Chem. Lett.* 1998, 8, 1883–1886.
74. Lee, J.-H.; Koo, T. H.; Yoon, H.; Jung, H. S.; Jin, H. Z.; Lee, K.; Hong, Y.-S.; Lee, J. J. Inhibition of NF-κB activation through targeting IκB kinase by celastrol, a quinone methide triterpenoid. *Biochem. Pharm.* 2006, 72, 1311–1321.
75. Sethi, G.; Ahn, K. S.; Pandey, M. K.; Aggarwal, B. B. Celastrol, a novel triterpene potentiates TNF-induced apoptosis and suppresses invasion of tumor cells by inhibiting

NF-κB-regulated gene products and TAK1-mediated NF-κB activation. *Blood* 2007, 109, 2727–2735.

76. Pinnaa, G. F.; Fiorucci, M.; Reimund, J.-M.; Taquet, N.; Arondel, Y.; Muller, C. D. Celastrol inhibits pro-inflammatory cytokine secretion in Crohn's disease biopsies. *Biochem. Biophys. Res. Commun.* 2004, 322, 778–786.
77. Zhang, D.-H.; Marconi, A.; Xu, L.-M.; Yang, C.-X.; Sun, G.-W.; Feng, X.-L.; Ling, C.-Q.; Qin, W.-Z.; Uzan, G.; d'Alessio, P. Tripterine inhibits the expression of adhesion molecules in activated endothelial cells. *J. Leukoc. Biol.* 2006, 80, 309–319.
78. Li, H.; Zhang, Y.-Y.; Huang, X.-Y.; Sun, Y.-N.; Jia, Y.-F.; Li, D. Beneficial effect of tripterine on systemic lupus erythematosus induced by active chromatin in BALB/c Mice. *Eur. J. Pharm.* 2005, 512, 231–237.
79. Xu, X.; Zhong, J.; Wu, Z.; Fang, Y.; Xu, C. Effects of tripterine on mRNA expression of TGF-β1 and collagen IV expression in BW F1 mice. *Cell Biochem. Funct.* 2007, 25, 501–507.
80. Traynor, B. J.; Bruijn, L.; Conwit, R.; Beal, F.; O'Neill, G.; Fagan, S. C.; Cudkowicz, M. E. Neuroprotective agents for clinical trials in ALS. A systematic assessment. *Neurology* 2006, 67, 20–27.
81. Jung, H. W.; Chung, Y. S.; Kim, Y. S.; Park, Y.-K. Celastrol inhibits production of nitric oxide and proinflammatory cytokines through MAPK signal transduction and NF-κB in LPS-stimulated BV-2 microglial cells. *Exp. Mol. Med.* 2007, 39, 715–721.
82. Cleren, C.; Calingasan, N. Y.; Chen, J.; Beal, M. F. Celastrol protects against MPTP- and 3-nitropropionic acid-induced neurotoxicity. *J. Neurochem.* 2005, 94, 995–1004.
83. Westerheide, S. D.; Bosman, J. D.; Mbadugha, B. N. A.; Kawahara, T. L. A.; Matsumoto, G.; Kim, S.; Gu, W.; Devlin, J. P.; Silverman, R. B.; Morimoto, R. I. Celastrol as inducers of the heat shock response and cytoprotection. *J. Biol. Chem.* 2004, 279, 56053–56060.
84. Wang, J.; Gines, S.; MacDonald, M. E.; Gusella, J. F. Reversal of a full-length mutant huntingtin neuronal cell phenotype by chemical inhibitors of polyglutamine-mediated aggregation. *BMC Neurosci.* 2005, 6,(1)1–12.
85. Chow, A. M.; Brown, I. R. Induction of heat shock proteins in differentiated human and rodent neurons by celastrol. *Cell Stress Chaperones* 2007, 12, 237–244.
86. Zhang, Y.-Q.; Sarge, K. D. Celastrol inhibits polyglutamine aggregation and toxicity through induction of the heat shock response. *J. Mol. Med.* 2007, 85, 1421–1428.
87. Allison, A. C.; Cacabelos, R.; Lombardi, V. R. M.; Alveraz, X. A.; Vigo, C. Celastrol, a potent antioxidant and anti-inflammatory drug, as a possible treatment for Alzheimer's disease *Prog. Neuro-Psychopharmacol. Biol. Psychiatry* 2001, 25, 1341–1357.
88. Paris, D.; Patel, N.; Quadros, A.; Linan, M.; Bakshi, P.; Ait-Ghezala, G.; Mullan, M. Inhibition of Aβ production by NF-κB inhibitors. *Neurosci. Lett.* 2007, 415, 11–16.
89. Powers, M. V.; Workman, P. Inhibitors of the heat shock response: biology and pharmacology. *FEBS Lett.* 2007, 581, 3758–3769.
90. Hieronymus, H.; Lamb, J.; Ross, K. N.; Peng, X. P.; Clement, C.; Rodina, A.; Nieto, M.; Du, J.; Stegmaier, K.; Raj, S. M.; Maloney, K. M.; Clardy, J.; Hahn, W. C.; Chiosis, G.; Golub, T. R. Gene expression signature-based chemical genomic prediction identifies a novel class of HSP90 pathway modulators. *Cancer Cell* 2006, 10, 321–330.

91. Yang, H.; Chen, D.; Cui, Q. C.; Yuan, X.; Dou, Q. P. Celatrol, a triterpene extracted from the Chinese "thunder of god vine" is a potent proteasome inhibitor and suppresses human prostate cancer growth in nude mice. *Cancer Res.* 2006, 66, 4758–4765.
92. Yang, H.; Landis-Piwowar, K. R.; Lu, D.; Yuan, P.; Li, L.; Reddy, G. P.-V.; Yuan, X.; Dou, Q. P. Pristemerin induces apoptosis by targeting the proteasome in prostate cancer cells. *J. Cell. Biochem.* 2008, 103, 234–244.
93. He, W.; Huang, F.-C.; Morytko, M.; Jariwala, N.; Yu, K.-T. Novel cytokine release inhibitors. Part II: steroids. *Bioorg. Med. Chem. Lett.* 1998, 8, 2825–2828.
94. He, W.; Huang, F.-C.; Gavai, A.; Chan, W. K.; Amato, G.; Yu, K.-T.; Zilberstein, A. Novel cytokine release inhibitors. Part III: truncated analogs of tripterine. *Bioorg. Med. Chem. Lett.* 1998, 8, 3659–3664.
95. Moujir, L.; Gutiérrez-Navarro, A. M.; González, A. G.; Ravelo, A. G.; Luis, J. G. The relationship between structure and antimicrobial activity in quinones from the Celastraceae. *Biochem. Syst. Ecol.* 1990, 18, 25–28.
96. González, A. G.; Alvarenga, N. L.; Ravelo, A. G.; Bazzocchi, I. L.; Ferro, E. A.; Navarro, A. G.; Moujir, L. M. Suctione, a new bioactive norquinonemethide triterpene from *Maytenus suctiodes* (Celastraceae). *Bioorg. Med. Chem.* 1996, 4, 815–820.
97. Sotanaphun, U.; Suttisri, R.; Lipipun, V.; Bavovada, R. Quinone-methide triterpenoids from *Glyptopetalum sclerocarpum*. *Phytochemistry* 1998, 49, 1749–1755.
98. Sotanaphun, U.; Lipipun, V.; Suttisri, R.; Bavovada, R. Antimicrobial activity and stability of tingenone derivatives. *Planta Med.* 1999, 65, 450–452.
99. Mean-Rejón, G. J.; Pérez-Espadas, A. R.; Moo-Puc, R. E.; Cedillo-Rivera, R.; Bazzocchi, I. L.; Jiménez-Diaz, I. A.; Quijano, L. Antigiardial activity of triterpenoids from root bark of *Hippocratea excelsa*. *J. Nat. Prod.* 2007, 70, 863–865.
100. Figueiredo, J. N.; Räz, B.; Séquin, U. Novel quinone methides from *Salacia kraussii* with *in vitro* antimalarial activity. *J. Nat. Prod.* 1998, 61, 718–723.
101. Thiem, D. A.; Sneden, A. T.; Khan, S. I.; Tekwani, B. L. Bionortriterpenes from *Salacia madagascariensis*. *J. Nat. Prod.* 2005, 68, 251–254.
102. Jeller, A. H.; Salva, D. H. S.; Lião, L. M.; Bolzani, V. S.; Burlan, M. Antioxidant phenolic and quinonemethide triterpenes from *Cheiloclinium cognatum*. *Phytochemistry* 2004, 65, 1977–1982.
103. Bai, J.-P.; Shi, Y.-L.; Fang, X.; Shi, Q.-X. Effects of demethylzeylasteral and celastrol on spermatogenic cell Ca^{2+} channels and progesterone-induced sperm acrosome reaction. *Eur. J. Pharm.* 2003, 464, 9–15.
104. Sun, H.; Liu, X.; Xiong, Q.; Shikano, S.; Li, M. Chronic inhibition of cardiac Kir2.1 and hERG potassium channel by celastrol with dual effects on both ion conductivity and protein trafficking. *J. Biol. Chem.* 2006, 281, 5877–5884.
105. Morikawa, T.; Kishi, A.; Pongpiriyadacha, Y.; Matsuda, H.; Yoshikawa, M. Structures of new friedelane-type triterpenes and eudesmane-type sesquiterpene and aldose reductase inhibitors from *Salacia chinensis*. *J. Nat. Prod.* 2003, 66, 1191–1196.
106. Murayama, T.; Eizuru, Y.; Yamada, R.; Sadanari, H.; Matsubara, K.; Rukung, G.; Tolo, F. M.; Mungai, G. M.; Kofi-Tsekpo, M. Anticytomegalovirus activity of pristimerin, a triterpenoid quinone methide isolated from *Maytenus heterophylla* (Eckl. & Zeyh.). *Antiviral Chem. Chemother.* 2007, 18, 133–139.

107. Avilla, J.; Teixidò, A.; Velásquez, C.; Alvarenga, N.; Ferro, E.; Canela, R. Insecticidal activity of *Maytenus* species (Celastraceae) nortriterpene quinone methide against codling moth, *Cydia pomonella* (L.) (Lepidoptera: Tortricidae). *J. Agric. Food Chem.* 2000, 48, 88–92.

108. Corsino, J.; de Carvalho, P. R. F.; Kato, M. J.; Latorre, L. R.; Oliveira, O. M. M. F.; Araújo, A. R.; Bolzani, V. S.; França, S. C.; Pereira, A. M. S.; Furlan, M. Biosynthesis of friedelane and quinonemethide triterpenoids is compartmentalized in *Maytenus aquifolium* and *Salacia campestris*. *Phytochemistry* 2000, 55, 741–748.

109. Oikawa, S.; Hirosawa, I.; Hirakawa, K.; Kawanishi, S. Site specificity and mechanism of oxidative DNA damage induced by carcinogenic catechol. *Carcinogenesis* 2001, 22, 1239–1245.

110. Schweigert, N.; Acero, J. L.; von Gunten, U.; Canonica, S.; Zehnder, A. J. B.; Eggen, R. I. L. DNA degradation by the mixture of copper and catechol is caused by DNA–copper-hydroperoxo complexes probably DNA–Cu(I)OOH. *Environ. Sci. Mol. Mutagen.* 2000, 36, 5–12.

111. Flowers, L.; Ohnishi, T.; Penning, T. M. DNA strand scission by polycylic hydrocarbon *o*-quinone: role of reactive oxygen species, Cu(II)/(I) redox cycling, and *o*-semiquinone anion radicals. *Biochemistry* 1997, 36, 8640–8648.

112. Zhou, Q.; Zuniga, M. A. Quinone methides formations in the Cu^{2+}-induced oxidation of a diterpenone catechol and concurrent damage on DNA. *Chem. Res. Toxicol.* 2005, 18, 382–388.

113. Zuniga, M. A.; Wehunt, M. P.; Zhou, Q. DNA oxidative damage by catechol analogs of terpene quinone methides in the presence of Cu(II) and/or NADH. *Chem. Res. Toxicol.* 2006, 19, 828–836.

114. Trott, A.; West, J. D.; Klaic, L.; Westerheide, S. D.; Silverman, R. B.; Morimoto, R. I.; Morano, K. A. Activation of heat shock and antioxidant responses by the natural product celastrol: transcriptional signatures of a thiol-targeted molecule. *Mol. Biol. Cell* 2008, 19, 1104–1112.

9

REVERSIBLE ALKYLATION OF DNA BY QUINONE METHIDES

STEVEN E. ROKITA

Department of Chemistry and Biochemistry, University of Maryland, College Park, MD 20742, USA

9.1 INTRODUCTION

The reversibility of DNA alkylation by quinone methides (QM), while logical and perhaps even predictable, only became obvious after investigating model systems based on deoxynucleosides. Equivalent studies with DNA were hampered by its high molecular weight since this necessitated its complete hydrolysis to the component deoxynucleosides before the alkylation products could be identified. The conditions and time required for this hydrolysis often allow or even promote degradation of labile products prior to their detection. Thus, reversible alkylation of DNA by QMs remained unexplored until recently. Besides, alkylation is not generally associated with reversibility. Imines, oximes, disulfides, and related types of bonds are instead most commonly associated with reversibility. This characteristic has provided the basis for self-selecting and assembling systems held by covalent bonds,[1,2] and now QM alkylation offers similar opportunities in addition to all of those described below and throughout this book.

The chemical and biological implications of compounds that support reversible but covalent processes are often lost in a common assumption that all covalent reactions of DNA are irreversible. Difficulty with reversible reactions is often encountered while attempting to isolate labile products as mentioned above and described more fully in Section 9.2. Yet, reversibility also has the potential to extend the effective lifetime of transient intermediates (Section 9.3.2) and support selective, target-promoted QM

Quinone Methides, Edited by Steven E. Rokita

transfer (Section 9.3.3). Whether or not this reversibility plays an important role in the cytotoxicity of a new series of QM-generating compounds now under preclinical investigation should become apparent in the near future.[3–5]

9.1.1 Reversible Alkylation of DNA

One of the first compounds to demonstrate reversible alkylation of DNA was published little more than 15 years ago and involved the antitumor antibiotic CC-1065 and related derivatives based on its central cyclopropylpyrrolindole core (Scheme 9.1).[6–8] Subsequent efforts explored a wide range of additional compounds containing this core to identify a relationship between chemical reactivity and cytotoxicity as well as to develop new therapeutic agents.[9,10] Interestingly, one report noted an increase in cytotoxicity with an increase in reversibility of alkylation.[11] The potential advantage of reversibility was recognized early when bizelesin, a conjugate of two cyclopropylpyrrolindoles, was found to migrate from kinetically favored sites within DNA that allowed for only monoalkylation to thermodynamically favored sites that allowed for the desired cross-linking (bis-alkylation).[12] Alternative reagents that act irreversibly would have been forever trapped at the initial site of monoalkylation and not have had the opportunity to cross-link DNA.

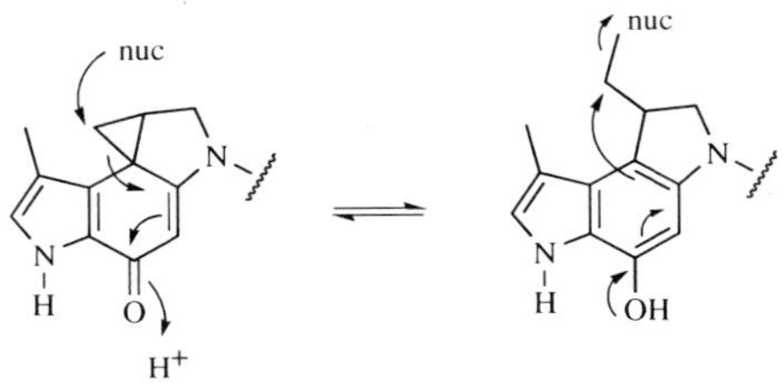

SCHEME 9.1 Reversible alkylation by a central cyclopropylpyrrolindole.

Another natural product, ecteinascidin 743 (Et 743), also demonstrated reversible alkylation of DNA through its reaction with the 2-amino group of guanine (Scheme 9.2). Again, migration was observed from sites favored kinetically to sites favored thermodynamically. The mechanism of this migration likely involved repetitive alkylation by and regeneration of Et 743.[13] Although unproductive and off-path reactions are not necessarily suppressed in this process, their impact is minimized since they only consume reactants transiently rather than terminally. Regeneration of the reactant may even play a significant role in maximizing its biological action since its effective lifetime may be dramatically extended in this manner (Section 9.3.2). Reversible reaction also provides a mechanism by which a DNA adduct may even evade DNA repair. Once a section of damaged DNA is excised from the chromosome and hydrolyzed to the deoxynucleotide level, the reactant may release and return to the chromosome to form a new adduct (Scheme 9.3).

Metabolism of nitrosamines from tobacco generates a series of electrophiles that alkylate DNA, and some of the resulting adducts can also regenerate electrophilic intermediates illustrated in Scheme 9.4. This oxonium ion can be terminally quenched

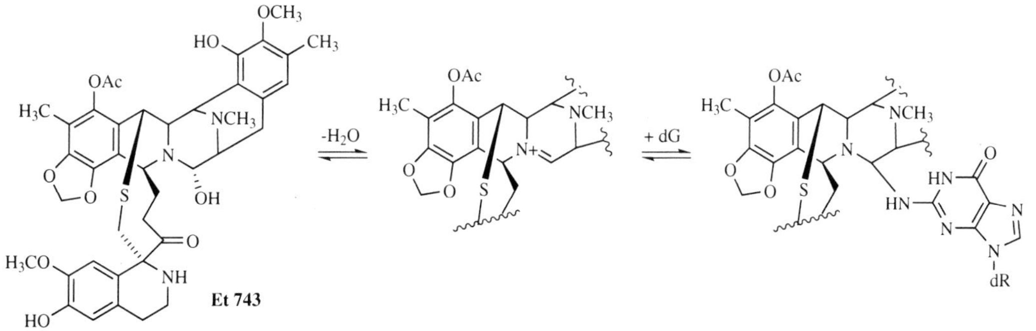

OAc

H_3C

HO

OCH_3

CH_3

NCH_3

N

S

O

O

OH

O

H_3CO

HO

NH

Et 743

$-H_2O$

OAc

H_3C

NCH_3

N+

S

O

O

+ dG

OAc

H_3C

NCH_3

N

S

O

O

HN

HN

O

N

N

N

dR

SCHEME 9.2 Reversible alkylation by ecteinascidin 743.

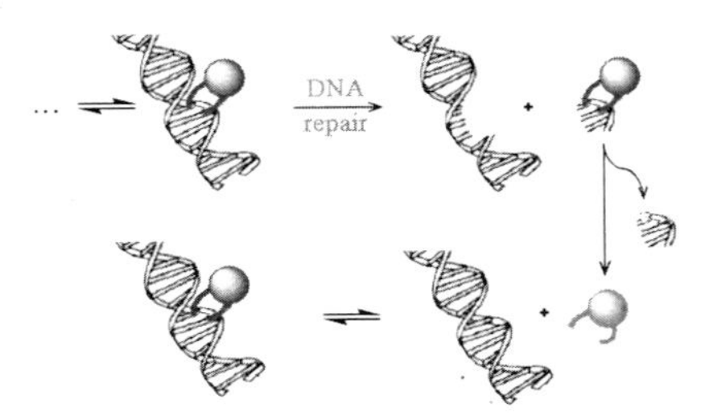

SCHEME 9.3 DNA adducts may reform after excision repair and return to react with DNA again.

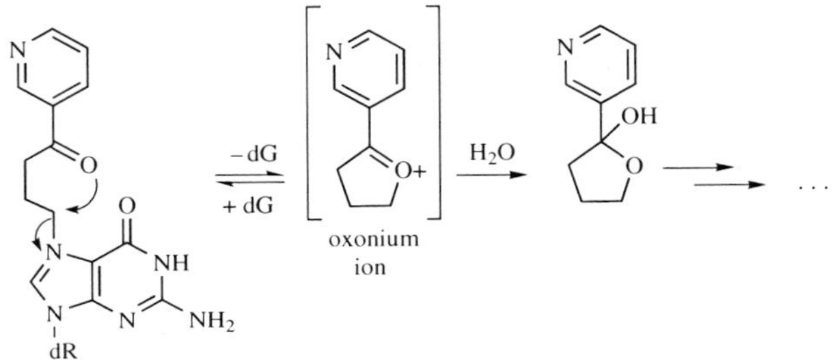

SCHEME 9.4 Reversible alkylation of DNA by an oxonium ion.

by water.[14] However, the oxonium ion can also alkylate DNA and regenerate itself repetitively. The potential biological relevance of this reversibility has not yet been examined but will likely contribute to the potency of such reactive intermediates.

The fascinating natural product leinamycin exhibits significant antitumor activity due at least in part to its ability to alkylate the N7 position of guanine via transient formation of an episulfonium ion.[15] Surprisingly, this intermediate can also reform after reversing its addition to guanine and either transfer back to DNA or be quenched by water (Scheme 9.5).[16] Another natural product azinomycin contains an epoxide and aziridine for DNA cross-linking, and its epoxide acts analogously to the episulfonium ion of leinamycin since both act reversibly (Scheme 9.6).[17] Both also alkylate the N7 position of dG and can be regenerated from the resulting dG N7 adducts. Additionally,

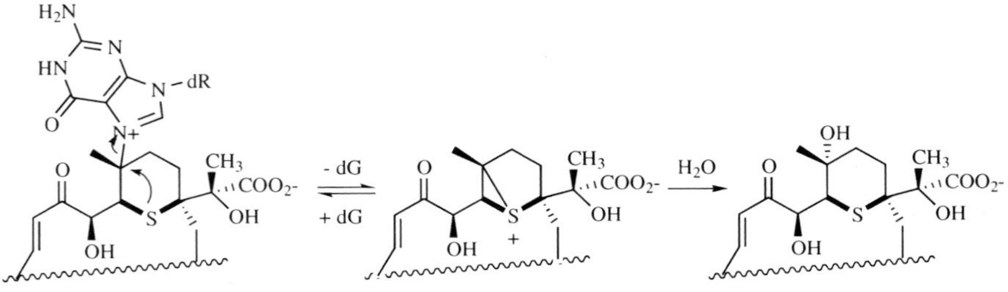

SCHEME 9.5 Reversible alkylation of DNA by leinamycin.

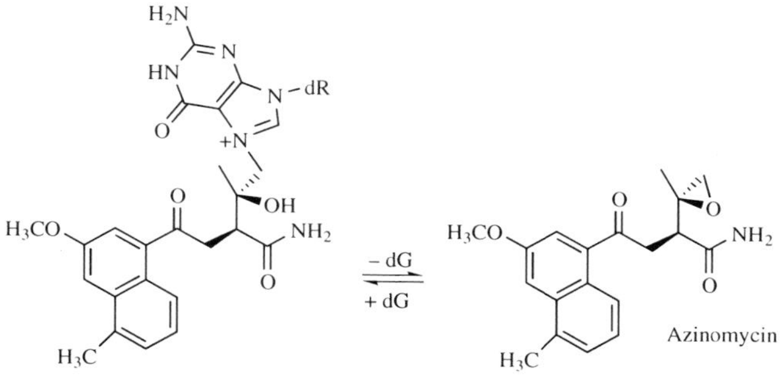

SCHEME 9.6 Reversible alkylation of DNA by azinomycin.

both guanine adducts are susceptible to depurination in competition with release of the episulfonium ion and epoxide groups associated with leinamycin and azinomycin, respectively. Whether the resulting DNA adducts or apurinic sites dominate the biological activity of these compounds has yet to be determined and remains a challenging question.[16,17]

Malondialdehyde does not alkylate DNA in the same direct manner as do QMs or the compounds described above, but its reversible reaction is worth including in this chapter on the basis of its significance in biology and extensive chemical and biochemical characterization. This electrophile appears to react initially with the 2-amino group of dG to form an 3-oxo-1-propenyl adduct that can subsequently cyclize with the N1 of dG to form a pyrimido[1,2-*a*]-purin-10(3*H*)-one derivative ($\mathbf{M_1G}$, Scheme 9.7).[18] Each step in this process is reversible, and the equilibrium between the cyclized and open-chain adducts depends on the neighboring structural environment.[19] *In vivo*, malondialdehyde originates in part from lipid peroxidation. Nucleobase propenals generated from DNA oxidation also provide a source of malondialdehyde through direct exchange of its 3-oxo-1-propenyl group.[20–22] Indeed, the base propenal appears to generate $\mathbf{M_1G}$ even more efficiently than malondialdehyde.[20] This latter observation is particularly pertinent to this chapter, since it introduces the importance of an endogenous carrier for extending and enhancing the effect of a reactive compound.

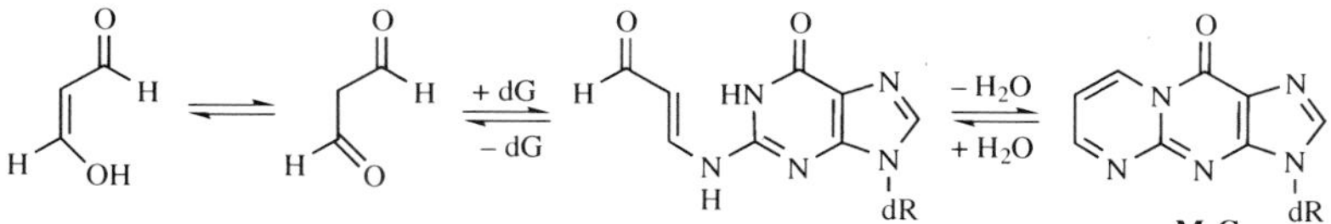

SCHEME 9.7 Reversible reaction between malondialdehyde and DNA.

9.1.2 Initial Reports of Reversible Alkylation by Quinone Methides

Earlier model studies designed to explore the action of anthracycline antitumor antibiotics first noted a lability of QM adducts. **QM1** was generated by oxidation of its precursor (**QMP1**) with silver oxide and shown to undergo reversible reaction

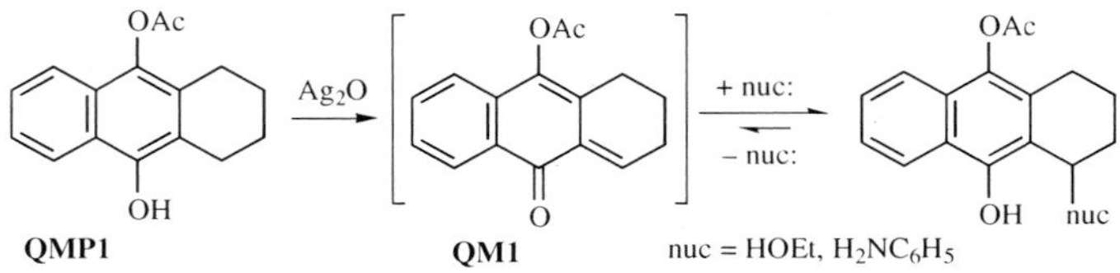

nuc = HOEt, $H_2NC_6H_5$

SCHEME 9.8 Reversible alkylation by a model anthracycline.

with ethanol and aniline (Scheme 9.8).[23] This report also described reaction between **QM1** and an adenosine derivative to form a product that was initially assigned as an adduct of adenosine N1.[23] Further study suggested an alternative adduct formed instead by alkylation of the *exo* 6-amino group, and likewise an equivalent adduct of the *exo* 2-amino group of a guanine derivative was also identified after reaction with this nucleobase.[24] Difficulties in isolating and assigning the N1 and 6-amino adducts of adenosine are quite understandable and in part such ambiguities instigated our own studies on the reversibility of QM alkylation as described in Section 9.2.1.

Further investigation of the original **QM1** also demonstrated that simple thiols could react reversibly although regeneration of **QM1** was very slow.[25] In contrast, a later model of anthracycline was shown to form stable adducts with a thiol despite the lability of corresponding adducts formed by oxygen and nitrogen nucleophiles.[26]

The first systematic study with deoxynucleotides (dNs) using a neutral aqueous system involved two QMs that form *in vivo* by P_{450} oxidation of the food preservative, 2,6,-di-*tert*-butyl-4-methylphenol (**BHT**). For experimental ease, however, the QMs of interest were most conveniently generated by oxidation of the appropriate precursors by silver oxide (Scheme 9.9). All dNs exhibited at least some reactivity with the resulting *para*-quinone methide **QM2**.[27] Even the relatively nonnucleophilic dT reacted, albeit weakly, to form its N3 adduct (Fig. 9.1). Adducts formed by reaction at the N3 and *exo* 4-amino groups of dC were also detected. The purines were alkylated with even greater efficiency than the pyrimidines and yielded adducts of dG attached to its N1, N7, and *exo* 2-amino groups alternatively and an adduct of dA attached to its *exo* 6-amino group. Kinetics of both adduct formation and decomposition were observed to be highly structure dependent. Although QM regeneration was not examined specifically in this effort, dG N7 and dC N3 adducts were observed to be quite labile.[27] The dG N7 adduct appeared to decompose by depurination, but the dC N3 adduct instead regenerated dC and formed the benzyl alcohol derivative of BHT, the product expected from elimination of the QM followed by trapping with

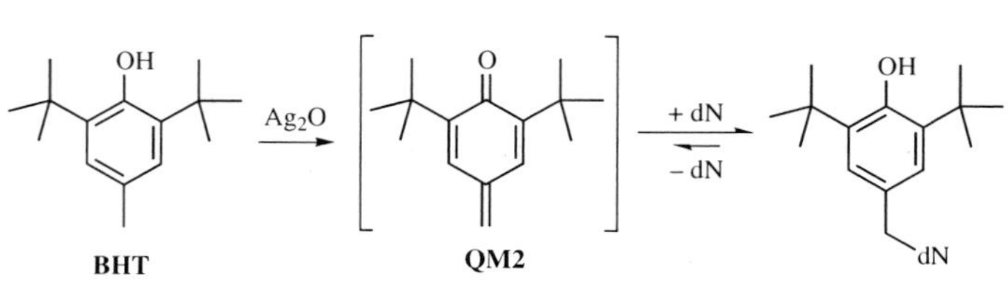

SCHEME 9.9 Reversible alkylation of deoxynucleotides by a metabolite of 2,6,-di-*tert*-butyl-4-methylphenol.

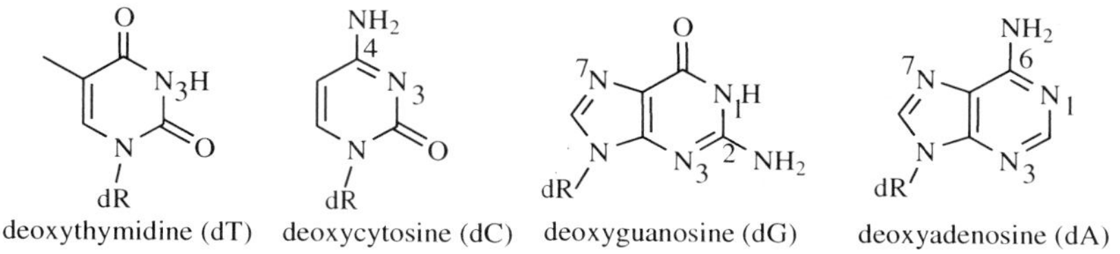

deoxythymidine (dT) deoxycytosine (dC) deoxyguanosine (dG) deoxyadenosine (dA)

FIGURE 9.1 Standard numbering of atoms used to identify sites of reaction within nucleobases.

water. The N1 adduct of dA might have also been formed but would likely have been too transient for detection under the assay conditions.

In contrast to the lability of certain dN adducts formed by the BHT metabolite above, amino acid and protein adducts formed by this metabolite were relatively stable.[28,29] The thiol of cysteine reacted most rapidly in accord with its nucleophilic strength and was followed in reactivity by the α-amine common to all amino acids. This type of amine even reacted preferentially over the ε-amine of lysine.[28] In proteins, however, the ε-amine of lysine and thiol of cysteine dominate reaction since the vast majority of α-amino groups are involved in peptide bonds. Other nucleophilic side chains such as the carboxylate of aspartate and glutamate and the imidazole of histidine may react as well, but their adducts are likely to be too labile to detect as suggested by the relative stability of QMs and the leaving group ability of the carboxylate and imidazole groups (see Section 9.2.3).

Reaction selectivity of the parent *ortho*-QM has also been explored with a variety of amino acid and related species.[30] In these examples, the rates of alkylation and adduct yields were quantified over a range of temperatures and pH values. The initial **QM3** was generated by exposing a quaternary benzyl amine (**QMP3**) to heat or ultraviolet radiation (Scheme 9.10). Reversible generation of **QM3** was implied by subsequent exchange of nucleophiles at the benzylic position under alternative photochemical or thermal activation.[30] Report of this work also included the first suggestion that the reversible nature of QM alkylation could be used for controlled delivery of a potent electrophile.

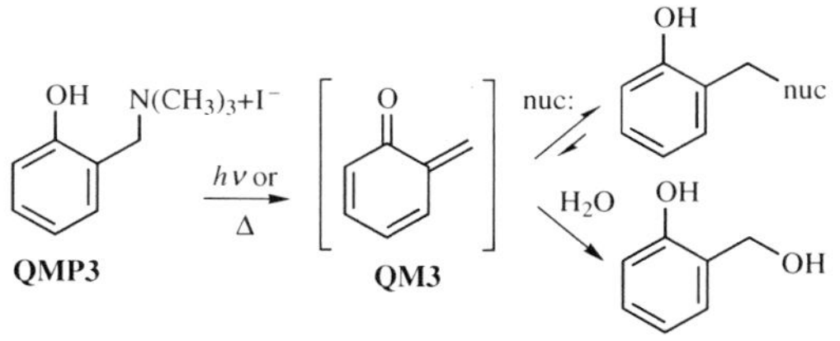

SCHEME 9.10 Generation and reversible reaction of an *ortho*-quinone methide.

9.2 REVERSIBLE ALKYLATION OF DEOXYNUCLEOSIDES BY A SIMPLE QUINONE METHIDE

Our laboratory did not at first fully appreciate the consequences of reversible alkylation by QM as our interest shifted toward investigations of this reactive intermediate with nucleic acids from earlier work related to its applications to mechanistic enzymology.[31–35] The apparent selectivity of certain QMs to alkylate the *exo* amino group of purines was surprising since these groups are not the most nucleophilic sites.[24,27] An electrophile is expected to demonstrate either selectivity based on the strength of its nucleophilic partner or no selectivity for any competing nucleophiles. To address the unusual observations suggesting a preference for weak, rather than strong, nucleophiles, our laboratory developed a model system to isolate and characterize each adduct formed by the parent *ortho*-QM (**QM3**) and quantify the yield of each from reaction of deoxynucleosides to duplex DNA.[36–38] A scheme for generating **QM3** was adopted based on fluoride-dependent release of a silyl-protected precursor (**QMP4**) to allow for easy application in simple chemical systems as well as more complex biochemical mixtures (Scheme 9.11). In general, our results from reactions extending for ~24 h confirmed the previous reports that DNA yielded adducts of the *exo* amino groups of purines. Intermolecular proton transfer during alkylation was proposed[37] to explain the apparent selectivity based on the limited data available at the time and the known activation of QMs by protonation.[39–42] The significance of this proposal diminished because an even more fundamental question became apparent as described below.

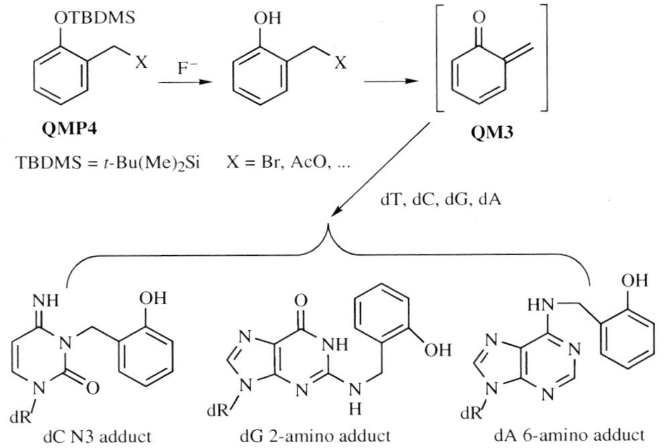

SCHEME 9.11 Generation of an *ortho*-quinone methide (**QM3**) under control of fluoride ion.

9.2.1 Quinone Methide Regeneration is Required for Isomerization between Its N1 and 6-Amino Adducts of dA

The nucleophiles participating in reaction with **QM3** are typically obvious from the QM linkage identified in the products of all deoxynucleosides except for dA. Direct reaction to form the observed adduct at the *exo* 6-amino group of dA was still considered surprising due to its weak nucleophilicity, and an alternate pathway involving initial reaction at the most nucleophilic dA N1 followed by a Dimroth rearrangement that interconverts the N1 to the 6-amino groups seemed more reasonable

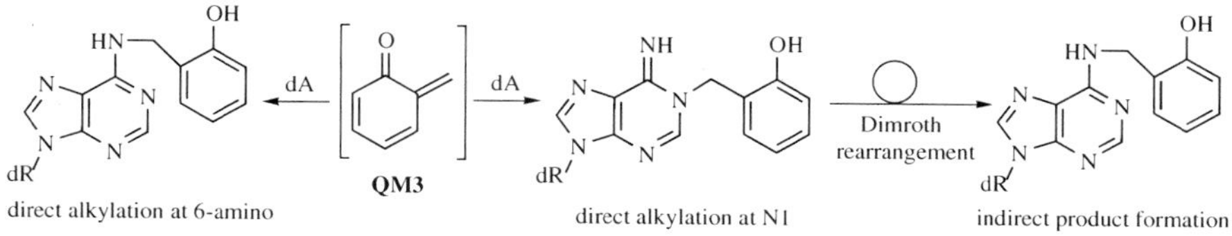

SCHEME 9.12 The **QM3** adduct attached to the 6-amino of dA may form through direct and indirect mechanisms.

(Scheme 9.12).[43–45] This alternative might also have explained the rapid decomposition of the dA N1 adduct prepared as a synthetic standard under mild conditions and the more forcing conditions required for accumulation of the 6-amino adduct.[46] Unexpectedly, no Dimroth rearrangement was evident when monitoring reaction between **QM3** and 6-[^{15}N]-dA.[46]

Intermediate formation of the N1 adduct still seemed most reasonable since it was found to be a kinetic product of alkylation with **QM3** and would initially dominate the product profile prior to detection of the 6-amino adduct.[46] Alternative mechanisms were thus explored. Intramolecular and intermolecular processes were differentiated by examining conversion of the N1 adduct of 6-[^{15}N]-dA to its 6-amino derivative in the presence of 10-fold excess nonlabeled dA. This classic isotope dilution experiment revealed a complete and statistical exchange with the nonlabeled dA and consequently indicated that isomerization was a dissociative process involving repeated regeneration and consumption of **QM3** until the kinetic product (N1 adduct) dissipated and the thermodynamic product (6-amino adduct) accumulated (Scheme 9.13).[46] Of course in hindsight, regeneration of **QM3** did not require great imagination since it was originally generated by elimination of a benzylic substituent, and the N1 of dA also acts as a good leaving group as indicated by the acidity of its conjugate acid.[47] However, the ability for this exchange to occur under aqueous conditions and avoid substantial trapping by water was not anticipated. Finally, the potential for the N1 adduct of dA to serve as a kinetic intermediate during QM alkylation was confirmed by the ability of both **QMP4** and the **QM3** adduct of dA N1 to produce equivalent profiles of adducts from an equimolar mixture of dA, dG, dC, and dT.[46]

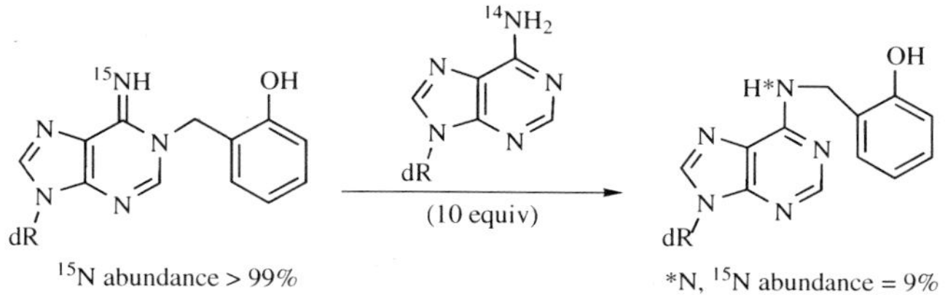

SCHEME 9.13 A dissociative mechanism releases the ^{15}N-labeled dA from its QM adduct.

9.2.2 Kinetic and Thermodynamic Adducts Formed by Quinone Methides

The reversibility of dA alkylation by **QM3** creates both problems and opportunities. For example, an accurate profile of products formed by initial exposure of DNA to QMs may not be easily determined since labile products such as the N1 adduct of dA would not persist through the typical procedures of enzyme digestion and analysis of the subsequent deoxynucleoside products by reverse phase HPLC. Thus, the full extent of DNA alkylation by QMs may be severely underestimated.

Deoxynucleosides in contrast to DNA can be analyzed rapidly, and thus alkylation of these monomeric components was used to determine the yield and structure of potential kinetic products. Indeed, adducts of the strong nucleophiles, dA N1 and dC N3, respectively, formed in approximately 7- and 10-fold greater efficiency relative to the corresponding adducts of weak nucleophiles in an equimolar mixture of dA, dC, dG, and dT during the initial 5 h after triggering the formation of **QM3** from **QMP4** (Fig. 9.2).[48] These results were very satisfying since the efficiency for QM addition was now shown to depend on the strength of the participating nucleophile. The apparent specificity noted previously for the *exo* amino groups was merely a function of product stability and not alkylation efficiency.

Lability of the resulting QM products revealed a logical correlation with the leaving group ability of the participating nucleophile. For example, the conjugate acid of dA N1 is stronger than the corresponding conjugate of dC N3, and hence the adduct of

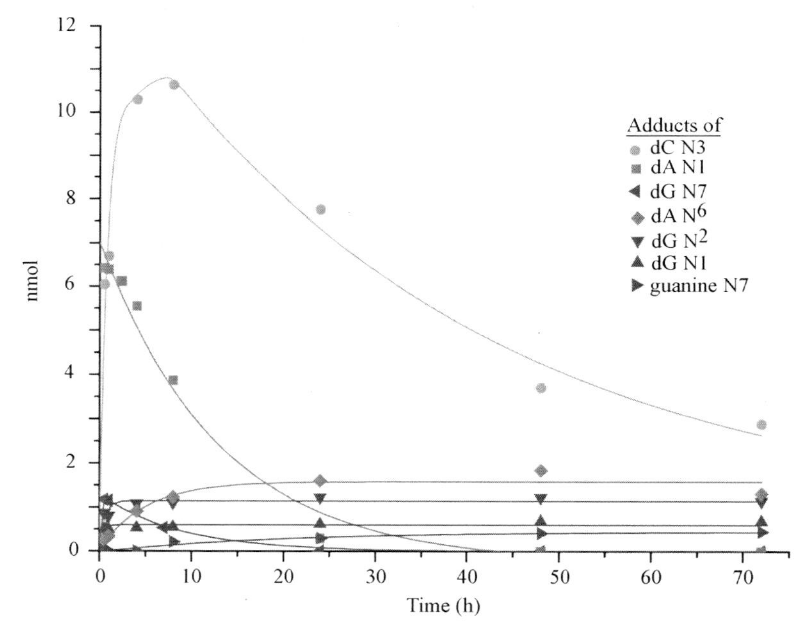

FIGURE 9.2 Time-dependent evolution of quinone methide adducts formed by an equimolar mixture of dA, dC, dG, and dT was monitored by reverse-phase chromatography. *Source*: Reproduced with permission from *Chem. Res. Toxicol.* **2005**, *18*, 1364–1370.[48] Copyright 2005, American Chemical Society.

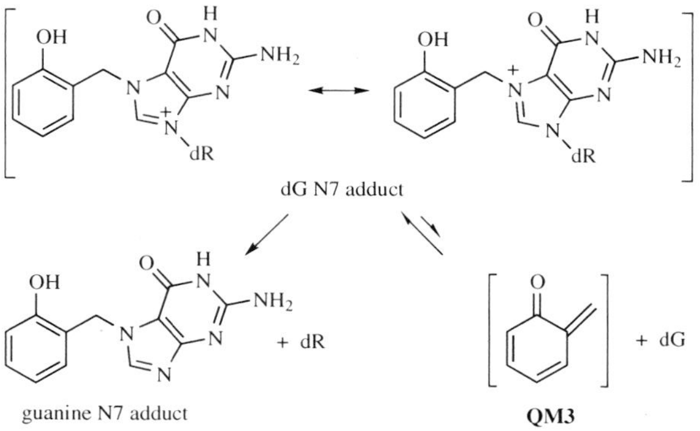

dG N7 adduct

guanine N7 adduct + dR

QM3 + dG

SCHEME 9.14 Partitioning of the dG N7 adduct between release of **QM3** and release of deoxyribose (deglycosylation).

dA N1 is more labile than the adduct of dC N3.[48] The *exo* amino groups form the weakest conjugate acids and are expectedly the most stable QM adducts. Perhaps by serendipity, the strongest nucleophiles of DNA are also stable leaving groups. The N7 of dG is also a strong nucleophile and a good leaving group, although its initial yield of QM capture is lower than expected from its nucleophilicity. This observation may in part be the result of competition between regeneration of **QM3** and the ability of this parent adduct to deglycosylate and form a stable N7 adduct of guanine (Scheme 9.14). The QM adduct of dG N7 is too labile to isolate for independent study,[38] but the adduct could be monitored *in situ*. The rate of deglycosylation was measured by formation of the guanine adduct, and the rate of QM release was measured by trapping this intermediate with phenylhydrazine. Under these conditions, both paths contributed equally to the decomposition of the dG N7 adduct.[48] The half-life of this adduct also decreased from approximately 6 to 2.5 h after phenylhydrazine was added. This suggests that **QM3** likely reacted with and was regenerated from the N7 position multiple times in the absence of phenylhydrazine. This type of reversible addition and elimination was also even evident during generation and consumption of a simple *para*-QM (**QM4**) and its precursor derived from a benzyl bromide.[49] QM reactivity may therefore be explained in all of these examples by rapid, efficient but reversible nucleophilic addition forming kinetic products that ultimately succumb to less efficient and irreversible nucleophilic addition leading to accumulation of thermodynamic products (Scheme 9.15).

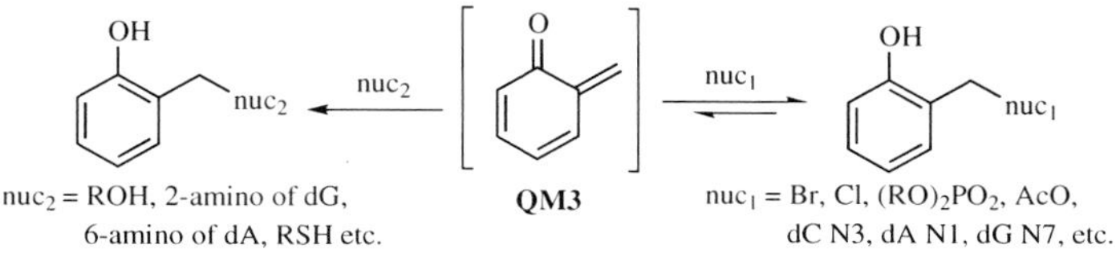

nuc_2 = ROH, 2-amino of dG, 6-amino of dA, RSH etc.

QM3

nuc_1 = Br, Cl, $(RO)_2PO_2$, AcO, dC N3, dA N1, dG N7, etc.

SCHEME 9.15 Product profiles formed by **QM3** depend on the fast and reversible addition of nuc_1 in competition with the slow and irreversible addition of nuc_2.

Detection of the dA N1 and dC N3 adducts may not in one sense be particularly important for DNA based on their central position within the helical conformation and hydrogen bonding network.[37,38] Still, the deoxynucleoside studies helped to focus attention on the reversibility of alkylation by QM and provided insight into the reactions of duplex DNA described below in Section 9.3. Reaction at the deoxynucleoside level also provided an essential system for developing a theoretical treatment of QM reaction.[50–52] Computations based on density functional theory well rationalized the published results on dA and correctly anticipated the results on dG and dC reviewed above and described in more detail in Chapter 2 (Freccero).

9.2.3 The Structure of Quinone Methides and Their Precursors Modulate the Reversibility of Reaction

The above section already introduced the influence of leaving groups at the benzylic position that eliminate to form and regenerate **QM3**, and the trend extends beyond adducts formed by the deoxynucleosides as expected. The standard benzylic acetate of **QMP4** eliminates completely from the deprotected phenol under neutral aqueous conditions and ambient temperature within approximately 20 h, while an equivalent benzyl bromide eliminates completely within 5 min.[48] Benzylic phosphates are also extremely labile, and, if the phosphate backbone of DNA is able to trap QM, the resulting products are likely to be too labile for standard detection.[53,54] In contrast, amines and thiols are much less susceptible to elimination from the benzylic position and require forcing conditions to regenerate the parent QM.[26,30] The benzylic alcohol derivative also appears stable under almost all thermal conditions and only eliminates routinely to form a QM after photochemical excitation.[55]

The ease of eliminating a benzylic substituent to (re)form a QM additionally depends on the electronic characteristics of its aromatic precursor. QMs are electron deficient and therefore subject to stabilization by electron-donating substituents and destabilization by electron-withdrawing substituents. This is well illustrated by a series of QM adducts formed by alkylation of dC N3.[56] The successive withdrawal of electron density from the QM by varying R from a methyl to hydrogen to carboxymethyl group suppressed the rate of initial QM generation as reflected by adduct formation (Fig. 9.3). This trend was observed for regenerating the transient QMs from their dC adducts. The carboxymethyl substituent suppressed QM regeneration sufficiently to prevent decomposition of its dC adduct. In contrast, the methyl substituent significantly enhanced the lability of its corresponding dC adduct. However, this substituent was still not sufficient to labilize the QM adducts formed by the *exo* amino groups of dA and dG. These purine adducts remained stable under all examined conditions.[56] A *N*-morpholino adduct (**QMP9**), however, did become labile when the electron-donating strength of *R* was increased further by replacing the methyl group with a methoxy group (see below).

The weak leaving group ability of the morpholino group provided an excellent opportunity to measure a far greater range of substituent effects than those possible with the dC N3 adduct. From this effort, the reversibility of QM alkylation was observed to vary over many orders of magnitude by the mere presence of a single substituent attached

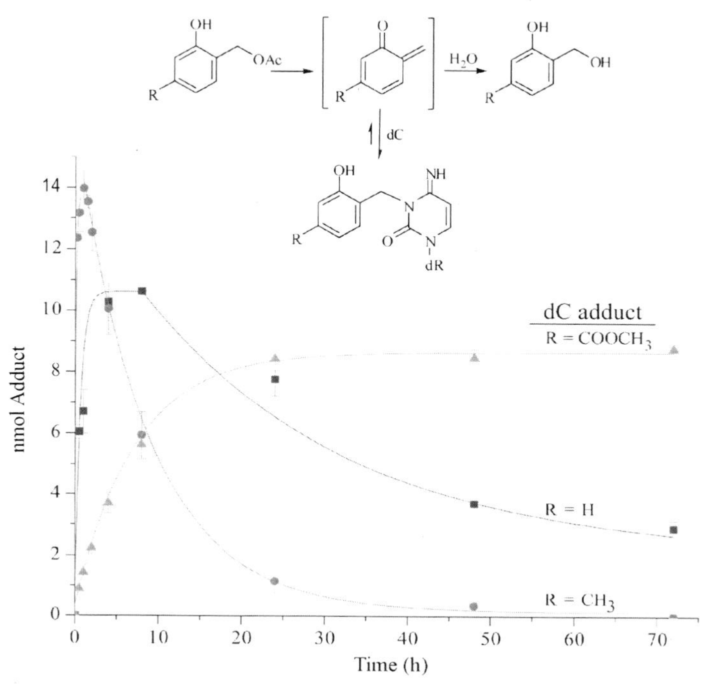

FIGURE 9.3 Substituent effects modulate the formation and lability of QM adducts. *Source*: Adapted from *J. Am. Chem. Soc.* **2006**, *128*, 11940–11947.[56]

directly to the conjugated system. For example, a carboxymethyl group positioned *para* to the nascent *exo* methylene rendered the *N*-morpholino adduct (**QMP5**) stable. No decomposition was detected even after sustained heating to 100°C (Table 9.1).[56] Replacing the carboxymethyl group with a methoxy group, in contrast, generated an equivalent derivative **QMP9** that readily forms its corresponding QM (**QM8**) under ambient temperature within minutes. A similar trend was observed for substitution *meta* to the nascent *exo* methylene (Y), but the overall magnitude of the effects was smaller than that of the corresponding *para* substituents (X) (Table 9.1).[56]

TABLE 9.1 Substituents Affect the Lability of Quinone Methide Adducts

QMP	X	Y	QM	*T* (°C)	$t_{1/2}$ of QMP (min) in H_2O
QMP5	$COOCH_3$	H	**QM5**	100	Stable
QMP6	H	$COOCH_3$	**QM6**	100	149
QMP7	H	H	**QM3**	50	115
QMP8	H	OCH_3	**QM7**	22	11
QMP9	OCH_3	H	**QM8**	22	4

Source: Adapted from *J. Am. Chem. Soc.* **2006**, *128*, 11940–11947.[56]

TABLE 9.2 Substituents Affect the Susceptibility of Quinone Methides to Nucleophilic Addition

QM	Y	k_2 ($M^{-1}s^{-1}$)
QM8	OCH_3	1.9
QM3	H	7.8
QM9	Cl	23
QM6	$COOCH_3$	360
QM10	CN	2000
QM11	NO_2	25,000

Source: Adapted from *J. Am. Chem. Soc.* **2006**, *128*, 11940–11947.[56]

The stabilization observed for the QMPs of Table 9.1 was primarily the result of destabilization of the intermediate QMs, and thus, substituents affect QM and QMP in a reciprocal manner. Just as the carboxymethyl substituent in **QMP6** greatly suppressed formation of **QM6**, the same substituent greatly enhanced subsequent reaction of **QM6** relative to the unsubstituted derivative **QM3**. Rate constants ranged over four orders of magnitude for water addition to a series of QM derivatives containing substituents ranging from electron-withdrawing to electron-donating groups on the position *meta* to the *exo* methylene group (Table 9.2).[56] Even greater effects can be predicted for substituents *para* to the *exo* methylene group. These results in general should help to explain and predict the characteristics of complex QMs that form in nature as well as guide the design of synthetic analogues with the desired balance of reactivity and stability.

9.3 REVERSIBLE ALKYLATION OF DNA BY QUINONE METHIDE BIOCONJUGATES

A satisfactory description of DNA alkylation by QMs such as **QM3** is complicated by our current inability to detect the transient products of alkylation. For some QMs, reversibility may not be a major concern but for others, the consequences of reversibility may be quite significant as illustrated below. Minimally, the product profiles generated by DNA may differ from those generated by deoxynucleosides as a result of the diminished accessibility of certain nucleophiles within the double helical structure. In addition, selective binding and localization of a QM within the helix can direct reaction to one specific nucleophile. For example, the selectivity of mitomycin C for alkylating the weakly nucleophilic *exo* amino group of dG is easily explained by the binding orientation of this drug in the minor groove of DNA.[57] An equivalent adduct generated by metabolism of 4-hydroxytamoxifen[58] may also result from a similar positioning of its quinone methide intermediate within the minor groove. In contrast, the simple **QM3** is unlikely to bind DNA selectively, and thus its apparent selectivity for the *exo* amino group of dG reflects a combination of nucleophile accessibility and product stability.[38]

The N7 of dG is one of the most nucleophilic sites within duplex DNA and is relatively accessible within the major groove.[59] However, QM-dependent alkylation of this site is not typically detected since its products are often reversible and thermodynamically unstable. When this reversibility is suppressed, participation of dG N7 becomes apparent. One mechanism for this suppression is spontaneous deglycosylation (Scheme 9.14) to release a guanine derivative after alkylation of its N7 position.[48] Such a process was previously detected during reaction between DNA and a *para*-QM derivative formed by metabolism of BHT.[27] Interstrand cross-linking can also highlight the role of dG N7 by effectively anchoring QM near to its nucleophilic partner as it continually reforms from reversible adduct formation. Similarly, bioconjugates with high affinity for DNA have the potential to maintain a QM proximal to a site of reversible reaction as well.

9.3.1 The Reversibility of Quinone Methide Reaction Does Not Preclude Its Use in Forming DNA Cross-Links

The presence of two appropriately substituted benzyl groups within a QMP such as **QMP10** provides an opportunity to generate two QM intermediates successively for cross-linking DNA (Scheme 9.16). Removal of the silyl protecting group from **QMP10** initiates QM formation and subsequent reaction with dA, dC, and dG residues within an oligodeoxynucleotide duplex.[60] Cross-linking appeared to form most readily at a 5′. . .CG. . .3′ sequence, and no reaction with DNA was observed prior to deprotection of **QMP10**. Direct displacement of the benzylic groups in **QMP10** rather than tandem QM formation is consequently unlikely. The specificity and yield of cross-linking also depends on the geometry and distance between the two benzylic groups involved in QM formation. These two groups need not even remain attached to the same aromatic ring or conjugated system (Fig. 9.4). Biphenol,[61,62] bipyridyl,[63] binaphthol,[64,65] and two phenol derivatives linked through a variety of spacer groups[66] have all demonstrated efficient QM-dependent cross-linking of DNA.

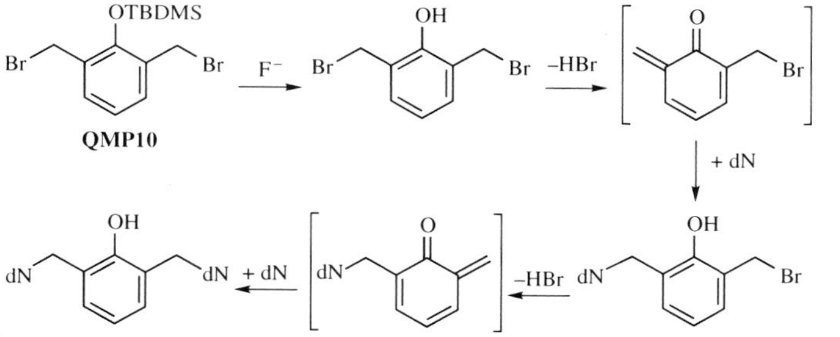

OTBDMS Br Br QMP10 F⁻ OH Br Br −HBr O Br + dN OH dN Br −HBr O dN + dN OH dN dN

SCHEME 9.16 DNA cross-linking by tandem quinone methide formation.

The reversibility of DNA cross-linking by the QMPs illustrated in Fig. 9.4 has not been examined directly but is presumably controlled as before by the leaving group strength of the DNA nucleophiles and the electronic characteristics of the QM intermediates. Unless both possible QMs form simultaneously, the cross-linking agents still remain bound to DNA at one site as the other is transiently released during

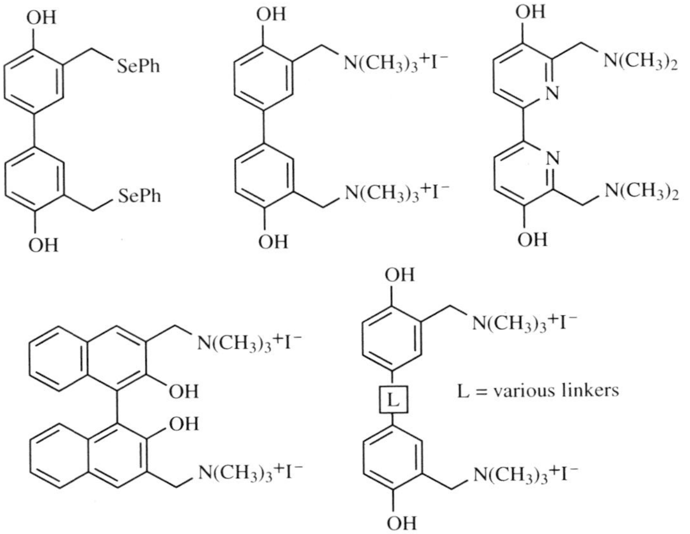

FIGURE 9.4 Representative quinone methide precursors for DNA cross-linking.[61–66]

QM regeneration. Such reversibility may ultimately enhance the efficiency of cross-linking by providing a mechanism to escape nonproductive trapping in regions of DNA that are not capable of stable cross-linking. Reversibility may also allow for efficient trapping of the QM through the formation of kinetic products and then subsequent isomerization to their thermodynamic products.

Conjugation of a DNA ligand such as acridine to a derivative of **QMP10** significantly increased its cross-linking efficiency by over 50-fold and appeared to direct reaction almost exclusively to dG N7.[67] Most importantly, the ratio of cross-linking to monoalkylation for this conjugate (**QMP11**, Scheme 9.17) was significantly greater than that for a similar conjugate based on a *N*-mustard.[67,68] This alternative cross-linking agent also reacts at the same dG N7 albeit irreversibly. The greater ratio of cross-linking to monoalkylation may be rationalized by the reversibility of QM reaction since its capture by dG N7 is not terminal and may regenerate QM for subsequent reaction. In contrast, the *N*-mustard reaction is terminal and excludes further sampling of potential sites for cross-linking. Although cross-linking by **QMP11** persists during gel electrophoresis under denaturing conditions, the bond between dG N7 and the nascent QM should exhibit a half-life in the range of 2–6 h as determined earlier for **QM3** and dG N7 (Scheme 9.14). Thus, cross-links formed by **QMP11** likely remain dynamic and should allow the QM to migrate along duplex

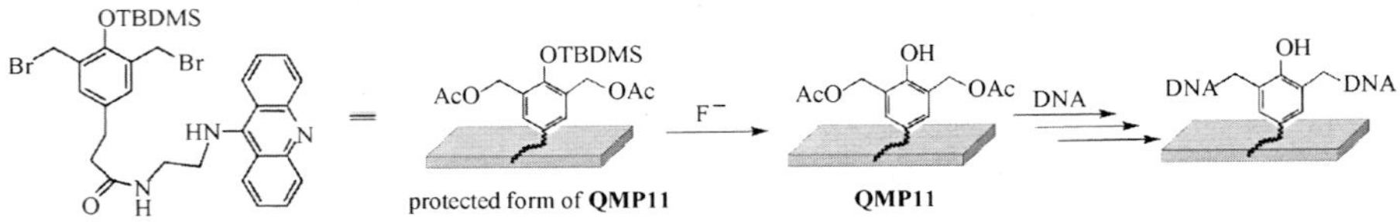

SCHEME 9.17 DNA cross-linking by a quinone methide–acridine conjugate.

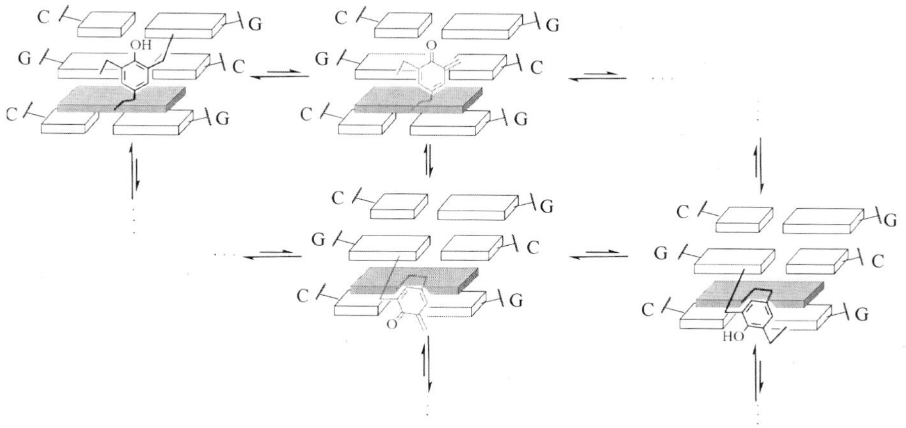

FIGURE 9.5 Migration of a dynamic cross-link in duplex DNA.

DNA from one reversibly acting nucleophile to the next (Fig. 9.5). This migration is currently under investigation in our laboratory, but the consequences of repetitive capture and release of a QM have already been observed from the extended lifetime of a QM in the presence of strong and reversibly acting nucleophiles as described below.

9.3.2 Repetitive Capture and Release of a Quinone Methide Extends Its Effective Lifetime

Standard alkylating and cross-linking agents such as dimethylsulfate or *N*-mustards, respectively, have only one opportunity to partition between various nucleophiles since their reactions are irreversible. In contrast, QMs have the potential to partition between nucleophiles multiple times as long as the resulting adducts are formed reversibly. Continual capture and release of QMs consequently can extend their effective lifetime almost indefinitely and is ultimately limited by only the competitiveness of possible irreversible reactions. For DNA, the strongest nucleophiles act reversibly so terminal quenching remains an infrequent event.

In the absence of reversible reaction, for example when water acts as the lone nucleophile, **QMP11** is consumed with a half-life of approximately 0.5 h as measured by its diminished ability to cross-link DNA (Scheme 9.18).[69] Elimination of acetate to form the first of two possible QM intermediates (**QM12**) is likely rate-determining in this process since subsequent addition by water is estimated to occur with a half-life in the millisecond range.[56] The resulting hydroxy substituent at the benzylic position does not eliminate and regenerate **QM12** under ambient conditions. Thus, water

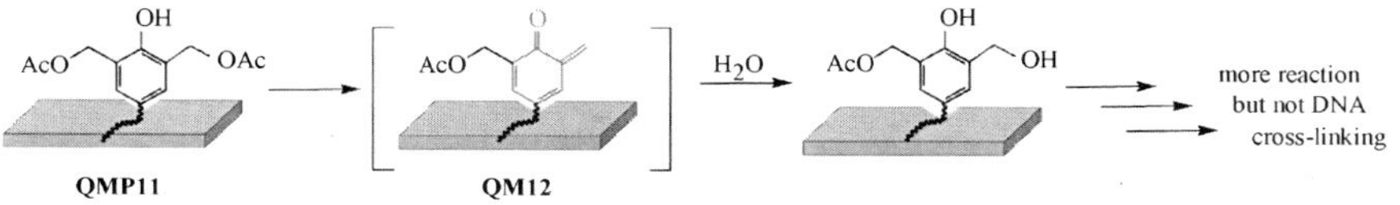

SCHEME 9.18 Quinone methides are quenched by water and prevented from cross-linking DNA.

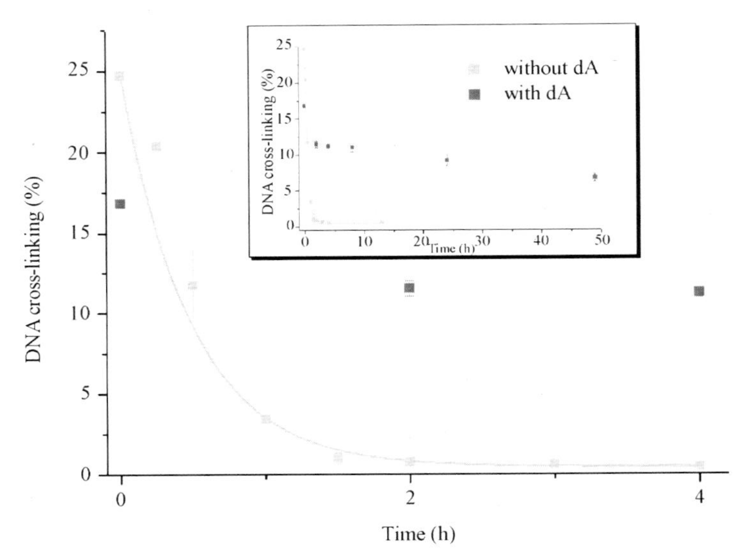

FIGURE 9.6 The loss of cross-linking activity of an aqueous solution of **QMP11** (100 μM) in the presence (black) and absence (gray) of dA (20 mM). Cross-linking activity was measured at the indicated times by addition of duplex DNA (3 μM) and subsequent analysis by denaturing gel electrophoresis. *Source*: Adapted from *Angew. Chem. Int. Ed.* **2008**, *47*, 1291–1293.[69]

addition to **QM12** is terminal, and cross-linking is consequently prohibited even though alkylation at the remaining QM site is still possible.

The added presence of dA in an aqueous solution of **QMP11** extends the effective lifetime for cross-linking DNA by 100-fold as indicated by a half-life of 50 h rather than the 0.5 h described above (Fig. 9.6).[69] This result is most likely explained by the ability of dA N1 to react reversibly and competitively with **QM12** and prevent irreversible quenching by water (Scheme 9.19). Neither the persistence of **QM12** nor **QMP11** should be directly affected by dA. Instead, capture and release of **QM12** by the N1 of dA is expected to occur multiple times during incubation prior to the addition of duplex DNA used for assessing the remaining capacity for cross-linking. Based on a maximum half-life for the dA N1 adduct of 4 h,[48] this adduct likely forms reversibly a minimum of 25 times on average to support the observed 100-fold increase in half-life

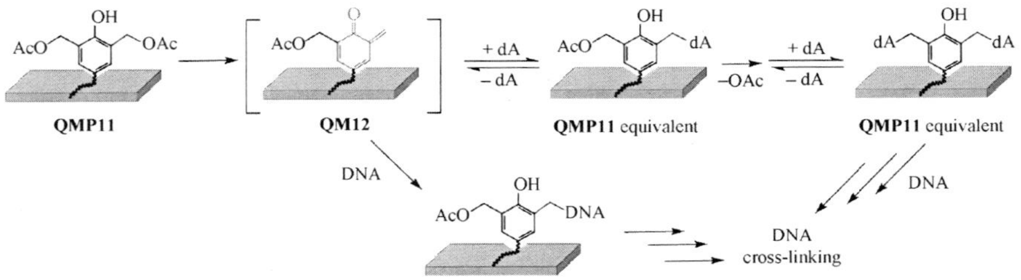

SCHEME 9.19 Reversible capture and release of quinone methides by dA extends their effective lifetime.

for retaining cross-linking activity. The effect of dA is also concentration dependent as expected and saturates at 3 mM when the concentration of **QMP11** is 100 μM.[69] Only 3 μM of an oligodeoxynucleotide duplex was used to detect cross-linking, but this was sufficient to compete for the QM in the presence of excess dA. Duplex DNA rather than dA generates the thermodynamically favored product by offering sites for both cross-linking of the QM and intercalation of the conjugated acridine.

Single-stranded oligodeoxynucleotides were even more effective than dA at preserving the reactivity of **QMP11** in an aqueous solution. A mere 3 μM of 5′-d (CTTGAGATCCTTTTTTCTGCGCGTAA) trapped sufficient quantities of **QMP11** (30 μM) to retain the ability to cross-link complementary DNA that was added to the reaction mixture a few days later.[69] This is remarkable when considering the half-life for **QMP11** in aqueous solution is only 0.5 h in the absence of additional nucleophiles. The final product of cross-linking is likely a culmination of individual strands (i) capturing both QM equivalents, (ii) annealing to complementary strands, and (iii) transferring one of the QM equivalents to the complementary strand (Scheme 9.20).

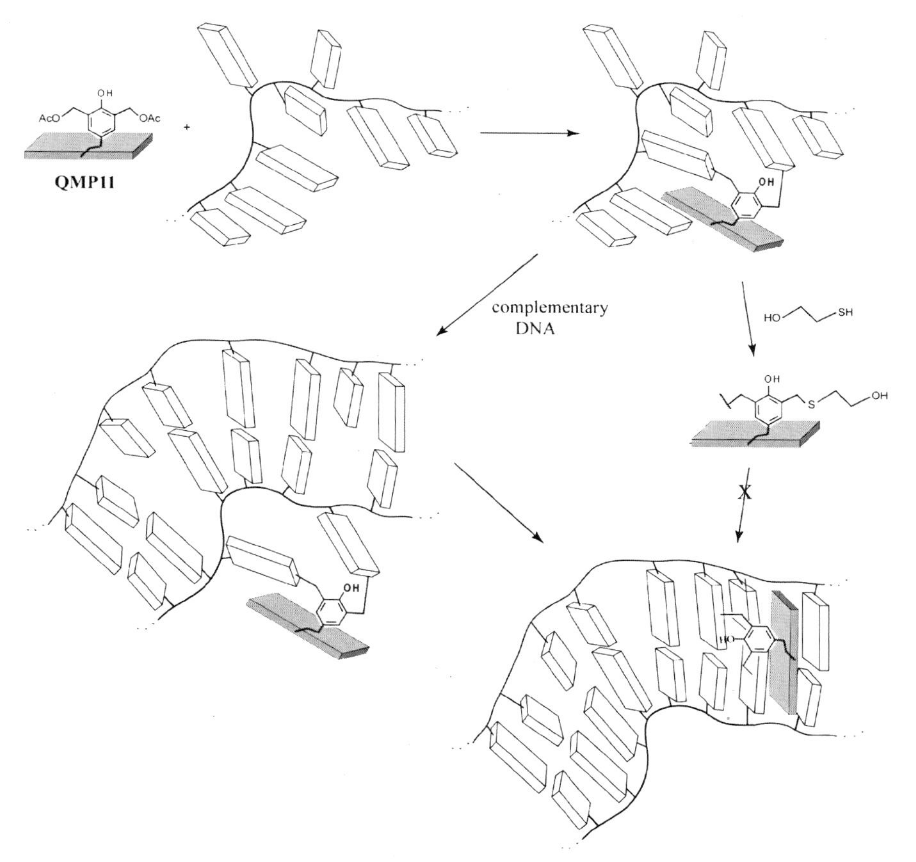

SCHEME 9.20 Single-stranded DNA has the ability to trap both quinone methide equivalents of **QMP11** and then transfer one of these equivalents to a complementary strand of DNA.

No cross-linking was observed when noncomplementary rather than complementary DNA was added to the mixture, and thus interstrand transfer of QM appears to be effective only from within a hybridized complex. Despite the obvious complications of annealing DNA that is constrained by capture of both QM equivalents of **QMP11**, such a species is suggested since both QM sites require protection from irreversible reaction over the many hours in which the cross-linking activity is preserved.

Regardless of the actual nature of the intermediates formed by **QMP11** and single-stranded DNA, their weak sensitivity to a strong and irreversible nucleophile is surprising. When a 1000-fold excess of 2-mercaptoethanol (3 mM) was added along with a complementary strand of DNA (**OD1**), cross-linking by an aged (24 h) mixture of **QMP11** and single-stranded DNA (**OD2**) was only suppressed by 50% (Fig. 9.7).[69] In contrast, 20 mM dA was completely ineffective at maintaining the ability of 100 μM **QMP11** to cross-link DNA in the presence of 5 mM 2-mercaptoethanol. Since the ultimate longevity of each QM species depends on the relative competition between nucleophiles that act reversibly versus those that act irreversibly, intramolecular capture of QM intermediates can be very efficient relative intermolecular capture.

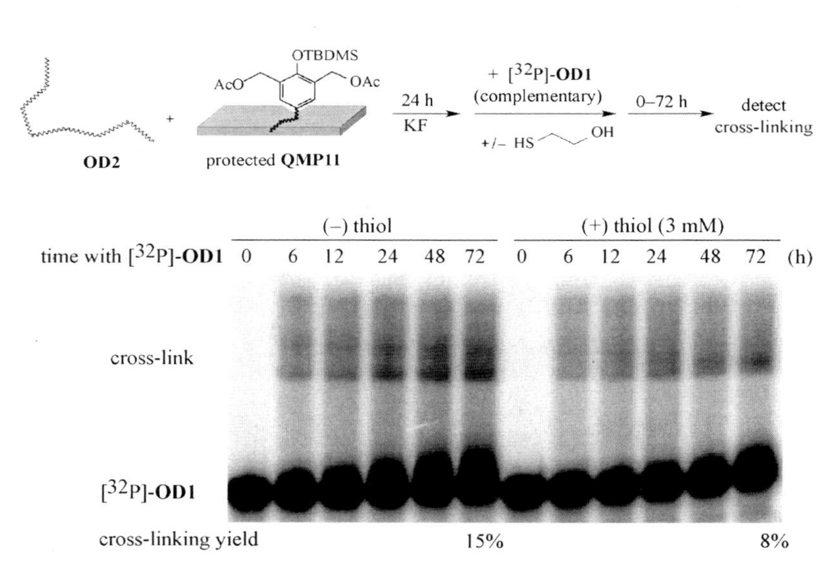

FIGURE 9.7 Interstrand transfer of a quinone methide equivalent is moderately protected within an oligodeoxynucleotide from quenching by a thiol. **QMP11** (30 μM) was generated *in situ* by addition of fluoride and incubated with a single-stranded oligodeoxynucleotide (**OD2**, 3 μM) for 24 h. The cross-linking activity that persists after 24 h was then measured by addition of a complementary strand (**OD1**, 3 μM) in the presence and absence of 2-mercaptoethanol as illustrated. *Source*: Adapted from *Angew. Chem. Int. Ed.* **2008**, *47*, 1291–1293.[69]

9.3.3 Intramolecular Capture and Release of a Quinone Methide Provides a Method for Directing Alkylation to a Chosen Sequence of DNA

The ability to alkylate a target sequence of DNA with an oligodeoxynucleotide–QMP conjugate such as **QMP12** (Scheme 9.21) should not be surprising particularly from the precedence of DNA cross-linking by the conjugates formed fortuitously between **QMP11** and **OD2** (Fig. 9.7).[69] However, this latter cross-linking was discovered only recently, whereas **QMP12** and related conjugates were first investigated almost two decades ago before the reversibility of QM reaction was examined. These earliest conjugates expressed an ability to alkylate their targets under conditions that were unexpected at the time.[34,70] The source of this activity became apparent only after the reversibility of DNA alkylation by QM was well documented. In retrospect, these initial observations can be rationalized by the presence of reversible self-adducts formed between the transient QM and the attached oligodeoxynucleotide, essentially an intramolecular analogue of the reversible trapping illustrated by dA in Section 9.3.2.

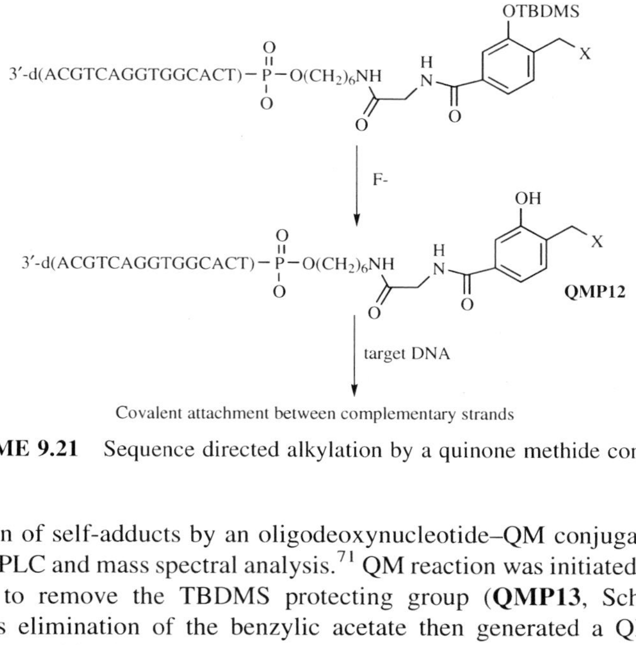

3′-d(ACGTCAGGTGGCACT)−P(O)(O)−O$(CH_2)_6$NH ... OTBDMS ... X

F-

3′-d(ACGTCAGGTGGCACT)−P(O)(O)−O$(CH_2)_6$NH ... OH ... X

QMP12

target DNA

Covalent attachment between complementary strands

SCHEME 9.21 Sequence directed alkylation by a quinone methide conjugate.

Formation of self-adducts by an oligodeoxynucleotide–QM conjugate was confirmed by HPLC and mass spectral analysis.[71] QM reaction was initiated by addition of fluoride to remove the TBDMS protecting group (**QMP13**, Scheme 9.22). Spontaneous elimination of the benzylic acetate then generated a QM that was efficiently trapped by intramolecular alkylation. This self-adduct could be aged for many days and still retain its ability to alkylate a complementary strand of DNA by transferring its reversible QM linkage. Such transfer was also characterized by HPLC and mass spectroscopy as well as by denaturing gel electrophoresis.[71] From the kinetics of QM reversibility, the self-adduct can be expected to form and decompose many times during the days of its incubation prior to addition of the complementary strand. Such continual generation of the QM provides multiple opportunities for

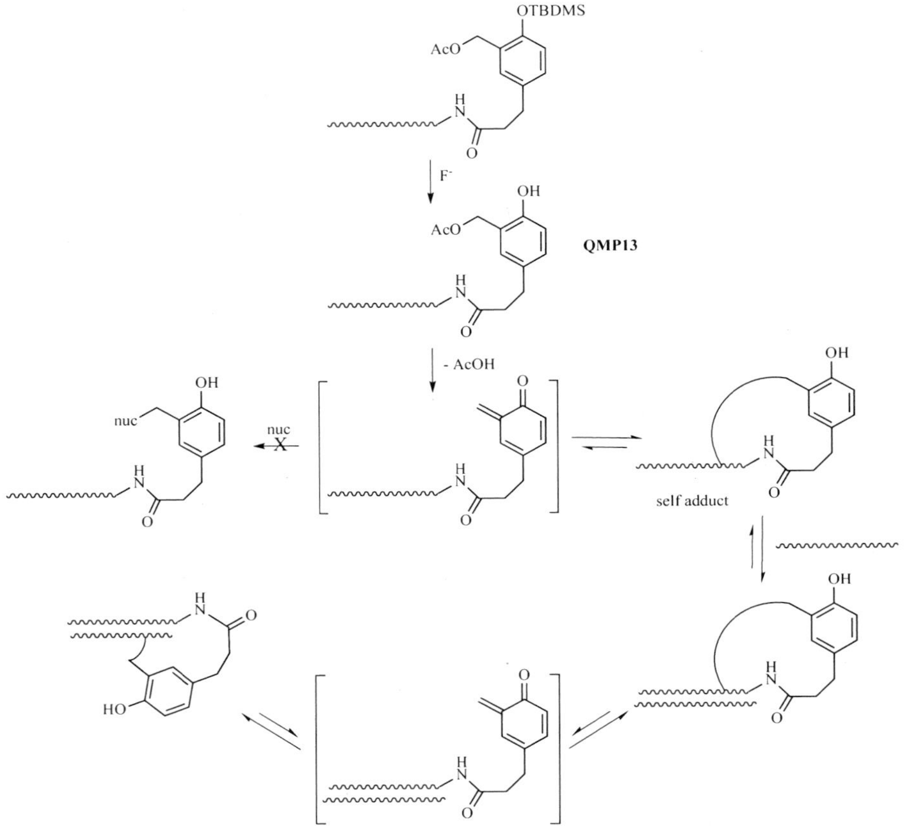

OTBDMS
AcO
H
N
O
F⁻
OH
AcO
QMP13
H
N
O
- AcOH
OH
nuc
nuc
H
N
O
O
H
N
O
OH
H
N
self adduct
O
OH
H
N
O
H
N
O
HO
O
H
N
O

SCHEME 9.22 Intramolecular trapping produces a self-adduct of the quinone methide conjugate.

irreversible trapping by addition of water (Scheme 9.22). However, HPLC and mass spectroscopy detected no water adduct. Instead, the slow (days) loss in the observed potency of the self-adduct for alkylating its target DNA is likely due to the eventual accumulation of self-adducts that form with the nucleophiles of DNA that act irreversibly rather than reversibly.[71]

Similar incubations of the self-adduct in the presence of excess noncomplementary DNA had no effect after eight days on subsequent target alkylation. The self-adduct (1.1 μM) was even resistant to quenching by 0.5 mM 2-mercaptoethanol during a six day incubation under ambient conditions. Once again, intrastrand trapping of the transient QM seems to be extremely efficient, yet remains sufficiently dynamic for subsequently transfer of its QM to a complementary sequence (Scheme 9.22). These results are also consistent with those observed with **QMP11** (Section 9.3.2).

Only after the oligodeoxynucleotide component of the conjugate was replaced with $d(T)_{10}$ could the adducts of external nucleophiles be observed. This alternative oligodeoxynucleotide was chosen based on the lack of detectable reaction between T and the parent *ortho*-**QM3**.[37] Aqueous incubation of the $d(T)_{10}$ QM conjugate in the absence and presence of 2-mercaptoethanol yielded the water and 2-mercaptoethanol adducts, respectively, as characterized by HPLC and mass spectroscopy.[71]

The same principle of reversible self-adduct formation has recently been used to guide platinum-based cross-linking to selected sequences of DNA.[72] However, not all self-adduct formation is beneficial as we learned in two critical ways. The bifunctional **QMP10** used to generate the acridine conjugate **QMP11** was also used to generate an oligodeoxynucleotide conjugate (**QMP14**) for recognition of duplex DNA through triplex formation (Scheme 9.23). No cross-linking of the target duplex was detected despite the ability of the equivalent acridine conjugate **QMP11** to cross-link the very same duplex efficiently.[67,73] Instead, strands were individually and alternatively alkylated by **QMP14**. The purine-rich strand reacted with a maximum yield of almost 30% after a 36 h treatment, but the pyrimidine strand reacted with a maximum yield of only 4% after the same period.[73] The primary site for alkyation in each strand was the first dG within the sequence 5′-d(. . .CGC. . .)/3′-d(. . .GCG. . .) that was just beyond the triplex binding region. This sequence should have supported cross-linking as shown in earlier studies with the nonconjugated precursor **QMP10**.[60] The surprising absence of cross-linking by **QMP14** may have been caused by the particular linker used in this conjugate since an equivalent *N*-mustard conjugate was also limited to monoalkylation rather than the expected cross-linking.[74] However, the ability of the QM conjugate to form a triplex structure could have been compromised by the structural constraints created by intramolecular reaction at both nascent QM sites (Scheme 9.23). Future cross-linking agents will

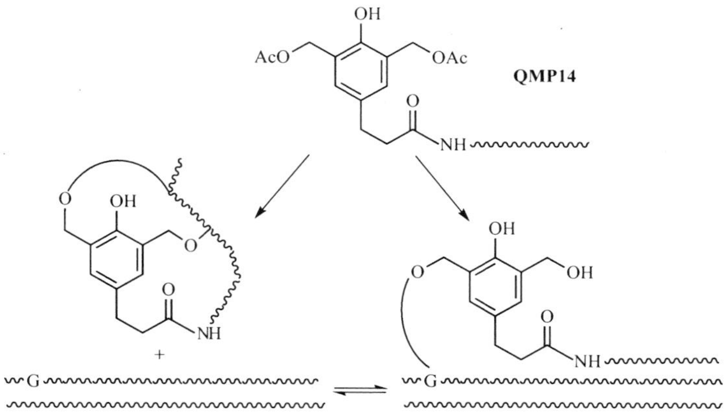

SCHEME 9.23 An oligodeoxynucleotide–bis(quinone methide) conjugate was designed for cross-linking duplex DNA by initial triplex formation but resulted at best in only monoalkylation of its target.

therefore be designed with single-QM equivalents attached to each terminus of a sequence-directing component.

As a first attempt to create a conjugate with potential application *in vivo*, the same QM used for cross-linking was also coupled to a β-hairpin forming pyrrole–imidazole polyamide that had previously been shown to associate in the minor groove of duplex DNA containing 5′-d(. . .AGCTGCT. . .)/3′-d(. . .TCGACGA. . .).[75] The resulting conjugate **QMP15** (Scheme 9.24) was capable of alkylating individual strands marginally better than a *N*-mustard equivalent when the recognition site was followed by 5′-d(. . .TATAT)/3′-d(. . .ATATA).[76] Cross-linking of both strands of duplex DNA by **QMP15** required a sequence of 5′-d(. . .CGCGC)/3′-d(. . .GCGCG) to follow the recognition site. However, yields of cross-linking by **QMP15** remained a low 4%.[76] The lack of efficient cross-linking became clear as soon as the fate of **QMP15** was identified. Within 1 h, **QMP15** was converted to self-adducts attached at both QM sites as indicated by HPLC and mass spectroscopy (Scheme 9.24). Such self-adducts are key to prolonging the effective lifetime of QMs, but only when they form reversibly. Unlike the major self-adducts of DNA, the self-adducts of pyrrole–imidazole polyamides were irreversible under ambient conditions.[76] Thus, the intramolecular reaction consumed QM and prevented its transfer to target DNA.

Pyrrole–imidazole polyamides may still be considered in future applications for a selective delivery of QMs despite their current problem. Reversibility of QM reaction responds to the simple rules of organic chemistry, and thus self-adducts may be adjusted for a desired balance of stability and lability. This can be accomplished by modifying the imidazole residues within the polyamide to strengthen their leaving group ability and/or adding electron-donating groups to the aromatic QMP (Section 9.2.3). Such modifications should overcome the kinetic barriers that limit reversibility and prevent accumulation of the desired product favored by thermodynamics. The benefits of overcoming these barriers are already evident by the efficiency and adaptability of DNA cross-linking as described in this section as well as Sections 9.3.1 and 9.3.2.

9.4 CONCLUSIONS AND FUTURE PROSPECTS

The ability to understand, control, and apply reversible alkylation by QMs offers significant opportunities in chemistry and biology. To date, reversibility has been examined in most detail using simple *ortho*-QMs and their conjugates with model nucleophiles and nucleic acids. Efforts to explain the origins of QM's unusual specificity for weak nucleophiles of DNA reaffirmed the importance of distinguishing between thermodynamic and kinetic products. The strongest nucleophiles of DNA do indeed react most rapidly but generate products that spontaneously regenerate QM for further reaction. Ultimately, products accumulate from the less efficient but irreversible reactions. The reversibility of alkylation additionally supports mechanisms for extending the effective life of QM and delivering QMs to chosen targets selectively. In both cases, repetitive capture and release of a QM by

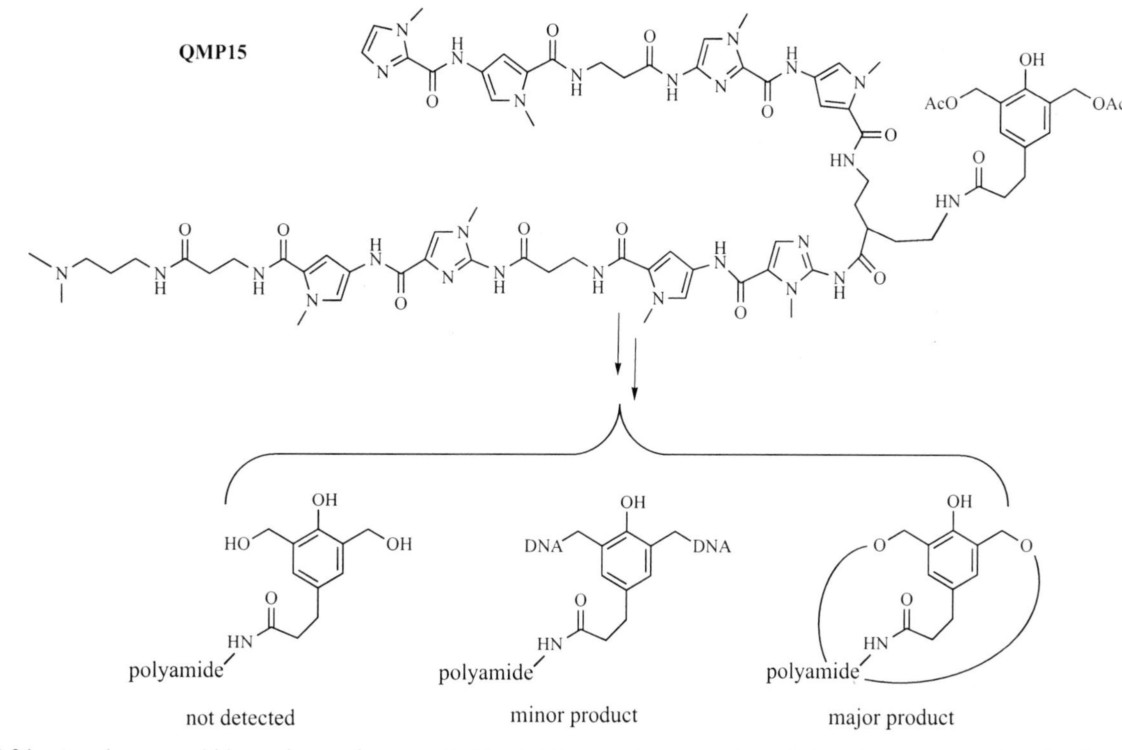

SCHEME 9.24 A quinone methide conjugate for cross-linking DNA through recognition of the minor groove primarily formed self-adducts irreversibly.

covalent addition protects against terminal quenching and other undesirable processes. The rates of both capture and release are sensitive to substituents and can be modulated in a predictable manner.

The reversibility of QM adducts also creates numerous challenges. For example, measuring the full burden of DNA alkylation by a QM can be obscured by the loss of its labile products during or before chemical identification can be completed. Results from a deoxynucleotide model system indicated that only a small fraction of the possible adducts could be measured after the interval required for analysis of DNA. Perhaps the kinetic products of QMs also contribute to the cellular activity of these intermediates although this has yet to be explored. QM equivalents can be envisioned to migrate from one reversible nucleophile such as the N1 of adenine in such cofactors as ATP to another until quenched by a compound such as glutathione that is present in cells as a defense against undesirable electrophiles.

Many opportunities conversely are supported by reversible reactions of QM despite the noted complications. One example includes the synthesis and chiral resolution of binaphthol derivatives by two cycles of QM formation and alkylation.[77] The reversibility of QM reaction may also be integrated in future design of self-assembling systems to provide covalent strength to the ultimate thermodynamic product. To date, QMs have already demonstrated great success in supporting the opposite process, spontaneous disassembly of dendrimers (Chapter 5).

The full potential for therapeutics based on covalent and reversible reactions has not been explored sufficiently. Typically, reversible interactions are associated with noncovalent forces and irreversible interactions are associated with covalent forces. QM may be placed between these extremes by forming strong covalent bonds yet maintaining a kinetic lability to escape from off-target reactions and respond to changes in its surrounding environment. In this manner, intramolecular capture and release of a reactive QM may be used to direct alkylation to a chosen target selectively for applications in biochemistry and medicine. Most likely, the first indication that QM reversibility may contribute to the potency of a therapeutic compound will derive from studies on a series of NO-generating aspirin derivatives undergoing preclinical evaluation.[3–5] These drug candidates were originally designed for esterase-dependent release of aspirin followed by spontaneous decomposition of a QM precursor to release nitrate as a NO precursor (Scheme 9.25). Interestingly, much of the biological activity is retained when nitrate is replaced by other leaving groups such as chloride. Thus, QM may instead be crucial for the activity of these types of compounds although its role as a cytoprotective versus cytotoxic agent remains to be determined.

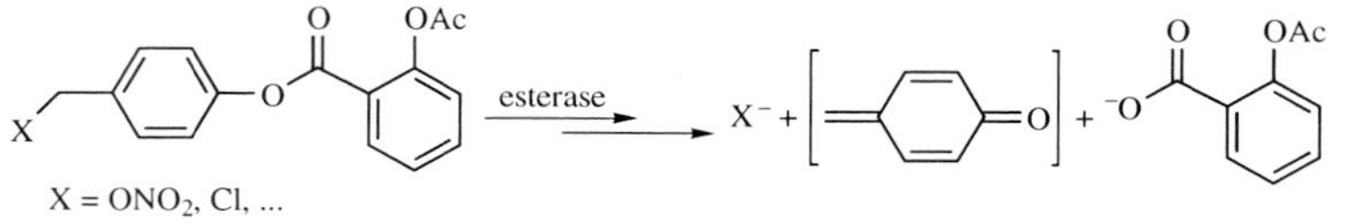

O OAc X O esterase X^- + O + $^-$O O OAc

$X = ONO_2$, Cl, ...

SCHEME 9.25 NO-generating aspirin derivatives are also a source of quinone methide.

ACKNOWLEDGMENTS

Progress on quinone methides in the Rokita laboratory is the culmination of efforts by many contributors. All undergraduate and graduate students, postdoctoral associates and collaborators participating in this effort are acknowledged for their dedication, enthusiasm, and hard work. Funding was kindly provided over the last two decades by the William J. and Florence M. Catacosinos Young Investigator Award for Cancer Research, the University Exploratory Research Program of The Procter and Gamble Company, the Center for Biotechnology at Stony Brook University in conjunction with the former Lederle Laboratories, the National Cancer Institute, and the National Science Foundation.

REFERENCES

1. Rowan, S. J.; Cantrill, S. J.; Cousins, G. R. L.; Sanders, J. K. M.; Stoddart, J. F. Dynamic covalent chemistry. *Angew. Chem. Int. Ed.* 2002, 41, 898–952.
2. Nitschke, J. R.; Lehn, J.-M. Self-organized by selection: generation of a metallosupramolecular grid architecture by selection of components in a dynamic library of ligands. *Proc. Natl. Acad. Sci. USA* 2003, 100, 11970–11974.
3. Hulsman, N.; Medema, J. P.; Bos, C.; Jongejan, A.; Luers, R.; Smit, M. J.; de Esch, I. J. P.; Richel, D.; Witjmans, M. Chemical insights in the concept of hybrid drugs: the antitumor effect of nitric oxide-donating aspirin involves a quinone methide but not nitric oxide nor aspirin. *J. Med. Chem.* 2007, 50, 2424–2431.
4. Kashfi, K.; Rigas, B. The mechanism of action of nitric oxide-donating aspirin. *Biochem. Biophys. Res. Commun.* 2007, 358, 1096–1101.
5. Dunlap, T.; Chandrasena, R. E. P.; Wang, Z.; Sinha, V.; Wang, Z.; Thatcher, G. R. J. Quinone formation as a chemoprevention strategy for hybrid drugs: balancing cytotoxity and cytoprotection. *Chem. Res. Toxicol.* 2007, 20, 1903–1912.
6. Warpehoski, M. A.; Harper, D. E.; Mitchell, M. A.; Monroe, T. J. Reversibility of the covalent reaction of CC-1065 and analogues with DNA. *Biochemistry* 1992, 31, 2502–2508.
7. Lee, C.-S.; Gibson, N. W. DNA interstrand cross-links induced by the cyclopropylpyrroloindole antitumor agent bizelesin are reversible upon exposure to alkali. *Biochemistry* 1993, 32, 9108–9144.
8. Boger, D. L.; Yun, W. Reversibility of the duocarmycin A and SA DNA alkylating reaction. *J. Am. Chem. Soc.* 1993, 115, 9872–9873.
9. Parrish, J. P.; Hughes, T. V.; Hwang, I.; Boger, D. L. Establishing the parabolic relationship between reactivity and activity for derivatives and analogues of the duocarmycin and CC-1065 alkylation subunits. *J. Am. Chem. Soc.* 2004, 126, 80–81.
10. Parrish, J. P.; Kastrinsky, D. B.; Wolkenberg, S. E.; Igarashi, Y.; Boger, D. L. DNA alkylation properties of yatakemycin. *J. Am. Chem. Soc.* 2003, 125, 10971–10976.
11. Asai, A.; Nagamura, S.; Saito, H.; Takahashi, I.; Nakano, H. The reversible DNA-alkylating activity of duocarmycin and its analogue. *Nucleic Acids Res.* 1994, 22, 88–93.

12. Lee, S.-J.; Seaman, F. C.; Sun, D.; Xiong, H.; Kelly, R. C.; Hurley, L. H. Replacement of the bizelesin ureadiyl linkage by a guanidinium moiety retards translocation from monoalkylation to cross-linking sites on DNA. *J. Am. Chem. Soc.* 1997, 119, 3434–3442.
13. Zewail-Foote, M.; Hurley, L. H. Differential rates of reversibility of ecteinascidin 743-DNA covalent adducts from different sequences lead to migration to favored bonding sites. *J. Am. Chem. Soc.* 2001, 123, 6485–6495.
14. Wang, M.; Cheng, G.; Sturla, S. J.; Shi, Y.; McIntee, E. J.; Villalta, P. W.; Upadhyaya, P.; Hecht, S. S. Identification of adducts formed by pyridyloxobutylation of deoxyguanosine and DNA by 4-(acetoxymethylnitrosamino)-1-(3-pyridyl)-1-butanone, a chemical activated form of tobacco specific carcinogens. *Chem. Res. Toxicol.* 2003, 16, 616–626.
15. Gates, K. S. Mechanisms of DNA damage by leinamycin. *Chem. Res. Toxicol.* 2000, 13, 953–956.
16. Nooner, T.; Dutta, S.; Gates, K. S. Chemical properties of the leinamycin-guanine adduct in DNA. *Chem. Res. Toxicol.* 2004, 17, 942–949.
17. David-Cordonnier, M.-H.; Casely-Hayford, M.; Kouach, M.; Briand, G.; Patterson, L. H.; Bailly, C.; Searcey, M. Stereoselectivity, sequence specificity and mechanism of action of the azinomycin epoxide. *ChemBiochem.* 2006, 7, 1658–1661.
18. Reddy, G. R.; Marnett, L. J. Mechanism of reaction of β-(aryloxy)acroleins with nucleosides. *Chem. Res. Toxicol.* 1996, 9, 12–15.
19. Riggins, J. N.; Daniels, S.; Rouzer, C. A.; Marnett, L. J. Kinetic and thermodynamic analysis of the hydrolytic ring-opening of the malondialdehyde-deoxyguanosine adduct, 3-(2′-deoxy-D-*erythro*-pentofuranosyl)pyrimido[1,2-α]purin-10(3*H*)-one. *J. Am. Chem. Soc.* 2004, 126, 8237–8243.
20. Dedon, P. C.; Plastaras, J. P.; Rouzer, C. A.; Marnett, L. J. Indirect mutagenesis by oxidative DNA damage: formation of the pyrimidopurinone adduct of deoxyguanosine by base propenal. *Proc. Natl. Acad. Sci. USA* 1998, 95, 11113–11116.
21. Plastaras, J. P.; Riggins, J. N.; Otteneder, M.; Marnett, L. J. Reactivity and mutagenicity of endogenous DNA oxopropenylating agents: base propenals, malondialdehyde, and N^{ϵ}-oxopreopenyllysine. *Chem. Res. Toxicol.* 2000, 13, 1235–1242.
22. Plastaras, J. P.; Dedon, P. C.; Marnett, L. J. Effects of DNA structure on oxopropenylation by the endogenous mutagens malondialdehyde and base propenal. *Biochemistry* 2002, 41, 5033–5042.
23. Angle, S. R.; Yang, W. Synthesis and chemistry of a quinone methide model for anthracycline antitumor antibiotics. *J. Am. Chem. Soc.* 1990, 112, 4524–4528.
24. Angle, S. R.; Yang, W. Nucleophilic addition of 2′-deoxynucleosides to the *o*-quinone methides 10-(acetyloxy) and 10-methoxy-3,4-dihydro-9(2*H*)-anthracenone. *J. Org. Chem.* 1992, 57, 1092–1097.
25. Angle, S. R.; Yang, W. pH-Dependent stability and reactivity of a thiol-quinone methide adduct. *Tetrahedron Lett.* 1992, 33, 6089–6092.
26. Angle, S. R.; Rainer, J. D.; Woytowicz, C. Synthesis and chemistry of quinone methide models for the anthracycline antitumor antibiotic. *J. Org. Chem.* 1997, 62, 5884–5892.
27. Lewis, M. A.; Yoerg, D. G.; Bolton, J. L.; Thompson, J. A. Alkylation of 2′-deoxynucleosides and DNA by quinone methides derived from 2,6-di-*tert*-butyl-4-methylphenol. *Chem. Res. Toxicol.* 1996, 9, 1368–1374.

28. Bolton, J. L.; Turnipseed, S. B.; Thompson, J. A. Influence of quinone methide reactivity on the alkylation of thiol and amino groups in proteins: studies utilizing amino acid and peptide models. *Chem. Biol. Interact.* 1997, 107, 185–200.
29. Lemercier, J.-N.; Meier, B.; Gomez, J. D.; Thompson, J. A. Inhibition of glutathione S-transferase P1-1 in mouse lung epithelial cells by the tumor promoted 2,6,di-*tert*-butyl-4-methylene-2,5-cylcohexadienone (BHT-quinone methide): protein adducts investigated by electrospray mass spectrometry. *Chem. Res. Toxicol.* 2004, 17, 1675–1683.
30. Modica, E.; Zanaletti, R.; Freccero, M.; Mella, M. Alkylation of amino acids and glutathione in water by *o*-quinone methides. Reactivity and selectivity. *J. Org. Chem.* 2001, 66, 41–52.
31. Woolridge, E. M.; Rokita, S. E. 6-(Difluoromethyl)tryptophan as a probe for substrate activation during the catalysis of tryptophanase. *Biochemistry* 1991, 30, 1852–1857.
32. Woolridge, E. M.; Rokita, S. E. The use of 6-(difluoromethyl)indole to study the activation of indole by tryptophan synthase. *Arch. Biochem. Biophys.* 1991, 286, 473–480.
33. Chatterjee, M.; Rokita, S. E. Sequence specific alkylation of DNA activated by an enzymatic signal. *J. Am. Chem. Soc.* 1991, 113, 5116–5117.
34. Li, T.; Rokita, S. E. Selective modification of DNA controlled by an ionic signal. *J. Am. Chem. Soc.* 1991, 113, 7771–7773.
35. Chatterjee, M.; Rokita, S. E. The role of a quinone methide in the sequence specific alkylation of DNA. *J. Am. Chem. Soc.* 1994, 116, 1690–1697.
36. Rokita, S. E.; Yang, J.; Pande, P.; Greenberg, W. A. Quinone methide alkylation of deoxycytidine. *J. Org. Chem.* 1997, 62, 3010–3012.
37. Pande, P.; Shearer, J.; Yang, J.; Greenberg, W. A.; Rokita, S. E. Alkylation of nucleic acids by a model quinone methide. *J. Am. Chem. Soc.* 1999, 121, 6773–6779.
38. Veldhuyzen, W.; Lam, Y.-F.; Rokita, S. E. 2-Deoxyguanosine reacts with a model quinone methide at multiple sites. *Chem. Res. Toxicol.* 2001, 14, 1345–1351.
39. Bolton, J. L.; Sevestre, H.; Ibe, B. O.; Thompson, J. A. Formation and reactivity of alternative quinone methides from butylated hydroxytoluene: possible explanation for species-specific pneumotoxicity. *Chem. Res. Toxicol.* 1990, 3, 65–70.
40. Bolton, J. L.; Valerio, L. G.; Thompson, J. A. The enzymatic formation and chemical reactivity of quinone methides correlate with alkylphenol-induced toxicity in rat hepatocytes. *Chem. Res. Toxicol.* 1992, 5, 816–822.
41. Richard, J. P. Mechanisms for the uncatalyzed and hydrogen ion catalyzed reactions of a simple quinone methide with solvent and halide ions. *J. Am. Chem. Soc.* 1991, 113, 4588–4595.
42. Zhou, Q.; Turnbull, K. D. Phosphodiester alkylation with a quinone methide. *J. Org. Chem.* 1999, 64, 2847–2851.
43. Brookes, P.; Lawley, P. D. The methylation of adenosine and adenylic acid. *J. Chem. Soc.* 1960, 539–545.
44. Qian, C.; Dipple, A. Different mechanisms of aralkylation of adenosine at the 1- and N^6-positions. *Chem. Res. Toxicol.* 1995, 8, 389–395.
45. Fujii, T.; Saito, T.; Terahara, N. The Dimroth rearrangement in the adenine series: a review updated. *Heterocycles* 1998, 48, 359–390.
46. Veldhuyzen, W. F.; Shallop, A. J.; Jones, R. A.; Rokita, S. E. Thermodynamic versus kinetic products of DNA alkylation as modeled by reaction of deoxyadenosine. *J. Am. Chem. Soc.* 2001, 123, 11126–11132.

47. Dawson, R. M. C.; Elliott, D. C.; Elliott, W. H.; Jones, K. M. *Data for Biochemical Research*, 3rd edition; Oxford University Press: New York, 1986; Chapter 5, pp 103–114.
48. Weinert, E. E.; Frankenfield, K. N.; Rokita, S. E. Time-dependent evolution of adducts formed between deoxynucleosides and a model quinone methide. *Chem. Res. Toxicol.* 2005, 18, 1364–1370.
49. Chiang, Y.; Kresge, A. J.; Zhu, Y. Flash photolytic generation and study of *p*-quinone methide in aqueous solution. An estimate of rate and equilibrium constants for heterolysis of the carbon–bromine bond in *p*-hydroxybenzyl bromide. *J. Am. Chem. Soc.* 2002, 124, 6349–6356.
50. Di Valentin, C.; Freccero, M.; Zanaletti, R.; Sarzi-Amadè, M. *o*-Quinone methide as alkylating agent of nitrogen, oxygen and sulfur nucleophiles. The role of H-bonding and solvent effects on the reactivity through DFT computational study. *J. Am. Chem. Soc.* 2001, 123, 8366–8377.
51. Freccero, M.; Di Valentin, C.; Sarzi-Amadè, M. Modeling H-bonding and solvent effects in the alkylation of pyrimidine bases by prototype quinone methide. A DFT study. *J. Am. Chem. Soc.* 2003, 125, 3544–3553.
52. Freccero, M.; Gandolfi, R.; Sarzi-Amadè, M. Selectivity of purine alkylation by a quinone methide. Kinetic or thermodynamic control? *J. Org. Chem.* 2003, 68, 6411–6423.
53. Zhou, Q.; Turnbull, K. D. Quinone methide phosphodiester alkylation under aqueous conditions. *J. Org. Chem.* 2001, 66, 7072–7077.
54. Ouyang, A.; Skibo, E. B. Design of a cyclopropyl quinone methide reductive alkylating agent. *J. Org. Chem.* 1998, 63, 1893–1900.
55. Wan, P.; Barker, B.; Diao, L.; Fischer, M.; Shi, Y.; Yang, C. Quinone methides: relevant intermediates in organic chemistry. *Can. J. Chem.* 1996, 74, 465–475.
56. Weinert, E. E.; Dondi, R.; Colloredo-Melz, S.; Frankenfield, K. N.; Mitchell, C. H.; Freccero, M.; Rokita, S. E. Substituents on quinone methides strongly modulate formation and stability of their nucleophilic adducts. *J. Am. Chem. Soc.* 2006, 128, 11940–11947.
57. Tomasz, M. Mitomycin C: small, fast and deadly (but very selective). *Chem. Biol.* 1995, 2, 575–579.
58. Marques, M. M.; Beland, F. A. Identification of tamoxifen–DNA adducts formed by 4-hydroxytamoxifen quinone methide. *Carcinogenesis* 1997, 18, 1949–1954.
59. Pullman, A.; Pullman, B. Molecular electrostatic potential of nucleic acids. *Q. Rev. Biophys.* 1981, 14, 289–380.
60. Zeng, Q.; Rokita, S. E. Tandem quinone methide generation for cross-linking DNA. *J. Org. Chem.* 1996, 61, 9080–9081.
61. Wang, P.; Liu, R.; Wu, X.; Ma, H.; Cao, X.; Zhou, P.; Zhang, J.; Weng, X.; Zhang, X. L.; Zhou, X.; Weng, L. A potent, water-soluble and photoinducible DNA cross-linking agent. *J. Am. Chem. Soc.* 2003, 125, 1116–1117.
62. Weng, X.; Ren, L.; Weng, L.; Huang, J.; Zhu, S.; Zhou, X.; Weng, L. Synthesis and biological studies of inducible DNA cross-linking agents. *Angew. Chem. Int. Ed.* 2007, 46, 8020–8023.
63. Verga, D.; Richter, S. N.; Palumbo, M.; Gandolfi, R.; Freccero, M. Bipyridyl ligands as photoactivatable mono- and bis-alkylating agents capable of DNA cross-linking. *Org. Biomol. Chem.* 2007, 5, 233–235.

64. Richter, S. N.; Maggi, S.; Mels, S. C.; Palumbo, M.; Freccero, M. Binol quinone methides as bisalkylating and DNA cross-linking agents. *J. Am. Chem. Soc.* 2004, 126, 13973–13979.

65. Doria, F.; Richter, S. N.; Nadai, M.; Colloredo-Melz, S.; Mella, M.; Palumbo, M.; Freccero, M. BINOL-amino acid conjugates as triggerable carriers of DNA-targeted potent photocytotoxic agents. *J. Med. Chem.* 2007, 50, 6570–6579.

66. Song, Y.; Tian, T.; Wang, P.; He, H.; Liu, W.; Zhou, X.; Cao, X.; Zhang, X.-L.; Zhou, X. Phenol quaternary ammonium derivatives: charge and linker effect on their DNA photoinducible cross-linking abilities. *Org. Biomol. Chem.* 2006, 4, 3358–3366.

67. Veldhuyzen, W. F.; Pande, P.; Rokita, S. E. A transient product of DNA alkylation can be stabilized by binding localization. *J. Am. Chem. Soc.* 2003, 125, 14005–14013.

68. Prakash, A. S.; Denny, W. A.; Gourdie, T. A.; Valu, K. K.; Woodgate, P. D.; Wakelin, L. P. G. DNA-directed alkylating ligands as potential antitumor agents: sequence specificity of alkylation by intercalating aniline mustards. *Biochemistry* 1990, 29, 9799–9807.

69. Wang, H.; Wahi, M. S.; Rokita, S. E. Immortalizing a transient electrophile for DNA crosslinking. *Angew. Chem. Int. Ed.* 2008, 47, 1291–1293.

70. Li, T.; Zeng, Q.; Rokita, S. E. Target promoted alkylation of DNA. *Bioconjug. Chem.* 1994, 5, 497–500.

71. Zhou, Q.; Rokita, S. E. A General strategy for target-promoted alkylation in biological systems. *Proc. Natl. Acad. Sci. USA* 2003, 100, 15452–15457.

72. Algueró, B.; López de la Osa, J.; González, C.; Pedroso, E.; Marchán, V.; Grandas, A. Selective platination of modified oligonucleotides and duplex cross-links. *Angew. Chem. Int. Ed.* 2006, 45, 8194–8197.

73. Zhou, Q.; Pande, P.; Johnson, A. E.; Rokita, S. E. Sequence-specific delivery of a quinone methide intermediate to the major groove of DNA. *Bioorg. Med. Chem.* 2001, 9, 2347–2354.

74. Reed, M. W.; Lukhtanov, E. A.; Gorn, V.; Dutyavin, I.; Gall, A.; Wald, A.; Meyer, R. B. Synthesis and reactivity of aryl nitrogen mustard-oligodeoxyribonucleotide conjugates. *Bioconjug. Chem.* 1998, 9, 64–71.

75. Wurtz, N. R.; Dervan, P. B. Sequence specific alkylation of DNA by hairpin pyrrole–imidazole polyamide conjugates. *Chem. Biol.* 2000, 7, 153–161.

76. Kumar, D.; Veldhuyzen, W. F.; Zhou, Q.; Rokita, S. E. Conjugation of a hairpin pyrrole–imidazole polyamide to a quinone methide for control of DNA cross-linking. *Bioconjug. Chem.* 2004, 15, 915–922.

77. Colloredo-Melz, S.; Dorr, R. T.; Verga, D.; Freccero, M. Photogenerated quinone methides as useful intermediates in the synthesis of chiral BINOL ligands. *J. Org. Chem.* 2006, 71, 3889–3895.

10

FORMATION AND REACTIONS OF XENOBIOTIC QUINONE METHIDES IN BIOLOGY

JUDY L. BOLTON[1] AND JOHN A. THOMPSON[2]

[1]*Department of Medicinal Chemistry and Pharmacognosy (M/C 781), College of Pharmacy, University of Illinois at Chicago, 833 S. Wood Street, Chicago, IL 60612-7231, USA*
[2]*Department of Pharmaceutical Chemistry, School of Pharmacy, University of Colorado, Denver, C238-L15, 12631 E. 17th Avenue, Aurora, CO 80045, USA*

10.1 INTRODUCTION

Quinone methides (QMs) are reactive metabolites of a variety of synthetic and natural phenols containing *ortho-* or *para*-alkyl substituents, and are most likely responsible for the cytotoxic/genotoxic effects of the parent phenols.[1–3] Like all reactive intermediates, the selectivity of a QM depends on both its rate of formation and reactivity. Due to positive charge density on the exocyclic methylene group (Scheme 10.1), QMs are much more reactive electrophiles than quinones, are not involved in redox cycling,[1,2] and generally react in biological systems via nonenzymatic Michael addition. With simple QMs, these reactions produce benzylic adducts with thiol or amino groups of peptides, proteins, and nucleic acids.[1,2,4] Theoretically, QMs can be characterized as resonance-stabilized carbocations[5,6] due to the contribution of a charged aromatic resonance form. Inter- and intramolecular factors that stabilize or destabilize this resonance structure, therefore, lead to wide variations in QM reactivities and selectivities with cellular nucleophiles.

Quinone Methides, Edited by Steven E. Rokita

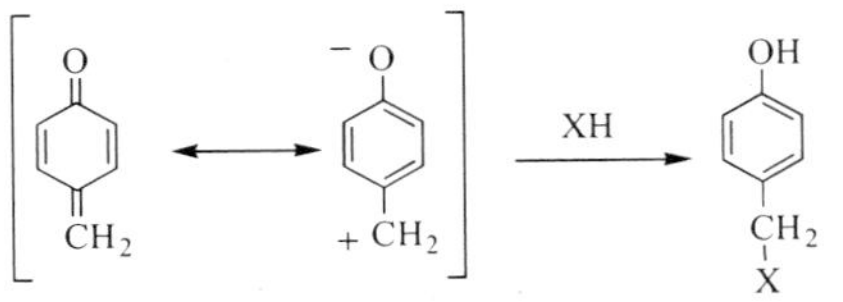

SCHEME 10.1 Resonance structures of a simple *p*-QM and the Michael addition product with nucleophile XH.

10.2 FORMATION OF QMs

The generation of QMs from a variety of synthetic and naturally occurring phenols can explain adverse biological effects of the parent compounds, including cytotoxicity and genotoxiciy. There are two major pathways by which these intermediates are formed *in vivo*, enzymatic oxidation and *o*-quinone isomerization. In the first case, the successive removal of two electrons (or alternatively, an electron and a hydrogen atom) from 4-alkyl-substituted phenols is typically catalyzed by cytochromes P450; however, in some cases a peroxidase has been shown to catalyze QM formation as well (Scheme 10.2a).[1] Oxidation of 2,6-di-*tert*-butyl-4-methylphenol (butylated hydroxytoluene, BHT) and 4-allyl-2-methoxyphenol (eugenol) represents the most extensively studied examples of QM formation by oxidative enzymes.[1] More recently, it has been shown that cytochrome P450 and peroxidases also convert selective estrogen receptor modulators (SERMs) such as tamoxifen, toremifene, and acolbifene to QMs.[7] Alternatively, 4-alkylcatechols can be oxidized initially to *o*-quinones that, depending on the acidity of the hydrogen on the 4-alkyl carbon, spontaneously isomerize to *o*-hydroxy-*p*-QMs (Scheme 10.2b). This oxidation–isomerization pathway occurs, for example, with hydroxychavicol[8,9] and catechol estrogens.[10]

Phenols with an appropriate leaving group in the benzylic position such as fluoride may form QMs by spontaneous hydrolysis, possibly catalyzed by a basic amino acid residue as shown in Scheme 10.2c. Evidence for this process was obtained with 4-(fluoromethyl)phenyl phosphate involving initial enzymatic hydrolysis of the phosphate followed by nonenzymatic formation of a QM.[11] Similarly, several lines of evidence demonstrate nonenzymatic QM formation from 4-trifluoromethylphenol under physiologic conditions.[12]

10.3 ALKYLPHENOLS

10.3.1 BHT and Related Alkylphenols: Historical Overview

The contributions of QMs to the toxicities of the food additive BHT and structurally related alkylphenols in rats and mice have been investigated extensively over the past 25 years. This work has generated considerable insight into the relationships between toxicity and the metabolic formation and reactivity of these electrophiles. Most studies have focused on liver and lung; damage is primarily observed in these organs due,

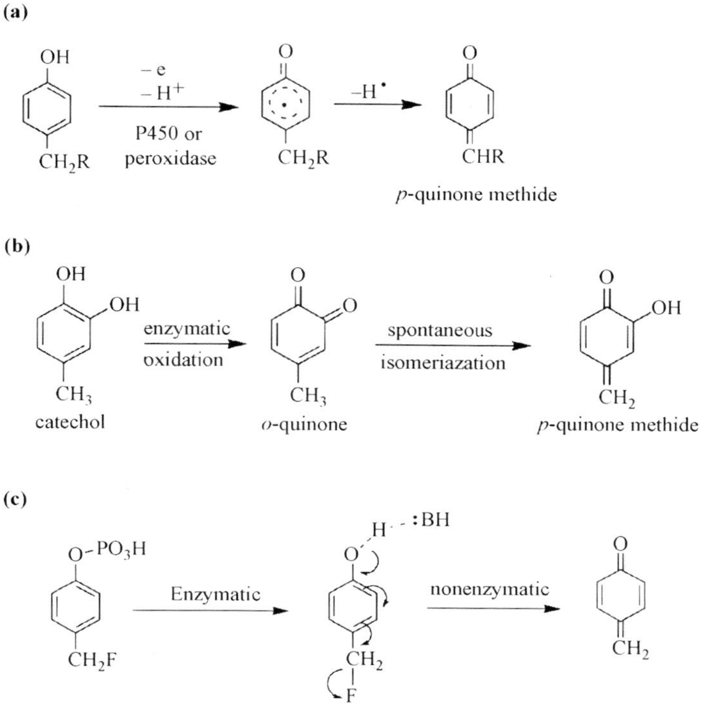

SCHEME 10.2 Common pathways of QM formation in biological systems. (a) Stepwise two-electron oxidation by cytochrome P450 or a peroxidase. (b) Enzymatic oxidation of a catechol followed by spontaneous isomerization of the resulting *o*-quinone. (c) Enzymatic hydrolysis of a phosphate ester followed by base-catalyzed elimination of a leaving group from the benzylic position.

presumably, to the presence of cytochrome P450 capable of oxidizing alkylphenols to QMs. BHT is a potent lung toxin and tumor promoter in mice. A single dose of BHT destroys pulmonary type 1 pneumocytes resulting in compensatory hyperplasia of type 2 cells and differentiation to replace damaged type 1 cells.[13,14] Lung damage has routinely been assessed by measuring increases in lung weight to body weight ratios. Chronic weekly injections of BHT following a single low (noncarcinogenic) dose of a tumor initiator such as urethane or 3-methylcholanthrene result in as much as a 12-fold increase in the number of lung tumors in susceptible strains of mice.[15–17] This two-stage model of carcinogenesis has proven to be a very useful system for studying mechanisms of lung tumor promotion in an animal model that develops pulmonary adenocarcinomas similar to humans.[18]

Witschi and colleagues[19] identified the requirement for metabolic activation of BHT in determining that radioactivity from ^{14}C-labeled BHT became covalently bound to proteins in mouse lung. Both toxicity and protein binding were prevented when mice were treated with cytochrome P450 inhibitors, thereby indicating the

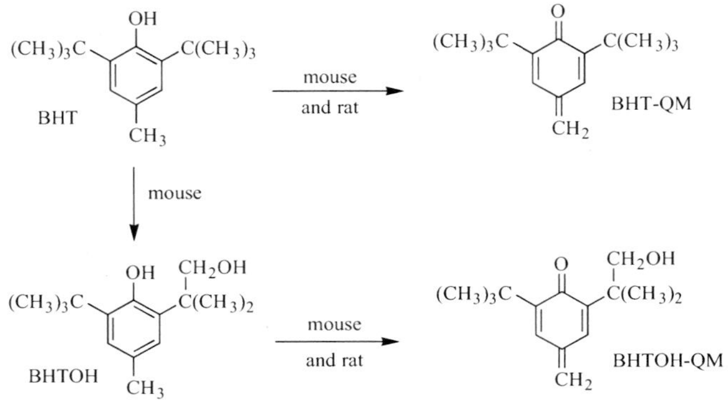

SCHEME 10.3 Direct oxidation of BHT to BHT–QM by rats and mice, and a mouse-specific BHT hydroxylation followed by QM formation.

formation of a reactive metabolite. Takahashi and Hiraga[20] trapped and identified the glutathione (GSH) conjugate of BHT–QM by incubating BHT with NADPH-fortified rat liver microsomes in the presence of GSH. More recent work[21,22] demonstrated that cytochrome P450 in mouse tissues hydroxylates a *tert*-butyl substituent of BHT-forming BHTOH that is subsequently oxidized to BHTOH–QM (Scheme 10.3). This hydroxylation pathway is a major contributor to BHT metabolism in mouse lung but not in liver or rat tissues. The latter QM is substantially more reactive than BHT–QM as discussed in the next section and is believed to mediate much of the lung damage resulting from BHT administration.

10.3.2 Relationships of QM Structure to Reactivity and Toxicity

10.3.2.1 Effects of Alkyl Substitution A structure–activity approach has proven useful for implicating QM intermediates in alkylphenol-induced toxicity; the toxic potency of an alkylphenol is compared with the reactivities and rates of formation of the corresponding QMs. Replacing the 4-methyl group of BHT by trideuteriomethyl decreases QM formation due to a kinetic isotope effect on C—H bond cleavage[23,24] and replacing the methyl by *tert*-butyl prevents QM formation. Both of these analogues (Fig. 10.1) have been employed to support the role of a QM intermediate in BHT-induced toxicity by decreasing or preventing toxicity, respectively.[23–25] Replacing the 4-methyl by a larger alkyl group such as ethyl or isopropyl also decreases toxicity and QM reactivity (Table 10.1) due to steric effects on nucleophilic attack at the exocyclic methylene.[26,27]

Substituents on the 2- and 6-positions of phenol rings greatly influence QM reactivity. Reaction rates for QMs derived from several of the phenols, shown in Fig. 10.1, were determined in methanolic or aqueous solutions and are listed in Table 10.1. Replacing a *tert*-butyl substituent of BHT by a methyl group (i.e., BDMP–QM) increased the rate of hydration by 60–70-fold at pH 7.4 and this

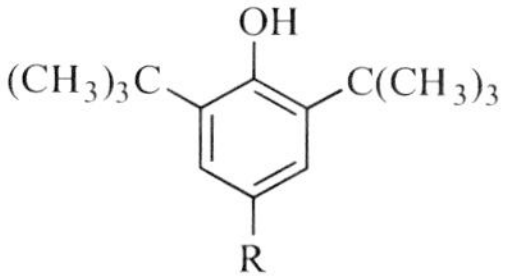

	R
BHT	CH_3
BHT-*d*$_3$	CH_3
Et-BHT	CH_2CH_3
***t*Bu-BHT**	$C(CH_3)_3$

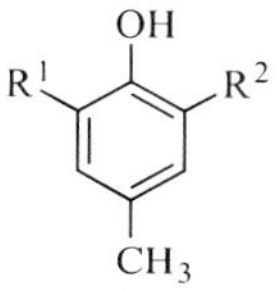

	R^1	R^2
DMP	H	CH_3
TMP	CH_3	CH_3
BMP	H	$C(CH_3)_3$
BDMP	CH_3	$C(CH_3)_3$
BHT	$C(CH_3)_3$	$C(CH_3)_3$
BHTOH	$C(CH_3)_3$	$C(CH_3)_3CH_2OH$
BPPOH	$C(CH_3)_3$	$C(CH_3)_2OH$

FIGURE 10.1 Structures and abbreviations of alkylphenols mentioned in the text.

rate effect was nearly doubled by replacing both *ortho* substituents by methyl groups (i.e., TMP–QM).[28] Evidence suggests that decreasing steric/hydrophobic shielding of the oxo group enhances hydrogen bonding interactions with water effectively stabilizing the charged aromatic resonance form and increasing positive charge density at the position of nucleophilic attack (Scheme 10.4). Removing an *ortho*-alkyl group as with DMP–QM and BMP–QM further increases reactivity and precludes the preparation of stable solutions of these compounds.[28]

TABLE 10.1 Half-Lives (s) for QM Solvolysis in Methanol and Water[a]

	Methanol		Water	
p-QM Derived from	Ref. 26	Ref. 31	Ref. 28[b]	Ref. 31[c]
BHT	7340	5676	3060	>7000
BHTOH	—	306	400	1602
BPPOH	—	20	—	130
BDMP	182	—	47	—
TMP	17	—	26	—
Et–BHT	12,545	—	—	—

[a] Rates of QM disappearance measured spectrophotometrically at 25°C.
[b] In phosphate buffer at pH 7.4.
[c] In water/acetonitrile 1:1.

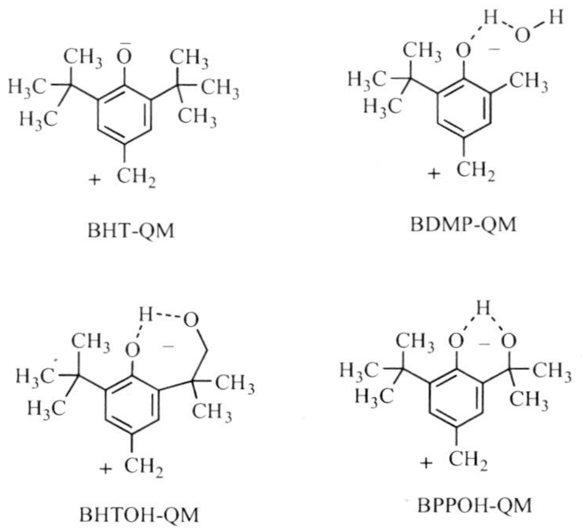

BHT-QM BDMP-QM

BHTOH-QM BPPOH-QM

SCHEME 10.4 Aromatic resonance forms of QMs indicating the stabilizing influences of inter- or intramolecular hydrogen bonding. Bulky alkyl groups shield the phenolate anion of BHT–QM but the anion of BDMP–QM is accessible to solvent. Analogues with a side-chain hydroxyl group can form intramolecular hydrogen bonds.

10.3.2.2 Hepatotoxicity Relationships of QM formation and reactivity to hepatotoxicity were investigated *in vitro* with isolated rat hepatocytes and rat liver slices containing active cytochromes P450. Hepatocytes were treated with a series of 4-methylphenols containing various alkyl substituents in the 2- and 6-positons.[28] GSH depletion and loss of cell viability correlated with both rates of QM formation (trapped as the GSH conjugates) and reactivity (Fig. 10.2). BHT, however, was not toxic to these cells, presumably due to the relatively sluggish reactivity of BHT–QM because of hydrophobic shielding of the carbonyl oxygen by bulky *tert*-butyl groups. On the

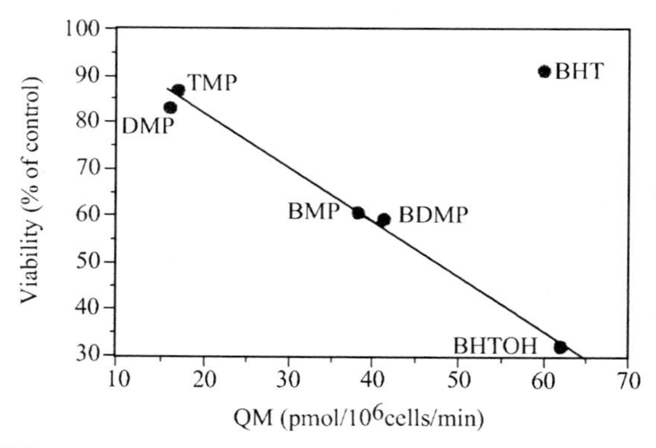

FIGURE 10.2 Effects of alkylphenols on cell viability compared to the amounts of *p*-QMs formed in isolated rat hepatocytes. *Source*: From Ref. 28, with permission from the American Chemical Society.

contrary, BHTOH was the most hepatotoxic compound in the group; this phenol is readily metabolized to BHTOH–QM that is considerably more reactive than BHT–QM (Table 10.1) because intramolecular hydrogen bonding stabilizes the resonance form shown in Scheme 10.4.

In rat liver slices, evidence also supports the roles of QMs in mediating the toxicity of a series of 4-methylphenols.[24] The potency correlates with rates of QM formation in the order 2-bromo-4-methylphenol > 4-methylphenol = DMP > TMP > 2-methoxy-4-methylphenol. None of these compounds contain two bulky *ortho* substituents, so as discussed earlier the corresponding QMs are expected to be highly reactive. The authors suggested that differences in the reactivities of these QMs determine their relative toxic potencies as electron-donating substituents on the ring stabilize the QM and thereby reduce its toxicity (e.g., 2-methoxy-4-methylphenol is less toxic than DMP) and conversely, electron-withdrawing substituents destabilize QMs and enhance toxicity (e.g., 2-bromo-4-methylphenol is more potent than DMP).

10.3.2.3 Lung Damage Several structure–activity studies implicate QMs in the adverse pulmonary effects observed in mice treated with BHT and related alkylphenols. The toxicity and the tumor-promoting activity of BHT were compared to the analogue with an ethyl group (Et–BHT) substituted for the 4-methyl resulting in a less reactive QM (Table 10.1). In contrast to BHT, Et–BHT did not increase lung/body weight ratios[27] and did not promote lung tumors in mice treated first with a tumor initiator.[29] As mentioned earlier, mice but not rats hydroxylate BHT and this metabolic pathway is especially important in lungs leading to the formation of BHTOH–QM.[22] Mouse lung epithelial cell lines examined in another study[30] do not contain cytochrome P450 for QM formation, so synthetic samples of BHT–QM and BHTOH–QM were directly added to the incubates and losses of cell viability were measured. The LC_{50} values determined in four epithelial cell lines were 10–20-fold lower for BHTOH–QM than BHT–QM, and toxicities of both compounds were enhanced substantially by depleting intracellular GSH. Relationships between QM reactivity and lung damage were investigated further by injecting mice with BHT, BHTOH, or BPPOH, three phenols that produce similar amounts of QMs in lungs, and measuring increases in lung/body weight ratios.[31] BPPOH–QM was found to be several-fold more reactive than BHTOH–QM (Table 10.1) due to more efficient intramolecular hydrogen bonding; a six- rather than a seven-membered ring is formed (Scheme 10.4). As predicted, the potency for both acute lung toxicity and lung tumor promotion correlates with the reactivities of the corresponding QMs in the order BPPOH > BHTOH > BHT.

10.3.3 Mechanisms of BHT Toxicity: Identification of Intracellular Targets

10.3.3.1 Gene Induction There is scarce information regarding the specific intracellular targets that mediate QM toxicities. Measurements of toxic end points (e.g., cell necrosis and oxidative stress *in vitro* or lung damage and tumor promotion *in vivo*) provide little mechanistic insight. There is evidence that BHT–QM is transcriptionally active. For example, employing the skin tumor promoter BHTOOH as a precursor to BHT–QM in cultured keratinocytes (Scheme 10.5), Guyton and

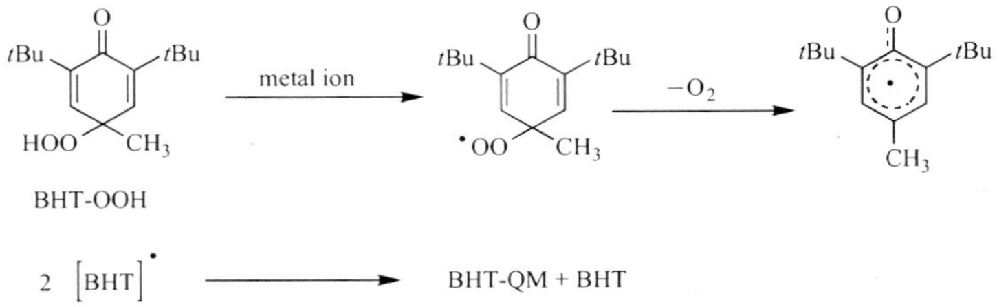

BHT-OOH —(metal ion)→ •OO peroxy radical —($-O_2$)→ BHT phenoxy radical

2 [BHT]• → BHT-QM + BHT

SCHEME 10.5 Proposed pathway for the nonenzymatic conversion of BHTOOH to BHT–QM in keratinocytes. BHTOOH is oxidized to a peroxy radical that spontaneously loses oxygen. Two BHT phenoxy radicals then undergo disproportionation.

colleagues[32] demonstrated induction of ornithine decarboxylase, a gene closely associated with tumor promotion. In addition, BHTOOH (through the QM) caused the activation of mitogen-activated protein kinase (MAPK) in murine skin indicating that MAPK is an important target for BHT–QM.[33] When recombinant HepG2 cells containing promoters or response elements for stress-related genes were exposed to BHT–QM and BHTOH–QM, gene expression was induced for metallothionein IIA, 70-kDa heat shock protein, glutathione *S*-transferase Ya, and xenobiotic response element.[34]

10.3.3.2 Reactivities of QMs with Cellular Nucleophiles Reactions with thiol groups on GSH or cysteines are generally very fast. For example the half-life under pseudo-first-order conditions for BHT–QM with cysteine is 90 s (versus 3060 s for water).[35] For the more reactive compound BDMP–QM, the $t_{1/2}$ is <3 s with cysteine and in the range 168–192 s with, respectively, the ε-amino of lysine and the imidazole group of histidine at pH 7.4. Clearly, thiol and amino groups can effectively compete with water for addition to QMs, however, thiols are preferred nucleophiles and partitioning depends on the pH of the protein microenvironment. Nonenzymatic conjugation with intracellular GSH occurs readily and limits QM binding to proteins. When GSH levels were depleted, for example by pretreatment with buthionine sulfoximine, covalent binding and toxicity of QMs were substantially enhanced in several cases.[28,30,36] Covalent binding to thiol or amino groups of peptides and proteins such as hemoglobin, myoglobin, and glutathione *S*-transferase has been demonstrated for BDMP–QM and BHT-derived QMs.[37–40] The extent of binding to a specific amino acid residue is influenced by neighboring residues that may assist in deprotonating the side-chain nucleophile or in protonating the oxo group of a QM and also provide a complimentary steric/hydrophobic environment to enhance QM interactions.[11,38] There is little or no evidence that simple QMs such as BHT–QM are genotoxic, and binding to nucleic acids has not been widely investigated. In one study, BHTOH–QM was substantially more reactive than BHT–QM toward 2′-deoxynucleosides, forming adducts mainly with the exocyclic amino groups of deoxyadenosine and deoxyguanosine.[41] Incubating BHTOH–QM with DNA generated the N6 adduct of deoxyadenosine as the main product.

10.3.3.3 Detection of QM–Protein Adducts Formed In Vitro Polyclonal antibodies were developed to the haptens BHT[42] and BHTOH[39] by linking their benzyl thioether derivatives formed with *N*-acetylcysteine to keyhole limpet hemocyanin (KLH, Fig. 10.3) and injecting rabbits with these immunogens. The specificities of the resulting antibodies were established by enzyme-linked immunosorbant assay (ELISA) and demonstrated that both thioether- and amine-linked BHT and BHTOH groups were detected and that antibodies raised to BHTOH–QM adducts also recognized BHT–QM adducts. Cytosolic proteins from pulmonary epithelial cells treated with BHT–QM or BHTOH–QM were separated by two-dimensional gel electrophoresis (2-DE), adducts detected by Western blotting (Fig. 10.3), and the adducted proteins identified by liquid chromatography–mass spectrometry (LC–MS) of the tryptic digests.[39] This approach led to the identification of 37 adducted proteins in several functional classes, including proteins related to stress (e.g., heat shock proteins and glutathione *S*-transferase P1), carbohydrate metabolism (e.g., α-enolase and triose phosphate isomerase), nucleic acid synthesis (e.g., nucleoside diphosphate kinase), and RNA and protein processing (e.g., histone H2B and translation elongation factor eEF-2). Cytoskeletal proteins including actins, tubulins, and calcyclin were targeted as well.

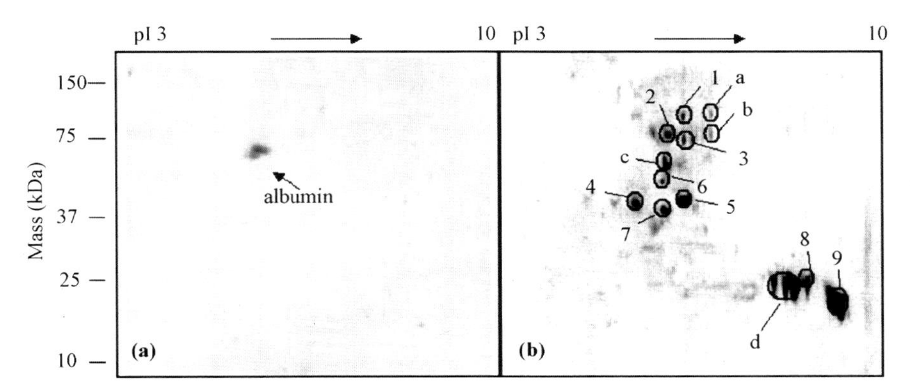

FIGURE 10.3 Structure of the BHTOH–KLH conjugate utilized for raising polyclonal antibodies. Western blots from 2D gels of cytosolic proteins isolated from mouse lung epithelial cells. The blot on the right is from cells incubated with BHT–QM and the blot on the left is from untreated cells. Immunoreactive proteins were identified by electrospray LC–MS analysis of the tryptic digests. *Source*: From Ref. 39, with permission from the American Chemical Society.

10.3.3.4 Glutathione S-Transferase P1 (GSTP1) Adduct An adduct of GSTP1, the predominant form of this enzyme in cultured mouse lung epithelial cells, was identified in the work mentioned earlier.[39] Incubating cells with BHT–QM decreased GST activity to 60–70% of control values and to only 35–45% of controls after first depleting intracellular GSH.[43] Human GSTP1 treated with a molar excess of BHT–QM combined with up to three molecules of the QM as shown in Fig. 10.4a.

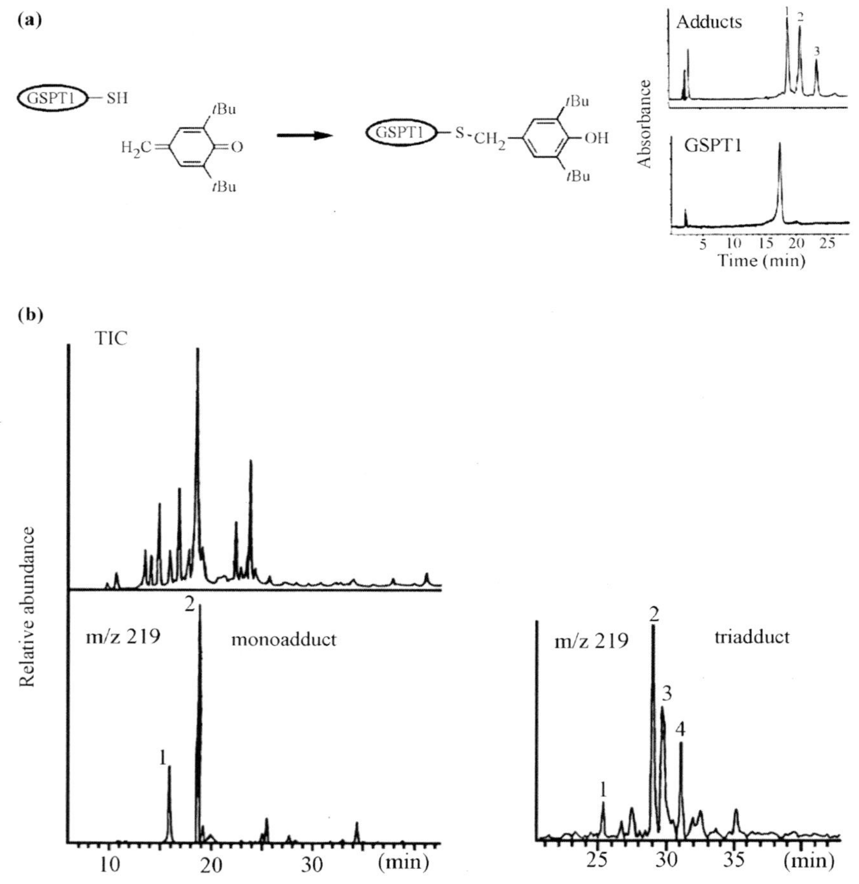

FIGURE 10.4 (a) BHT–QM adducts with human GSTP1. Intact proteins were analyzed by LC–MS. The total ion chromatograms (TIC) of untreated (lower) and QM-treated GSTP1 (upper) demonstrate the incorporation of one, two, or three molecules of BHT–QM. (b) LC–MS analyses of tryptic digests of mono- and triadducted GSTP1. The total ion chromatogram is shown for the monoadduct along with selected ion chromatograms for *m/z* 219, the mass of the BHT benzyl cation. *Source*: From Ref. 43, with permission from the American Chemical Society.

Comparisons of adduct formation with enzyme activity after treatment with increasing amounts of BHT–QM demonstrated that at least two molecules of BHT–QM must bind for significant inhibition to occur. LC–MS/MS analysis of tryptic digests revealed that Cys101 is alkylated first, followed by Cys47, and finally by Cys14. Cys47 is located in the GSH-binding site and is essential for activity, thereby explaining GSTP1 inactivation. During LC–MS work, it was discovered that thiol adducts of cysteine-containing peptides are labile, partially decomposing in the ion transport region between the electrospray source and the mass analyzer to produce the nonadducted peptide together with small amounts of the BHT benzyl cation at *m/z* 219. The detection of adducted peptides in tryptic digests, therefore, was facilitated by monitoring *m/z* 219 as shown in Fig. 10.4b, and the coeluting peptides identified from their MS/MS spectra. Alkylation and inhibition of GSTP1 *in vivo* may have adverse consequences including compromised protection from hydroperoxides and α,β-unsaturated aldehydes formed during lipid peroxidation, and a reduced ability to participate in the regulation of signal transduction.[44]

10.3.3.5 Protein Adducts Formed In Vivo Adducts formed in lungs of mice injected with BHT were detected utilizing 2-DE separations, immunoblotting, and LC–MS methodology as described above. Eight proteins were found to be alkylated reproducibly in four to six separate groups of mice; peroxiredoxin 6, Cu, Zn superoxide dismutase, carbonyl reductase, selenium-binding protein, tropomyosin 5, apolipoprotein A1, annexin A3, and β-actin.[40] Of particular importance are the antioxidant proteins peroxiredoxin 6 and Cu, Zn superoxide dismutase. Both were inhibited when the purified forms were exposed to BHT–QM. Increases in lipid peroxidation, hydrogen peroxide, and superoxide were observed in lung homogenates treated with the BHT–QM suggesting that these enzymes are also inhibited *in vivo*. A tryptic digest of BHT–QM-treated human peroxiredoxin 6 analyzed by LC–MS demonstrated that Cys47 and Cys91 were alkylated. The former residue is essential for activity, thereby explaining inhibition. Treating bovine Cu, Zn superoxide dismutase with BHT–QM generated predominantly a monoadduct that was detected by matrix-assisted laser desorption ionization-time of flight (MALDI-TOF) MS of the intact protein (Fig. 10.5). Analyses of peptides in the resulting digest revealed a QM-modified peptide containing His78, believed to be the main site of alkylation. The mechanism of inhibition, however, was not identified in this case. A third, potentially important QM, target in murine lung is carbonyl reductase. Inhibition of this enzyme is expected to enhance intracellular levels of lipid-derived α,β-unsaturated aldehydes leading to protein and DNA damage through adduct formation. These data suggest that alkylation and inhibition of several protective enzymes by QMs may contribute to inflammation and tumor promotion occurring in the lungs of mice treated with BHT.

10.3.3.6 ortho-Alkylphenols The very high reactivity of *o*-QM with thiols and amines under aqueous conditions was investigated by Modica et al.[45] This electrophile

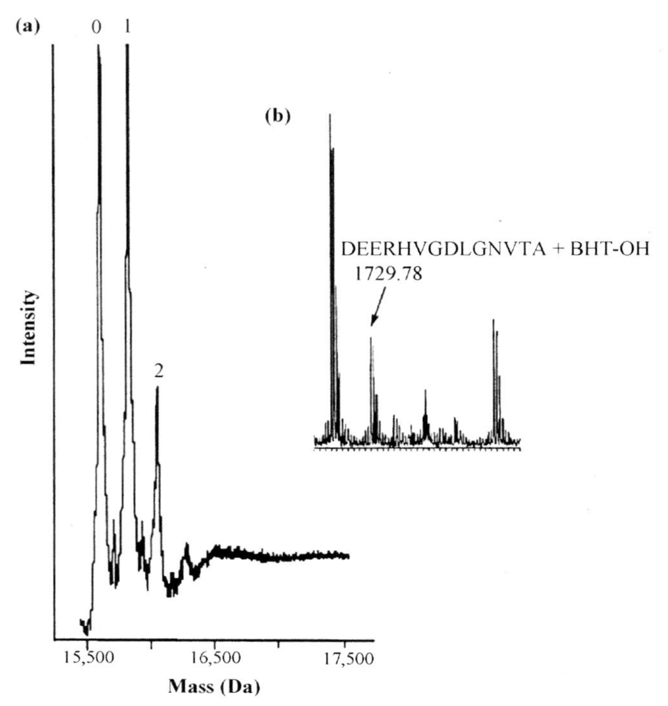

FIGURE 10.5 (a) MALDI-TOF MS analysis of the apo-form of bovine Cu, Zn superoxide dismutase after treatment with BHT–QM. A mixture of intact proteins bound to either zero, one, or two molecules of BHT–QM. (b) Portion of an Asp-N digest of the BHT–QM-treated protein showing the mass range corresponding to a His78-adducted peptide. *Source*: From Ref. 40, with permission from the American Chemical Society.

was generated by thermal or photochemical decomposition of the trimethylammonium salt as shown in Scheme 10.6a, and rates of reaction with water, amino acids, GSH, and amines were measured at several pH values. *o*-QM was found to react 21,000-fold faster than the BDMP–QM with water and 39-fold faster with cysteine. *o*-QMs undergo Michael additions to amino groups of peptides more readily than *p*-QMs, however, both react preferentially with the α-amino versus the ε-amino group of lysine and may be expected to alkylate *N*-terminal amino groups of peptides and proteins.[35] There are few examples in which an *o*-QM has been proposed to mediate alkylphenol toxicity, however, the severe hepatotoxicity of troglitazone, an oral diabetic drug that is no longer on the market, may be due in part to such an intermediate.[46] Incubating the drug with NADPH-fortified human liver microsomes and GSH yielded several conjugates including the one derived from the *o*-QM shown in Scheme 10.6b. Several P450 isoforms such as 3A4, 2C9, and 2C19 catalyze the formation of this *o*-QM and it was suggested that this reactive metabolite might contribute to troglitazone-induced liver damage.

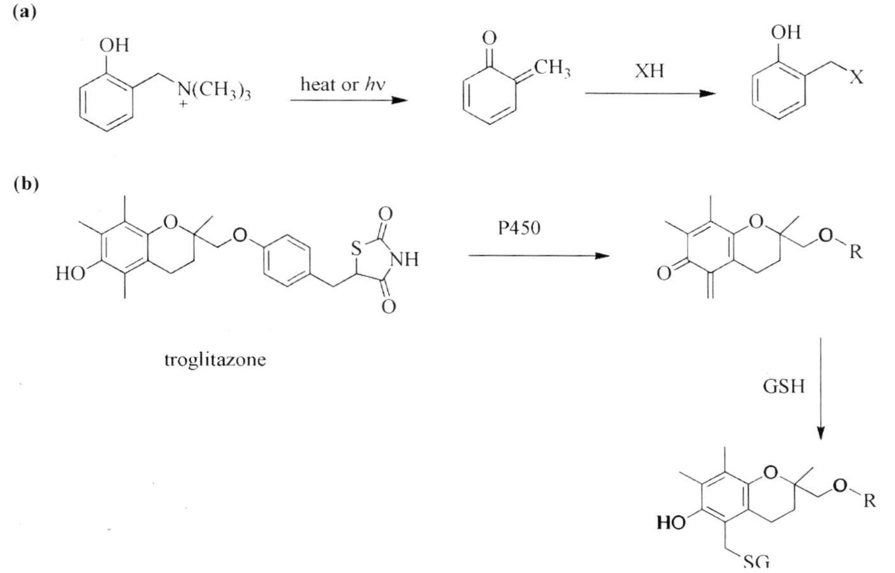

(a) OH, $N(CH_3)_3^+$, heat or $h\nu$, O, CH_3, XH, OH, X

(b) HO, O, O, S, NH, O, O, P450, O, O, O, R

troglitazone

GSH

HO, O, O, R, SG

SCHEME 10.6 (a) Formation of *o*-QM. (b) Oxidation of troglitazone by human cytochrome P450 to an *o*-QM and trapping by GSH.

10.4 METHOXYPHENOLS AND CATECHOLS

10.4.1 Methoxyphenols

As for BHT, QM formation from eugenol is also well characterized (Scheme 10.7). Human exposure occurs through its use as an analgesic and also from smoking clove cigarettes.[47] The toxicity of eugenol in isolated rat hepatocytes is believed to involve QM formation and subsequent covalent binding to proteins and/or DNA.[48,49] Incubating rat hepatocytes with eugenol depletes intracellular GSH before the onset of cell death, and this effect is accompanied by the formation of a eugenol–GSH conjugate and covalent binding to cellular protein.[50] These findings indicate that

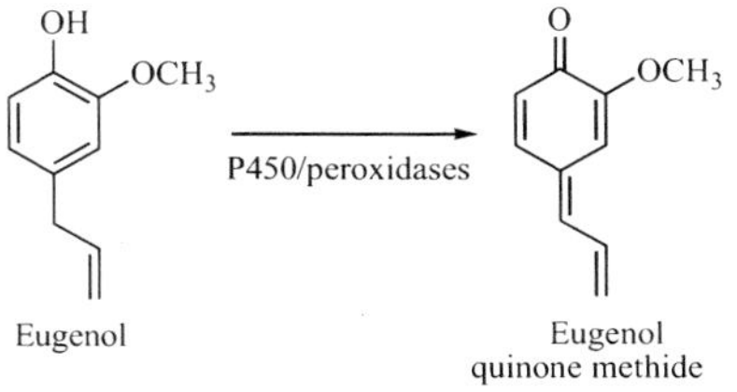

OH, OCH_3, P450/peroxidases, O, OCH_3

Eugenol

Eugenol quinone methide

SCHEME 10.7 Two-electron oxidation of eugenol.

eugenol–QM is formed in hepatocytes and that this reactive metabolite is responsible for eugenol-mediated cytotoxicity.

Analogues of eugenol with various alkyl substituents in the *para* position are converted to QMs in rat liver microsomes and hepatocytes.[50] The general trend seen in this work involves modest increases in cytochrome P450-catalyzed QM formation with larger alkyl substituents, however, the amounts differed by a factor of only three. In contrast to relatively small differences in the rates of QM formation, these electrophiles exhibited a large (3300-fold) range of reactivities (Table 10.2). For example, similar amounts of the QMs from eugenol and 2-methoxy-4-propylphenol were produced, suggesting that the lack of hepatotoxicity for the latter phenol in mice depleted of GSH[51,52] may be due to the extreme reactivity of the corresponding QM and rapid detoxification by hydrolysis. These data suggest that there may be a window of reactivities for QMs derived from 4-alkyl-2-methoxyphenols that is optimal for cytotoxicity. In support of this hypothesis, a plot of LC_{50} values versus QM hydrolysis rates yields a reasonable parabolic correlation for these compounds.[53] This relationship seems to indicate that phenols that are oxidized to highly reactive QMs (e.g., the metabolite from 2-methoxy-4-methylphenol) will cause little cellular damage since nucleophiles on protein side chains cannot compete with solvent. In addition, the formation of QMs from phenols that are stabilized by extended π-conjugation and electron-releasing substituents (e.g., the QM from 2,6-dimethoxy-4-allylphenol) is less cytotoxic due to an increased selectivity for sulfur nucleophiles such as GSH.

TABLE 10.2 Relative Reactivities of Quinone Methides

Phenolic Precursor	Half-Life (s) H_2O	References
Methoxy phenols		
4-Ally-2,6-dimethoxyphenol	4332	50
Eugenol	408	50
4-Benzyl-2-methoxyphenol	299	53
4-Isopropyl-2-methoxyphenol	87	50
4-Propyl-2-methoxyphenol	6	50
4-Ethyl-2-methoxyphenol	6	50
4-Methyl-2-methoxyphenol	1.3	50
Catechols		
4-Cinnamylcatechol	2700	9
Hydroxychavicol	336	8
SERMs		
Tamoxifen	10,800	59
Toremifene	3600	59
Acolbifene	32	64
DMA di-QM	15	96
Raloxifene di-QM	<1	95

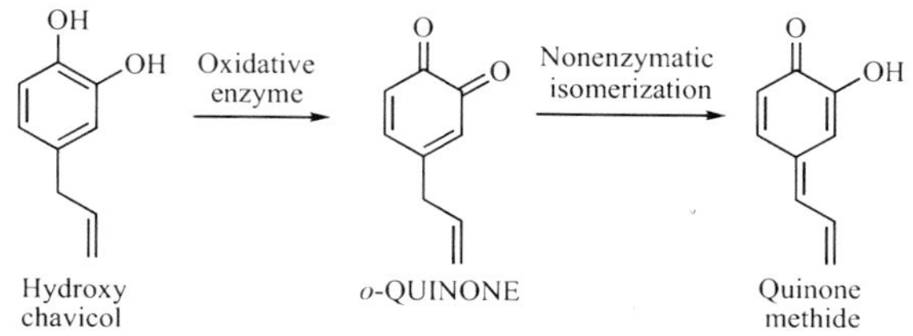

SCHEME 10.8 Hydroxychavicol oxidizes to an *o*-quinone that isomerizes to a *o*-hydroxy-*p*-QM.

10.4.2 Catechols

Hydroxychavicol (Scheme 10.8) is a major component of the Indian betel leaf, which is consumed by millions of people every year. Chewing areca quid that contains betel leaf has been implicated as a major risk factor for the development of oral squamous-cell carcinoma.[54] In addition, hydroxychavicol is a major metabolite of the hepatocarcinogen safrole[55] as well a minor metabolite of eugenol.[56] Hydroxychavicol is readily oxidized by a variety of oxidative enzymes including cytochrome P450 and peroxidases, forming a relatively stable *o*-quinone ($t_{1/2} = 9$ min, pH 7.4) that is readily trapped by thiol nucleophiles including GSH.[8] However, in the absence of thiol-trapping agents hydroxychavicol isomerizes to the *p*-QM, which is fully characterized as the GSH conjugates. These data suggest that reactive, potentially toxic metabolites of hydroxychavicol, and structurally related catechols include both the corresponding redox active/electrophilic *o*-quinones and *p*-QMs, which are more potent alkylating agents.

The effects of changing π-conjugation at the 4-position on both the rate of isomerization of the initially formed *o*-quinones to QMs and the reactivity of the quinoids formed from 4-propylcatechol, 2,3-dihydroxy-5,6,7,8-tetrahydronaphthalene (2-THNC), hydroxychavicol, and 4-cinnamylcatechol were studied (Fig. 10.6).[9] These catechols were selectively oxidized to the corresponding *o*-quinones or QMs and trapped with GSH. Microsomal incubations with the parent catechols produced only *o*-quinone–GSH conjugates. However, if GSH was added after an initial incubation period both *o*-quinone– and QM–GSH conjugates were observed. The results indicate that the extended π-conjugation at the *para* position enhances the rate

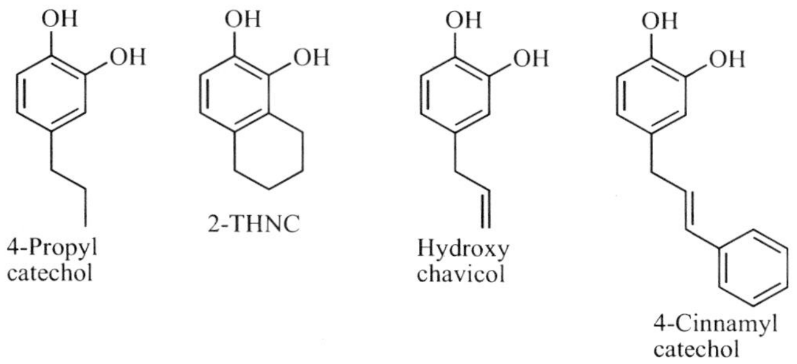

FIGURE 10.6 Structures of alkylcatechols.

of isomerization of an *o*-quinone to the QM. The half-lives, therefore, of *o*-quinones derived from the following catechols decreased in the order 4-propylcatechol > 2-THNC > hydroxychavicol > 4-cinnamylcatechol. AM1 semiempirical calculations showed the same trend, that is, an increase in QM versus *o*-quinone stability by extending π-conjugation at the 4-position. Finally, kinetic studies showed that the reactivity of the QMs with water increases with decreasing π-conjugation, similar to the results obtained with QMs from *o*-methoxyphenols (Table 10.2).[50,57] These data suggest that alkyl substituents at the 4-position of *o*-quinones attenuate both their ability to isomerize and the electrophilicity of the resulting QM tautomers. The dependence of the *o*-quinone/QM pathway on structure may provide a means of designing pharmacologically active compounds with selective modes of action.

10.5 QUINONE METHIDES FROM SERMs

A relatively stable QM is produced by initial P450-catalyzed aromatic hydroxylation of the SERM tamoxifen to yield 4-hydroxytamoxifen, followed by a cytochrome P450-catalyzed direct two-electron oxidation (Scheme 10.9).[7,58] This QM is extremely long lived at physiological pH and temperature ($t_{1/2} = 3$ h, Table 10.2),[59] most likely due to stabilization imposed by the two aryl substituents and the π-system of the additional vinyl group. The vinyl substituent alone can decrease the reactivity of *o*-methoxy–QMs by a factor of 100[60] and an aryl substituent leads to a 230-fold increase in stability relative to QMs with an unsubstituted exocyclic methylene group.[4] This π-stabilization, in addition to steric factors, completely changes the chemistry of the 4-hydroxytamoxifen QM. Most QMs react instantaneously with GSH, whereas the 4-hydroxytamoxifen QM has a half-life in the presence of GSH of approximately 4 min.[59] This reaction with GSH is reversible as the GSH conjugates slowly decompose to regenerate the QM.[59] Tamoxifen–GSH conjugates were detected

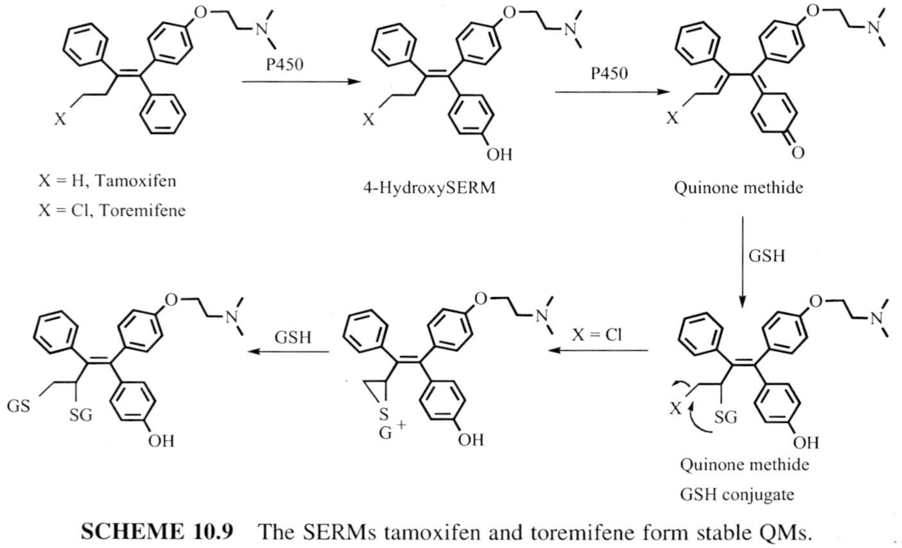

SCHEME 10.9 The SERMs tamoxifen and toremifene form stable QMs.

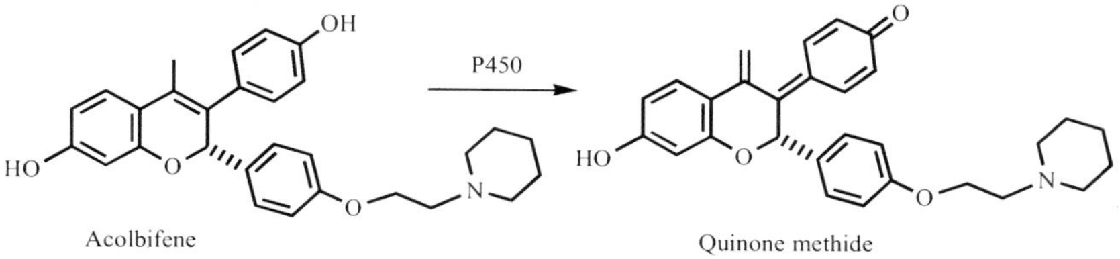

P450

Acolbifene Quinone methide

SCHEME 10.10 Metabolism of acolbifene to a QM.

in liver microsomal incubations with 4-hydroxytamoxifen, however, none were observed in incubations with breast cancer cells (MCF-7). Finally, although the tamoxifen QM does not react with deoxynucleosides,[59] DNA adducts have been reported both *in vitro* and *in vivo*.[58,61–63] This is likely a very minor pathway for DNA adduct formation compared to those produced via carbocation formation,[62] however, it has been reported that the tamoxifen QM DNA adducts are considerably more mutagenic leading to GC- > AT transitions in human Ad293 cells.[61]

Toremifene also undergoes oxidative metabolism to form a QM.[59] The 4-hydroxytoremifene QM has a half-life of 1 h at physiological pH and temperature (Table 10.2), while its half-life in the presence of GSH is approximately 6 min.[59] The 4-hydroxytoremifene QM reacts with two molecules of GSH and loses chlorine to yield the corresponding di-GSH conjugate (Scheme 10.9). This reaction mechanism likely involves an electrophilic episulfonium ion intermediate, which could contribute to the potential cytotoxicity of toremifene.

Acolbifene is also metabolized to a QM (Scheme 10.10)[64] formed by oxidation at the C-17 methyl group. This QM is considerably more reactive compared to the tamoxifen quinone methide, which indicates that the acolbifene quinone methide is an electrophile of intermediate stability (Table 10.2). In addition, the acolbifene QM was determined to react with deoxynucleosides, with one of the major adducts resulting from reaction with the exocyclic amino group of adenine.[64]

10.6 QUINONE METHIDES FROM ESTROGENS

The molecular mechanisms of steroidal estrogen carcinogenesis are highly complex and ambiguous.[65–69] Malignant phenotypes arise as a result of a series of mutations, most likely in genes associated with tumor suppressor, oncogene, DNA repair, or endocrine functions.[70] One major pathway considered to be important is the extensively studied hormonal pathway, by which estrogen stimulates cell proliferation through nuclear ER-mediated signaling pathways, thus resulting in an increased risk of genomic mutations during DNA replication.[70–73] An additional pathway involves estrogen metabolism, mediated by cytochrome P450, that generates catechol estrogens, which can be oxidized by virtually any oxidative enzyme or metal ion giving *o*-quinones.[10,65,68] The *o*-quinone formed from 2-hydroxyestrone has a half-life of 47 s, whereas the 4-hydroxyestrone-*o*-quinone is considerably longer lived ($t_{1/2} = 12$ min).[74] In the absence of nucleophilic-trapping agents, both *o*-quinones isomerize to QMs (Scheme 10.11), although the relative importance and biological

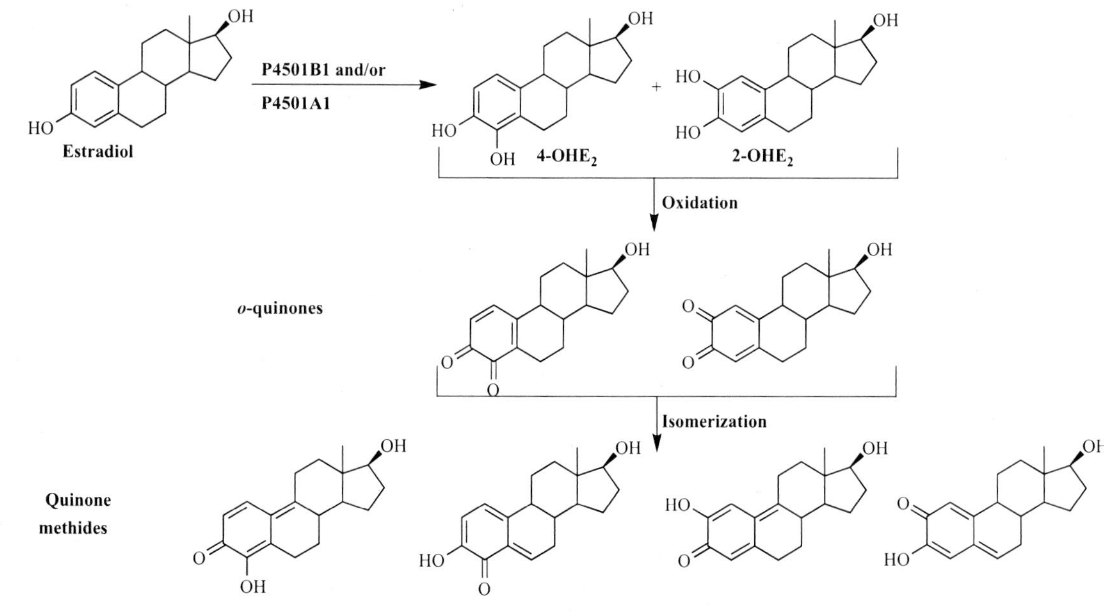

SCHEME 10.11 Estrone and estradiol are oxidized to catechols and *o*-quinones, which isomerizes to *p*-QMs.

targets of these potentially highly electrophilic intermediates have not been explored in detail.[74–76] Estrogen quinoids can directly damage cellular DNA leading to genotoxic effects.[68,77–82] Cavalieri's group has shown that the major DNA adducts produced from 4-hydroxyestradiol-*o*-quinone are depurinating N7-guanine and N3-adenine adducts resulting from 1,4-Michael addition both *in vitro* and *in vivo*.[68,76,81,83–85] In contrast, the considerably more rapid isomerization of the 2-hydroxyestradiol-*o*-quinone to the corresponding QMs results in 1,6-Michael addition products with the exocyclic amino groups of adenine and guanine.[76,86] Unlike N3 and N7 purine DNA adducts, these adducts are stable, which may alter their rate of repair and relative mutagenicity *in vivo*. The mutagenic properties of 2-hydroxyestrogen QM-derived stable DNA adducts have been evaluated using oligonucleotides containing site-specific adducts transfected into simian kidney (COS-7) cells where G $\rightarrow$ T and A $\rightarrow$ T mutations were observed.[87] It is important to mention that stable DNA adducts have been detected by ^{32}P postlabeling in Syrian hamster embryo cells treated with estradiol and its catechol metabolites.[88] The rank order of DNA adduct formation, which correlated with cellular transformation was 4-OHE$_2$ > 2-OHE$_2$ > estradiol. Finally, stable bulky adducts of 4-hydroxyestrone and 4-OHE$_2$ corresponding to alkylation of guanine have been detected in human breast tumor tissue.[89] These data suggest that the relative importance of *o*-quinone depurinating adducts versus stable QM DNA adducts in catechol estrogen carcinogenesis remains unclear.

10.7 NONCLASSICAL QUINONE METHIDES

Quercetin is a naturally occurring flavonoid with both antioxidant and prooxidant activities (Scheme 10.12).[90] It has been demonstrated in a variety of bacterial and mammalian mutagenicity experiments that quercetin has mutagenic properties that could be related to quinoid formation.[91,92] Quercetin is initially oxidized to an *o*-quinone, which rapidly isomerizes to di-QMs that could also be called extended

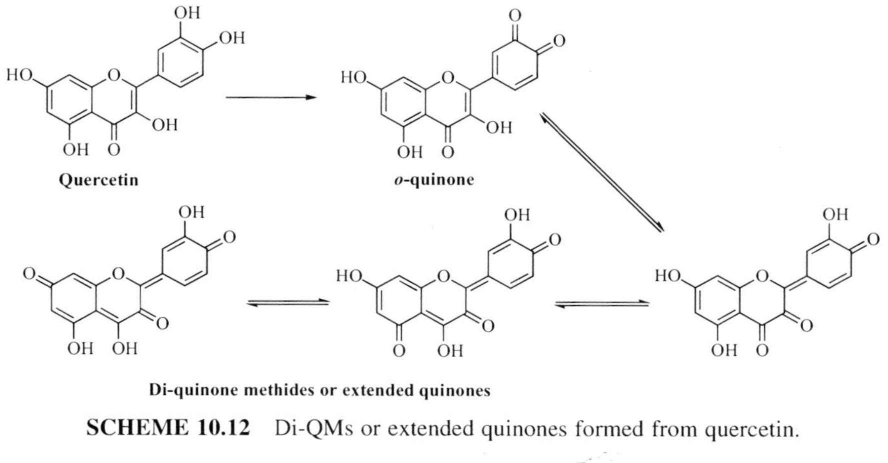

Quercetin

***o*-quinone**

Di-quinone methides or extended quinones

SCHEME 10.12 Di-QMs or extended quinones formed from quercetin.

quinones.[90,93] These di-QMs can be trapped with GSH although the GSH conjugates are unstable and equilibrate over time producing an isomeric mixture of both GSH conjugates. Protein and DNA adducts have also been observed in Caco-2 and HepG2 cells exposed to ^{14}C-labeled quercetin although these adducts were also unstable.[94] The transient nature of the quercetin di-QM adducts may have consequences for extrapolating quercetin genotoxicity to carcinogenicity *in vivo*.

Di-QM (extended quinones) formation also occurs with the SERMs raloxifene (Scheme 10.13),[95] arzoxifene,[96] and acolbifene.[64] The raloxifene di-QM is relatively short-lived, with a half-life of less than 1 s at physiological pH and temperature (Table 10.1).[95] This highly reactive metabolite has the potential to contribute to cytotoxicity through alkylation of proteins *in vivo*. For example, raloxifene has been associated with decreases in cytochrome P450 aromatase activity in human colon carcinoma cells[97] and irreversible inhibition of P450 3A4.[98,99] Furthermore, raloxifene–GSH conjugates were detected in incubation mixtures derived from a raloxifene di-QM. To identify microsomal proteins covalently modified by raloxifene metabolites, a novel raloxifene covert oxidatively activated tag (COATag) in which the SERM was linked to biotin was synthesized.[100] The raloxifene COATag allowed identification of covalently modified proteins by immunoblotting and LC–MS/MS analysis. Four major bands were observed in the blots from microsomal incubations, for which peptide maps were obtained using in-gel digestion, followed by MALDI-TOF or ESI mass spectral analysis of the resulting peptide mixtures. Cytosolic glucose-regulated protein (78 kDa, GRP78/BiP), protein disulfide isomerase isozyme A4 precursor (72 kDa, ERp72), protein disulfide isomerase isozyme A1 (57 kDa, PDIA1), protein disulfide isomerase isozyme A3 (58 kDa, ER-60), and microsomal glutathione *S*-transferase (17 kDa, mGST1) were identified as targets.[100] These data show that raloxifene produces a highly reactive intermediate that modifies tissue microsomal proteins with a low degree of selectivity, which might be an expected feature of a reactive intermediate with a relatively short lifetime.

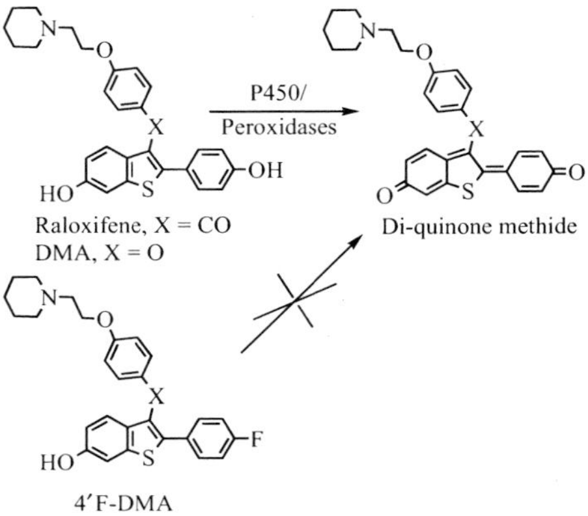

SCHEME 10.13 Di-QMs formed from raloxifene and DMA. 4-Fluoro substitution prevents di-QM formation.

Desmethyl arzoxifene (DMA) is a metabolite of arzoxifene and a structural analogue of raloxifene in which the carbonyl group of raloxifene has been replaced by an ether linkage (Scheme 10.13). DMA can be oxidized to a di-QM in the presence of rat or human liver microsomes.[96] The half-life of the DMA di-QM was found to be 15 s,[96] which indicates that the electron-donating ether linkage stabilizes the di-QM relative to the electron-withdrawing substituent in the raloxifene di-QM (Table 10.2). Since the uterus is another major target tissue of estrogens and antiestrogens, it was of interest to determine if quinoids could be formed from SERMs in uterine tissue potentially producing cytotoxic effects.[101] Incubations with rat uterine microsomes showed that both raloxifene and DMA can be oxidized to electrophilic di-QMs that were trapped as the corresponding GSH conjugates. Interestingly, the metabolism of raloxifene and DMA in rat uterine microsomes was not NADPH-dependent and could be inhibited by cyanide and catalase or enhanced by H_2O_2. Incubations of raloxifene and DMA COATags with rat uterine microsomes showed several modified proteins by Western blot analysis. The protein modification could be enhanced by the addition of H_2O_2 and decreased by the addition of NADPH, which suggests that unlike liver metabolism the formation of quinoids in the uterus could be mediated by uterine peroxidases.

Given the need to develop new SERMs with attenuated toxicity and increased bioavailability while maintaining their beneficial effects, a fluorinated DMA derivative (4′F-DMA) was synthesized, which is incapable of forming a di-QM (Scheme 10.13). 4′F-DMA showed similar ER-binding affinity compared to that of DMA, whereas the antiestrogenic activity was 10-fold lower than that of DMA, however, comparable to that of raloxifene.[96] No GSH conjugates were detected in microsomal incubations with 4′F-DMA in the presence of GSH or in cryopreserved rat hepatocytes. Furthermore, DMA significantly decreased the GSH levels in these cells within 30 min, whereas 4′F-DMA had no effect on GSH levels.[96] These data suggest that 4′F-DMA represents a promising SERM with comparable antiestrogenic activity to raloxifene, but improved metabolic stability and an attenuated potential for toxicity compared to the current benzothiophene SERMs. In addition, these experiments illustrate the general point that small structural modifications can prevent QM formation while maintaining the efficacy of drugs.

10.8 CONCLUSIONS AND FUTURE PROSPECTS

These are several examples of both structurally simple and complex phenols for which data strongly implicate QM intermediates as mediators of toxicity and/or carcinogenesis. These electrophiles could be considerably more important to the metabolism and biological properties of synthetic and naturally occurring phenols than is currently recognized. QMs are formed both enzymatically and nonenzymatically, but the details of these processes and relationships to the structures of phenolic compounds are just beginning to emerge. As Michael acceptors, QMs are unique because of a stabilized ionic resonance form. Variations in the contributions of this form modulate QM reactivity over a wide range (10^4–10^5, Table 10.1) suggesting substantial differences in the intracellular effects of QMs. It is clear that binding to both proteins and DNA competes with detoxification and that QMs are capable of inducing cytotoxic and possibly genotoxic responses. Future studies will seek to clarify relationships between reactivities and biological actions of these electrophiles and to gain insight into the mechanisms involved in cell damage. The data obtained will assist in clarifying the complex biological properties of phenolic compounds and provide new information on intracellular targets as a function of electrophile reactivity, which may be applicable to other types of electrophilic intermediates.

ACKNOWLEDGMENTS

Work cited from the authors' laboratories was supported by NIH Grants CA130037, CA79870, and CA041248.

REFERENCES

1. Thompson, D. C.; Thompson, J. A.; Sugumaran, M.; Moldeus, P. Biological and toxicological consequences of quinone methide formation. *Chem.-Biol. Interact.* 1993, 86, 129–162.
2. Peter, M. G. Chemical modifications of bio-polymers by quinones and quinone methides. *Angew. Chem. Int. Ed.* 1989, 28, 555–570.
3. Monks, T. J.; Jones, D. C. The metabolism and toxicity of quinones, quinonimines, quinone methides, and quinone-thioethers. *Curr. Drug Metab.* 2002, 3, 425–438.
4. Monks, T. J.; Hanzlik, R. P.; Cohen, G. M.; Ross, D.; Graham, D. G. Quinone chemistry and toxicity. *Toxicol. Appl. Pharmacol.* 1992, 112, 2–16.
5. Richard, J. P.; Amyes, T. L.; Bei, L.; Stubblefield, V. The effect of beta-fluorine substituents on the rate and equilibrium-constants for the reactions of alpha-substituted 4-methoxybenzyl carbocations and on the reactivity of a simple quinone methide. *J. Am. Chem. Soc.* 1990, 112, 9513–9519.
6. Hulbert, P. B.; Grover, P. L. Chemical rearrangement of phenol-epoxide metabolites of polycyclic aromatic hydrocarbons to quinone-methides. *Biochem. Biophys. Res. Commun.* 1983, 117, 129–134.
7. Dowers, T. S.; Qin, Z. H.; Thatcher, G. R.; Bolton, J. L. Bioactivation of selective estrogen receptor modulators (SERMs). *Chem. Res. Toxicol.* 2006, 19, 1125–1137.
8. Bolton, J. L.; Acay, N. M.; Vukomanovic, V. Evidence that 4-allyl-*o*-quinones spontaneously rearrange to their more electrophilic quinone methides: potential bioactivation mechanism for the hepatocarcinogen safrole. *Chem. Res. Toxicol.* 1994, 7, 443–450.
9. Iverson, S. L.; Hu, L. Q.; Vukomanovic, V.; Bolton, J. L. The influence of the *para*-alkyl substituent on the isomerization of *o*-quinones to *p*-quinone methides: potential bioactivation mechanism for catechols. *Chem. Res. Toxicol.* 1995, 8, 537–544.
10. Bolton, J. L.; Thatcher, G. R. Potential mechanisms of estrogen quinone carcinogenesis. *Chem. Res. Toxicol.* 2008, 21, 93–101.

11. Born, T. L.; Myers, J. K.; Widlanski, T. S.; Rusnak, F. 4-(Fluoromethyl)phenyl phosphate acts as a mechanism-based inhibitor of calcineurin. *J. Biol. Chem.* 1995, 270, 25651–25655.

12. Thompson, D. C.; Perera, K.; London, R. Spontaneous hydrolysis of 4-trifluoromethylphenol to a quinone methide and subsequent protein alkylation. *Chem.-Biol. Interact.* 2000, 126, 1–14.

13. Witschi, H.; Lock, S. Toxicity of butylated hydroxytoluene in mouse following oral administration. *Toxicology* 1978, 9, 137–146.

14. Witschi, H.; Malkinson, A. M.; Thompson, J. A. Metabolism and pulmonary toxicity of butylated hydroxytoluene (BHT). *Pharmacol. Ther.* 1989, 42, 89–113.

15. Malkinson, A. M.; Koski, K. M.; Evans, W. A.; Festing, M. F. Butylated hydroxytoluene exposure is necessary to induce lung tumors in BALB mice treated with 3-methylcholanthrene. *Cancer Res.* 1997, 57, 2832–2834.

16. Bauer, A. K.; Dwyer-Nield, L. D.; Keil, K.; Koski, K.; Malkinson, A. M. Butylated hydroxytoluene (BHT) induction of pulmonary inflammation: a role in tumor promotion. *Exp. Lung Res.* 2001, 27, 197–216.

17. Bauer, A. K.; Dwyer-Nield, L. D.; Hankin, J. A.; Murphy, R. C.; Malkinson, A. M. The lung tumor promoter, butylated hydroxytoluene (BHT), causes chronic inflammation in promotion-sensitive BALB/cByJ mice but not in promotion-resistant CXB4 mice. *Toxicology* 2001, 169, 1–15.

18. Malkinson, A. M. Primary lung tumors in mice: an experimentally manipulable model of human adenocarcinoma. *Cancer Res.* 1992, 52, 2670s–2676s.

19. Kehrer, J. P.; Witschi, H. Effects of drug metabolism inhibitors on butylated hydroxytoluene-induced pulmonary toxicity in mice. *Toxicol. Appl. Pharmacol.* 1980, 53, 333–342.

20. Takahashi, O.; Hiraga, K. 2,6-Di-*tert*-butyl-4-methylene-2,5-cyclohexadienone: a hepatic metabolite of butylated hydroxytoluene in rats. *Food Cosmet. Toxicol.* 1979, 17, 451–454.

21. Bolton, J. L.; Sevestre, H.; Ibe, B. O.; Thompson, J. A. Formation and reactivity of alternative quinone methides from butylated hydroxytoluene: possible explanation for species-specific pneumotoxicity. *Chem. Res. Toxicol.* 1990, 3, 65–70.

22. Bolton, J. L.; Thompson, J. A. Oxidation of butylated hydroxytoluene to toxic metabolites: factors influencing hydroxylation and quinone methide formation by hepatic and pulmonary microsomes. *Drug Metab. Dispos.* 1991, 19, 467–472.

23. Mizutani, T.; Yamamoto, K.; Tajima, K. Isotope effects on the metabolism and pulmonary toxicity of butylated hydroxytoluene in mice by deuteration of the 4-methyl group. *Toxicol. Appl. Pharm.* 1983, 69, 283–290.

24. Thompson, D. C.; Perera, K.; London, R. Studies on the mechanism of hepatotoxicity of 4-methylphenol(*p*-cresol)—effects of deuterium labeling and ring substitution. *Chem.-Biol. Interact.* 1996, 101, 1–11.

25. Guyton, K. Z.; Bhan, P.; Kuppusamy, P.; Zweier, J. L.; Trush, M. A.; Kensler, T. W. Free radical-derived quinone methide mediates skin tumor promotion by butylated hydroxytoluene hydroperoxide: expanded role for electrophiles in multistage carcinogenesis. *Proc. Natl. Acad. Sci. USA* 1991, 88, 946–950.

26. Filar, L. J.; Winstein, S. Preparation and behavior of simple quinone methides. *Tetrahedron Lett.* 1960, 25, 9–16.

27. Mizutani, T.; Ishida, I.; Yamamoto, K.; Tajima, K. Pulmonary toxicity of butylated hydroxytoluene and related alkylphenols: structural requirements for toxic potency in mice. *Toxicol. Appl. Pharmacol.* 1982, 62, 273–281.

28. Bolton, J. L.; Valerio, L. G. J.; Thompson, J. A. The enzymatic formation and chemical reactivity of quinone methides correlate with alkylphenol-induced toxicity in rat hepatocytes. *Chem. Res. Toxicol.* 1992, 5, 816–822.

29. Thompson, J. A.; Carlson, T. J.; Sun, Y.; Dwyer-Nield, L. D.; Malkinson, A. M. Studies using structural analogs and inbred strain differences to support a role for quinone methide metabolites of butylated hydroxytoluene (BHT) in mouse lung tumor promotion. *Toxicology* 2001, 160, 197–205.

30. Sun, Y.; Dwyer-Nield, L. D.; Malkinson, A. M.; Zhang, Y. L.; Thompson, J. A. Responses of tumorigenic and non-tumorigenic mouse lung epithelial cell lines to electrophilic metabolites of the tumor promoter butylated hydroxytoluene. *Chem.-Biol. Interact.* 2003, 145, 41–51.

31. Kupfer, R.; Dwyer-Nield, L. D.; Malkinson, A. M.; Thompson, J. A. Lung toxicity and tumor promotion by hydroxylated derivatives of 2,6-di-*tert*-butyl-4-methylphenol (BHT) and 2-*tert*-butyl-4-methyl-6-iso-propylphenol: correlation with quinone methide reactivity. *Chem. Res. Toxicol.* 2002, 15, 1106–1112.

32. Guyton, K.; Dolan, P. M.; Kensler, T. W. Quinone methide mediates *in vitro* induction of ornithine decarboxylase by the tumor promoter butylated hydroxytoluene hydroperoxide. *Carcinogenesis* 1994, 15, 817–821.

33. Guyton, K. Z.; Gorospe, M.; Kensler, T. W.; Holbrook, N. J. Mitogen-activated protein kinase (MAPK) activation by butylated hydroxytoluene hydroperoxide: implications for cellular survival and tumor promotion. *Cancer Res.* 1996, 56, 3480–3485.

34. Desjardins, J. P.; Beard, S. E.; Mapoles, J. E.; Gee, P.; Thompson, J. A. Transcriptional activity of quinone methides derived from the tumor promoter butylated hydroxytoluene in HepG2 cells. *Cancer Lett.* 1998, 131, 201–207.

35. Bolton, J. L.; Turnipseed, S. B.; Thompson, J. A. Influence of quinone methide reactivity on the alkylation of thiol and amino groups in proteins: studies utilizing amino acid and peptide models. *Chem.-Biol. Interact.* 1997, 107, 185–200.

36. Guyton, K. Z.; Thompson, J. A.; Kensler, T. W. Role of quinone methide in the *in vitro* toxicity of the skin tumor promoter butylated hydroxytoluene hydroperoxide. *Chem. Res. Toxicol.* 1993, 6, 731–738.

37. Bolton, J. L.; Le Blanc, J. C. Y.; Siu, K. W. M. Reaction of quinone methides with proteins: analysis of myoglobin adduct formation by electrospray mass spectrometry. *Biol. Mass Spectrom.* 1993, 22, 666–668.

38. McCracken, P. G.; Bolton, J. L.; Thatcher, G. R. J. Covalent modification of proteins and peptides by the quinone methide from 2-*tert*-butyl-4,6-dimethylphenol: selectivity and reactivity with respect to competitive hydration. *J. Org. Chem.* 1997, 62, 1820–1825.

39. Meier, B. W.; Gomez, J. D.; Zhou, A.; Thompson, J. A. Immunochemical and proteomic analysis of covalent adducts formed by quinone methide tumor promoters in mouse lung epithelial cell lines. *Chem. Res. Toxicol.* 2005, 18, 1575–1585.

40. Meier, B. W.; Gomez, J. D.; Kirichenko, O. V.; Thompson, J. A. Mechanistic basis for inflammation and tumor promotion in lungs of 2,6-di-*tert*-butyl-4-methylphenol-treated mice: electrophilic metabolites alkylate and inactivate antioxidant enzymes. *Chem. Res. Toxicol.* 2007, 20, 199–207.

41. Lewis, M. A.; Yoerg, D. G.; Bolton, J. L.; Thompson, J. A. Alkylation of 2′-deoxynucleosides and DNA by quinone methides derived from 2,6-di-*tert*-butyl-4-methylphenol. *Chem. Res. Toxicol.* 1996, 9, 1368–1374.

42. Reed, M.; Thompson, D. C. Immunochemical visualization and identification of rat liver proteins adducted by 2,6-di-*tert*-butyl-4-methylphenol (BHT). *Chem. Res. Toxicol.* 1997, 10, 1109–1117.

43. Lemercier, J. N.; Meier, B. W.; Gomez, J. D.; Thompson, J. A. Inhibition of glutathione *S*-transferase P1-1 in mouse lung epithelial cells by the tumor promoter 2,6-di-*tert*-butyl-4-methylene-2,5-cyclohexadienone (BHT-quinone methide): protein adducts investigated by electrospray mass spectrometry. *Chem. Res. Toxicol.* 2004, 17, 1675–1683.

44. Adler, V.; Yin, Z.; Fuchs, S. Y.; Benezra, M.; Rosario, L.; Tew, K. D.; Pincus, M. R.; Sardana, M.; Henderson, C. J.; Wolf, C. R.; Davis, R. J.; Ronai, Z. Regulation of JNK signaling by GSTp. *EMBO J.* 1999, 18, 1321–1334.

45. Modica, E.; Zanaletti, R.; Freccero, M.; Mella, M. Alkylation of amino acids and glutathione in water by *o*-quinone methide. Reactivity and selectivity. *J. Org. Chem.* 2001, 66, 41–52.

46. Kassahun, K.; Pearson, P. G.; Tang, W.; McIntosh, I.; Leung, K.; Elmore, C.; Dean, D.; Wang, R.; Doss, G.; Baillie, T. A. Studies on the metabolism of troglitazone to reactive intermediates *in vitro* and *in vivo*. Evidence for novel biotransformation pathways involving quinone methide formation and thiazolidinedione ring scission. *Chem. Res. Toxicol.* 2001, 14, 62–70.

47. *IARC Monographs;* International Agency for Research on Cancer: Lyon, France, 1985; Vol. 36, pp 75–97.

48. Thompson, D. C.; Constantin, T. D.; Moldeus, P. Metabolism and cytotoxicity of eugenol in isolated rat hepatocytes. *Chem.-Biol. Interact.* 1991, 77, 137–147.

49. Bodell, W. J.; Ye, Q.; Pathak, D. N.; Pongracz, K. Oxidation of eugenol to form DNA adducts and 8-hydroxy-2′-deoxyguanosine: role of quinone methide derivative in DNA adduct formation. *Carcinogenesis* 1998, 19, 437–443.

50. Bolton, J. L.; Comeau, E.; Vukomanovic, V. The influence of 4-alkyl substituents on the formation and reactivity of 2-methoxy-quinone methides: evidence that extended π-conjugation dramatically stabilizes the quinone methide formed from eugenol. *Chem.-Biol. Interact.* 1995, 95, 279–290.

51. Mizutani, T.; Satoh, K.; Nomura, H. Hepatotoxicity of eugenol and related compounds in mice depleted of glutathione: structural requirements for toxic potency. *Res. Commun. Chem. Pathol. Pharmacol.* 1991, 73, 87–95.

52. Mizutani, T.; Satoh, K.; Nomura, H.; Nakanishi, K. Hepatotoxicity of eugenol in mice depleted of glutathione by treatment with DL-buthionine sulfoximine. *Res. Commun. Chem. Pathol. Pharmacol.* 1991, 71, 219–230.

53. Thompson, D. C.; Perera, K.; Krol, E. S.; Bolton, J. L. *o*-Methoxy-4-alkylphenols that form quinone methides of intermediate reactivity are the most toxic in rat liver slices. *Chem. Res. Toxicol.* 1995, 8, 323–327.

54. *Betel-Quid and Areca Nut Chewing (IARC Monograph);* International Agency for Research on Cancer: Lyon, France, 1985; Vol. 37, pp 141–291.

55. Benedetti, M. S.; Malnoe, A.; Broillet, A. L. Absorption, metabolism and excretion of safrole in the rat and man. *Toxicology* 1977, 7, 69–83.

56. Sakano, K.; Inagaki, Y.; Oikawa, S.; Hiraku, Y.; Kawanishi, S. Copper-mediated oxidative DNA damage induced by eugenol: possible involvement of *O*-demethylation. *Mutat. Res.* 2004, 565, 35–44.

57. Bolton, J. L.; Pisha, E.; Shen, L.; Krol, E. S.; Iverson, S. L.; Huang, Z.; van Breemen, R. B.; Pezzuto, J. M. The reactivity of *o*-quinones which do not isomerize to quinone methides correlates with alkylcatechol-induced toxicity in human melanoma cells. *Chem.-Biol. Interact.* 1997, 106, 133–148.

58. Potter, G. A.; McCague, R.; Jarman, M. A mechanistic hypothesis for DNA adduct formation by tamoxifen following hepatic oxidative metabolism. *Carcinogenesis* 1994, 5, 439–442.

59. Fan, P. W.; Zhang, F.; Bolton, J. L. 4-Hydroxylated metabolites of the antiestrogens tamoxifen and toremifene are metabolized to unusually stable quinone methides. *Chem. Res. Toxicol.* 2000, 13, 45–52.

60. Bolton, J. L.; Comeau, E.; Vukomanovic, V. The influence of 4-alkyl substituents on the formation and reactivity of 2-methoxy-quinone methides: evidence that extended pi-conjugation dramatically stabilizes the quinone methide formed from eugenol. *Chem.-Biol. Interact.* 1995, 95, 279–290.

61. McLuckie, K. I.; Routledge, M. N.; Brown, K.; Gaskell, M.; Farmer, P. B.; Roberts, G. C.; Martin, E. A. DNA adducts formed from 4-hydroxytamoxifen are more mutagenic than those formed by alpha-acetoxytamoxifen in a shuttle vector target gene replicated in human Ad293 cells. *Biochemistry* 2002, 41, 8899–8906.

62. Beland, F. A.; McDaniel, L. P.; Marques, M. M. Comparison of the DNA adducts formed by tamoxifen and 4-hydroxytamoxifen *in vivo*. *Carcinogenesis* 1999, 20, 471–477.

63. Marques, M. M.; Beland, F. A. Identification of tamoxifen–DNA adducts formed by 4-hydroxytamoxifen quinone methide. *Carcinogenesis* 1997, 18, 1949–1954.

64. Liu, J.; Liu, H.; van Breemen, R. B.; Thatcher, G. R.; Bolton, J. L. Bioactivation of the selective estrogen receptor modulator acolbifene to quinone methides. *Chem. Res. Toxicol.* 2005, 18, 174–182.

65. Yager, J. D.; Davidson, N. E. Estrogen carcinogenesis in breast cancer. *N. Engl. J. Med.* 2006, 354, 270–282.

66. Russo, J.; Hu, Y. F.; Yang, X.; Russo, I. H. Developmental, cellular, and molecular basis of human breast cancer. *J. Natl. Cancer Inst. Monogr.* 2000, 27, 17–37.

67. Jefcoate, C. R.; Liehr, J. G.; Santen, R. J.; Sutter, T. R.; Yager, J. D.; Yue, W.; Santner, S. J.; Tekmal, R.; Demers, L.; Pauley, R.; Naftolin, F.; Mor, G.; Berstein, L. Tissue-specific synthesis and oxidative metabolism of estrogens. *J. Natl. Cancer Inst. Monogr.* 2000, 27, 95–112.

68. Cavalieri, E.; Chakravarti, D.; Guttenplan, J.; Hart, E.; Ingle, J.; Jankowiak, R.; Muti, P.; Rogan, E.; Russo, J.; Santen, R.; Sutter, T. Catechol estrogen quinones as initiators of breast and other human cancers: implications for biomarkers of susceptibility and cancer prevention. *Biochim. Biophys. Acta* 2006, 1766, 63–78.

69. Russo, J.; Russo, I. H. The role of estrogen in the initiation of breast cancer. *J. Steroid Biochem. Mol. Biol.* 2006, 102, 89–96.

70. Henderson, B. E.; Feigelson, H. S. Hormonal carcinogenesis. *Carcinogenesis* 2000, 21, 427–433.

71. Feigelson, H. S.; Henderson, B. E. Estrogens and breast cancer. *Carcinogenesis* 1996, 17, 2279–2284.

72. Nandi, S.; Guzman, R. C.; Yang, J. Hormones and mammary carcinogenesis in mice, rats, and humans: a unifying hypothesis. *Proc. Natl. Acad. Sci. USA* 1995, 92, 3650–3657.

73. Flototto, T.; Djahansouzi, S.; Glaser, M.; Hanstein, B.; Niederacher, D.; Brumm, C.; Beckmann, M. W. Hormones and hormone antagonists: mechanisms of action in carcinogenesis of endometrial and breast cancer. *Horm. Metab. Res.* 2001, 33, 451–457.

74. Iverson, S. L.; Shen, L.; Anlar, N.; Bolton, J. L. Bioactivation of estrone and its catechol metabolites to quinoid-glutathione conjugates in rat liver microsomes. *Chem. Res. Toxicol.* 1996, 9, 492–499.

75. Bolton, J. L.; Shen, L. *p*-Quinone methides are the major decomposition products of catechol estrogen *o*-quinones. *Carcinogenesis* 1996, 17, 925–929.

76. Stack, D. E.; Byun, J.; Gross, M. L.; Rogan, E. G.; Cavalieri, E. L. Molecular characteristics of catechol estrogen quinones in reactions with deoxyribonucleosides. *Chem. Res. Toxicol.* 1996, 9, 851–859.

77. Bolton, J. L.; Yu, L.; Thatcher, G. R. Quinoids formed from estrogens and antiestrogens. *Methods Enzymol.* 2004, 378, 110–123.

78. Prokai-Tatrai, K.; Prokai, L. Impact of metabolism on the safety of estrogen therapy. *Ann. N.Y. Acad. Sci.* 1052, 2005, 243–257.

79. Liehr, J. G. Role of DNA adducts in hormonal carcinogenesis. *Regul. Toxicol. Pharmacol.* 2000, 32, 276–282.

80. Russo, J.; Russo, I. H. Genotoxicity of steroidal estrogens. *Trends Endocrinol. Metab.* 2004, 15, 211–214.

81. Li, K. M.; Todorovic, R.; Devanesan, P.; Higginbotham, S.; Kofeler, H.; Ramanathan, R.; Gross, M. L.; Rogan, E. G.; Cavalieri, E. L. Metabolism and DNA binding studies of 4-hydroxyestradiol and estradiol-3,4-quinone *in vitro* and in female ACI rat mammary gland *in vivo*. *Carcinogenesis* 2004, 25, 289–297.

82. Chakravarti, D.; Mailander, P. C.; Li, K. M.; Higginbotham, S.; Zhang, H. L.; Gross, M. L.; Meza, J. L.; Cavalieri, E. L.; Rogan, E. G. Evidence that a burst of DNA depurination in SENCAR mouse skin induces error-prone repair and forms mutations in the H-ras gene. *Oncogene* 2001, 20, 7945–7953.

83. Cavalieri, E.; Frenkel, K.; Liehr, J. G.; Rogan, E.; Roy, D. Estrogens as endogenous genotoxic agents—DNA adducts and mutations. *J. Natl. Cancer Inst. Monogr.* 2000, 27, 75–93.

84. Saeed, M.; Rogan, E.; Fernandez, S. V.; Sheriff, F.; Russo, J.; Cavalieri, E. Formation of depurinating N3-adenine and N7-guanine adducts by MCF-10F cells cultured in the presence of 4-hydroxyestradiol. *Int. J. Cancer* 2007, 120, 1821–1824.

85. Zahid, M.; Kohli, E.; Saeed, M.; Rogan, E.; Cavalieri, E. The greater reactivity of estradiol-3,4-quinone vs. estradiol-2,3-quinone with DNA in the formation of depurinating adducts: implications for tumor-initiating activity. *Chem. Res. Toxicol.* 2006, 19, 164–172.

86. Debrauwer, L.; Rathahao, E.; Jouanin, I.; Paris, A.; Clodic, G.; Molines, H.; Convert, O.; Fournier, F.; Tabet, J. C. Investigation of the regio- and stereoselectivity of deoxyguanosine linkage to deuterated 2-hydroxyestradiol by using liquid chromatography/ESI-ion trap mass spectrometry. *J. Am. Soc. Mass Spectrom.* 2003, 14, 364–372.

87. Terashima, I.; Suzuki, N.; Shibutani, S. Mutagenic properties of estrogen quinone-derived DNA adducts in simian kidney cells. *Biochemistry* 2001, 40, 166–172.

88. Hayashi, N.; Hasegawa, K.; Barrett, J. C.; Tsutsui, T. Estrogen-induced cell transformation and DNA adduct formation in cultured Syrian hamster embryo cells. *Mol. Carcinog.* 1996, 16, 149–156.
89. Embrechts, J.; Lemiere, F.; Dongen, W. V.; Esmans, E. L.; Buytaert, P.; van Marck, E.; Kockx, M.; Makar, A. Detection of estrogen DNA-adducts in human breast tumor tissue and healthy tissue by combined nano LC–nano ES tandem mass spectrometry. *J. Am. Soc. Mass Spectrom.* 2003, 14, 482–491.
90. Rietjens, I. M.; Boersma, M. G.; van der Woude, H.; Jeurissen, S. M.; Schutte, M. E.; Alink, G. M. Flavonoids and alkenylbenzenes: mechanisms of mutagenic action and carcinogenic risk. *Mutat. Res.* 2005, 574, 124–138.
91. MacGregor, J. T.; Jurd, L. Mutagenicity of plant flavonoids: structural requirements for mutagenic activity in *Salmonella typhimurium*. *Mutat. Res.* 1978, 54, 297–309.
92. Brown, J. P. A review of the genetic effects of naturally occurring flavonoids, anthraquinolines and related compounds. *Mutat. Res.* 1980, 75, 243–277.
93. Boersma, M. G.; Vervoort, J.; Szymusiak, H.; Lemanska, K.; Tyrakowska, B.; Cenas, N.; Segura-Aguilar, J.; Rietjens, I. M. Regioselectivity and reversibility of the glutathione conjugation of quercetin quinone methide. *Chem. Res. Toxicol.* 2000, 13, 185–191.
94. Walle, T.; Vincent, T. S.; Walle, U. K. Evidence of covalent binding of the dietary flavonoid quercetin to DNA and protein in human intestinal and hepatic cells. *Biochem. Pharmacol.* 2003, 65, 1603–1610.
95. Yu, L.; Liu, H.; Li, W.; Zhang, F.; Luckie, C.; van Breemen, R. B.; Thatcher, G. R.; Bolton, J. L. Oxidation of raloxifene to quinoids: potential toxic pathways via a diquinone methide and *o*-quinones. *Chem. Res. Toxicol.* 2004, 17, 879–888.
96. Liu, H.; Liu, J.; van Breemen, R. B.; Thatcher, G. R.; Bolton, J. L. Bioactivation of the selective estrogen receptor modulator desmethylated arzoxifene to quinoids: 4′-fluoro substitution prevents quinoid formation. *Chem. Res. Toxicol.* 2005, 18, 162–173.
97. Fiorelli, G.; Picariello, L.; Martineti, V.; Tonelli, F.; Brandi, M. L. Estrogen synthesis in human colon cancer epithelial cells. *J. Steroid Biochem. Mol. Biol.* 1999, 71, 223–230.
98. Chen, Q.; Ngui, J. S.; Doss, G. A.; Wang, R. W.; Cai, X.; DiNinno, F. P.; Blizzard, T. A.; Hammond, M. L.; Stearns, R. A.; Evans, D. C.; Baillie, T. A.; Tang, W. Cytochrome P450 3A4-mediated bioactivation of raloxifene: irreversible enzyme inhibition and thiol adduct formation. *Chem. Res. Toxicol.* 2002, 15, 907–914.
99. Baer, B. R.; Wienkers, L. C.; Rock, D. A. Time-dependent inactivation of P450 3A4 by raloxifene: identification of Cys239 as the site of apoprotein alkylation. *Chem. Res. Toxicol.* 2007, 20, 954–964.
100. Liu, J.; Li, Q.; Yang, X.; van Breemen, R. B.; Bolton, J. L.; Thatcher, G. R. Analysis of protein covalent modification by xenobiotics using a covert oxidatively activated tag: raloxifene proof-of-principle study. *Chem. Res. Toxicol.* 2005, 18, 1485–1496.
101. Liu, H.; Qin, Z.; Thatcher, G. R.; Bolton, J. L. Uterine peroxidase-catalyzed formation of diquinone methides from the selective estrogen receptor modulators raloxifene and desmethylated arzoxifene. *Chem. Res. Toxicol.* 2007, 20, 1676–1684.

11

QUINONE METHIDES AND AZA-QUINONE METHIDES AS LATENT ALKYLATING SPECIES IN THE DESIGN OF MECHANISM-BASED INHIBITORS OF SERINE PROTEASES AND β-LACTAMASES

Michèle Reboud-Ravaux[1] and Michel Wakselman[2]

[1] *Enzymologie Moléculaire et Fonctionnelle, FRE 2852, CNRS-Université Paris 6 UPMC, T43, Institut Jacques Monod, 2 Place Jussieu, 75251 Paris Cedex 05, France*
[2] *Institut Lavoisier de Versailles, UMR 8180, CNRS-Université Versailles Saint-Quentin, 45 Avenue Des Etats Unis, F-78035 Versailles, France*

11.1 INTRODUCTION

The starting point for much of the work described in this article is the idea that quinone methides (QMs) are the electrophilic species that are generated from *ortho*-hydroxybenzyl halides during the relatively selective modification of tryptophan residues in proteins. Therefore, a series of suicide substrates (a subtype of mechanism-based inhibitors) that produce quinone or quinonimine methides (QIMs) have been designed to inhibit enzymes. The concept of mechanism-based inhibitors was very appealing and has been widely applied. The present review will be focused on the inhibition of mammalian serine proteases and bacterial serine β-lactamases by suicide inhibitors. These very different classes of enzymes have however an analogous step in their catalytic mechanism, the formation of an acyl-enzyme intermediate. Several studies have examined the possible use of quinone or quinonimine methides as the latent

Quinone Methides, Edited by Steven E. Rokita

electrophilic species unmasked by the catalytic action of the enzyme through the formation of the acyl-enzyme.

11.1.1 Mechanism-Based Inhibitors/Suicide Substrates

The design of mechanism-based inhibitors is based on knowledge of the structure and mechanism of action of an enzyme and the principles of organic chemical reactivity.[1–5] An initial molecule is modified by the catalytic action of an enzyme leading to the formation of an intermediate that forms a stable covalent bond with the enzyme. Krantz pointed out the need for multiple designations to stipulate the type of activity directly responsible for the inhibition.[4] Catalytic processing by the enzyme may convert inhibitors to "(1) tight binding active-site complements (mechanism-based inhibitors/transition-state analogues); (2) reactive intermediates that combine irreversibly with the enzyme in a step that lies outside normal catalysis (mechanism-based inhibitors/reactive intermediates or suicide substrates); (3) compounds that are converted to stable analogues that are unable to proceed to product because they lack the requisite functionality for further processing (dead-end inhibitors); (4) stable analogues that have the potential for conversion to product, but are trapped in potential energy wells (alternate substrate inhibitors)." This review uses this classification with subtypes.

We will be mainly concerned with suicide substrates that are analogues of β-lactamase or serine protease substrates but also possess a latent electrophilic function that is unmasked during one step of the catalytic cycle of the target enzyme. The resulting electrophile reacts with a nucleophilic group in the enzyme active site. The formation of a covalent acyl-enzyme will maintain part of the inhibitor in the active site facilitating the enzyme inactivation.[2,3] These inhibitors are likely to be extremely selective *in vitro* and *in vivo* since their inhibitory activity requires discrimination in the binding steps, the catalytic activation by the enzyme, and the irreversible modification of the active site. Furthermore, from a mechanistic point of view, suicide substrates can reveal the nature and reactivity of the trapped functional groups present in the enzyme active site.

11.1.2 Reactivity of Quinone Methides

Quinone methides are cyclic vinylogs of α,β-unsaturated ketones. They are good Michael acceptors that react with many nucleophiles, including amines, thiols, indoles, and phenols;[6,7] they can thus be the electrophile generated during an enzyme turnover. They can be formed by a dissociative 1–4 or 1–6 elimination mechanism from an *ortho*- or *para*-hydroxybenzyl starting compound possessing a good leaving group such as a halide anion or a neutral sulfide.[8] Unconjugated *ortho*-QMs are generally more reactive than their *p*-QM isomers.[9] The influence of substituents on the reactivity of these electrophilic species has been studied recently.[10]

One example of a reactive *ortho*-hydroxybenzyl derivatives is the Koshland reagent I (Fig. 11.1).[11–13] Its hydrolysis half-life in water (pH 3.5 and 25 °C) is

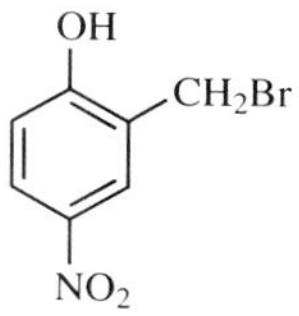

OH
CH_2Br
NO_2

FIGURE 11.1 Structure of Koshland reagent I.

less than 30 s and 2-hydroxy-5-nitrobenzylbromide is relatively selective for the indole nucleus of protein tryptophan residue. Other nucleophiles such as the thiol function of cysteine, the imidazole ring of histidine, or the phenolate anion of tyrosine residues can also react. In general, substitution at the benzylic position of an *o*- or *p*-hydroxybenzyl compound occurs by an elimination–addition ($D_N + A_N$) mechanism.

11.2 SERINE PROTEASES: MINIMAL SCHEMES; CATALYTIC MECHANISMS; SUICIDE INHIBITION

Proteinases (=peptidases) whose catalytic activity depends upon the hydroxyl group of a serine residue acting as the nucleophile that attacks the peptide bond are termed serine proteinases (serine peptidases). The families of serine proteases (about 50) have been grouped into nine clans by comparing their tertiary structures and the order of the catalytic residues in their sequences.[14] The catalytic machinery usually involves a proton donor (or general base) in addition to the nucleophilic serine (often called active serine). The proton donor is a histidine residue that is part of a catalytic triad in the enzymes of clans PA(S), SB, SC, SH, SK, and SN. Clans and families are groups of homologous peptidases; a clan contains one or more families that appear to have come from a unique origin of peptidases. The third residue (usually an aspartate except in clan SH where it is a histidine) is believed to be implicated in the orientation of the imidazolium ring. A lysine residue is a proton donor in clans SE, SF, and SM, and a third catalytic residue is not needed. Some peptidases have a Ser/His dyad (clans SF and SM). The well-known pancreatic chymotrypsin belongs to clan PA(S) whose members have a double barrel fold; in contrast, subtilisin (clan SB) has an α,β-fold with a parallel β-sheet. Clan PA(S) and clan SB enzymes were the two major known groups for a long time but gene cloning has revealed a variety of other serine proteases that do not fit into these two clans. The enzymes of clans PA(S), SB, and SC all act via the same mechanism whereas those of clans SE, SF, SH, and PB have different mechanisms of action.[15] Fig. 11.2a and b shows the basic features of the reaction scheme and mechanism of action of serine proteases.

The hydrolysis of esters, amides, and peptides catalyzed by serine proteases involves the nucleophilic attack by the serine hydroxyl group on the carbonyl group of the substrate. The addition reaction is catalyzed by the imidazole group of the active site histidine acting as a general base leading to a tetrahedral intermediate and an imidazolium group. This charged group catalyzes the intermediate breakdown to the acyl-enzyme, imidazole, and the product alcohol or amine (elimination reaction).

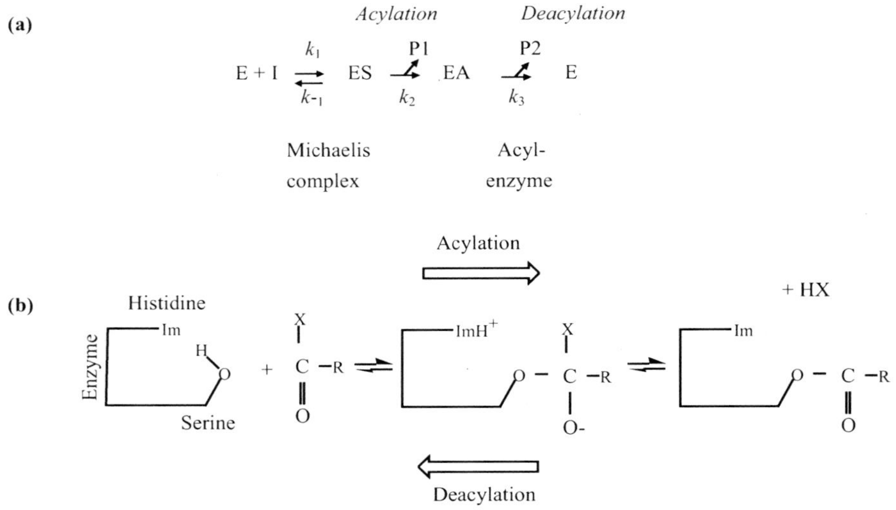

(a)

Acylation *Deacylation*

$$E + I \underset{k_{-1}}{\overset{k_1}{\rightleftharpoons}} ES \xrightarrow[k_2]{P1} EA \xrightarrow[k_3]{P2} E$$

Michaelis complex Acyl-enzyme

(b)

Acylation

Histidine Im, H, O, Enzyme, Serine + X–C(=O)–R ⇌ ImH$^+$ O–C(X)(O-)–R ⇌ Im O–C(=O)–R + HX

Deacylation

FIGURE 11.2 Hydrolysis of esters and peptides by serine proteases: reaction scheme (a) and mechanism of action (b) (after Polgàr[15]). (a) ES, noncovalent enzyme–substrate complex (Michaelis complex); EA, the acyl-enzyme; P1 and P2, the products. (b) X = OR or NHR (acylation); X = OH (deacylation).

Deacylation is the reverse of acylation in which a water molecule is the attacking nucleophile (Fig. 11.2b).

The specificities of serine proteases are exceedingly diverse.[16] Occupation of the S rather S′ subsites is important (terminology from Schechter and Berger;[17] Fig. 11.3).

Serine proteases usually show primary specificity (occupation of subsite S_1) for positively charged arginine or lysine (trypsin, plasmin, plasminogen activators, thrombin), large hydrophobic side chains of phenylalanine, tyrosine, and tryptophan (chymotrypsin, cathepsin G, chymase, and subtilisin), or small aliphatic side chains (elastases). Nevertheless, there are a large number of variations and in many cases, other subsites like S_2 and S_3 are more discriminating while maintaining the

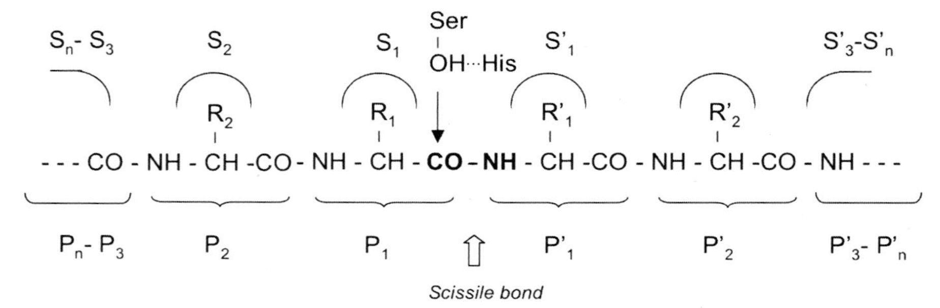

FIGURE 11.3 Representation of extended substrate binding site of a serine protease according to Schechter and Berger.[17]

specificity of S_1.[18] When designing protease inhibitors, modulation of the peptide chain of the inhibitor is a classical way to improve the selectivity (see below).

Serine proteases are involved in numerous biological processes in mammals such as digestion (chymotrypsin, trypsin, and pancreatic elastase), phagocytosis (leukocyte elastase, proteinase 3, and cathepsin G), and hemostasis and fibrinolysis (thrombin, plasmin, plasminogen activators, and factor Xa). When the control of proteolysis is deficient, the uncontrolled proteolytic action may have deleterious effects explaining why serine proteases are implicated in a large variety of diseases, including pulmonary emphysema, inflammation, tumor invasion and cancer, and thrombosis disorders. Serine proteases have also important roles in microorganisms, for example, subtilisin Carlsberg (*Bacillus licheniformis*), lon protease and signal peptidase I (*Escherichia coli*), and viruses like hepacivirin (hepatitis C virus), flavivirin (yellow fever virus), and cytomegalovirus assembling (human cytomegalovirus). These enzymes are thus important targets and developing inhibitors of them may lead to clinically useful drugs. The characteristics of their mechanism of action make them suitable for inhibition by mechanism-based inhibitors, especially suicide substrates.

The general kinetic outline for the inactivation of serine proteases by suicide substrates can be described by Eq. 11.1:[19]

$$\mathrm{E+I} \underset{k_{-1}}{\overset{k_1}{\rightleftharpoons}} \underset{\text{Encounter complex}}{\mathrm{EI}} \xrightarrow{k_2} \underset{\text{Acyl-enzyme}}{\mathrm{E{-}I}} \xrightarrow{k_4} \underset{\text{Inactivated enzyme}}{\mathrm{E{-}I'}} \qquad \mathrm{E{-}I} \xrightarrow{k_3} \mathrm{E+P} \tag{11.1}$$

where EI is the enzyme–inhibitor encounter complex (Michaelis complex) that is converted to an acyl-enzyme E-I, which may break down by two pathways, to give either the inactivated enzyme E-I′ or the product P and free enzyme E. The partition ratio r, defined as [P]/[E-I′], or k_3/k_4 is a key parameter of this mechanism-based process. It represents the number of catalytic turnovers per inactivation step.[20]

When examining the first moments of the reaction, the kinetic constant k_3 is usually small enough to be neglected. If the enzyme is inactivated, the acyl-enzyme cannot be kinetically distinguished from the Michaelis complex. Thus, the minimum kinetic scheme for inactivation is described by Eq. 11.2:

$$\mathrm{E+I} \overset{K_I}{\rightleftharpoons} \mathrm{E^{*}I} \xrightarrow{k_i} \mathrm{E{-}I'} \tag{11.2}$$

$E^{*}I$ is a kinetic chimera; K_I and k_i are the constants characterizing the inactivation process: k_i is the first-order rate constant for inactivation at infinite inhibitor concentration and K_I is the counterpart of the Michaelis constant. The k_i/K_I ratio is an index of the inhibitory potency. The parameters K_I and k_i are determined by analyzing the data obtained by using the incubation method or the progress curve method. In the incubation method, the pseudo-first-order constants k_{obs} are determined from the slopes of the semilogarithmic plots of remaining enzyme activity versus time (Eq. 11.3) where $[E]/[E]_0$ is the amount of remaining activity at time t. The pseudo-first-order constants k_{obs} are related to K_I and k_i by Eq. 11.4.

$$\ln[E]/[E]_0 = k_{obs} \times t, \tag{11.3}$$

$$k_{obs} = (k_i \times [I]_0)/(K_I + [I]_0). \tag{11.4}$$

The ratio k_i/K_I is obtained as $k_{obs}/[I]$ at low inhibitor concentrations. With efficient inhibitors, parameters K_I and k_i can be obtained using the progress curve method in which the enzyme substrate competes with the inhibitor as described for example in Ref. 21.

11.2.1 Linear Versus Cyclized Suicide Substrates

An acyloxybenzyl derivative **a** (Scheme 11.1) could be a precursor of the required hydroxybenzyl compound. Enzymatic cleavage of the ester function could generate the acyl-enzyme and the intermediate **b**, and then rapidly the corresponding QM. Chemical activation of the benzylic function will result from the change of the poor electron-donating ester substituent (σ_p^+ $OCOCH_3 = -0.19$[22]) to the donor group OH ($\sigma_p^+ = -0.92$) and the very strong electron releasing substituent O^- ($\sigma_p^+ = -2.30$).

However, diffusion of the reactive QM out of the enzyme active site is a major concern. For instance, a 2-acyloxy-5-nitrobenzylchloride does not modify any nucleophilic residue located within the enzyme active site but becomes attached to a tryptophan residue proximal to the active site of chymotrypsin or papain.[23,24] The lack of inactivation could also be due to other factors: the unmasked QM being poorly electrophilic, active site residues not being nucleophilic enough, or the covalent adduct being unstable. Cyclized acyloxybenzyl molecules of type **a′** could well overcome the diffusion problem. They will retain both the electrophilic hydroxybenzyl species **b′**, and then the tethered QM, in the active site throughout the lifetime of the acyl-enzyme (Scheme 11.1). This reasoning led us to synthesize functionalized

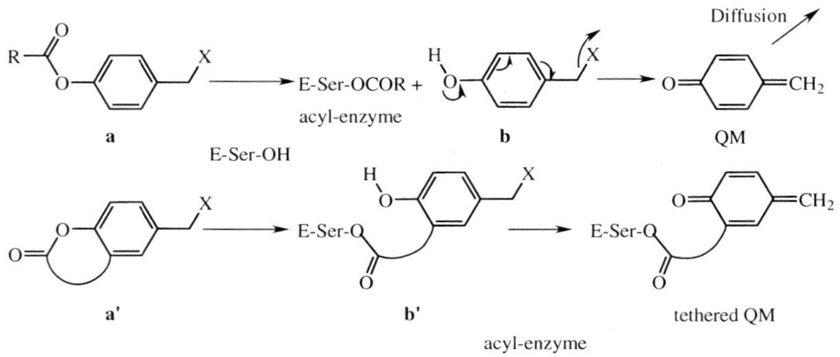

SCHEME 11.1 Comparison between linear (**a**) and cyclized (**a′**) acyloxybenzyl compounds.

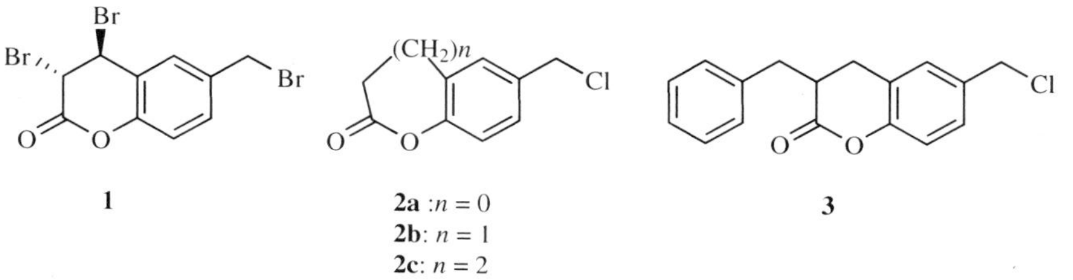

FIGURE 11.4 Structures of some functionalized lactones **1–3** that are general suicide substrates of proteases.

lactones such as 3,4-dihydrocoumarins (chroman-2-ones) **1–3** (Fig. 11.4) that have a latent electrophilic function.[25–29]

11.3 INHIBITOR SYNTHESIS

The synthesis of these inhibitors is not always straightforward because some hydroxybenzyl derivatives that have very good leaving groups (like their aminobenzyl analogues) are unstable. Hence, more stable precursors are often used, such as phenyl ethers[30–32] or silyl ethers.[33–35] Other functional groups present in the molecule must be sometimes protected (*N-tert*-butoxycarbonylation of amides for example).[36,37]

11.4 PROTEASES: NEUTRAL DIHYDROCOUMARINS

6-Bromomethyl-3,4-dibromo-3,4-dihydrocoumarin **1** (Fig. 11.4) and its chloromethylated analogue **2b** rapidly and progressively inactivate α-chymotrypsin and also the activities of a series of trypsin-like proteases.[25,38,39] A benzyl substituent characteristic of good substrates of α-chymotrypsin was introduced at the 3-position to make inhibition more selective. This substituted dihydrocoumarin **3** irreversibly inhibited α-chymotrypsin and other proteases.[27–29,38–41] These functionalized six-membered aromatic lactones, and their five- and seven-membered counterparts, 3*H*-benzofuran-2-ones **2a**[26] and 4,5-dihydro-3*H*-benzo[**b**]oxepin-2-ones **2c**,[27] were the first efficient suicide inhibitors of serine proteases. Their postulated mechanism of action is shown in Scheme 11.2.

After the nucleophilic attack by the hydroxyl function of the active serine on the carbonyl group of the lactone, the formation of the acyl-enzyme unmasks a reactive hydroxybenzyl derivative and then the corresponding QM. The cyclic structure of the inhibitor prevents the QM from rapidly diffusing out of the active center. Substitution of a second nucleophile leads to an irreversible inhibition. The second nucleophile was shown to be a histidine residue in α-chymotrypsin[28] and in urokinase.[39] Thus, the action of a functionalized dihydrocoumarin results in the cross-linking of two of the most important residues of the protease catalytic triad.

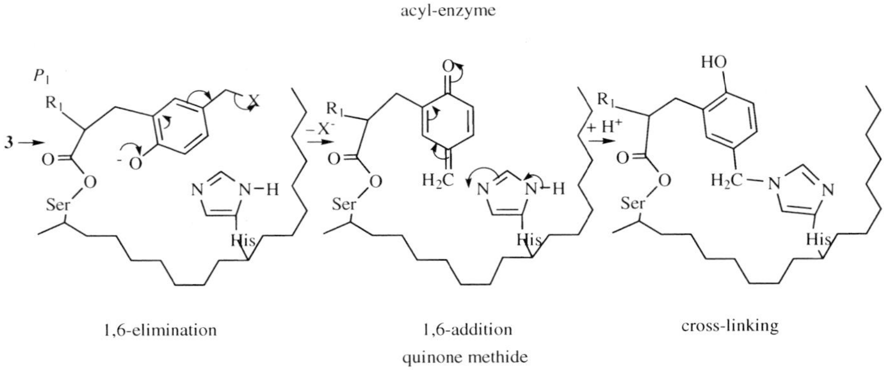

acyl-enzyme

1,6-elimination   1,6-addition   cross-linking

quinone methide

SCHEME 11.2 Postulated mechanism of inactivation of a serine protease by a functionalized dihydrocoumarin such as molecule **3** (R_1 = benzyl).[28]

However, there are two problems with these unconjugated lactones: lack of selectivity and limited stability of the inhibitor in biological buffers. Coumarin carboxylates have been developed to improve selectivity toward a given serine protease (Section 11.4.1). On the other hand, the amide bond is chemically and enzymatically much more stable than the ester one. This raised the question of whether a starting functionalized lactam behaved like the previous lactones and generated *in situ* a quinonimine methide, the aza-analogue of the quinone methide (Section 11.5).

11.4.1 Proteases: Coumarincarboxylates

Coumarins are more stable than dihydrocoumarins because of the "aromatic" character of their heterocyclic ring, and their rates of alkaline hydrolysis is much slower. Therefore, a series of esters and amides of 6-(chloromethyl)-2-oxo-2*H*-1-benzopyran-3-carboxylic acid (series **4**, Fig. 11.5) was designed that were simple to synthesize and should hopefully be more efficient and selective.[42] These molecules have a strong electron-withdrawing group at the 3-position, which should increase the electrophilicity of the conjugated lactonic carbonyl group, thus facilitating the nucleophilic attack by active serine. A discrimination between different classes of serine proteases was expected by varying the nature of R′.

Coumarincarboxylate derivatives are versatile, efficient, low molecular weight, nonpeptidic protease inhibitors. Both esters and amides behave as time-dependent inhibitors of α-chymotrypsin but the esters are clearly more efficient than the corresponding amides. The criteria for a suicide mechanism are met. The presence of a latent alkylating function at the 6-position (chloromethyl group) is required to produce to inactivation by a suicide mechanism (Scheme 11.3, pathway a). Aryl esters, in particular the *meta*-substituted phenyl esters are the best inhibitors. Thus, *m*-chlorophenyl 6-(chloromethyl)-2-oxo-2*H*-1-benzopyran-3-carboxylate is one of the well-known inactivator of α-chymotrypsin ($k_i/K_I = 760{,}000 M^{-1}s^{-1}$ at pH 7.5 and 25 °C, Table 11.1).

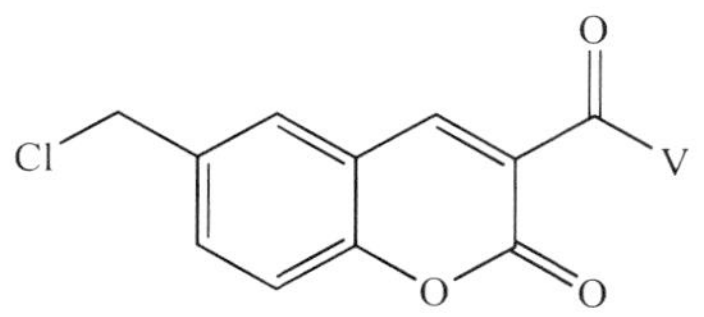

4
V = O–R′, NH–R′
R′ = alkyl, aryl

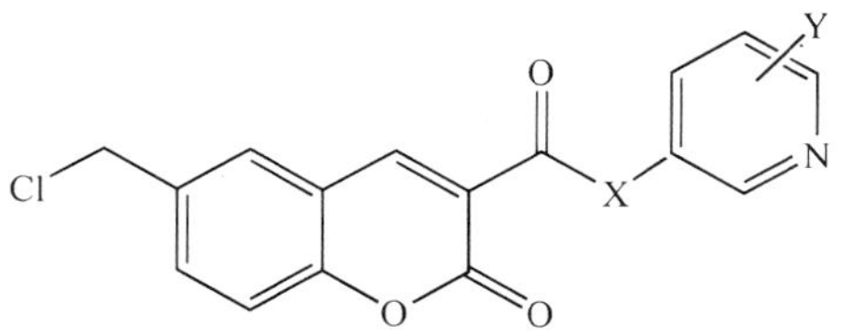

5
X = O or NH
Y = H, alkyl, halide, nitro, or methoxy

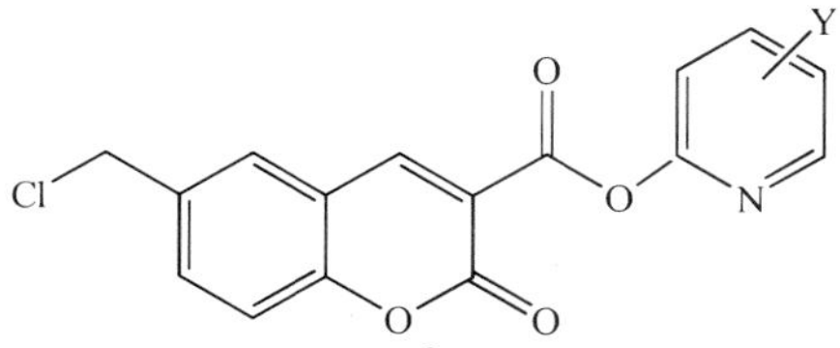

6
Y = H, alkyl or halide

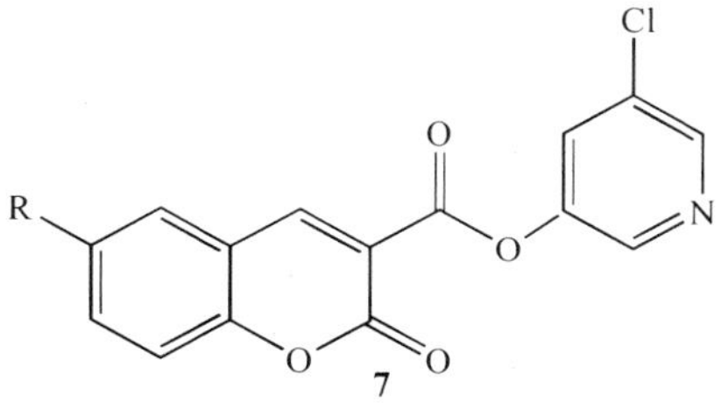

7
R = H, alkyl, halomethyl, or acyloxymethyl

FIGURE 11.5 General structure of coumarincarboxamides and alkyl, aryl and heteroaryl coumarincarboxylates **4**–**7**.

Irreversible inhibition is probably due to the alkylation of a histidine residue.[43] Chymotrypsin is selectively inactivated with no or poor inhibition of human leukocyte elastase (HLE) with a major difference: the inactivation of HLE is transient.[42,43] The calculated intrinsic reactivity of the coumarin derivatives, using a model of a nucleophilic reaction between the ligand and the methanol–water pair, indicates that the inhibitor potency cannot be explained solely by differences in the reactivity of the lactonic carbonyl group toward the nucleophilic attack.[43] Studies on pyridyl esters of 6-(chloromethyl)-2-oxo-2*H*-1-benzopyran-3-carboxylic acid (**5** and **6**, Fig. 11.5) and related structures having various substituents at the 6-position (**7**, Fig. 11.5) revealed that compounds **5** and **6** are powerful inhibitors of human leukocyte elastase and α-chymotrypsin; thrombin is inhibited in some cases whereas trypsin is not inhibited.[21]

The 6-chloromethyl substituent (series **5** and **6**) is required for the inactivation of α-chymotrypsin. Nevertheless, there is only a transient inactivation of HLE and thrombin through the formation of a stable acyl-enzyme in spite of the presence of this group as demonstrated by the spontaneous or hydroxylamine-accelerated reactivation of the treated enzymes (Scheme 11.3, pathway b).[21] HLE is specifically inhibited when such an alkylating function is absent (series **7**), always through the formation of a transient acyl-enzyme (Table 11.2).

Finally, coumarin derivatives may act as general inhibitors of serine proteases or as specific inhibitors of human leukocyte elastase, depending on the nature of the substituents, through two distinct mechanisms, suicide substrates (α-chymotrypsin)

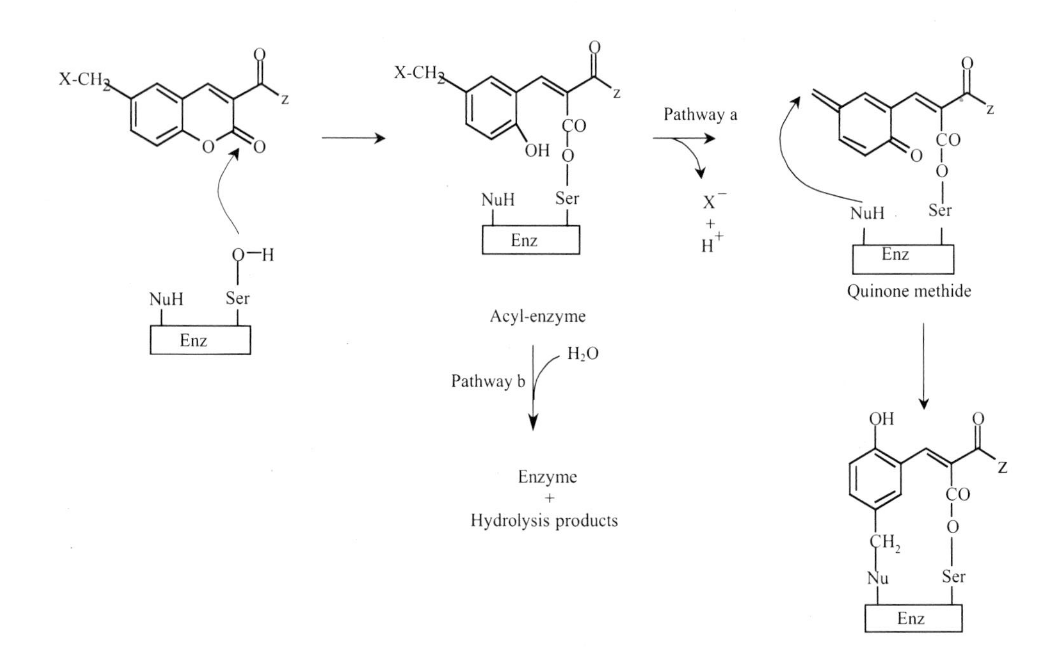

SCHEME 11.3 Postulated mechanisms for the inhibition of serine proteases by coumarin derivatives. NuH: nucleophile. Pathway a: suicide-type inactivation (suicide substrate). Pathway b: transient inactivation by formation of a stable acyl-enzyme (alternate substrate-inhibitor).

TABLE 11.1 Inhibition of α-Chymotrypsin and Human Leukocyte Elastase by Phenolic Esters of 6-(Chloromethyl)-2-oxo-2*H*-1-Benzopyran-3-Carboxylic Acid[42]

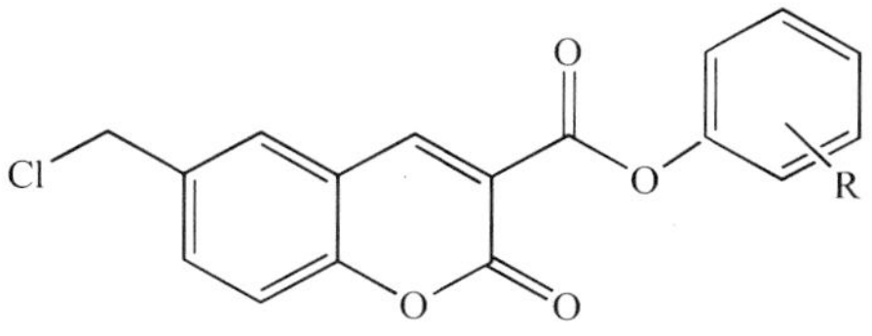

R	α-Chymotrypsin k_i/K_I ($M^{-1}s^{-1}$)	Human Leukocyte Elastase $k_{obs}/[I]$ ($M^{-1}s^{-1}$)
o-CH_3	11,4000	ni
m-CH_3	71,750	ni
p-CH_3	685	ni
o-Cl	48,100	23
m-Cl	762,700	630
o-I	80,000	85
m-I	72,120	85
p-I	1775	ni
m-NO_2	11,300	500
m-CH_3-p-Cl	7550	90

ni, No inhibition.

TABLE 11.2 Inhibition of Human Leukocyte Elastase, α-Chymotrypsin and Thrombin by 5′-Chloropyrid-3′-yl Derivatives at pH 8.0 and 25°C[21]

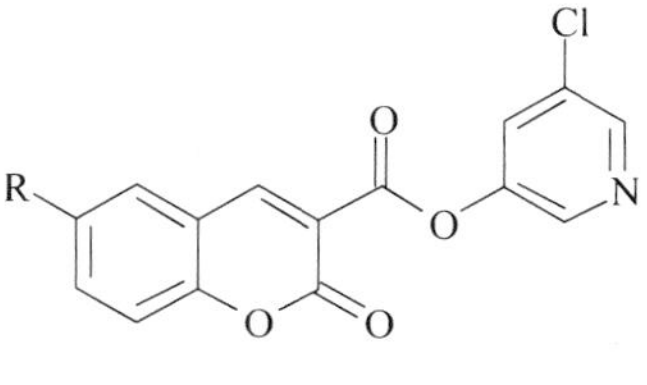

	k_i/K_I ($M^{-1}s^{-1}$)		
Substituent at the 6-Position (R)	Human Leukocyte Elastase	α-Chymotrypsin	Thrombin
CH_2Cl	64,600	30,970	≅200
H	32,850	ni	ni
CH_3	32,500	ni	ni
CH_2OCOCH_3	107,000	ni	ni
$CH_2OCOCH\ C_2H_5$	62,000	ni	ni
$CH_2OCOCH\ (CH_3)_2$	45,000	ni	ni
$CH_2OCO(CH_2)_2CH_3$	72700	ni	ni
$CH_2OCOC\ (CH_3)_3$	95,500	ni	ni
CH_2NH_2,HCl	9100	ni	ni
Br	58,000	ni	ni

ni, No inhibition.

or alternate substrate inhibitors (human leukocyte elastase).[21] Chloromethyl derivatives of coumarin esters were also found to inhibit thrombin more selectively than the coagulation factor Xa or trypsin. Treatment with hydrazine induced only a partial reactivation of the enzyme.[44] The selectivity toward thrombin is improved by introducing of a 2-(*N*-ethyl-2′-oxoacetamide)-5′-chlorophenyl ester side chain.[45]

Several aryl esters of 6-chloromethyl-2-oxo-2*H*-1-benzopyran-3-carboxylic acid act as human Lon protease inhibitors (alternate substrate inhibitors)[46] without having any effect on the 20S proteasome. Proteasomes are the major agents of protein turnover and the breakdown of oxidized proteins in the cytosol and nucleus of eukaryotic cells,[47] whereas Lon protease seems to play a major role in the elimination of oxidatively modified proteins in the mitochondrial matrix. The coumarin derivatives are potentially useful tools for investigating the various biological roles of Lon protease without interfering with the proteasome inhibition.

11.5 AZA-QUINONE METHIDES/QUINONIMINE METHIDES

Like the hydroxyl group, the amino substituent ($\sigma_p^+ = -1.30$)[22] is strongly electron donating and good benzylic leaving groups are eliminated when this substituent is present in the *ortho* or *para* position of the benzene ring. Elimination leads to the unstable aza-analogue of QM, the QIM.[48] Model reactions of *p*-dimethylaminobenzylbromide with a series of nucleophiles in water at room temperature showed that this type of molecules is very reactive.[49] The enzymatic hydrolysis of an amidobenzyl precursor could place the QIM within the enzyme active site in a manner analogous to that described above for the acyloxybenzyl derivatives **a′** (Scheme 11.1).

11.5.1 α-Chymotrypsin: Inactivity of Five- and Six-Membered Lactams

An aza-analogue of the functionalized dihydrocoumarin **2b**, the 3-benzyl-3,4-dihydro-1*H*-quinolin-2-one **8b**, and its five-membered analogue, a substituted indolin-2-one **8a**, were synthesized (Fig. 11.6).

These molecules do not inhibit α-chymotrypsin, probably because the protease is unable to open lactam rings in which the amide function has a *cis* configuration.[50]

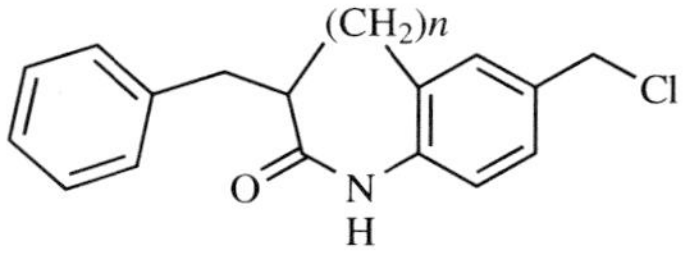

$(CH_2)n$ Cl O N H

8a: $n = 0$
8b: $n = 1$

FIGURE 11.6 Structures of the functionalized five- and six-membered lactams **8**.

11.5.2 Proteases: Functionalized Cyclopeptides

The resistance of small cyclic peptides to hydrolysis by proteases is attributed to the *cis* configuration of their amide bond. We considered synthesizing functionalized cyclopeptides large enough to have the scissible peptide bond $P_1-P'_1$ (Fig. 11.3) in a *trans* configuration.[50] It should also be possible to introduce $P_n \cdots P_2$ sequence having a good affinity for the $S_n \cdots S_2$ secondary subsites of the target enzyme into a cyclopeptide, and consequently discriminate among proteases of different specificities (see above).

Any suicide substrate must be a substrate of the enzyme, even if it is not a very good one. The selective hydrolysis of the $P_1-P'_1$ bond was checked by preparing unfunctionalyzed cyclopeptides likely to be substrates of serine proteases. Two series of compounds were synthesized: c(Gly$_n$-P$_1$-*o*Aba) and c(Gly$_n$-P$_1$-*m*Aba) in which P$_1$ was Phe or Arg and *o*Aba and *m*Aba were ring-methylated *ortho*- and *meta*-aminobenzoic acid residues. The enzymic hydrolysis of these cyclopeptides catalyzed by α-chymotrypsin (P$_1$ = Phe) and bovine trypsin and human urokinase (P$_1$ = Arg) occurred specifically at the P$_1$-Aba bond. The enzymic efficiencies k_{cat}/K_m were generally higher for the *meta*-compounds than for their *ortho*-counterparts.[51,52] An NMR conformational analysis suggested a greater conformational mobility of the more reactive *meta*-analogues.[53] We therefore concentrated on the functionalized 3-aminobenzoic derivatives *m*Aba(5-CH$_2$X) by preparing a series of cyclopeptides of type c[Phe-*m*Aba(5-CH$_2$X)-Gly$_n$] (Fig. 11.7). Molecules possessing a poor benzylic leaving group X such as OC_6H_5 or $OCOCH_3$ behaved only as substrates of α-chymotrypsin whereas the cyclopeptides with X = Br or Cl irreversibly inhibited the target protease.[30] Another molecule, a sulfonium salt having a latent sulfide leaving group c[Arg-*m*Aba(5-CH$_2$S$^+$MePh)-Gly$_4$, CF$_3$CO$_2^-$] was synthesized and studied as a potential mechanism-based inhibitor for plasminogen activators and plasmin (Table 11.3). This cyclopeptide preferentially inactivates urokinase-type plasminogen activator (u-PA; $k_{inact} = 0.021\,s^{-1}$, $K_I = 9\,\mu M$ at pH 7.5 and 25 °C). It inactivates plasmin and tissue plasminogen activator (*t*-PA) 40-fold and 2330-fold less efficiently than u-PA (Table 11.3). The criteria for a suicide inhibition are fulfilled: the reaction is a first-order and irreversible process, show saturation kinetics, and the enzyme is protected by substrate.[31,54] Thrombin is also very weakly inhibited.

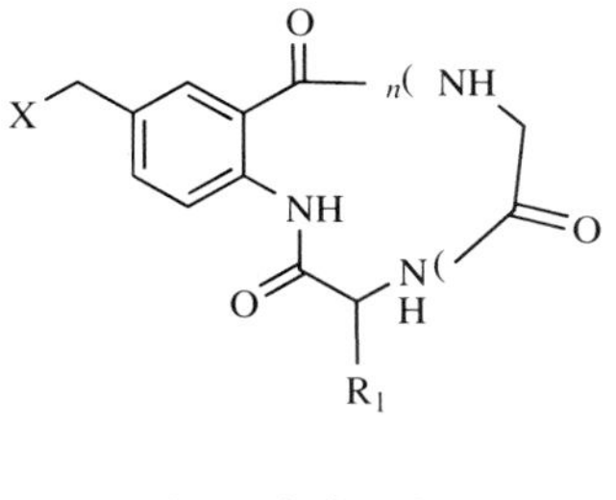

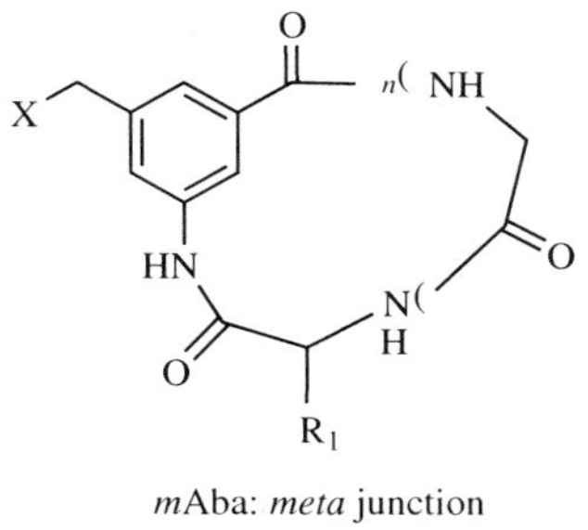

*o*Aba: *ortho* junction | *m*Aba: *meta* junction

FIGURE 11.7 Structure of cyclopeptides c(Gly$_n$-P$_1$-*o*Aba-5-CH$_2$X) (**9**) and c(Gly$_n$-P$_1$-*m*Aba-5-CH$_2$X) (**10**).

TABLE 11.3 Kinetic Parameters for the Inactivation of Proteases by Cyclopeptides c(Gly$_n$-P$_1$-*o*Aba-5-CH$_2$X) 9 and c(Gly$_n$-P$_1$-*m*Aba-5-CH$_2$X) 10 (Fig. 11.7) at pH 7.5 and 25°C

			Bovine Trypsin			Urokinase			Plasmin	*t*-PA	Thrombin
Junction	P_1	X	k_i (s^{-1})	$10^3 K_I$ (M)	k_i/K_I ($M^{-1} s^{-1}$)	k_i (s^{-1})	$10^3 K_I$ (M)	k_i/K_I ($M^{-1} s^{-1}$)	k_i/K_I ($M^{-1} s^{-1}$)	k_i/K_I ($M^{-1} s^{-1}$)	k_i/K_I ($M^{-1} s^{-1}$)
o	Lys	Br			ni			ni	ni	ni	ni
m	Lys	Br	0.02	0.08	250			5	ni	ni	ni
o	Arg	Br			≈5			14	ni	ni	ni
m	Arg	Br	0.031	0.055	560	0.017	0.092	185	≈ 2	ni	ni
m	Lys	$^+S(CH_3)_2$			ni			≈ 1	ni	ni	ni
m	Arg	$^+S(CH_3)_2$			≈1	0.014	0.041	341	ni	ni	ni
o	Lys	$^+S(CH_3)C_6H_5$			ni			≈ 25	ni	ni	ni
m	Lys	$^+S(CH_3)C_6H_5$	0.0075	0.0087	862	0.077	0.5	154	75	ni	ni
o	Arg	$^+S(CH_3)C_6H_5$			50			14	ni	ni	ni
m	Arg	$^+S(CH_3)C_6H_5$	0.011	0.012	916	0.021	0.09	2330	40	≈1	≈12

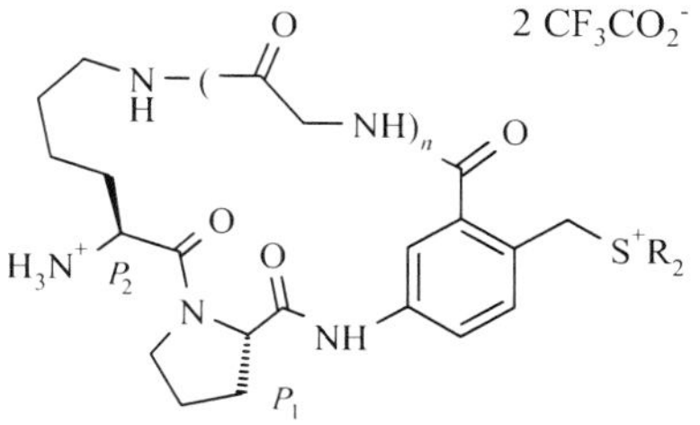

FIGURE 11.8 Structure of the cyclopeptides **11** c[($^{\alpha}H_2N^+$)-Lys-Pro-*m*Aba(5-$CH_2S^+R_2$)-Gly$_n$], 2 $CF_3CO_2^-$.

By introducing the suitable peptide sequence D-Phe-Pro-Arg, an efficient and selective thrombin substrate was obtained: c(Gly$_2$-D-Phe-Pro-Arg-*m*Aba) with $k_{cat}/K_M = 1.6 \times 10^5 M^{-1}s^{-1}$. Its functionalization led to an efficient suicide substrate of this enzyme: c(Gly$_2$-D-Phe-Pro-Arg-*m*Aba-5-$CH_2S^+Me_2$, $CF_3CO_2^-$) with $k_{inact}/K_i = 3500 M^{-1}s^{-1}$ at pH 7.5 and 25 °C.[55]

Another interesting target for this type of inhibitors is the dipeptidyl peptidase IV (DPP IV). This exodipeptidase, which can cleave peptides behind a proline residue is important in type 2 diabetes as it truncates the glucagon-like peptide 1. Taking into account the P_2-P_1(Pro)-P'_1 cleavage and the requirement for a free terminal amine, the synthesis of a suicide inhibitor was planned. It looked as if the the ε-amino group of a P_2 lysine residue could be cyclized because of the relative little importance of the nature of the P_2 residue on the rate of enzymatic hydrolysis of known synthetic substrates. Therefore, a new series of cyclopeptides **11** was synthesized (Fig. 11.8).

Molecules **11** rapidly and irreversibly inhibit the DPP IV activity of the CD26 antigen, with IC_{50} values in the nanomolar range. Cycle enlargement ($n = 4$ instead of 2) improves inhibitory activity, whereas increasing the alkyl chain length on the sulfur atom R has no apparent effect. Other aminopeptidases are not inhibited. Molecules **11** inhibit DPP IV-β, an isoform of DPP IV, much more poorly.[32,56]

As for the above cyclopeptides **10**, the first step in inactivation process is a selective ring opening. Formation of the acyl-enzyme is then followed by a fast 1,6-elimination and the unmasking of a reactive tethered QIM (Scheme 11.4).

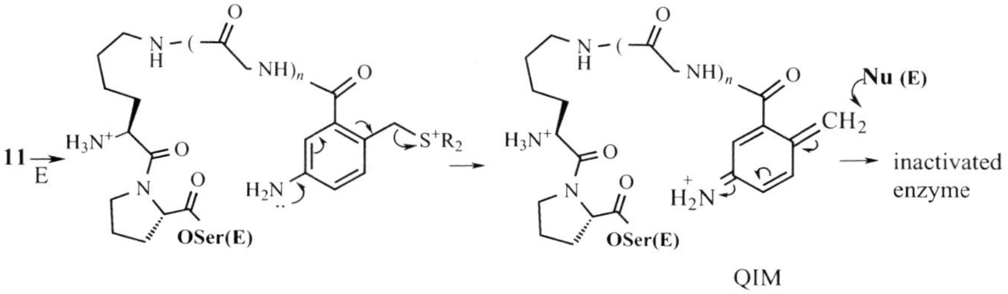

SCHEME 11.4 Postulated mechanism for inactivation of DPP IV by a functionalized cyclopeptide **11**.[32]

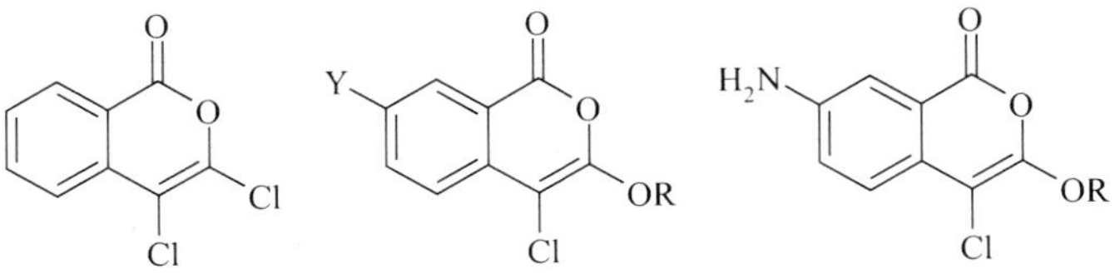

12 (DCI) **13** **14**

FIGURE 11.9 Structure of several isocoumarins **12–14**.

11.5.3 Proteases: Isocoumarins

Many isocoumarins, such as the 4-chloro-3-alkoxy-isochromen-1-ones (Fig. 11.9), are heterocyclic inhibitors of serine proteases.[57] Inhibition occurs by opening of the isocoumarin ring by the active site serine residue to form an acyl-enzyme that is quite often stable. For example, 3,4-dichloroisocoumarin (or DCI, **12**) is a general serine protease inhibitor that also acts on the proteasome (threonine protease).[58] The 3-alkoxy-4-chloro-7-substituted derivatives **13** are more selective acting on human leukocyte elastase, porcine pancreatic elastase (PPE), proteinase 3, cathepsin G, chymotrypsin, tryptase, granzyme, and the trypsin-like enzymes that are involved in blood coagulation and the complement cascade. The 3-alkoxy group, which interacts with the S_1 subsite of the enzyme, provides selectivity for certain proteases.[57]

Some of these heterocycles can also be suicide inhibitors if a new reactive structure is unmasked during acylation and this reactive species can further react with an active site nucleophile. For example, the presence of a 7-amino substituent (compounds **14**, Fig. 11.9) makes the formation of a QIM possible.

The acyl-enzyme can eliminate the 4-chlorine atom to generate this reactive intermediate that can then react with a nearby nucleophile such as His57 to give an alkylated acyl-enzyme derivative in which the inhibitor moiety is bound to the enzyme by two covalent bonds (Scheme 11.5). Inhibition is irreversible.[59] The mechanism has been confirmed by X-ray structural analysis of protease–isocoumarin complexes. There is a cross-link between the inhibitor and the Ser195 and His57 residues of PPE.[60] Human leukocyte elastase is also very efficiently inactivated.[61]

Some new 3-alkoxy-7-amino-4-chloroisocoumarins have been found recently to be poor inhibitors of α-chymotrypsin, trypsin, caspase 3, and HIV protease and to

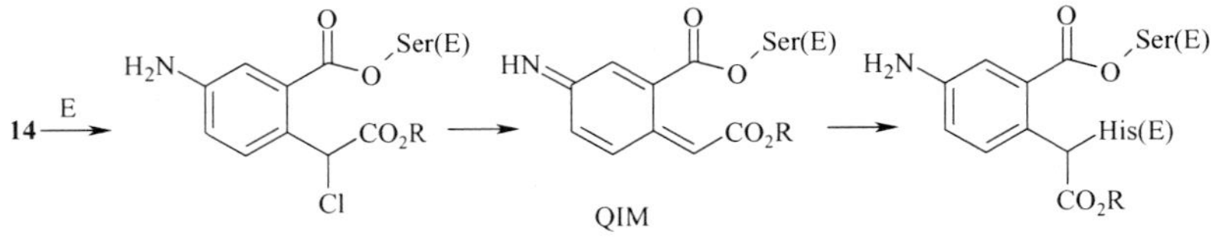

SCHEME 11.5 Mechanism of the inactivation of a protease with a 3-alkoxy-7-amino-4-chloro-isochromen-1-one **14**.[57]

prevent the γ-secretase-mediated production of the Aβ 40/42 amyloid peptides known to be involved in Alzheimer's disease.[62]

11.6 β-LACTAMASES: FUNCTIONALIZED ARYL PHENACETURATES

The antibiotic activity of certain β-lactams depends largely on their interaction with two different groups of bacterial enzymes. β-Lactams, like the penicillins and cephalosporins, inhibit the DD-peptidases/transpeptidases that are responsible for the final step of bacterial cell wall biosynthesis.[63] Unfortunately, they are themselves destroyed by the β-lactamases,[64] which thereby provide much of the resistance to these antibiotics. Class A, C, and D β-lactamases and DD-peptidases all have a conserved serine residue in the active site whose hydroxyl group is the primary nucleophile that attacks the substrate carbonyl. Catalysis in both cases involves a double-displacement reaction with the transient formation of an acyl-enzyme intermediate. The major distinction between β-lactamases and their evolutionary parents the DD-peptidase residues is the lifetime of the acyl-enzyme: it is short in β-lactamases and long in the DD-peptidases.[65–67]

Aryl phenylacetylglycinates (aryl phenaceturates) **15** (Fig. 11.10) with a carboxylate substituent *ortho*, *meta*, or *para* to the oxygen atom of the phenoxide leaving group are β-lactamase substrates.[68,69] The structure of these molecules lends itself to the preparation of derivatives **16** (Fig. 11.10) possessing a latent *ortho*- or *para*-quinone methide electrophile. The electrostatic attraction between the carboxylate group of the phenol leaving group and the positive potential of the β-lactamase or DD-peptidase active site may ensure that the product phenol is retained in the active site long enough for the electrophilic QM to be generated and to react within the active site.[37]

The functionalized phenaceturates **16** (Fig. 11.10) are substrates of class A and C β-lactamases, especially the class C enzymes, as observed with the parent unfunctionalized phenaceturates **15**. They are also modest inhibitors of these enzymes and the serine DD-peptidase of *Streptomyces* R61. The inhibition of class C β-lactamases is turnover dependent, as expected for a mechanism-based inhibitor. Inhibition is not very dependent on the nature of the leaving group, suggesting that the QM is generated in solution after the product phenol has been released from the active site. It therefore

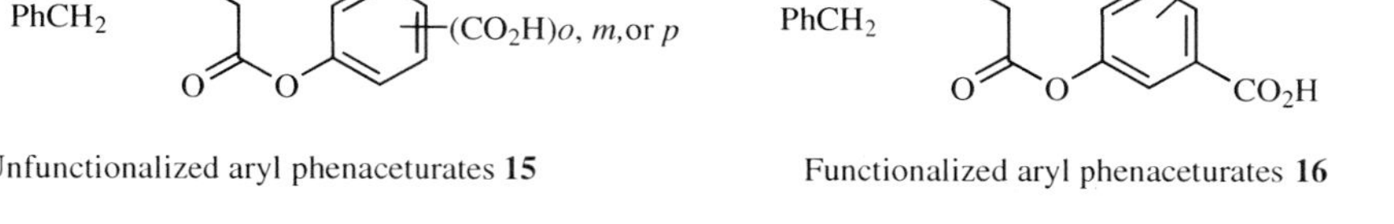

PhCH$_2$ ... (CO_2H)*o*, *m*, or *p*

Unfunctionalized aryl phenaceturates **15**

PhCH$_2$ ... (CH_2X)*o* or *p* ... CO_2H

Functionalized aryl phenaceturates **16**

X = Br, Cl, OAc, $OCOC_6H_3(CF_3)_2$, or $S^+(Et)Me,BF4^-$

FIGURE 11.10 Structures of the unfunctionalized (**15**) and functionalized (**16**) phenaceturates.

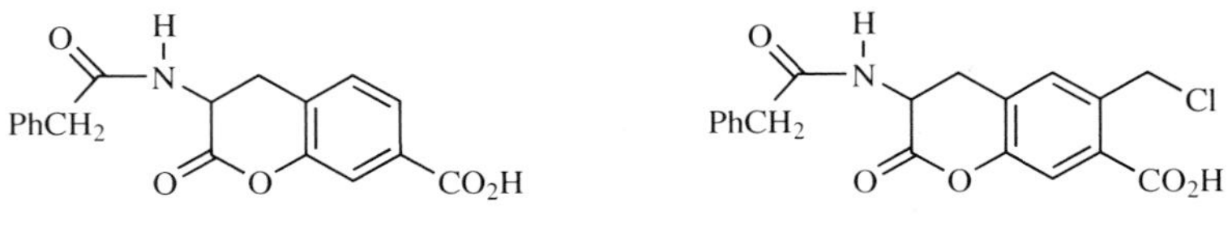

Unfunctionalized acidic dihydrocoumarin **17**   Functionalized acidic dihydrocoumarin **18**

FIGURE 11.11 Structure of the unfunctionalized (**17**) and functionalized (**18**) acidic dihydrocoumarins.

seems likely that some intramolecular tethering is needed to produce an effective β-lactamase or DD-peptidase suicide substrate.

11.6.1 β-Lactamases: Acidic Dihydrocoumarins

A new series of unfunctionalized (**17**) and functionalized (**18**) 3,4-dihydrocoumarins (Fig. 11.11) was synthesized. These new benzopyranones differ from the previous neutral dihydrocoumarins, which are suicide substrates of proteinases (Section 11.4, in that they have a carboxy substituent on their aromatic ring and a side chain phenylacetamido group at the 3-position of the heterocycle instead of a benzyl one in compound **3**.

All these 3,4-dihydro-2*H*-1-benzopyran-2-ones **17** and **18** are substrates of class A and class C β-lactamases. They are thus the first δ-lactones that are hydrolyzed by β-lactamases. The k_{cat} values for these substrates are generally smaller than those of the analogous acyclic phenaceturates suggesting that the tethered leaving group obstructs the attack of water on the acyl-enzyme. Despite the apparent advantage of the long-lived acyl-enzymes, the irreversible inhibition by the functionalized compounds is no better than that of acyclic molecules **16**. Thus, even the tethered QM cannot efficiently trap a second nucleophile at the β-lactamase active site, at least as placed as dictated by the structure of compounds **18**.[70]

11.6.2 β-Lactamases: N-Aryl Azetidinones

N-(*o*, *m*, or *p*-carboxy)phenyl azetidin-2-ones **19** (Fig. 11.12), which have two characteristic features of a β-lactamase substrate, a β-lactam ring and a carboxy

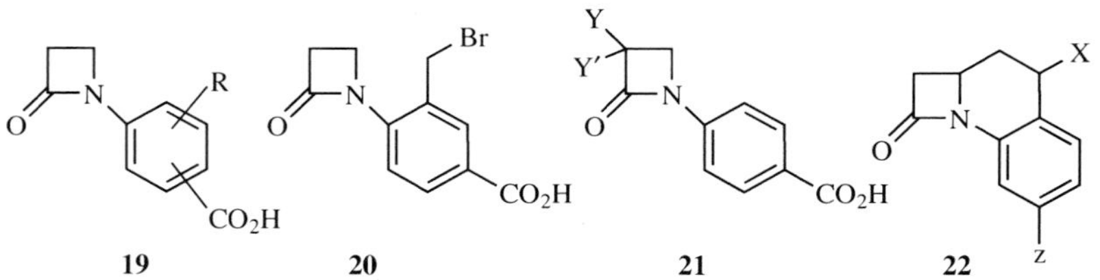

19   **20**   **21**   **22**

Y,Y′ = H, F, or Br   X = Cl or F
Z = H, CO_2R, or CO_2H

FIGURE 11.12 Structures of the azetidinones **19–22**.

substituent, are competitive inhibitors of a series of β-lactamases. However, they are not detectable substrates of the enzymes and their bromomethylated derivatives such as compound **20** are not mechanism-based inhibitors. Ring opening is a prerequisite for a suicide substrate of that type.[71]

To increase the reactivity of the β-lactam ring toward enzymatic ring opening, two ways were considered: halogen substitution in position α to the carbonyl to give compounds **21** (Fig. 11.12), and cyclization to give strained tricyclic β-lactams **22**. The presence of one or two halogen substituents should increase the polarization of the carbonyl group. Fluorine that introduces a little steric hindrance is particularly attractive.

11.6.3 β-Lactamases: Acidic Dihalogenoazetidinones and Benzocarbacephems

The β-lactam nucleus of these halogenoazetidin-2-ones **21** (Fig. 11.12) is potentially highly reactive, as indicated by the high ν C = O IR frequencies (1765–1790 cm^{-1}), an indicator of β-lactam ring reactivity. However, these azetidinones are only competitive inhibitors and are not substrates of the enzymes.[72]

The benzocarbacephems **22** (Fig. 11.12), with (X = Cl or F) or without (X = H) a halogen leaving group, were synthesized by a copper-mediated intramolecular aromatic substitution as the main step.[73,74] These tricyclic β-lactams are also competitive inhibitors of β-lactamases and not substrates.

Thus, neither halogen substitution nor ring strain induces enzymatic hydrolysis. Molecules **21** and **22** may be bound in such a way that the β-lactam carbonyl lies too far away from the catalytic serine hydroxyl group.[72]

11.7 ELASTASES: NEUTRAL DIHALOGENOAZETIDINONES

Human leukocyte elastase is a protease that degrades elastin and other connective tissue components. It is implicated in the pathogenesis of pulmonary emphysema and other inflammatory diseases such as rheumatoid arthritis and cystic fibrosis. Porcine pancreatic elastase has often been used as a model for HLE. Both enzymes have a small primary binding site S_1.

The neutral azetidin-2-one **23** (Fig. 11.13), which has neither the latent leaving group nor the carboxy substituent of the previous acidic azetidinone **21**, is a substrate of both HLE and PPE. Its functionalized analogue **24a** inactivates these proteases by an

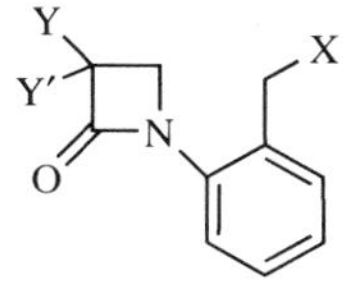

Y, Y′, X, N, O

23: Y=Y′=F, X=H
24a: Y=Y′=F, X=Cl

FIGURE 11.13 Structures of the neutral unfunctionalized (**23**) and functionalized (**24a**) dihalogenoazetidinones.

TABLE 11.4 Kinetic Parameters for the Inhibition of Elastases by Dihalogenoazetidinones 24 and 25[34]

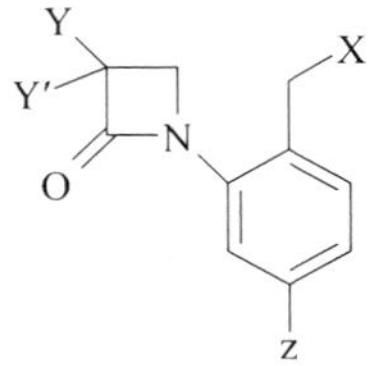

24a : Y=Y′=F, X=Cl, Z=H
24b : Y=Y′=F, X=Br, Z=H
24c : Y=Y′=F, X=Br, Z= CO_2tBu
24d : Y=Y′=F, X=F, Z=H
24e : Y=Y′=Cl, X=Cl, Z=H
24f: Y=Y′=Br, X=Cl, Z=H

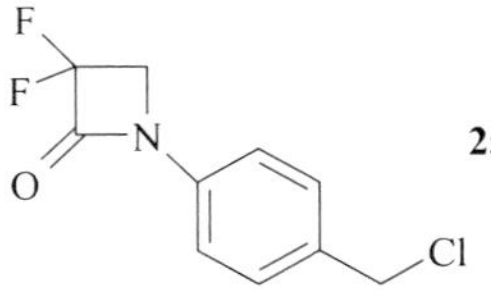

25

	HLE				PPE			
Compound	$10^3\ k_i$ (s^{-1})	K_I (μM)	k_i/K_I ($M^{-1}\ s^{-1}$)	r	$10^3 k_i$ (s^{-1})	K_I (μM)	k_i/K_I ($M^{-1}\ s^{-1}$)	r
24a	35	120	292	18	80	270	296	11
25			26				117	
24b	57	107	533	16	69	94	740	10
24c			11				58	
24d			29				<1	
24e	52	127	409	9			62	19
24f	53	162	327	4			50	9

enzyme-mediated process.[75] This molecule efficiently prevents the breakdown of lung elastin fibers induced by elastase[76] and intradermal microvascular hemorrhage.[77,78]

Several modifications of the structure of the azetidinone were made to improve the inhibitory properties.

The isomeric β-lactam **25** with the functionalized methylene group in the *para* position instead of the *ortho* position is a poorer inhibitor than the previous molecule **24a** (Table 11.4).[34] Compounds belonging to the *ortho* series were therefore preferentially synthesized.

The effects of varying the nature of halogen substituents at C3 and that of the potential leaving group at the benzylic position were then examined.

The 3,3-*gem*-dihalogeno group that favors the opening of the β-lactam ring by lowering the pK_a of the conjugated acid of the aniline leaving group also activates the carbonyl group for the nucleophilic attack of the active serine. The conformational analyses and energy mechanics of the interaction of HLE with compound **24a** agreed with the possible attack of the β-lactam carbonyl atom on the α-face of the β-lactam ring by the hydroxyl of enzyme Ser195, and with alkylation by the benzylic carbon of the imidazole nitrogen Nε2 of His57 leading to the inactivated enzyme (*bis*-adduct form).[79]

A series of substituents *meta* to the nitrogen atom and *para* to the functionalized methylene were then introduced. The hydrophobic electron-attracting ester substituents (**26**: Z = CO_2tBu, $CO_2C_6H_{13}$, $CO_2(CH_2)_3$ or CO_2tBu; Fig. 11.14) decreased the inhibitory efficiency. The acid ester **26** (Z = $CO_2(CH_2)_3CO_2H$) was prepared in the hope that the anionic charge of the corresponding carboxylate ion would interact with

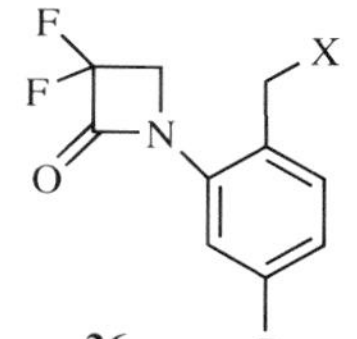

26

Z = CO_2tBu, $CO_2C_6H_{13}$,$CO_2(CH_2)_3CO_2tBu$, $CO_2(CH_2)_3CO_2H$, or OCH_3

FIGURE 11.14 Structure of ring-substituted azetidinones **26**.

Arg217 located in the S_4–S_5 subsites but this did not occur. The electron-donating methoxy substituent was introduced to enhance the departure of the benzylic leaving group X. But there was no enhancement.[35]

For a given set of substituents at position C3, the efficiency of the inhibition depends on the nature of the benzylic leaving group X (Table 11.4). No significant inactivation is found with a fluorine atom at this position. The brominated azetidinone (**24b**) is a better inactivator than its chlorinated counterpart (**24a**).

The *gem*-dichloro and *gem*-dibromo analogues (**24e** and **24f**) are more potent inhibitors of HLE compared to PPE, whereas the *gem*-difluoro β-lactam **24a** is equally efficient against the two enzymes (Table 11.4).[34]

With a 3,3-heterodihalogeno substitution of the β-lactam ring, a selective interaction of each enantiomer of the chiral azetidinone with the enzyme active site is expected. The enantiomer 3R of the 3F, 3Br derivative indeed has a more favorable kinetic parameter k_i/K_I than the enantiomer 3S.[33] The partition ratio k_{cat}/k_i ($=k_3/k_4$, Eq. 11.1) for the inactivation is also higher. Therefore, enantiomer 3R is a better suicide substrate for HLE since a lower partition ratio corresponds to a better suicide substrate.[20]

The reported 3,3-dihalogenoazetidinones **24** fulfill the criteria expected for mechanism-based inhibitors of HLE: irreversibility; time-dependent and *pseudo*-first-order inactivation; saturation kinetics; protection by a chromogenic substrate against inactivation; and no enzyme reactivation upon treatment with hydroxylamine. Furthermore, the hydrolysis of model compounds lacking the potential benzylic leaving group showed that HLE catalyzed the opening of the β-lactam ring and suggested that this would occur with molecules functionalized by a chlorine or bromine atom at that position. The inactivation is very selective: these compounds do not inhibit chymotrypsin-like (α-chymotrypsin, cathepsin G) or trypsin-like (trypsin itself, plasmin, thrombin, and urokinase) proteinases.

11.8 CONCLUSION AND FUTURE PROSPECTS

The studies described above show that a quinone methide or its aza-analogue quinonimine methide incorporated as a latent electrophilic species into a cyclic lactone or lactam precursor can modify a second nucleophilic residue within the enzyme active site after formation of the acyl-enzyme. Very efficient suicide substrates of proteases and an aminodipeptidase have been obtained (neutral 3,4-dihydrocoumarins, coumarincarboxylates) and some of these suicide substrates are very selective (functionalized cyclopeptides, neutral dihalogenoazetidinones, and coumarincarboxylates).

Coumarin derivatives appear to be a core structure suitable for designing selective inhibitors of serine proteases acting by different mechanisms (suicide or alternate substrate inhibitors). This mechanistic versatility may be very useful pharmacologically depending on the nature of the target enzyme: favoring a massive action in the treatment of acute diseases, or continuous, less aggressive action in chronic diseases. The great advantage of suicide substrates over affinity labels resides in the presence of a latent reactive group. The molecule administrated has no "warhead" and hence fewer side effects. The development of suicide substrates and alternate substrate inhibitors is an inspiring strategy for newly identified targets such as viral enzymes and mammalian proteasomes (threonine proteases).

ACKNOWLEDGMENTS

We are most grateful for the intellectual and experimental contributions of all our coworkers; their names appear in the publications from our laboratories. The English text was edited by Dr. Owen Parkes.

REFERENCES

1. Wolfenden, R. Transition state analogs and enzyme catalysis. *Ann. Rev. Biophys. Bioeng.* 1976, 5, 271–306.
2. Walsh, C. T. Suicide substrates: mechanism-based enzyme inactivators: recent developments. *Ann. Rev. Biochem.* 1984, 53, 493–535.
3. Silverman, R. B.; *Mechanism-Based Enzyme Inactivation: Chemistry and Enzymology;* CRC Press: Boca Raton, FL, 1988; Vol. 1, pp 3–22.
4. Krantz, A. A classification of enzyme inhibitors. *Bioorg. Med. Chem. Lett.* 1992, 2, 1327–1334.
5. Pratt, R. F. On the definition of mechanism-based enzyme inhibitors. *Bioorg. Med. Chem. Lett.* 1992, 2, 1323–1326.
6. Freccero, M. Quinones methides as alkylating and cross-linking agents. *Mini-Reviews Org. Chem.* 2004, 1, 403–415.
7. Van De Water, R. W.; Pettus, T. R. R. *o*-Quinone methides: intermediates underdeveloped and underutilized in organic chemistry. *Tetrahedron* 2002, 58, 5367–5405.
8. Wakselman, M. 1,4 and 1,6-Eliminations from hydroxy- and amino-substituted benzyl systems: chemical and biochemical applications. *Nouv J. Chim.* 1983, 7, 439–447.
9. Chiang, Y.; Kresge, A. J.; Zhu, Y. Flash photolysis generation and study of *p*-quinone methide in aqueous solution. An estimate of rate and equilibrium constants for heterolysis of the carbon–bromine bond in *p*-hydroxybenzyl bromide. *J. Amer. Chem. Soc.* 2002, 124, 6349–6356.

10. Weinert, E. E.; Dondi, R.; Colleredo-Melz, S.; Frankenfield, K. N.; Mitchell, C. H.; Feccero, M.; Rokita, S. E. Substituents on quinone methides strongly modulate formation and stability of their nucleophilic adducts. *J. Amer. Chem. Soc.* 2006, 128, 11940–11947, and references therein.

11. Horton, H. R.; Koshland, D. E. Jr. A highly reactive coloured reagent with selectivity for the tryptophan residue in proteins: 2-hydroxy-5-nitrobenzyl bromide. *J. Amer. Chem. Soc.* 1965, 87, 1126–1132.

12. Horton, H. R.; Koshland, D. E. Jr. Reaction with reactive alkyl halides. *Methods Enzymol.* 1967, 11, 556–565.

13. Lundblad, A. L.; Noyes, C. M. *Chemical Reagents for Protein Modification;* CRC Press: Boca Raton, FL, 1984; Vol. 2, pp 59–65.

14. Rawlings, N. D.; Barrett, A. J.; Introduction: serine peptidases and their clans. In *Handbook of Proteolytic Enzymes*, 2nd edition; Barrett, A. J.; Rawlings, N. D.; Woessner, J. F., Eds.; Elsevier Academic Press: Amsterdam, Boston, Heidelberg, London, New York, Oxford, Paris, San Diego, San Francisco, Singapore, Sydney, Tokyo 2004, pp 1417–1439.

15. Polgàr, L.; Catalytic mechanisms of serine and threonine peptidases. In *Handbook of Proteolytic Enzymes*, 2nd edition; Barrett, A. J.; Rawlings, N. D.; Woessner, J. F., Eds.; Elsevier Academic Press: Amsterdam, Boston, Heidelberg, London, New York, Oxford, Paris, San Diego, San Francisco, Singapore, Sydney, Tokyo 2004, pp 1441–1448.

16. Barrett, A. J.; *Proteinases Inhibitors;* Barrett, A. J.; Salvesen, G., Eds.; Elsevier: Amsterdam, New York, Oxford, 1986; pp 3–22.

17. Schechter, I.; Berger, A. On the size of the active site in proteases. I. Papain. *Biochem. Biophys. Res. Commun.* 1967, 27, 157–162.

18. Lorand, L., *Proteolytic Enzymes. Part C. Methods in Enzymology;* Academic Press: New York, 1981; p 80.

19. Waley, S. G. Kinetic of suicide substrates. *Biochem. J.* 1980, 185, 771–773.

20. Waley, S. G. Kinetic of suicide substrates, practical procedures for determining parameters. *Biochem. J.* 1985, 227, 843–849.

21. Doucet, C.; Pochet, L.; Thierry, N.; Pirotte, B.; Delarge, J.; Reboud-Ravaux, M. 6-Substituted 2-oxo-2*H*-1-benzopyran-3-carboxylic acid as a core structure for specific inhibitors of human leukocyte elastase. *J. Med. Chem.* 1999, 42, 4161–4171.

22. Hantsch, C.; Leo, A.; Taft, W. A survey of Hammett substituent constant and resonance and field parameters. *Chem. Rev.* 1991, 91, 165–195.

23. Horton, H. R.; Young, G. 2-Acetoxy-5-nitrobenzyl chloride. A reagent designed to introduce a reporter group near the active site of chymotrypsin. *Biochem. Biophys. Acta.* 1969, 94, 272–278.

24. Mole, J. E.; Horton, H. R. A kinetic analysis of the enhanced catalytic efficiency of papain modified by 2-hydroxy-5-nitrobenzylation. *Biochemistry* 1973, 12, 5278–5285.

25. Béchet, J.-J.; Dupaix, A.; Yon, J.; Wakselman, M.; Robert, J.-C.; Vilkas, M. Inactivation of α-chymotrypsin by a bifunctional reagent, 3,4-dihydro-3,4-dibromo-6-bromomethylcoumarin. *Eur. J. Biochem.* 1973, 35, 527–539.

26. Wakselman, M.; Hamon, J.-F.; Vilkas, M. Lactones phénoliques halométhylées, inhibiteurs bifonctionnels de protéases II. Préparation de dérivés halométhylés de la dihydro-3,4-coumarine et de la benzofuranone-2. *Tetrahedron* 1974, 30, 4069–4078.

27. Nicolle, J.-P.; Hamon, J.-F.; Wakselman, M. Lactones phénoliques halométhylées, inhibiteurs bifonctionnels de protéases IV. Préparation de dérivés chlorométhylés de la dihydro-3,4-benzyl-3-coumarine et de la tétrahydro-2,3,4,5-benzoxepinone-2. *Bull. Soc. Chim. Fr.* 1977, 83–88.

28. Béchet, J.-J.; Dupaix, A.; Blagoeva, I. Inactivation of α-chymotrypsin by new bifunctional reagents: halomethylated derivatives of dihydrocoumarins. *Biochimie* 1977, 59, 231–239.

29. Béchet, J.-J.; Dupaix, A.; Roucous, C.; Bonamy, A.-M. Inactivation of proteases and esterases by halomethylated derivatives of dihydrocoumarins. *Biochimie* 1977, 59, 241–246.

30. Wakselman, M.; Mazaleyrat, J.-P.; Xie, J.; Montagne, J.-J.; Vilain, A. C.; Reboud-Ravaux, M. Cyclopeptidic inactivators for chymotrypsin-like proteases. *Eur. J. Med. Chem.* 1991, 26, 699–707.

31. Wakselman, M.; Xie, J.; Mazaleyrat, J.-P.; Boggetto, N.; Vilain, A. C.; Montagne, J.-J.; Reboud-Ravaux, M. New mechanism-based inactivators of trypsin-like proteinases. Selective inactivation of urokinase by functionalized cyclopeptides incorporating a sulfoniomethyl-substituted *meta*-aminobenzoic acid residue. *J. Med. Chem.* 1993, 36, 1539–1547.

32. Nguyen, C.; Blanco, J.; Mazaleyrat, J.-P.; Krust, B.; Callebaut, C.; Jacotot, E.; Hovanessian, A. G.; Wakselman, M. Specific and irreversible cyclopeptide inhibitor of dipeptidyl peptidase IV activity of the T-cell activation antigen CD26. *J. Med. Chem.* 1998, 41, 2100–2110.

33. Doucet, C.; Vergely, I.; Reboud-Ravaux, M.; Guilhem, J.; Kobaiter, R.; Joyeau, R.; Wakselman, M. Inhibition of human leucocyte elastase by functionalized *N*-aryl azetidin-2-ones. Stereospecific synthesis and chiral recognition of disymmetrically C_3-substituted β-lactams. *Tetrahedron: Asymmetry* 1997, 8, 739–751.

34. Vergely, I.; Boggetto, N.; Okochi, V.; Golpayegani, S.; Reboud-Ravaux, M.; Kobaiter, R.; Joyeau, R.; Wakselman, M. Inhibition of human leucocyte elastase by functionalized *N*-aryl azetidin-2-ones: substituent effects at C-3 and benzylic positions. *Eur. J. Med. Chem.* 1995, 30, 199–208.

35. Joyeau, R.; Felk, A.; Guillaume, S.; Vergely, I.; Doucet, C.; Boggetto, N.; Reboud-Ravaux, M. Synthesis and inhibition of human leucocyte elastase by functionalized *N*-aryl azetidin-2-ones: effect of different substituents on the aromatic ring. *J. Pharm. Pharmacol.* 1996, 48, 1218–1230.

36. Cabaret, D.; Liu, J.; Wakselman, M. An efficient synthesis of aryl phenaceturates using acid catalyzed dicyclohexylcarbodiimide esterification and transient *N-tert*-butoxycarbonylation. *Synthesis* 1994, 480–483.

37. Cabaret, D.; Liu, J.; Wakselman, M.; Pratt, R. F.; Xu, Y. Functionalized depsipeptides, substrates and inhibitors of β-lactamases and DD-peptidases. *Bioorg. Med. Chem.* 1994, 2, 757–771.

38. Reboud-Ravaux, M.; Desvages, G.; Chapeville, F. Irreversible inhibition and peptide mapping of urinary plasminogen activator urokinase. *FEBS Lett* 1982, 140, 58–62.

39. Reboud-Ravaux, M.; Desvages, G. Inactivation of human high- and low-molecular-weight urokinases. Analysis of their active site. *Biochim. Biophys. Acta* 1984, 791, 333–341.

40. Mor, A.; Reboud-Ravaux, M.; Mazaleyrat, J.-P.; Wakselman, M. Susceptibility of plasminogen activators to suicide inactivation. *Thrombosis Res. Suppl VIII* 1988, 35–44.

41. Mor, A.; Maillard, J.; Favreau, C.; Reboud-Ravaux, M. Reaction of thrombin and proteinases of the fibrinolytic system with a mechanism-based inhibitor, 3,4-dihydro-3-benzyl-6-chloromethyl-coumarin. *Biochim. Biophys. Acta* 1990, 1038, 158–163.

42. Pochet, L.; Doucet, C.; Schynts, M.; Thierry, N.; Boggetto, N.; Pirotte, B.; Jiang, K. Y.; Masereel, B.; de Tullio, P.; Delarge, J.; Reboud-Ravaux, M. Esters and amides of 6-(chloromethyl)-2-oxo-2*H*-1-benzopyran-3-carboxylic acid as inhibitors of α-chymotrypsin: significance of the "aromatic" nature of the novel ester-type coumarin for strong inhibitory activity. *J. Med. Chem.* 1996, 39, 2579–2585.

43. Pochet, L.; Doucet, C.; Dive, G.; Wooters, J.; Masereel, B.; Reboud-Ravaux, M.; Pirotte, B. Coumarinic derivatives as mechanism-based inhibitors of α-chymotrypsin and human leukocyte elastase. *Bioorg. Med. Chem.* 2000, 8, 1489–1501.

44. Frédérick, R.; Robert, S.; Charlier, C.; de Ruyck, J.; Wouters, J.; Pirotte, B.; Masereel, B.; Pochet, L. 3,6-Substituted coumarins as mechanism-based inhibitors of thrombin and factor Xa. *J. Med. Chem.* 2005, 48, 7592–7603.

45. Frédérick, R.; Robert, S.; Charlier, C.; de Ruyck, J.; Wouters, J.; Masereel, B.; Pochet, L. Mechanism-based thrombin inhibitors: design, synthesis, and molecular docking of a new selective 2-oxo-2*H*-1-benzopyran derivative. *J. Med. Chem.* 2007, 50, 3645–3650.

46. Bayot, A.; Basse, N.; Lee, I.; Gareil, M.; Pirotte, B.; Bulteau, A. L.; Friguet, B.; Reboud-Ravaux, M. Towards the control of intracellular protein turnover: mitochondrial Lon protease inhibitors versus proteasome inhibitors. *Biochimie* 2008, 90, 260–269.

47. Papapostolou, D.; Reboud-Ravaux, M. Proteasome and proteolysis. *J. Soc. Biol.* 2004, 198, 263–278.

48. Wojciechowski, K. Aza-*ortho*-xylylenes in organic synthesis. *Eur. J. Org. Chem.* 2001, 3587–3605.

49. Domé, M.; Wakselman, M. Alkylations en milieu aqueux par les halogénures benzyliques aminés ou amidés II. Réactions avec quelques nucléophiles. *Bull. Soc. Chim. Fr.* 1975, 576–582.

50. Decodts, G.; Wakselman, M. Suicide inhibitors of proteases. Lack of activity of halomethyl derivatives of some aromatic lactams. *Eur. J. Med. Chem. Chim. Ther.* 1983, 18, 107–111.

51. Mazaleyrat, J.-P.; Reboud-Ravaux, M.; Wakselman, M. Synthesis and enzymic hydrolysis of cyclic peptides containing an anthranilic acid residue. *Int. J. Peptide Protein Res.* 1987, 30, 622–633.

52. Reboud-Ravaux, M.; Convert, O.; Mazaleyrat, J.-P.; Wakselman, M. Structural factors in the enzymic hydrolysis of cyclopeptides containing an *ortho*- or *meta*-amino benzoic acid residue. *Bull. Soc. Chim. Fr.* 1988, 267–271.

53. Convert, O.; Mazaleyrat, J.-P.; Wakselman, M.; Morize, I.; Reboud-Ravaux, M. NMR conformational analysis of cyclopeptidic substrates of serine proteases, containing an *ortho* or *meta* aminobenzoic acid residue. *Biopolymers* 1990, 30, 583–591.

54. Reboud-Ravaux, M.; Vilain, A. C.; Boggetto, N.; Maillard, J.-L.; Favreau, C.; Xie, J.; Mazaleyrat, J.-P.; Wakselman, M. A cyclopeptidic suicide substrate preferentially inactivates urokinase-type plasminogen activator. *Biochem. Biophys. Res. Commun.* 1991, 178, 352–359.

55. Wakselman, M.; Mazaleyrat, J.-P.; Lin, R. C.; Xie, J.; Vigier, B.; Vilain, A. C.; Fesquet, S.; Boggetto, N.; Reboud-Ravaux, M.; Design, synthesis and study of a selective cyclopeptidic mechanism-based inhibitor of human thrombin. In *Peptides: Chemistry, Structure*

and Biology. Proceedings of the13th American Peptide Symposium, Hodges, R. S.; Smith, J. A., Eds.; Escom: Leiden, 1994; pp 646–648.

56. Blanco, J.; Nguyen, C.; Callebaut, C.; Jacotot, E.; Krust, B.; Mazaleyrat, J.-P.; Wakselman, M.; Hovanessian, A. G. Dipetidyl-peptidase IV-β. Further characterization and comparison to dipetidyl-peptidase IV activity of CD26. *Eur. J. Biochem.* 1998, 256, 369–378.
57. Powers, J. C.; Asgian, J. L.; Dogan Ekici, O.; Ellis James, K. Irreversible inhibitors of serine, cysteine, and threonine proteases. *Chem. Rev.* 2002, 102, 4639–4750.
58. Orlowski, M.; Cardozo, C.; Eleutri, A. M.; Kohanski, R.; Kam, C. M.; Powers, J. C. Reactions of [^{14}C]-3,4-dichloroisocoumarin with subunits of pituitary and spleen multicatalytic proteinase complexes (proteasomes). *Biochemistry* 1997, 36, 13946–13953.
59. Wade Harper, J.; Powers, J. C. Reaction of serine proteases with substituted 3-alkoxy-4-chloroisocoumarins and 3-alkoxy-7-amino-4-chloroisocoumarins: new reactive mechanism-based inhibitors. *Biochemistry* 1985, 24, 7200–7213.
60. Vijayalakshmi, J.; Meyer, E. F.; Kam, C.-M.; Powers, J. C. Structural study of porcine pancreatic elastase complexed with 7-amino-3-(2-bromoethoxy)-4-chloroisocoumarin as a nonreactivable doubly covalent enzyme–inhibitor complex. *Biochemistry* 1991, 30, 2175–2183.
61. Kerrigan, J. E.; Oleksyszyn, J.; Kam, C.-M.; Selzler, J.; Powers, J. C. Mechanism-based isocoumarin inhibitors of human leucocyte elastase. Effects of the 7-amino substituent and the 3-alkoxy group in 3-alkoxy-7-amino-4-chloroisocoumarins on inhibitory potency. *J. Med. Chem.* 1995, 38, 544–552.
62. Bihel, F.; Quéléver, G.; Lelouard, H.; Petit, A.; Alvès de Costa, C.; Pourqié, O.; Checler, F.; Thellend, A.; Pierre, P.; Kraus, J.-L. Synthesis of new 3-alkoxy-7-amino-4-chloroisocoumarin derivatives as new beta-amyloid peptide production inhibitors and their activities on various classes of protease. *Bioorg. Med. Chem.* 2003, 11, 3141–3152.
63. Frère, J. M.; Nguyen-Distèche, M.; Coyette, J.; Joris, B. Mode of action. In *The Chemistry of β-Lactams;* Page, M. I., Ed.; Blackie Academic and Professional: London, UK, 1992; pp 148–197.
64. Waley, S. G.; β-Lactamase: mechanism of action. In *The Chemistry of β-Lactams;* Page, M. I., Ed.; Blackie Academic and Professional: London, UK 1992; pp 198–228.
65. Pratt, R. F. Functional evolution of the serine β-lactamase active site. *J. Chem. Soc. Perkin Trans. 2* 2002, 851–861.
66. Pratt, R. F.; β-Lactamases inhibitors. In *The Design of Enzyme Inhibitors as Drugs;* Sandler, M.; Smith, H. J., Eds.; Oxford University Press: Oxford, UK 1989; pp 596–619.
67. Pratt, R. F. β-Lactamase: inhibition. In *The Chemistry of β-lactams;* Page, M. I., Ed.; Blackie Academic and Professional: London, UK 1992; pp 229–231.
68. Pratt, R. F.; Govardhan, C. P. β-Lactamase-catalyzed hydrolysis of acyclic depsipeptides and acyl transfer to specific amino acid acceptors. *Proc. Natl. Acad. Sci. USA* 1984, 81, 1302–1306.
69. Govardhan, C. P.; Pratt, R. F. Kinetics and mechanism of the serine β-lactamase catalyzed hydrolysis of depsipeptides. *Biochemistry* 1987, 26, 3385–3395.
70. Cabaret, D.; Adediran, S. A.; Garcia Gonzalez, M. J.; Pratt, R. F.; Wakselman, M. Synthesis and reactivity with β-lactamases of "penicillin-like" cyclic depsipeptides. *J. Org. Chem.* 1999, 64, 713–720.
71. Zrihen, M.; Labia, R.; Wakselman, M. *N*-aryl azétidinones substituées, inhibiteurs potentiels de β-lactamases. *Eur. J. Med. Chem. Chim. Ther.* 1983, 18, 307–314.

72. Joyeau, R.; Molines, H.; Labia, R.; Wakselman, M. *N*-aryl 3-halogenated azetidin-2-ones and benzocarbacephems, inhibitors of β-lactamases. *J. Med. Chem.* 1988, 31, 370–374.
73. Joyeau, R.; Dugenet, Y.; Wakselman, M. Synthesis of fused β-lactams by copper promoted intramolecular aromatic substitution. *J. Chem. Soc. Chem. Commun.* 1983, 431–432.
74. Joyeau, R.; Yadav, L. D. S.; Wakselman, M. Synthesis of benzocarbacephem and benzocarbapenem derivatives by copper-promoted intramolecular aromatic substitution. *J. Chem. Soc. Perkin Trans. 1* 1987, 1899–1907.
75. Wakselman, M.; Joyeau, R.; Kobaiter, R.; Boggetto, N.; Vergely, I.; Maillard, J.-L.; Okochi, V.; Montagne, J.-J.; Reboud-Ravaux, M. Functionalized *N*-arylazetidinones as novel mechanism-based inhibitors of neutrophil elastase. *FEBS Lett.* 1991, 282, 377–381.
76. Maillard, J.-L.; Favreau, C.; Reboud-Ravaux, M.; Kobaiter, R.; Joyeau, R.; Wakselman, M. Biological evaluation of the inhibition of neutrophil elastase by a synthetic β-lactam derivative. *Eur. J. Cell. Biol.* 1990, 52, 213–218.
77. Maillard, J.-L.; Favreau, C.; Vergely, I.; Reboud-Ravaux, M.; Joyeau, R.; Kobaiter, R.; Wakselman, M. Protection of vascular basement membrane and microcirculation from elastase-induced damage with a fluorinated β-lactam derivative. *Clin. Chim. Acta* 1992, 213, 75–86.
78. Reboud-Ravaux, M.; Vergely, I.; Maillard, J.-L.; Favreau, C.; Kobaiter, R.; Joyeau, R.; Wakselman, M. Prevention of some types of inflammatory damage using AA 231-1, a fluorinated β-lactam. *Drugs Exp. Clin. Res.* 1992, 13, 159–162.
79. Vergely, I.; Laugaâ, P.; Reboud-Ravaux, M. Interaction of human leukocyte elastase with a *N*-aryl azetidinone suicide substrate; conformational analyses based on the mechanism of action of serine proteinases. *J. Mol. Graphics* 1996, 14, 158–167.

12

QUINONE METHIDES IN LIGNIFICATION

John Ralph[1,2,3], Paul F. Schatz[3], Fachuang Lu[1], Hoon Kim[1], Takuya Akiyama[3,4], and Stephen F. Nelsen[5]

[1]*Department of Biochemistry and Great Lakes Bioenergy Research Center, University of Wisconsin, Madison, WI, USA*
[2]*Department of Biological Systems Engineering, University of Wisconsin, Madison, WI, USA*
[3]*Dairy Forage Research Center, USDA-Agricultural Research Service, Madison, WI, USA*
[4]*RIKEN Plant Science Center, Suehiro, Tsurumi, Yokohama, Kanagawa 230-0045, Japan*
[5]*Department of Chemistry, University of Wisconsin, Madison, WI, USA*

12.1 INTRODUCTION

Reactive quinone methide (QM) intermediates are implicated in lignin and lignan biosynthesis, as well as in degradation reactions of lignins such as those that occur during alkaline pulping of wood.[1] Dehydrodimeric lignans are plant-extractive products derived from radical coupling reactions of monlignols,[2,3] primarily the two hydroxycinnamyl alcohols coniferyl **M**G (Fig. 12.1) and sinapyl alcohol **M**S—here we shall largely ignore the reactions of *p*-coumaryl alcohol, typically a minor monolignol. Lignins are the polymers produced by combinatorial radical coupling reactions of primarily the monolignols **M** with the growing polymer **P** (Fig. 12.2);[4–8] dehydrodimerization reactions are involved in chain initiation. Although lignin dimers may have the same structures as their lignan analogues and are produced from the same monolignols, it is generally considered that lignan formation and lignin biosynthesis are well separated in time and space. Importantly, lignans are invariably chiral[2] whereas

Quinone Methides, Edited by Steven E. Rokita

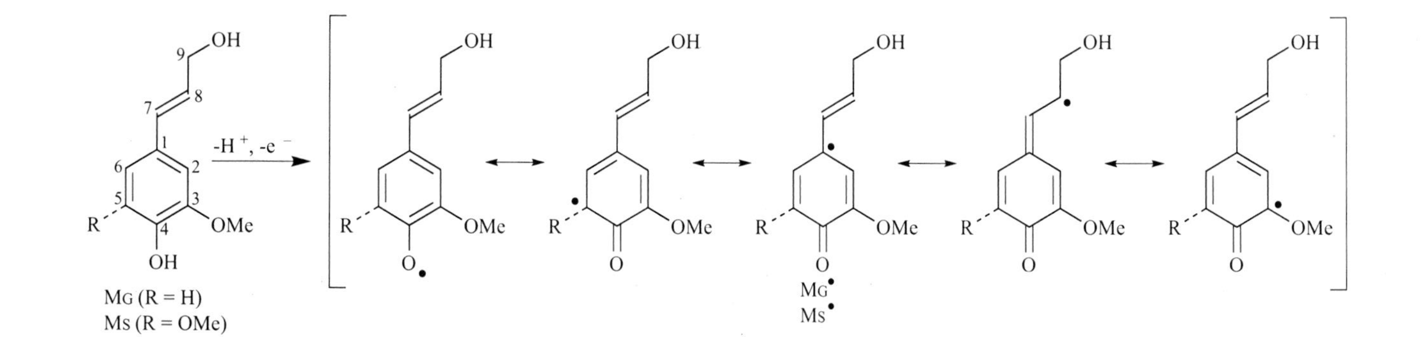

FIGURE 12.1 Radicals from monolignols. One-electron oxidation of coniferyl alcohol **MG** or sinapyl alcohol **MS** produces the phenolic radicals **MG**• or **MS**•. The resonance forms help to rationalize the combinatorial coupling products produced in coupling and cross-coupling reactions. Coupling involving the 8- and 4-*O*-positions, and 5-position for **MG** only, are known, although monolignol radicals favor coupling at their 8-positions. Coupling reactions at the substituted 3- and 1-positions are not evidenced for **MG** or **MS** nor at the 5-position for **MS**, although 1-coupling in seen from 8-*O*-4-ether units (Fig. 12.2).

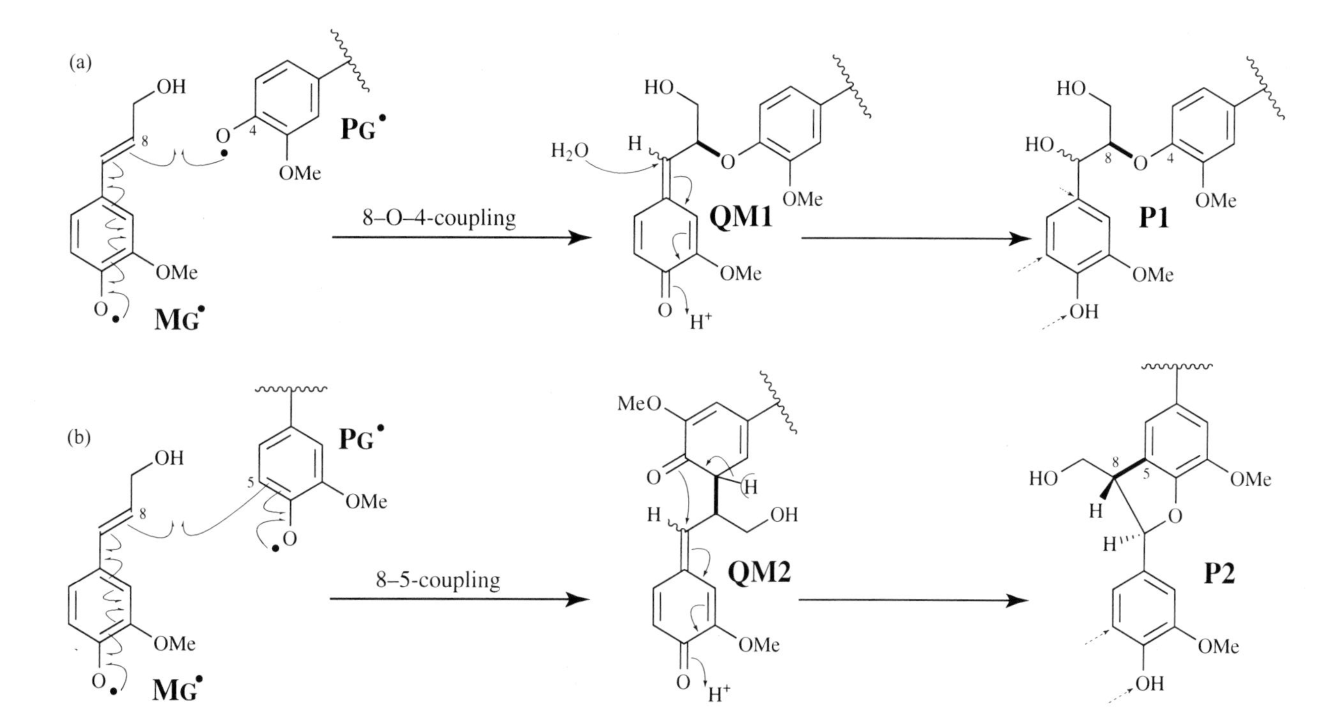

FIGURE 12.2 Coupling and cross-coupling reactions and rearomatization (illustrated using only the radical from the monolignol, coniferyl alcohol **M**G, and using only coupling with guaiacyl units, except for Fig. 12.2e). (a) 8-*O*-4-Coupling or cross-coupling produces a quinone methide **QM1** that rearomatizes by (proton-assisted) nucleophilic attack at C7, primarily by water, to give the so-called β-guaiacyl ether or 8-*O*-4-guaiacyl ether products **P1**. (b) 8-5-Coupling or cross-coupling produces a quinone methide **QM2**; rearomatization of the "top" ring produces a phenol/phenolate that internally traps the quinone methide moiety, yielding *trans*-phenylcoumaran structures **P2**. (c) 8-8-Coupling occurs only between monomers; no 8-coupling is possible on any of the dimers or oligomers. Each moiety of the bis-quinone

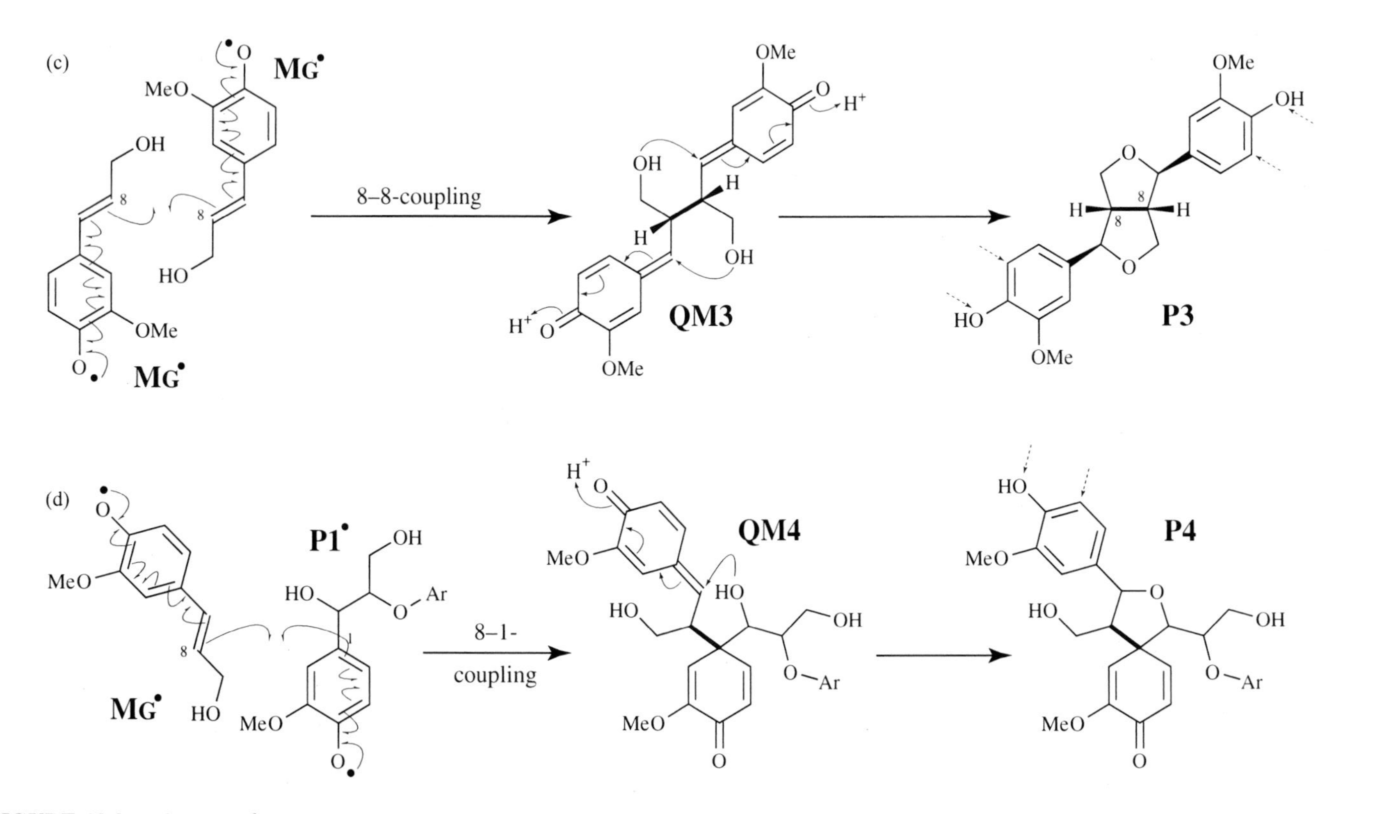

FIGURE 12.2 (*Continued*)

methide **QM3** produced rearomatizes via internal trapping by each 9-OH to produce pinoresinol structures **P3**. (d) A more unusual coupling is 8-1-coupling, only evidenced to date between a monolignol (coupling at its 8-position) and an 8-*O*-4-ether unit **P1**. The resulting quinone methide **QM4** is also a dienone; internal trapping of the quinone methide moiety, by the 7-OH, rearomatizes that ring,

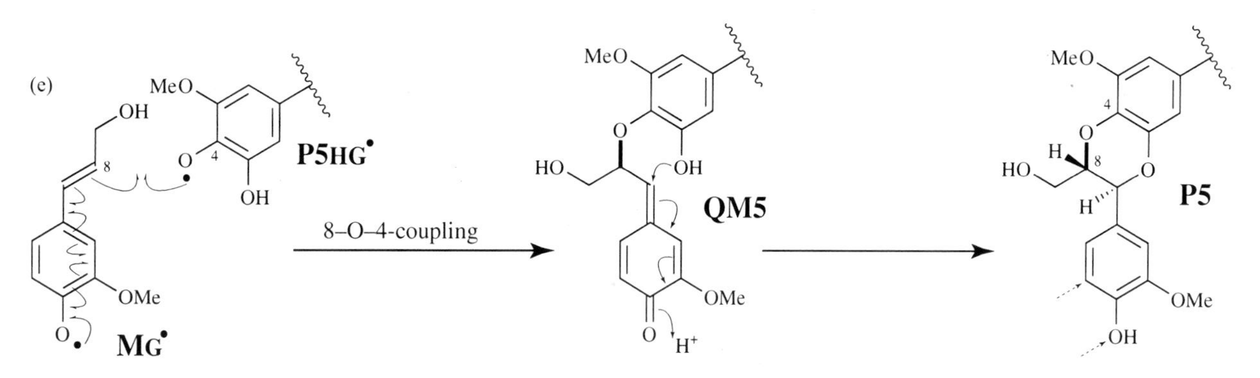

FIGURE 12.2 (*Continued*)

also creating a stable spirodienone unit **P4** that remains in the lignin polymer. (e) 8-*O*-4-Cross-coupling with a 5-hydroxyguaiacyl unit **P5HG** (that results in lignins in COMT-deficient plants due to the incorporation of 5-hydroxyconiferyl alcohol as a monomer substitute) produces a quinone methide **QM5** that is rearomatized via internal trapping by the phenolic 5-OH. Benzodioxane structures **P5**, mainly *trans*-isomers, are produced in the lignins. Dashed arrows on each of the final structures **P** indicate the sites of further possible radical coupling.

lignins are completely optically inactive.[9] As a result, lignification, the process of polymerization from the monomers, is theorized to involve only chemical reactions without the directing influence of proteins or enzymes.[5–7,10] Quinone methides are the initial products of radical coupling reactions that involve a monomer coupling at its 8-position (also often designated as the β-position). Subsequent rearomatization reactions produce the structures that are traditionally characterized in lignins.

This chapter does not purport to be a comprehensive review of quinone methides in lignification. It covers aspects of quinone methide chemistry that our laboratory has been involved in over the past 30 years and is therefore heavily weighted toward our own studies and those with which we have interacted.

12.2 QUINONE METHIDES FROM RADICAL COUPLING IN LIGNIFICATION

As has been recently reviewed,[6,7] radicals for lignification are generated primarily via peroxidase–H_2O_2 oxidation. One-electron oxidation of a monolignol **M** produces a phenolic radical[11–13] (Fig. 12.1). (The monolignol radicals and their coupling reactions seem at some variance with observations of other phenolic radicals. Thus, although it has been traditional to write a phenoxy radical with the *O*-centered resonance structure, it is considered to exist principally as the cyclic oxo-pentadienyl radical, as shown by its ESR spectrum,[11] which indicates that 92% of its π-spin density resides at the *ortho*- and *para*-carbons, and its photoelectron spectrum and calculations of its structure, as pointed out by Dewar in 1980,[12] as well as the coupling patterns of substituted phenoxy radicals. The majority of the coupling occurs between carbon atoms, although coupling products involving one oxygen center are also detected as minor products. ESR studies of actual monolignol radicals indicate considerable unpaired electron density on the phenolic oxygen, increasing from *p*-coumaryl to coniferyl to sinapyl alcohol radicals.[13] In view of the predominance of 4-*O*-coupling, and for convenience, we will continue to show *O*-centered radicals here.)

The combinatorial coupling and cross-coupling reactions are readily understood from examination of the resonance forms, as shown in Fig. 12.1 for coniferyl alcohol. Although radical coupling can, in principle, occur at sites 4-*O*, 5, 1, 3, and 8, the coniferyl alcohol radical is not found to couple at its 1- or 3-positions (except perhaps reversibly) nor does the sinapyl alcohol radical couple at its 5-position. In addition, in dehydrodimerization reactions, one monolignol invariably couples at its 8-position; the other may couple at either its 8-, 5-, or 4-*O*-positions (Fig. 12.2). For coniferyl alcohol **M**G, the three modes, 8-*O*-4-, 8-5-, and 8-8-dehydrodimerization, occur in roughly equal proportions under biomimetic aqueous conditions;[14,15] the coupling propensity can be markedly altered by changing solvent conditions. For sinapyl alcohol **M**S, dehydrodimerization is almost entirely via 8–8-coupling under biomimetic conditions.[15]

More importantly for lignification, where the main reaction is chain extension by cross-coupling of a monolignol with the (phenolic end of the) growing polymer, the monolignol **M** (coniferyl or sinapyl alcohol) invariably couples at its 8-position, with the polymer **P** coupling at its 4-*O*-position or, in the case of a guaiacyl unit **P**G only, at its 5-position. (During lignification, guaiacyl units derive from coniferyl alcohol and syringyl units from sinapyl alcohol.) There is another coupling mode that is seen in lignification accounting for a few percent of the linkages, 8-1 coupling (Fig. 12.2d). The mode has only been firmly established as a product of coupling between a monolignol **M**, at its 8-position, and a preformed 8-*O*-4-aryl ether end unit **P1** on the polymer.[7] As seen below, such 8-1-coupling eventually results in novel spirodienone structures **P4** in the lignin. The four types of quinone methides, **QM1–QM4**, generated from each of the four coupling modes are shown in Fig. 12.2. Each is rearomatized as discussed in Section 12.3.

Lignin monomers do not have any chiral centers. During lignification, however, each coupling reaction involving a monolignol coupling at its 8-position results in a quinone methide with a new optical center at carbon-8. As reviewed,[9,16] in a variety of studies over decades, using a variety of methods, fragments released from the lignin structure (under conditions that have been proven to retain optical activity) have never been found to be optically active. And entire isolated polymer fractions have no detectable optical activity. It is therefore logical that the radical coupling reactions involved in lignification are not controlled by proteins or enzymes; that is, they are simple (combinatorial) chemical reactions.[7,10]

12.3 REAROMATIZATION OF LIGNIN QUINONE METHIDES

As illustrated in Fig. 12.2, the various quinone methides **QM1–QM4** produced by radical coupling of a monolignol **M** (at its 8-position) with the growing polymer **P** or another monolignol are subsequently rearomatized by proton-assisted nucleophilic addition reactions. The 8-*O*-4-coupled product **QM1** in Fig. 12.2a is generally rearomatized by simple water addition. Again, the quinone methide quenching is theorized to be under simple chemical control as the ratio of *erythro*/*threo* (*anti*/*syn*) isomers is apparently kinetically controlled. Thus, 4-*O*-guaiacyl ethers (such as the one shown in Fig. 12.2a) produce ~50:50 mixtures of isomers, whereas 4-*O*-syringyl ethers produce the final product as ~75:25 *erythro*/*threo* mixture; in both cases, the thermodynamic distribution (in refluxing dioxane–water) is ~50:50.[17,18] The isomer ratios therefore correlate well with syringyl/guaiacyl ratios in dicots.[19] These ratios are noted *in vivo* and *in vitro*, suggesting that there are no obvious proteinaceous control mechanisms operating. In addition, even though a new optical center is created at carbon-7 by water addition to the quinone methide, the products (both the *erythro*- and *threo*-isomers) are again racemic.[20] The quinone methide **QM2** produced by 8–5-coupling rearomatizes (not necessarily in the single concerted step shown in Fig. 12.2b) by internal trapping with the available phenol to produce *trans*-phenylcoumaran units **P2**.[4] Both moieties of the bis-quinone methide

QM3 produced by 8–8-coupling are also internally trapped, this time by the hydroxyls at C9, to produce the bicyclic resinol structures **P3**. As further evidence for kinetic control of isomer ratios during lignification, 8–8-coupled structures (resinols) in lignins, both natural and synthetic, are essentially in a single form whereas the equilibrium ratio of pinoresinol to *epi*-pinoresinol is ~1:1.[21] Finally, the special case of 8–1-coupling between a monolignol **M** and a 8-*O*-4-ether end unit **P1** results in a quinone methide **QM4** that is internally trapped by the 7-OH on the original 8-*O*-4-ether unit, to form novel spirodienone structures **P4** in the lignin. Such units, recently authenticated via NMR spectroscopy, are more prevalent in high-syringyl lignins.[7,22–25]

There are indications that slow rearomatization of quinone methides can limit polymerization. Obviously, until the phenol is regenerated (by rearomatization), further radical coupling reactions cannot occur. A recent study illustrates that synthetic lignins from sinapyl alcohol have higher degree of polymerization if the reaction is aided by efficient trapping of the quinone methide, in this case by azide.[26] In addition to the stable quinone methides that have been isolated (see Section 12.4), this observation suggests that quinone methide rearomatization needs to receive more attention and that radical generation and radical coupling are not necessarily the rate-limiting reactions.

Lignification is readily perturbed by up- and downregulating genes in the monolignol biosynthetic pathway.[6,7] By downregulating an *O*-methyltransferase, the so-called caffeic acid *O*-methyltransferase although the substrate *in planta* is now recognized to be primarily 5-hydroxyconiferaldehyde,[27–29] substantial amounts of the novel monomer 5-hydroxyconiferyl alcohol can be incorporated into lignins during lignification.[30–39] The resulting 5-hydroxyguaiacyl groups **P5HG** appearing during lignification provide another pathway toward cyclic structures, benzodioxanes **P5**, via internal trapping of the quinone methide intermediate **QM5** (Fig. 12.2e).

Quinone methides formed during, for example, alkaline pulping reactions may have other mechanisms for rearomatization. Most commonly, in 8-*O*-4-, 8–5-, and 8–8-quinone methides **QM1–QM3** (Fig. 12.2), retro-aldol elimination of formaldehyde to give styryl aryl ethers or stilbenes is common.[40] Retro-aldol reactions using a strong base, for example, diazabicycloundecene (DBU) in CH_2Cl_2 can also provide these compounds conveniently at room temperatures.[41–43]

12.4 QUINONE METHIDES DERIVED FROM ACYLATED MONOLIGNOLS

Many lignins have various groups acylating the 9-hydroxyl, often at high levels.[44] For example, kenaf bast fiber lignins are over 50% 9-acetylated.[44,45] Grass lignins have 9-*p*-coumarate substituents on up to ~10% of their units; a mature maize isolated lignin had a 17% content by weight.[46] Similarly, various poplar/aspen, palm, and willow varieties have 9-*p*-hydroxybenzoate substituents.[47–53]

Examination of the postcoupling reactions available to the quinone methide intermediates **QM3** (Fig. 12.3)[54] provides compelling evidence that lignin acylation derives from the incorporation of acylated monolignols into lignification (as opposed to arising from postlignification events). 9-Acylated sinapyl alcohols **Ms′** predominantly undergo 8–8-coupling, analogously to sinapyl alcohol **Ms** itself. And cross-coupling between sinapyl alcohol **Ms** and its 9-acylated analogue **Ms′** is also an efficient reaction. The ratio of the three products (Fig. 12.3) from reacting a 1:1 molar mixture of sinapyl alcohol and sinapyl *p*-hydroxybenzoate, for example, is essentially 1:2:1, suggesting that coupling reactions are insensitive to the acylation of the 9-OH.[55] As shown in Fig. 12.3, the non-, mono-, or di-acylated bis-quinone methide intermediates **QM3**, **QM3′**, and **QM3″** are structural analogues, but the postcoupling rearomatization reactions are distinctly different.[54,55] Sinapyl alcohol radical 8–8-coupling produces an intermediate bis-quinone methide **QM3** (Fig. 12.3), which rearomatizes to afford syringaresinol **P3** by internal trapping via the two 9-OH groups. When the 9-OH group is acylated, it is obviously incapable of trapping the quinone methide in the resulting bis-quinone methide **QM3″**. Consequently, one quinone methide rearomatizes by external water addition, as is typically seen following 8-*O*-4-dehydrodimerization (Fig. 12.2a). At this point there is an internal hydroxyl capable of trapping the other quinone methide moiety. The resulting product is therefore not a dehydrodimer, but the product **P3″**, with a molecular mass 16 units higher. The cross-coupling reaction between **Ms** and **Ms′** also produces an intermediary bis-quinone methide **QM3′**, but one such moiety is internally trapped by the single 9-OH. Rearomatization of the other requires water addition and produces product **P3′**. Finding the cross-coupling *p*-hydroxybenzoate product **P3′** (R = *p*-hydroxybenzoyl) in poplar is compelling evidence that sinapyl *p*-hydroxybenzoate must therefore be an authentic "monomer" in the lignin (or lignan) pathway.[55] And the various acetylated, *p*-hydroxybenzoylated, and *p*-coumaroylated tetrahydrofuran structures **P3** are found in the lignin polymers of the appropriate plant materials.[44,56]

One of the more intriguing and unexpected chemical aspects of the homocoupling reaction using sinapyl *p*-hydroxybenzoate **Ms′** in peroxidase–H_2O_2 at pH 5 to independently generate compound **P3″** was that the intermediary quinone methide **QM3″** was isolable and stable.[55] It precipitated directly from the reaction mixture and was in fact difficult to dissolve in common organic solvents. Although certain syringyl and guaiacyl quinone methides were sufficiently stable in solution to allow NMR spectra to be recorded in 1983,[57] they have generally been considered unstable, although crystalline bromo analogues have been long known;[58] a crystal structure determination is available.[59] Solid quinone methide **QM3″** (R = *p*-hydroxybenzoyl), structurally authenticated by 1D and 2D NMR in DMSO-d_6, has not degraded after over 2 years in a freezer. Regrettably, it has not yet been possible to obtain sufficiently good crystals for an X-ray crystal structure. Extracting the reaction product (from dehydrodimerization of sinapyl *p*-hydroxybenzoate **Ms′**) into EtOAc and washing with saturated aqueous NH_4Cl acidified with HCl produces mainly compound **P3″** (along with a tetrahydronaphthalene).

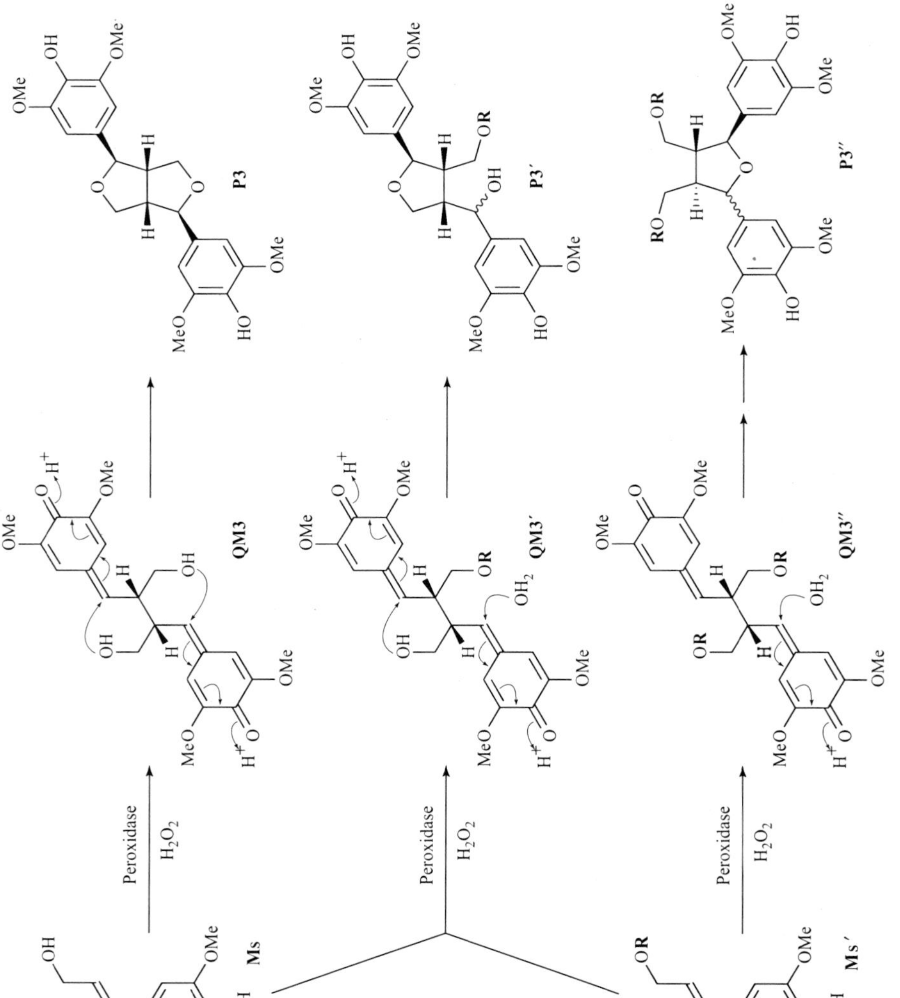
Peroxidase
H_2O_2
Ms
Ms′
QM3
QM3′
QM3″
P3
P3′
P3″

12.5 SYNTHESIS OF QUINONE METHIDES

Quinone methides for lignin studies have become readily accessible, primarily via two general approaches (Fig. 12.4). The first is to generate phenolics with a good leaving group at C7 so that simple base elimination generates the quinone methide (Fig. 12.4a). For 8-*O*-4-aryl ether quinone methides **QM1**, this is essentially a lower energy route to the quinone methides produced during alkaline pulping, for example, where temperatures of ~130°C are sufficient to effect the elimination from the native 7-hydroxy compound.[40] In fact, for phenylcoumaran structures **P2**, there is already a good leaving group at C7; reacting the free-phenolic benzyl-aryl ether in 1 M base will generate the quinone methide **QM2** readily at 10°C (Fig. 12.4b).[40,60] However, the quinone methide so generated is only present in low steady-state concentrations and is not readily characterized.

The second method is to generate the quinone methide via radical coupling reactions, analogously to the way they are produced via radical coupling during lignification (Fig. 12.4c) or to exploit radical disproportionation reactions (Fig. 12.4d).

12.5.1 From Free-Phenolic Units with a Good 7-Leaving Group

As shown in Fig. 12.4a for certain dimeric 8-*O*-4-ether quinone methides **P1**, simple elimination reactions readily produce the required quinone methides **QM1**, the solutions of which can be quite stable. The issue is in converting the 7-OH to a good leaving group. The simplest way, in a single step, directly from the 8-*O*-4-ether model (as long as it is only a dimer with only a single benzyl alcohol and no allyl (or cinnamyl) alcohol), is via bromination. The utilization of Me_3SiBr to effect this bromination[57,61] was a significant advance over the earlier use of HBr,[62] allowing the benzyl bromides **P1**–Br to be generated in a matter of minutes. If the reaction is carried out in a water-immiscible solvent such as CH_2Cl_2 or $CHCl_3$, treatment with aqueous $NaHCO_3$ is sufficient to readily generate the quinone methide **QM1** in that solvent. For NMR studies, the reaction is readily run in $CDCl_3$. Quinone methides generated this way became the first lignin-relevant quinone methides to be characterized by NMR (see Section 12.6, Fig. 12.6).

An alternative method, as long as acetylation is not a concern elsewhere in the molecule, is to produce free-phenolic 7-acetates **P1**–Ac (Fig. 12.4a, lower scheme). The quinone methides **QM1** or **QM1**–Ac can then be generated *in situ* by treatment

◀

FIGURE 12.3 8-8-Coupling and cross-coupling of acylated monolignols, and the fate of the 8-8-bis-quinone methides. The quinone methides **QM3**, **QM3′**, and **QM3″**, the various acylated coupling products derived from sinapyl alcohol **Ms** and its 9-acylated analogues **Ms′**, rearomatize via logical pathways to produce a variety of tetrahydrofuran products, **P3**, **P3′**, and **P3″**. Internal quinone methide trapping occurs via 9-OHs on the quinone methide when formed, or via the 7-OH produced by water addition/rearomatization. Quinone methide **QM3″** (R = *p*-hydroxybenzoyl) is readily isolated and is stable in the solid state. R = acetyl, *p*-hydroxybenzoyl, *p*-coumaroyl.

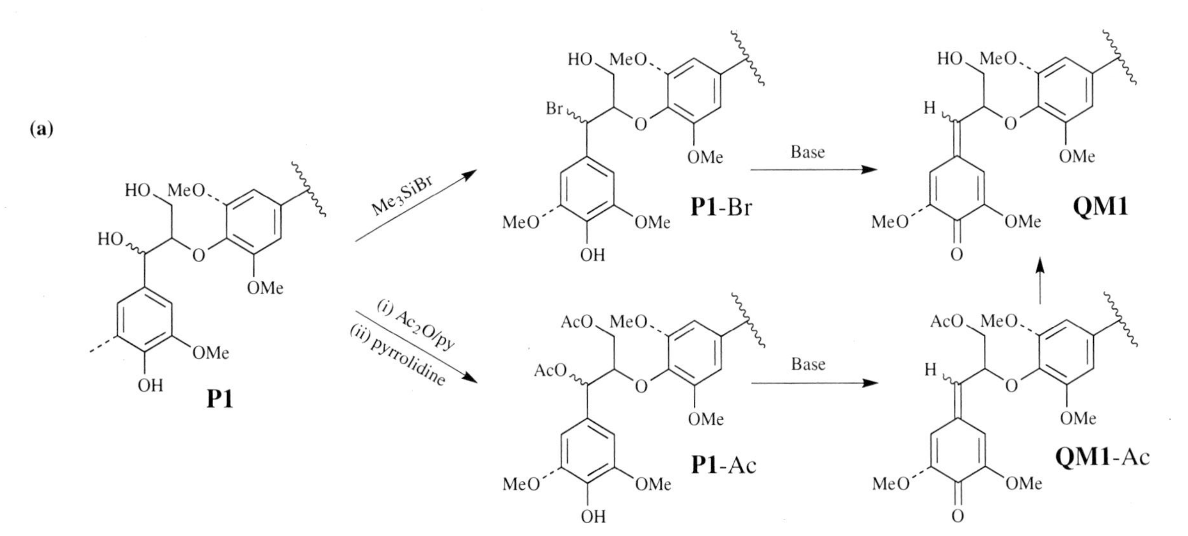

FIGURE 12.4 Syntheses of quinone methides. (a) Derivatization of 8-*O*-4-ether structures **P1** with good 7-leaving groups (bromide or acetate) allows the quinone methide **QM1** to be generated by base treatment. (b) Phenylcoumarans **P2** are readily opened in base to form quinone methides **QM2**, but only in low equilibrium concentrations. (c) Quinone methides can be generated directly by radical coupling reactions; in the case of the syringyl analogue of isoeugenol shown here, the 8-*O*-4-coupling reaction is in high yield and stops at the quinone methide **QM6**. (d) *p*-Hydroxyphenyl-CH_2-R units **7** (i.e., with benzylic methylene groups) undergo a formal double one-electron oxidation, for example, with silver(I) oxide, to produce the quinone methide **QM7**; in fact the quinone methide is likely formed from disproportionation of the phenoxy radical. Compounds such as eugenol **8** and its syringyl analogue form vinylogous quinone methides **QM8**.

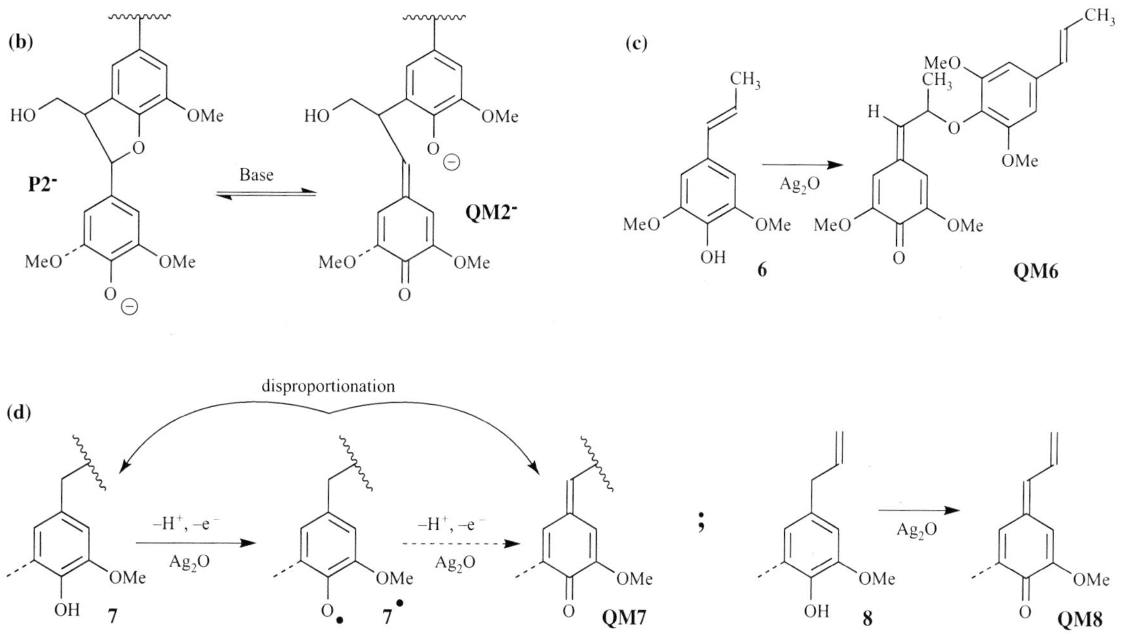

FIGURE 12.4 (*Continued*)

with a base. Such 7-acetate derivatives **P1**–Ac are produced from phenolic model compounds **P1** simply by first peracetylating and then selectively removing the phenolic acetate, for example, by treatment with neat pyrrolidine,[63] or from acetylating phenol-benzylated models and then debenzylating.[64] The free-phenolic 7-acetate **P1**–Ac can then be isolated, purified, and characterized before using it in base reactions. As the generation of the quinone methide is more rapid than base hydrolysis of the benzylic acetate, the reactions are very clean—an example is given in Section 12.5.3.

The quinone methide can also be generated *in situ*, at least in aqueous NaOH, directly from the peracetate, as hydrolysis of the phenolic acetate is faster than the benzylic acetate (see an example in Section 12.5.3). This method was used to demonstrate the addition of anthrahydroquinone (AHQ) and anthranol to (actual polymeric) lignin quinone methides in studies elucidating the anthraquinone (AQ)-catalyzed 8-*O*-4-aryl ether cleavage mechanisms in alkaline pulping.[64–66]

12.5.2 Via Radical Coupling

In a more biomimetic approach, quinone methides can be generated directly via radical coupling or by exploiting radical disproportionation reactions. Zanarotti generated the 8-*O*-4-quinone methide **QM6**, as shown in Fig. 12.4c, in high yields using silver(I) oxide (Ag_2O) as the oxidant, directly from the syringyl analogue **6** of isoeugenol.[67] The reaction is not quite so clean from hydroxycinnamyl alcohols ("real" monolignols), but metal oxidations are useful for preparing various lignin model dimers,[68] if not the quinone methides directly. Finally, an interesting avenue toward quinone methides comes from *p*-hydroxyphenyl-CH_2-R units **7** (Fig. 12.4d) using one-electron oxidants such as silver(I) oxide.[69–71] Although the reaction appears to be formally via two single-electron oxidations, the quinone methide **QM7** is likely produced by disproportionation of the phenoxy radical.[10] Although such reactions are only described for *p*-hydroxyphenyl-CH_2-R structures, there is the possibility that *p*-hydroxyphenyl-CH(OH)-R moieties may undergo analogous reactions producing the benzylic ketones; such structures in lignins were thought to arise from the harsh milling conditions usually required to isolate lignins, but these oxidized structures also appear in synthetic lignins derived from monolignols using metal oxidants on peroxidase–H_2O_2, particularly in syringyl lignins; syringyl units are easier to oxidize.[25,72,73]

12.5.3 Quinone Methides Generated *In Situ*

Rather than generating quinone methides as relatively stable entities in solution, generating them *in situ* allows various lignin model compounds to be synthesized. For example, the synthesis of dibenzodioxocin model compounds **P12** is shown in Fig. 12.5. Dibenzodioxocins are relatively newly discovered eight-membered ring heterocycles produced during lignification via the coupling of a monolignol with a 5–5-linked (biphenyl) unit in the growing polymer;[74–76] 5–5-linked structures

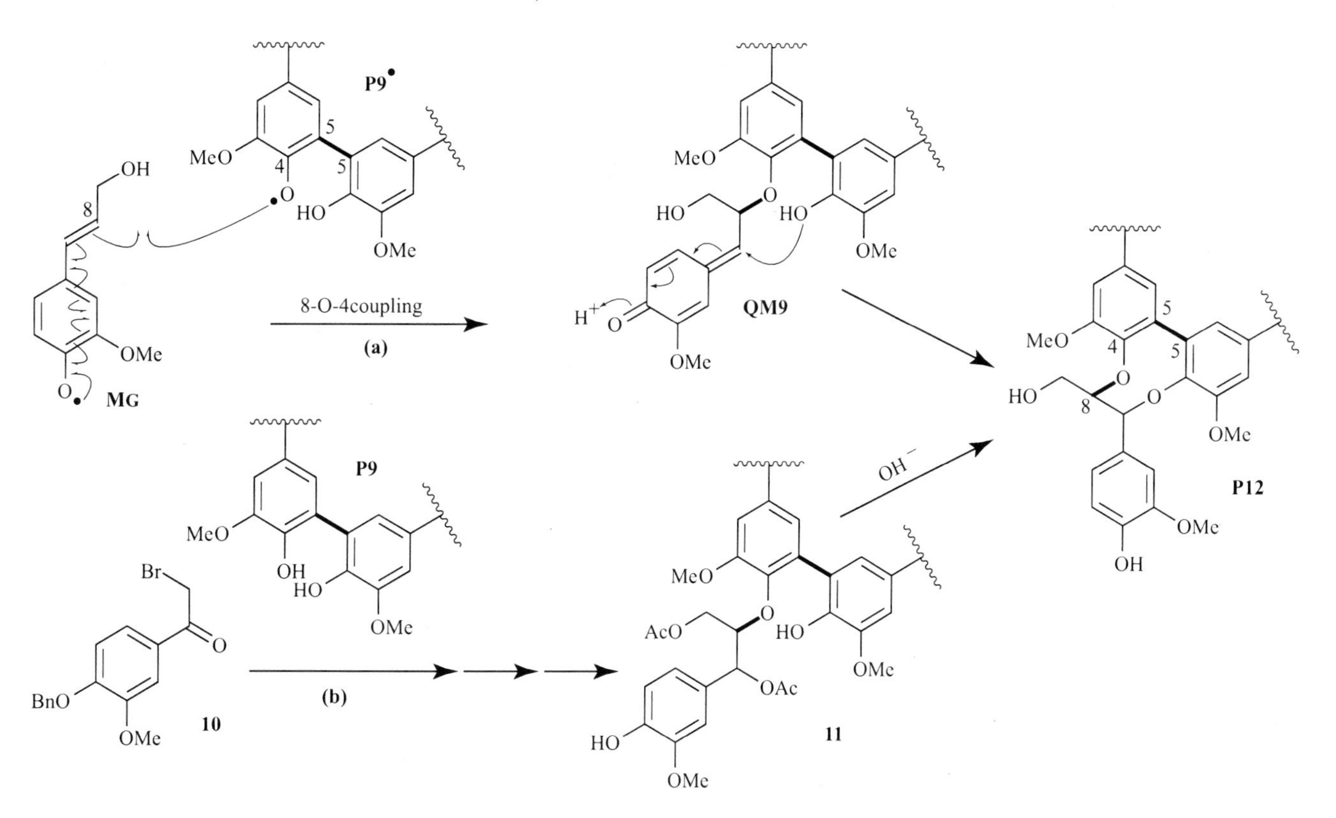

FIGURE 12.5 *In situ* generation of quinone methides to produce dibenzodioxocin model compounds. (See text for details.)

are themselves formed when guaiacyl (or *p*-hydroxyphenyl) dimers or higher oligomers couple—there is little evidence that coniferyl alcohol produces 5–5-coupled units, although the resultant dimer (synthesized from 5–5-diferulate) is particularly unstable,[77] see compound #2058 in the "NMR database of lignin and cell wall model compounds,"[78] *p*-coumaryl alcohol may 5–5-couple.[79]

Dibenzodioxocins **P12** (Fig. 12.5) can be synthesized by two basic approaches,[76] both involving the generation of quinone methide intermediates. The first way (Fig. 12.5a) is to generate the quinone methide **QM9** directly and biomimetically via radical coupling between a monolignol, for example, coniferyl alcohol **MG**, and the 5–5-model, **P9**. It is difficult to obtain yields above about 60% (based on the 5–5-unit) this way. A higher yielding and more traditional synthetic approach is to first create the 8-*O*-4-ether via usual methods (Fig. 12.5b),[80] which includes adding formaldehyde to complete the side chain, and then peracetylating and specifically deacetylating the phenols to produce the free-phenolic diacetate quinone methide precursor **11** shown, cf. the approach in Fig. 12.4a described in Section 12.5.1. Generating the quinone methide *in situ* is accomplished via a base. The trapping by the phenol on the biphenyl moiety spontaneously generates the dibenzodioxocin, which is isolated as compound **P12** following the hydrolysis of the 9-acetate that subsequently occurs in base. This approach can produce the dibenzodioxocin **P12** essentially quantitatively from the free-phenolic diacetate precursor **11**.[80]

12.6 STRUCTURE AND ISOMERISM

As might be expected, but not widely anticipated originally, there are two geometric isomers of the guaiacyl quinone methides, one in which the side chain is *syn* to the methoxyl and one in which it is *anti*.[57,81,82] NMR, which readily distinguishes the two isomers, shows that the ratio is typically ~70:30 *syn/anti* (Fig. 12.6); the carbon assignments in the original paper[57] have since been corrected.[83] The energy barriers are such that interconversion at room temperature is unlikely.[82] It has also been shown that each isomer rearomatizes by nucleophilic addition to C7 at a different rate.[81]

12.7 REACTIVITY WITH NUCLEOPHILES

Quinone methides are electron-deficient at C7, as readily understood via the resonance forms of **QM1** shown in Fig. 12.7. They are therefore susceptible to nucleophilic attack at that position. Although reactions during high-temperature pulping demonstrate that 8-*O*-4-aryl ether quinone methides **QM1** are rearomatized by attack with hard nucleophiles such as HO^- and HS^-,[84] these reactions do not readily occur at ambient temperatures.[41,85] Thus, HO^- will not add to quinone methide **QM1** under any conditions that we have tried (including with cosolvents, and using phase-transfer conditions). Of course water will add to quinone methides under acidic conditions

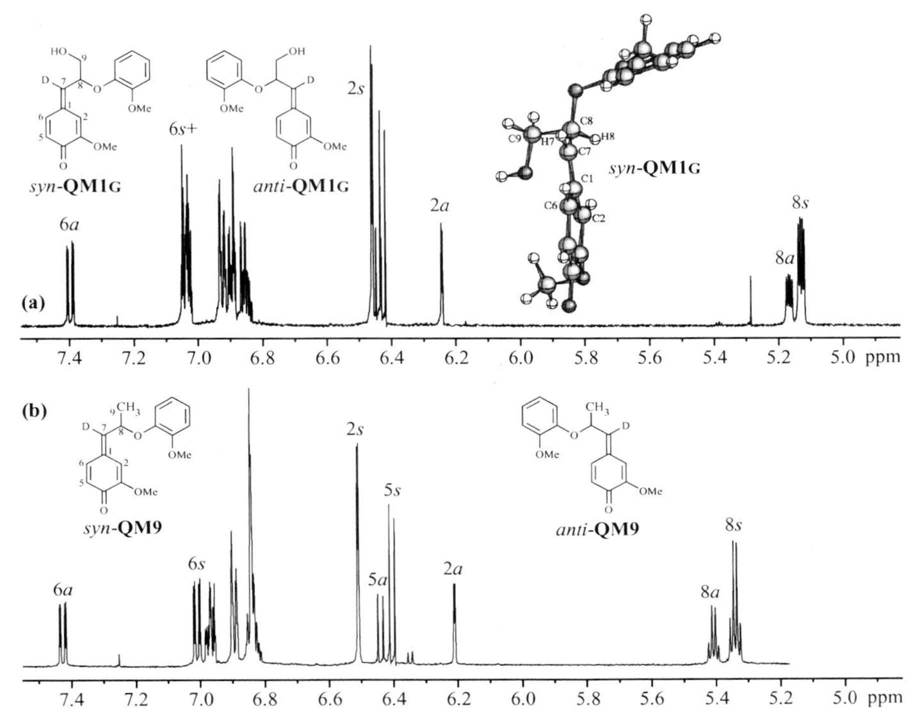

FIGURE 12.6 Proton NMR spectra of guaiacyl quinone methides. As the 7-proton overlaps with others, strategically 7-deutero-labeled quinone methides have simpler spectra in which the geometric isomers are readily distinguished. (a) Quinone methide **QM1G** in $CDCl_3$. (b) The 9-methyl analogue **QM9**, showing the superior dispersion of the 5-proton signals. The noninterconverting isomers are formed in an approximately 70:30 *syn/anti* ratio and rearomatize at different rates.

(Fig. 12.7a,e); it is important to realize that a protonated quinone methide **QM1H**$^+$ is actually a benzylic carbocation (Fig. 12.7a). Water will also add to the quinone methide under fairly neutral conditions.[86–88] The isomer distribution of the resulting compounds, **P1** can be determined directly from ^{1}H or ^{13}C (or 2D ^{13}C/^{1}H correlation) NMR[89,90] by making various cyclic derivatives, such as acetals **14**[61] or phenylboronates **15**[61,91] from the 1,3-diol, or via ozonolysis in Ac-OH-H_2O-MeOH to produce erythronic acid **16** and threonic acid **17**[92,93] (Fig. 12.7f–h). The latter method has been particularly useful for determining *erythro/threo* ratios of 8-*O*-4-ether units in lignins.[19] Claims for the room temperature addition of HO^- to quinone methides have, in several cases, been shown to be erroneous—it has been demonstrated that the quinone methide **QM1** remained in the reaction mixture and was quenched (via the carbocation **QM1H**$^+$) during the acidifying workup. If an alkaline reaction involving quinone methides is worked up by simply extracting the product without prior acidification, any unreacted quinone methide will extract into solvents such as

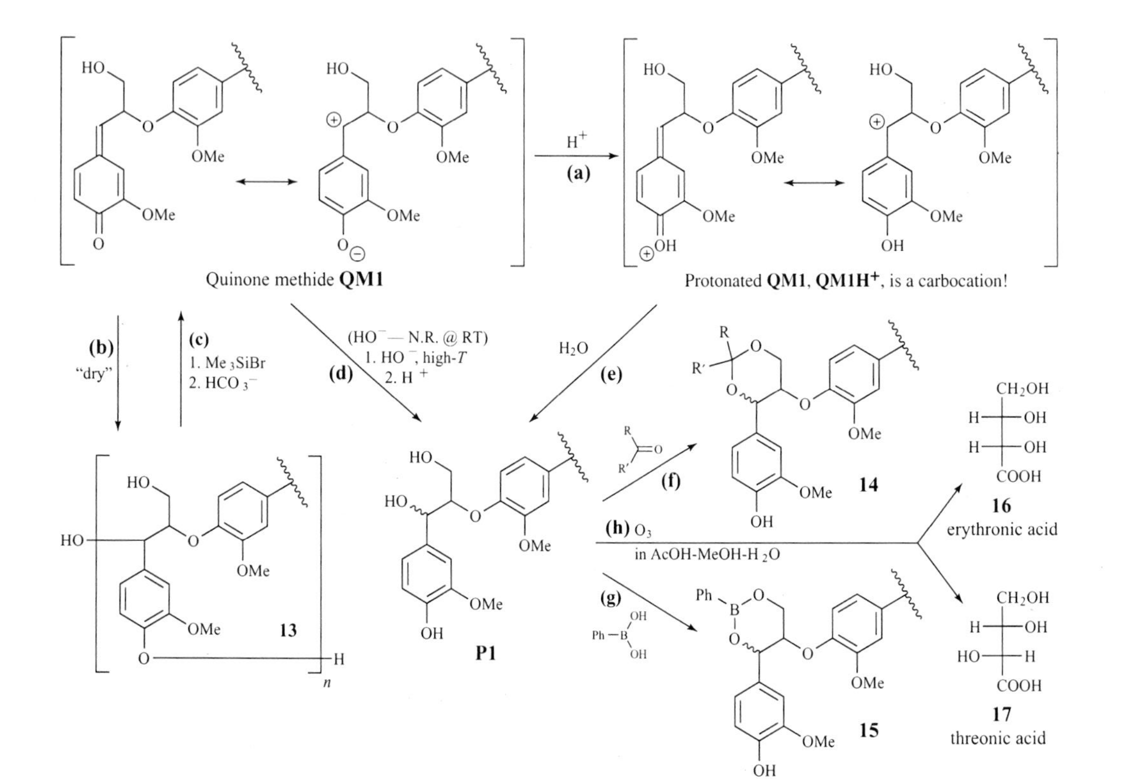

FIGURE 12.7 Quinone methide protonation and reactions. Quinone methides **QM1** are electron-deficient at carbon-7 as readily seen by drawing resonance structures. (a) Protonation of **QM1** produces a carbocation **QM1H**$^+$ that is even more reactive toward nucleophilic attack at C7. (b) If quinone methide **QM1** solutions are dried, a 7-*O*-aryl ether polymer **13** results. (c) Polymer **13** can be cleaved and the quinone methide **QM1** regenerated readily via the 7-bromide. (d) **QM1** will not nucleophilically add HO^- at room temperature under any conditions we have tried. At higher temperature, evidence is that addition does occur as cooking a single isomer of **P1** in 1 M NaOH at 130°C will cause isomerization.[40] (e) Water readily adds to the more reactive carbocation **QM1H**$^+$. (f–g) Isomer ratios of 8-*O*-4-ethers **P1** are readily determined via acetal **14** or phenylboronate **15** derivatives, or (h) via ozonolysis to threonic and erythronic acids **16–17**.

methylene chloride, chloroform, ethyl acetate, and ether. Once the solvent is evaporated, a 4-*O*-7-oligomer **13** is obtained (Fig. 12.7b).[41] This oligomer can be converted back to the original quinone methide **QM1** in the same way as the model **P1**, that is, by treatment with Me_3SiBr and then with bicarbonate.

Quinone methides **QM1** do, however, react extremely rapidly with softer nucleophiles at room temperature or hard nucleophiles at high temperature (Figs. 12.7d and 12.8) without requiring protonation. Primary and secondary amines add cleanly (Fig. 12.8c) producing predominantly *threo*-isomers of adducts **21**, as can be readily determined, in the case of primary amine products, by making cyclic derivatives **22** via formaldehyde addition.[57] Anthranol and anthrahydroquinone in NaOH also add rapidly and cleanly, producing *threo*-adducts (e.g., **18**, Fig. 12.8a) that can be readily isolated.[64–66,94–97] In the case of AHQ adducts **18**, Grob fragmentation at slightly elevated temperatures cleaves the 8-*O*-4-ether bond (Fig. 12.8a). The addition/fragmentation mechanism explains the efficacy of catalytic amounts of AQ in the caustic soda pulping process—AQ is reduced to AHQ by carbohydrate reducing end groups during pulping.

Cleavage of 8-*O*-4-ethers in alkaline pulping is also facilitated by HS^- as used in kraft pulping.[98] The major mechanisms (Fig. 12.8b) are via addition to the quinone methide **QM1** to give adduct **19**, followed by anchimerically assisted fragmentation via a thioepoxide **20**.

Nucleophilic addition of H^- (e.g., by treating a **QM1** solution in CH_2Cl_2 with $NaBH_4/SiO_2$) produces reduced structures **23**[41,99] (Fig. 12.8d). Biological reductants such as NADPH may also be capable of such reductions under physiological conditions even under the oxidizing conditions occurring during peroxidase–H_2O_2-catalyzed lignification (see Section 12.8).

12.8 REDUCTION OF QUINONE METHIDES DURING LIGNIFICATION?

One of the puzzles from lignin structural analysis is the existence of reduced structures in lignins, particularly in gymnosperms (softwoods). Dihydroconiferyl alcohol units are readily explained—they arise from the incorporation of the monomer dihydroconiferyl alcohol into lignins.[90,100–105] The source of the dihydroconiferyl alcohol, however, remains somewhat puzzling, even in a novel CAD-deficient pine where it is particularly prevalent. However, it is 8-8-dimeric units such as secoisolariciresinol units that are more mysterious. (Secoisolariciresinol, not shown, is the structure that would be obtained by hydride attack on the 8-8-bis-quinone methide **QM3** (Fig. 12.2) analogously to that shown in Fig. 12.8d for production of the reduced 8-*O*-4-ether **23** from quinone methide **QM1**.) One controversial review[106] suggested that such reduced components (including dihydroconiferyl alcohol) are due to contamination by lignans, a claim not borne out by the evidence.[90,100–104,107] Thus, although the lignan secoisolariciresinol isolated from various softwoods is optically active,[2] the units released from lignins (by methods shown to retain stereochemistry) are racemic.[7] It is therefore hypothesized that the

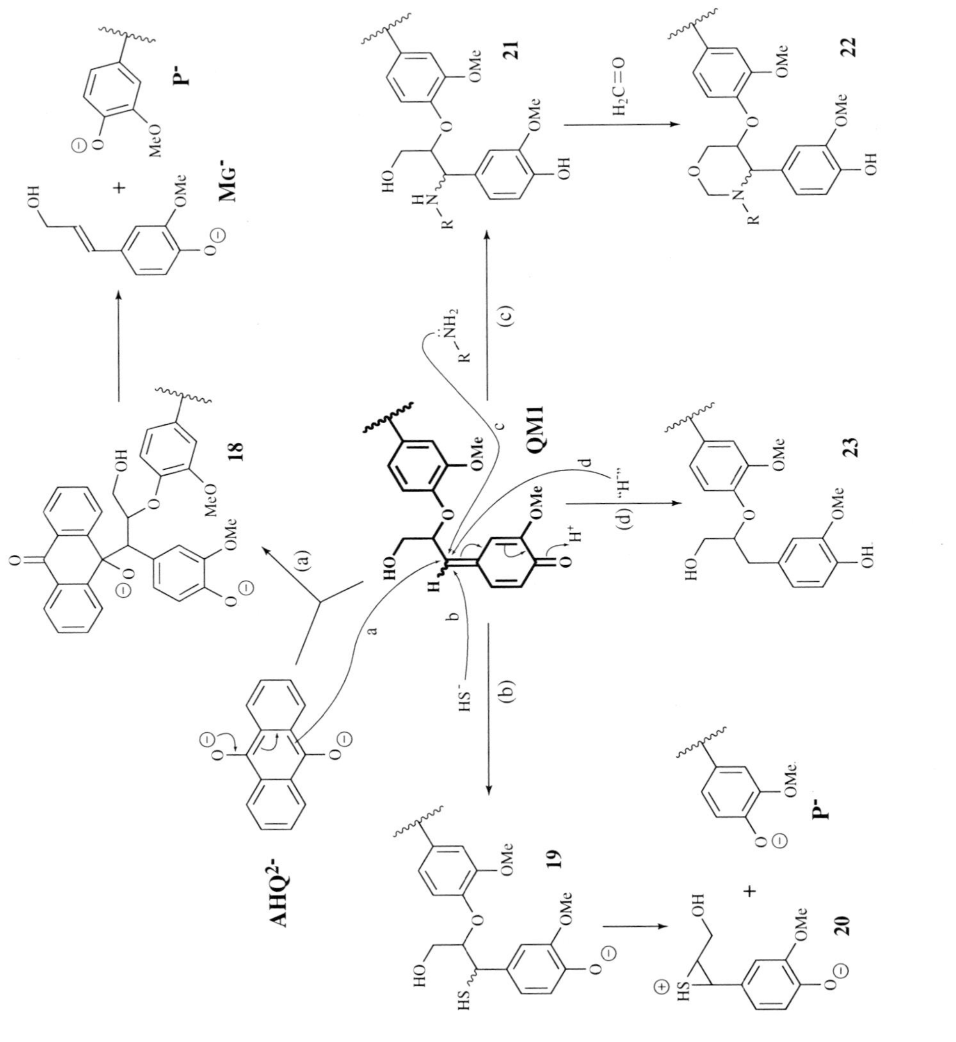
AHQ2-
18
MG-
P-
QM1
19
20
21
22
23
(a)
(b)
(c)
(d)
HS-
R–NH2
"H"
H2C=O

(racemic) dimer secoisolariciresinol is present during lignification and is therefore an authentic lignin precursor. It is also logical to hypothesize that it might arise from (nonenzymatically catalyzed) reduction of the 8–8-coupling product, quinone methide **QM3** (Fig. 12.2), using a biological reductant such as NADPH. Attempts to demonstrate this *in vitro* have not so far been successful.[108] However, reductive trapping of the 8-*O*-4-ether quinone methide **QM1** was achieved *in vitro* using NADPH (pathway d, Fig. 12.8). This was thought to have little relevance to actual lignification as reduced structures of the type **23** (Fig. 12.8) had not been detected in softwood lignins. We shall shortly report, however, fairly compelling NMR evidence for traces of such structures in softwood lignins (Kim, 2008, unpublished data). Whether an analogous reductive trapping of intermediate 8-8-quinone methides **QM3** can occur during lignification to produce racemic secoisolariciresinol (and lariciresinol) for incorporation into lignins remains to be elucidated.

12.9 ISSUES WITH THE RADICAL COUPLING MECHANISM, AND A SOLUTION?

The radical coupling mechanism for lignification is a rather well-accepted theory.[10] Nevertheless, some difficulties remain. Chemists express a general expectation that radical–radical coupling might not favor ether formation, yet ethers predominantly result during lignification. There is also the realization that the described process is not a typical radical polymerization (in the sense of, for example, styrene polymerization, which is a chain reaction), instead involving only radical–radical coupling reactions that would normally be considered termination reactions.

A major issue is how to oxidize the phenolic end group on the growing polymer. Peroxidases (primarily) can oxidize a monolignol **M** to its phenolic radical (Fig. 12.1). Evidence suggests that the oxidation occurs in the active site since the monolignols generate their radicals at distinctly different rates.[109–112] For most peroxidases studied, the rate of radical formation is *p*-coumaryl > coniferyl >sinapyl alcohol; in fact, many peroxidases oxidize sinapyl alcohol **MS** considerably more slowly. It is generally accepted, however, that even a modestly large oligomer could not easily diffuse back to the peroxidase to be oxidized directly, and it is difficult to imagine that the polymer can be accommodated in the narrow substrate channel of

---

FIGURE 12.8 QM reactions with nucleophiles. (a) Reaction of anthrahydroquinone with quinone methide **QM1** produces an intermediate adduct *threo*-**18** that undergoes Grob fragmentation at higher temperature, underlying the reason for anthraquinone's catalysis of alkaline delignification. (b) At least at elevated temperatures, HS^- adds to **QM1**; subsequent anchimerically assisted ether cleavage explains the effectiveness of kraft pulping over the older soda process. (c) Amines add rapidly to **QM1** to produce adducts **21**, mainly *threo*-isomers as can be determined via cyclic derivatives **22**. (d) Hydride addition, even possibly via biological reductants such as NADPH, produces reduced structures **23**.

peroxidases. As the oligomer must generate its phenolic radical again after every extension, every extension is a termination reaction. Speculation has therefore arisen that a major mechanism for oxidation of the growing oligomer/polymer must be via some type of radical transfer, that is, via a carrier. Small, easily diffusible species such as Mn(III) have been suggested,[113] and the oxidized monolignols themselves may mediate the dehydrogenation of the polymer.[114–119] Such transfer readily occurs from coniferyl alcohol to sinapyl alcohol,[117,120–122] but recently the transfer to polymeric lignin was shown to be inefficient.[121] The same authors isolated a peroxidase isoenzyme (CWPO-C) that, in contrast to horseradish peroxidase (HRP), was able to oxidize polymeric lignin directly. The synthetic lignin generated with this peroxidase had a higher proportion of 8-*O*-4-linkages, as determined by thioacidolysis, and a higher molecular weight than a synthetic lignin generated with horseradish peroxidase, probably due to a lower proportion of dehydrodimerization.[110,121] The fact that this enzyme can also oxidize ferrocytochrome *c* suggests an electron transport chain from the active site to the surface of the molecule. Further evidence for radical transfer comes from observing the increased coupling rate of sinapyl alcohol in the presence of certain rapidly oxidizable carriers. For example, the addition of traces of *p*-coumarate esters (including the monolignol conjugate, sinapyl *p*-coumarate) will vastly enhance sinapyl alcohol dimerization rates.[115,116,119,123,124] As none of the *p*-coumarate is found to react until all of the sinapyl alcohol has been consumed,[119] it seems clear that the *p*-coumarate radical is undergoing radical transfer with sinapyl alcohol; that is, *p*-coumarate radical is the oxidant producing sinapyl alcohol radicals.

Although researchers have occasionally attempted to discern pathways by which the oxidized monolignol might react with the ground-state oligomer, none of the hypotheses have withstood testing. Here we add the following to that list.

The most logical solution to the problem of having to oxidize both the monolignol and the phenolic end of the oligomer by one-electron oxidation is to instead oxidize only the monolignol by a two-electron oxidation, and have that oxidized entity react directly with the native oligomer. In fact, peroxidases are capable of executing two single-electron oxidation reactions taking the so-called peroxidase compound I via the so-called peroxidase compound II to the ground-state peroxidase.[125,126] But what would that oxidized monolignol look like? Figure 12.9 illustrates a hypothetical pathway to the carbocation $\mathbf{M}^+$. Such a product might spontaneously cyclize to its epoxide **QM24**, as shown in the figure. Alternatively, it could be trapped by an aldehyde or a ketone to produce an acetal **QM25**, the idea being that something as simple and prevalent as a reducing sugar (even a polysaccharide end group) could provide the necessary protection for a peroxidase-generated carbocation. Either of these products are quinone methides that represent latent carbocations (formally at the 8-position). The hypothesis can therefore be tested as the 8,9-epoxide or 8,9-acetal quinone methides **QM24** or **QM25** can be readily synthesized (Schatz, 2008, unpublished data). In fact, quinone methides **QM24** are sufficiently stable to obtain good NMR spectra in a variety of solvents. We synthesized them via silver(I) oxide oxidation of the benzylic-CH_2-R parent (cf. Fig. 12.4d). Attempts

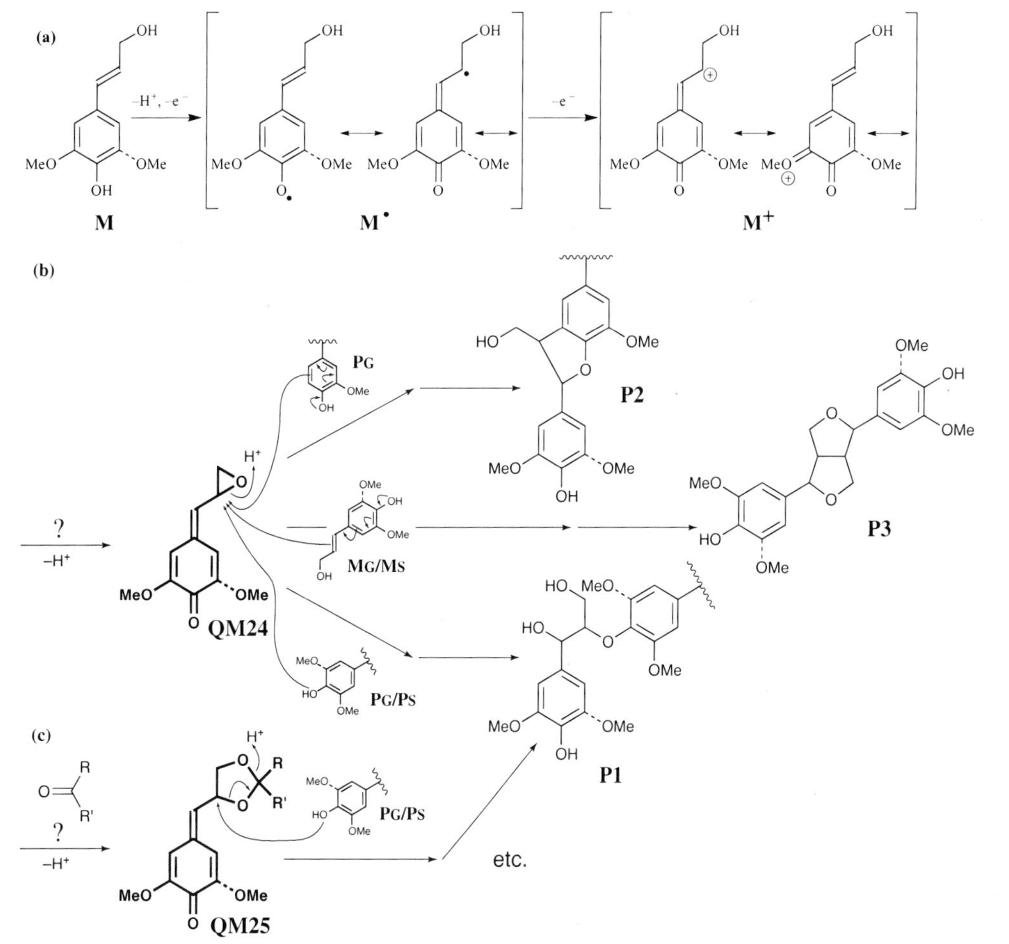

FIGURE 12.9 Hypothetical two-electron oxidation of monolignols and coupling reactions of potential quinone methides. (See text for details.)

to synthesize quinone methide **QM25** analogously resulted in a quinone methide but with apparent acetal partial opening.

To date, effecting the subsequent reaction chemistry (as hypothesized in Fig. 12.9) has not been possible in our laboratories. There have been indications that reaction of the epoxide quinone methide **QM24** in chloroform with acetic acid may have opened the epoxide before rearomatizing the quinone methide moiety, but this is not well authenticated. However, all attempts to add phenols to this epoxide, particularly aimed at producing 8-*O*-4-ethers **P1**, have been unsuccessful. Similar reactions using **QM25** have also failed.

Currently, the intriguing hypothesis that two-electron (or double one-electron) oxidation of a monolignol might produce an oxidized monomer that could couple with the phenolic end of the growing polymer appears to remain problematic. There are of course other issues with trying to move away from the theory of radical coupling. For example, it is rather well established that hydroxycinnamaldehydes and hydroxycinnamate esters can also cross-couple with monolignols and oligomers, *in vitro* and *in vivo* (see Section 12.11). In the latter case, these oxidized analogues of the monolignols become incorporated into the lignin polymer. The problem is that it is difficult to envision forming the 8-carbocation from such compounds—the 8-position is already electron-deficient in the neutral parent. Thus, one is forced to concede that even if two-electron reactions were responsible for some of the coupling reactions occurring during lignification, radical coupling reactions would be hard to dismiss for reactions involving, for example, ferulate or coniferaldehyde. We are beginning to feel, therefore, that this interesting hypothesis has little chance of ever rising to theory status, but include it here in case it prompts further ideas from interested researchers.

12.10 COUPLING REACTIONS WITH 9-OXIDIZED ANALOGUES OF THE MONOLIGNOLS

The hydroxycinnamyl alcohols are not the only monomers that enter into lignification. Among others, hydroxycinnamaldehydes are always found incorporated into lignins, particularly in angiosperms deficient in CAD (cinnamyl alcohol dehydrogenase),[127] and ferulates acylating arabinoxylans are incorporated into grass lignins.[119] The coupling and cross-coupling reactions, reviewed recently[33,119,128] so not detailed here, are analogous to those of the hydroxycinnamyl alcohol monolignols (Fig. 12.2); that is, 8-*O*-4-, 8–5-, 8–8-, and even 5–5-coupling reactions have all been documented. Apart from the different coupling propensities, the major differences result from postcoupling pathways available to the quinone methide intermediates.[128] The acidity of the quinone methide 8-proton when it is alpha to a carbonyl (ester or aldehyde) means that its elimination is an alternative mechanism to rearomatize the quinone methide. An example involving 8-*O*-4-cross-coupling of a ferulate ester **26** is shown in Fig. 12.10a. Elimination of the acidic 8-proton from the resulting quinone methide **QM26** regenerates an (unsaturated) hydroxycinnamate ester in the final 8-*O*-4-ether product **P26**.[42] The other coupling pathways produce other variants of the final product. In particular, 8-8-coupling of ferulates produces at least four final dehydrodimeric products, three having been documented,[129] and another having just been discovered in our laboratories (Lu, 2007, unpublished data).

A recently documented variant is the pathway by which ferulic acid incorporates into lignins, particularly in grasses and in angiosperms deficient in CCR (cinnamoyl–CoA reductase) (Fig. 12.10b).[130,131] 8-*O*-4-Cross-coupling of ferulic acid with a lignin phenolic end unit produces the anticipated quinone methide **QM27**. This quinone methide has yet another option for rearomatization, the elimination of CO_2. Even more intriguing is that with its truncated side chain, the radical derived from the product **P27** is capable of further 8-*O*-4-coupling. A second coupling reaction can produce the bis-ether quinone methide **QM28** and, following (proton catalyzed) water addition, the bis-ether **P28**. Such structures were recently authenticated in lignins by their diagnostic thioacidolysis products and directly by NMR.[130] The logical mechanisms and the demonstration of products **P28** in lignins confirmed that ferulic acid can also be a lignin monomer, in direct disagreement with a recent study refuting such contentions.[132]

12.11 CONCLUSIONS AND FUTURE PROSPECTS

Quinone methides play an important role in lignification. They are produced directly, as intermediates, when lignin monomers, be they hydroxycinnamyl alcohols, hydroxycinnamaldehydes, or hydroxycinnamates, couple or cross-couple at their 8- positions. A variety of postcoupling quinone methide rearomatization reactions leads to an array of structures in the complex lignin polymer (Fig. 12.2).

Lignification, and the reactions of the intermediary quinone methides in particular, is increasingly seen as crucial to understand due to the key role of lignin in limiting the efficient conversion of lignocellulosic biomass in a variety of natural and industrial processes. These include digestibility by ruminant animals;[133,134] the isolation, by solvolytic methods, of cellulose for pulp and paper production,[1] or as a feedstock for enzymatic saccharification;[135] and the enzymatic saccharification itself to provide simple sugars for fermentation to biofuels.[136] The stability and reactivity of quinone methides during lignification, *in planta*, remain largely unexplored. For example, in recent metabolomics studies, 7-*O*-methyl-8-*O*-4-ether products (7-*O*-methyl ether analogues of compound **P1**, Fig. 12.2) can be identified in methanol extracts of actively lignifying tissues (Morreel, 2007, unpublished data). Subjecting the normal 8-*O*-4-dimers **P1** to the same conditions does not produce such methylated products, suggesting that the quinone methide **QM1** itself was present in the plant tissues at the time of extraction; that is, it was not until during the extraction that it was rearomatized by (acid catalyzed) methanol addition. The stability and reactivity of quinone methides, under the conditions of cell wall lignification, presumably dictate the partitioning between rearomatization by water addition, the major pathway (Fig. 12.2a), versus the more minor pathway that possibly has a significant impact on biomass utilization—the addition of other cell

(a)

26 $\xrightarrow[H_2O_2]{POD}$ [26•] $\xrightarrow{\text{8-O-4-coupling}}$ QM26 $\xrightarrow{-H+,+H+}$ P26

P•

(b)

27 $\xrightarrow[H_2O_2]{POD}$ [27•] $\xrightarrow{\text{8-O-4-coupling}}$ QM27 $\xrightarrow{-CO_2}$ P27

P27 $\xrightarrow[H_2O_2]{POD}$ [P27•] $\xrightarrow{\text{8-O-4-coupling}}$ QM28 $\xrightarrow{+H_2O}$ P28

wall nucleophiles, including polysaccharides.[137] It is assumed that the resulting lignin–polysaccharide cross-linking is responsible for much of the encountered recalcitrance, yet the linkages in the wall have not yet been sufficiently well characterized, validated, or quantified.

Another process that needs further study is the reduction of quinone methides, which appears to be implicated in lignification, as covered in Section 12.9. Although the analogous lignans have defined pathways, the derivation of the racemic products, such as the secoisolariciresinol units in gymnosperm lignins, remains unexplained. If the pathway is through reduction of quinone methides via biological reductants such as NADH or NADPH, the conditions to effectively result in quinone methide reduction remain elusive, except for the case of 8-*O*-4-quinone methides **QM1** where NADPH reduction (to products **23**, Fig. 12.8) has been demonstrated *in vitro*.[108]

One of the major remaining challenges in the lignification step is to understand how the polymer becomes oxidized for the subsequent chain extension reaction. That is, if the polymerization is indeed by radical coupling reactions, between a monomer radical and the radical derived from the growing oligomer, it is unlikely that the increasingly massive polymer can continue to diffuse to peroxidases to accomplish the oxidation of the phenolic end directly. Is that oxidation via a radical-transfer agent? And, if so, are these simple organic agents such as the monolignols themselves or an inorganic carrier?[8,113] Our examination, in Section 12.9 and Fig. 12.9, of an alternative pathway involving quinone methide carbocation intermediates derived from two-electron oxidation of the monomer, and not requiring oxidation of the polymer, seems to be an attractive mechanism to obviate many of the conceptual problems encountered with lignification. However, the testing we have been able to conduct to date suggests that such processes are not involved.

Clearly, quinone methide chemistry and biochemistry remain at the forefront of biosynthetic process producing one of the world's most abundant classes of terrestrial biopolymers, the lignins that are required by plant cell walls for structural integrity, water transport, and defense. At the same time, the formation and subsequent reactions of the quinone methide intermediates produced during lignification are the key to understanding limitations to the efficient utilization of plant cell walls in a variety of natural and industrial processes. Although quinone methide intermediates in lignification are now rather well understood, as touched on in this chapter, significant areas of research into their properties and reactivity require further scrutiny.

FIGURE 12.10 Analogous reactions with ferulate and ferulic acid. (b) Ferulate 8-*O*-4-cross-coupling gives a quinone methide intermediate **QM26** that rearomatizes by elimination of the acidic 8-proton (rather than by water addition) to give the 8-substituted cinnamate **P26**. (b) Lignification with ferulic acid as a monomer is unusual for two reasons. First, the 8-*O*-4-quinone methide **QM27** rearomatizes by CO_2 elimination. Second, the truncated side chain of the resultant product **P27** allows a second 8-*O*-4-coupling reaction, producing the bis-8-*O*-4-ether quinone methide **QM28** and eventual product **P28** in the lignin.

REFERENCES

1. Sarkanen, K. V.; Ludwig, C. H. *Lignins, Occurrence, Formation, Structure and Reactions;* Wiley-Interscience: New York, 1971.
2. Umezawa, T. Diversity in lignan biosynthesis. *Phytochem. Rev.* 2004, 2, 371–390.
3. Lewis, N. G.; Davin, L. B. Lignans: biosynthesis and function. In *Comprehensive Natural Products Chemistry;* Barton, D. H. R.; Nakanishi, K.,Eds.; Elsevier: New York, 1999; Vol. 1.
4. Freudenberg, K.; Neish, A. C. *Constitution and Biosynthesis of Lignin;* Springer-Verlag: Berlin, 1968.
5. Gierer, J.; Opara, A. E. Enzymic degradation of lignin. Action of peroxidase and laccase on monomeric and dimeric model compounds. *Acta Chem. Scand.* 1973, 27, 2909–2922.
6. Boerjan, W.; Ralph, J.; Baucher, M. Lignin biosynthesis. *Annu. Rev. Plant Biol.* 2003, 54, 519–549.
7. Ralph, J.; Lundquist, K.; Brunow, G.; Lu, F.; Kim, H.; Schatz, P. F.; Marita, J. M.; Hatfield, R. D.; Ralph, S. A.; Christensen, J. H.; Boerjan, W. Lignins: natural polymers from oxidative coupling of 4-hydroxyphenylpropanoids. *Phytochem. Rev.* 2004, 3, 29–60.
8. Ralph, J.; Brunow, G.; Boerjan, W.; *Encyclopedia of Life Sciences* Rose, F.; Osborne, K., Eds.; John Wiley & Sons, Ltd.: Chichester, UK, 2007, pp. 1–10.
9. Ralph, J.; Peng, J.; Lu, F.; Hatfield, R. D. Are lignins optically active? *J. Agric. Food Chem.* 1999, 47, 2991–2996.
10. Ralph, J.; Brunow, G.; Harris, P. J.; Dixon, R. A.; Schatz, P. F.; Boerjan, W. Lignification: Are lignins biosynthesized via simple combinatorial chemistry or via proteinaceous control and template replication? In *Advances in Polyphenols Research;* Daayf, F.; El Hadrami, A.; Adam, L.; Ballance, G. M.,Eds.; Blackwell Publishing: Oxford, UK, 2008, pp. 36–66.
11. Lloyd, R. V.; Wood, D. E. Free-radicals in an adamantane matrix. 8. EPR and INDO study of benzyl, anilino, and phenoxy radicals and their fluorinated derivatives. *J. Am. Chem. Soc.* 1974, 96, 659–665.
12. Dewar, M. J. S.; David, D. E. Ultraviolet photoelectron-spectrum of the phenoxy radical. *J. Am. Chem. Soc.* 1980, 102, 7387–7389.
13. Russell, W. R.; Forrester, A. R.; Chesson, A.; Burkitt, M. J. Oxidative coupling during lignin polymerization is determined by unpaired electron delocalization within parent phenylpropanoid radicals. *Arch. Biochem. Biophys.* 1996, 332, 357–366.
14. Katayama, Y.; Fukuzumi, T. Enzymic synthesis of three lignin-related dimers by an improved peroxidase-hydrogen peroxide system. *Mokuzai Gakkaishi* 1978, 24, 664–667.
15. Tanahashi M.; Takeuchi H.; Higuchi T. Dehydrogenative polymerization of 3,5-disubstituted *p*-coumaryl alcohols. *Wood Res.* 1976, 61, 44–53.
16. Akiyama, T.; Magara, K.; Matsumoto, Y.; Meshitsuka, G.; Ishizu, A.; Lundquist, K. Proof of the presence of racemic forms of arylglycerol-β-aryl ether structure in lignin: studies on the stereo structure of lignin by ozonation. *J. Wood Sci.* 2000, 46, 414–415.
17. Brunow, G.; Karlsson, O.; Lundquist, K.; Sipilä, J. On the distribution of the diastereomers of the structural elements in lignins: the steric course of reactions mimicking lignin biosynthesis. *Wood Sci. Technol.* 1993, 27, 281–286.

18. Bardet, M.; Robert, D.; Lundquist, K.; von Unge, S. Distribution of *erythro* and *threo* forms of different types of β-*O*-4 structures in aspen lignin by carbon-13 NMR using the 2D inadequate experiment. *Magn. Reson. Chem.* 1998, 36, 597–600.

19. Akiyama, T.; Goto, H.; Nawawi, D. S.; Syafii, W.; Matsumoto, Y.; Meshitsuka, G. *Erythro/threo* ratio of β-*O*-4-structures as an important structural characteristic of lignin. Part 4: variation in the *erythro/threo* ratio in softwood and hardwood lignins and its relation to syringyl/guaiacyl ratio. *Holzforschung* 2005, 59, 276–281.

20. Akiyama, T.; Magara, K.; Matsumoto, Y.; Ishizu, A.; Meshitsuka, G.; Lundquist, K. Studies on the optical activities of arylglycerol-β-aryl ether structures in lignin. In *11th International Symposium of Wood and Pulping Chemistry, Nice, France, Association Technique de l'Industrie Papetière (ATIP), Paris;* 2001; Vol. III, pp 537–540.

21. Lundquist, K.; Stomberg, R. On the occurrence of structural elements of the lignan type (β-β structures) in lignins. The crystal structures of (+)-pinoresinol and (±)-*trans*-3,4-divanillyltetrahydrofuran. *Holzforschung* 1988, 42, 375–384.

22. Ämmälahti, E.; Brunow, G.; Bardet, M.; Robert, D.; Kilpeläinen, I. Identification of side-chain structures in a poplar lignin using three-dimensional HMQC-HOHAHA NMR spectroscopy. *J. Agric. Food Chem.* 1998, 46, 5113–5117.

23. Zhang, L.; Gellerstedt, G. NMR observation of a new lignin structure, a spiro-dienone. *J. Chem. Soc., Chem. Commun.* 2001, 2744–2745.

24. Zhang, L.; Gellerstedt, G.; Ralph, J.; Lu, F. NMR studies on the occurrence of spirodienone structures in lignins. *J. Wood Chem. Technol.* 2006, 26, 65–79.

25. Ralph, J.; Landucci, L. L. NMR of lignins. In *Lignins;* Heitner, C.; Dimmel, D. R.,Eds.; Marcel Dekker: New York, in press.

26. Tobimatsu, Y.; Takano, T.; Kamitakahara, H.; Nakatsubo, F. Azide ion as a quinone methide scavenger in the horseradish peroxidase catalyzed polymerization of sinapyl alcohol. *J. Wood Sci.* 2008, 54, 87–89.

27. Humphreys, J. M.; Hemm, M. R.; Chapple, C. Ferulate 5-hydroxylase from*Arabidopsis* is a multifunctional cytochrome P450-dependent monooxygenase catalyzing parallel hydroxylations in phenylpropanoid metabolism. *Proc. Natl. Acad Sci. USA* 1999, 96, 10045–10050.

28. Osakabe, K.; Tsao, C. C.; Li, L.; Popko, J. L.; Umezawa, T.; Carraway, D. T.; Smeltzer, R. H.; Joshi, C. P.; Chiang, V. L. Coniferyl aldehyde 5-hydroxylation and methylation direct syringyl lignin biosynthesis in angiosperms. *Proc. Natl. Acad. Sci. USA* 1999, 96, 8955–8960.

29. Li, L.; Popko, J. L.; Umezawa, T.; Chiang, V. L. 5-Hydroxyconiferyl aldehyde modulates enzymatic methylation for syringyl monolignol formation, a new view of monolignol biosynthesis in angiosperms. *J. Biol. Chem.* 2000, 275, 6537–6545.

30. Van Doorsselaere, J.; Baucher, M.; Chognot, E.; Chabbert, B.; Tollier, M.-T.; Petit-Conil, M.; Leplé, J.-C.; Pilate, G.; Cornu, D.; Monties, B.; Van Montagu, M.; Inzé, D.; Boerjan, W.; Jouanin, L. A novel lignin in poplar trees with a reduced caffeic acid/5-hydroxyferulic acid *O*-methyltransferase activity. *Plant J.* 1995, 8, 855–864.

31. Lapierre, C.; Tollier, M. T.; Monties, B. A new type of constitutive unit in lignins from the corn *bm3* mutant. *C. R. Biol.* 1988, 307, 723–728.

32. Lapierre, C.; Pollet, B.; Petit-Conil, M.; Toval, G.; Romero, J.; Pilate, G.; Leple, J. C.; Boerjan, W.; Ferret, V.; De Nadai, V.; Jouanin, L. Structural alterations of lignins in transgenic poplars with depressed cinnamyl alcohol dehydrogenase or caffeic acid

O-methyltransferase activity have an opposite impact on the efficiency of industrial kraft pulping. *Plant Physiol.* 1999, 119, 153–163.

33. Ralph, J.; Lapierre, C.; Marita, J.; Kim, H.; Lu, F.; Hatfield, R. D.; Ralph, S. A.; Chapple, C.; Franke, R.; Hemm, M. R.; Van Doorsselaere, J.; Sederoff R. R.; O'Malley, D. M.; Scott, J. T.; MacKay, J. J.; Yahiaoui, N.; Boudet, A.-M.; Pean, M.; Pilate, G.; Jouanin, L.; Boerjan, W. Elucidation of new structures in lignins of CAD- and COMT-deficient plants by NMR. *Phytochemistry* 2001, 57, 993–1003.
34. Marita, J. M.; Ralph, J.; Lapierre, C.; Jouanin, L.; Boerjan, W. NMR characterization of lignins from transgenic poplars with suppressed caffeic acid *O*-methyltransferase activity. *J. Chem. Soc., Perkin Trans.* 2001, 1, 2939–2945.
35. Ralph, J.; Lapierre, C.; Lu, F.; Marita, J. M.; Pilate, G.; Van Doorsselaere, J.; Boerjan, W.; Jouanin, L. NMR evidence for benzodioxane structures resulting from incorporation of 5-hydroxyconiferyl alcohol into lignins of *O*-methyl-transferase-deficient poplars. *J. Agric. Food Chem.* 2001, 49, 86–91.
36. Marita, J. M.; Ralph, J.; Hatfield, R. D.; Guo, D.; Chen, F.; Dixon, R. A. Structural and compositional modifications in lignin of transgenic alfalfa down-regulated in caffeic acid 3-*O*-methyltransferase and caffeoyl coenzyme A 3-*O*-methyltransferase. *Phytochemistry* 2003, 62, 53–65.
37. Marita, J. M.; Vermerris, W.; Ralph, J.; Hatfield, R. D. Variations in the cell wall composition of maize *brown midrib* mutants. *J. Agric. Food Chem.* 2003, 51, 1313–1321.
38. Morreel, K.; Ralph, J.; Lu, F.; Goeminne, G.; Busson, R.; Herdewijn, P.; Goeman, J. L.; Van der Eycken, J.; Boerjan, W.; Messens, E. Phenolic profiling of caffeic acid *O*-methyltransferase-deficient poplar reveals novel benzodioxane oligolignols. *Plant Physiol.* 2004, 136, 4023–4036.
39. Jouanin, L.; Gujon, T.; Sibout, R.; Pollet, B.; Mila, I.; Leplé, J. C.; Pilate, G.; Petit-Conil, M.; Ralph, J.; Lapierre, C. Comparison of the consequences on lignin content and structure of COMT and CAD downregulation in poplar and *Arabidopsis thaliana*. In *Plantation Forest Biotechnology in the 21st Century;* Walter, C.; Carson, M.,Eds.; Research Signpost: Kerala, India, 2004; pp. 219–229.
40. Miksche, G. E. Lignin reactions in alkaline pulping processes (rate processes in soda pulping). In *Chemistry of Delignification with Oxygen, Ozone, and Peroxides;* Nakano, J.; Singh, R. P.,Eds.; Uni Publishers Co: Tokyo, Japan, 1980; pp. 107–120.
41. Ralph, J. Reactions of lignin model quinone methides and NMR studies of lignins. Ph. D. thesis, University of Wisconsin—Madison, University Microfilms #DA 82-26987. 1982.
42. Ralph, J.; Quideau, S.; Grabber, J. H.; Hatfield, R. D. Identification and synthesis of new ferulic acid dehydrodimers present in grass cell walls. *J. Chem. Soc., Perkin Trans.* 1994, 1, 3485–3498.
43. Kim, H.; Ralph, J.; Yahiaoui, N.; Pean, M.; Boudet, A.-M. Cross-coupling of hydroxycinnamyl aldehydes into lignins. *Org. Lett.* 2000, 2, 2197–2200.
44. del Rio, J. C.; Marques, G.; Rencoret, J.; Martinez, A. T.; Gutierrez, A. Occurrence of naturally acetylated lignin units. *J. Agric. Food Chem.* 2007, 55, 5461–5468.
45. Ralph J. An unusual lignin from kenaf. *J. Nat. Prod.* 1996, 59, 341–342.
46. Ralph, J.; Hatfield, R. D.; Quideau, S.; Helm, R. F.; Grabber, J. H.; Jung, H.-J. G. Pathway of *p*-coumaric acid incorporation into maize lignin as revealed by NMR. *J. Am. Chem. Soc.* 1994, 116, 9448–9456.

47. Smith, D. C. C. *p*-Hydroxybenzoates groups in the lignin of aspen (*Populus tremula*). *J. Chem. Soc.* 1955, 2347.

48. Nakano, J.; Ishizu, A.; Migita, N. Studies on lignin. XXXII. Ester groups of lignin. *Tappi* 1961, 44, 30–32.

49. Landucci, L. L.; Deka, G. C.; Roy, D. N. A ^{13}C NMR study of milled wood lignins from hybrid *Salix* clones. *Holzforschung* 1992, 46, 505–511.

50. Sun, R. C.; Fang, J. M.; Tomkinson, J. Fractional isolation and structural characterization of lignins from oil palm trunk and empty fruit bunch fibres. *J. Wood Chem. Technol.* 1999, 19, 335–356.

51. Meyermans, H.; Morreel, K.; Lapierre, C.; Pollet, B.; De Bruyn, A.; Busson, R.; Herdewijn, P.; Devreese, B.; Van Beeumen, J.; Marita, J. M.; Ralph, J.; Chen, C.; Burggraeve, B.; Van Montagu, M.; Messens, E.; Boerjan, W. Modifications in lignin and accumulation of phenolic glucosides in poplar xylem upon down-regulation of caffeoyl-coenzyme A *O*-methyltransferase, an enzyme involved in lignin biosynthesis. *J. Biol. Chem.* 2000, 275, 36899–36909.

52. Li, S.; Lundquist, K. Analysis of hydroxyl groups in lignins by ^{1}H NMR spectrometry. *Nordic Pulp Pap. Res. J.* 2001, 16, 63–67.

53. Lu, F.; Ralph, J. Non-degradative dissolution and acetylation of ball-milled plant cell walls; high-resolution solution-state NMR. *Plant J.* 2003, 35, 535–544.

54. Lu, F.; Ralph, J. Preliminary evidence for sinapyl acetate as a lignin monomer in kenaf. *J. Chem. Soc., Chem. Commun.* 2002, 90–91.

55. Lu, F.; Ralph, J.; Morreel, K.; Messens, E.; Boerjan, W. Preparation and relevance of a cross-coupling product between sinapyl alcohol and sinapyl *p*-hydroxybenzoate. *Org. Biomol. Chem.* 2004, 2, 2888–2890.

56. Lu, F.; Ralph, J. Novel β–β-structures in natural lignins incorporating acylated monolignols. In *Thirteenth International Symposium on Wood, Fiber, and Pulping Chemistry, Auckland, New Zealand, APPITA, Australia;* 2005; Vol. 3, pp 233–237.

57. Ralph, J.; Adams, B. R. Determination of the conformation and isomeric composition of lignin model quinone methides by NMR. *J. Wood Chem. Technol.* 1983, 3, 183–194.

58. Zinke, T.; Hahn, O. Über die Einwirkung von Brom und von Chlor auf Phenole: Substitutionsproducted, Pseudobromide and Pseudochloride; IX. Über die Einwirkung von Brom auf Isoeugenol. *Justus Liebigs Ann. Chem.* 1903, 329, 1–36.

59. Li, S. M.; Lundquist, K.; Soubbotin, N.; Stomberg, R. 2-Bromo-4-[2-bromo-(*E*)-propylidene]-6-methoxy-2,5-cyclohexadien-1-one. *Acta Crystallogr.* 1995, 51, 2366–2369.

60. Ralph, J.; Ede, R. M.; Robinson, N. P.; Main, L. Reactions of β-aryl lignin model quinone methides with anthrahydroquinone and anthranol. *J. Wood Chem. Technol.* 1987, 7, 133–160.

61. Ralph, J.; Young, R. A. Stereochemical aspects of addition reactions involving lignin model quinone methides. *J. Wood Chem. Technol.* 1983, 3, 161–181.

62. Johansson, B.; Miksche, G. E. Über die Benzyl-arylätherbindung im Lignin. II. Versuche an Modellen. *Acta Chem. Scand.* 1972, 26, 289–308.

63. Lu, F.; Ralph, J. Facile synthesis of 4-hydroxycinnamyl *p*-coumarates. *J. Agric. Food Chem.* 1998, 46, 2911–2913.

64. Landucci, L. L.; Ralph, J. Adducts of anthrahydroquinone and anthranol with lignin model quinone methides. 1. Synthesis and characterization. *J. Org. Chem.* 1982, 47, 3486–3495.

65. Landucci, L. L. Formation of carbon-linked anthrone-lignin and anthrahydroquinone-lignin adducts. *J. Wood Chem. Technol.* 1981, 1, 61–74.

66. Ralph J.; Landucci, L. L. Adducts of anthrahydroquinone and anthranol with lignin model quinone methides. 9,10-^{13}C labeled anthranol-lignin adducts; examination of adduct formation and stereochemistry in the polymer. *J. Wood Chem. Technol.* 1986, 6, 73–88.

67. Zanarotti, A. Synthesis and reactivity of lignin model quinone methides. *J. Org. Chem.* 1983, 50, 941–945.

68. Quideau, S.; Ralph, J. A biomimetic route to lignin model compounds *via* silver(I) oxide oxidation. 1. Synthesis of dilignols and non-cyclic benzyl aryl ethers. *Holzforschung* 1994, 48, 12–22.

69. Zanarotti, A. Synthesis and reactivity of vinyl quinone methides. *J. Org. Chem.* 1985, 50, 941–945.

70. Zanarotti, A. Synthesis and reactivity of lignin model quinone methides. Biomimetic synthesis of 8.0.4′ neolignans. *J. Chem. Res., Synop.* 1983, 306–307.

71. Zanarotti, A. Preparation and reactivity of 2,6-dimethoxy-4-allylidene-2,5-cyclohexa-dien-1-one (vinyl quinone methide). A novel synthesis of sinapyl alcohol. *Tetrahedron Lett*, 1982, 23, 3815–3818.

72. Marita, J.; Ralph, J.; Hatfield, R. D.; Chapple, C. NMR characterization of lignins in *Arabidopsis* altered in the activity of ferulate-5-hydroxylase. *Proc. Natl. Acad. Sci. USA* 1999, 96, 12328–12332.

73. Hu, W.-J.; Lung, J.; Harding, S. A.; Popko, J. L.; Ralph, J.; Stokke, D. D.; Tsai, C.-J.; Chiang, V. L. Repression of lignin biosynthesis in transgenic trees promotes cellulose accumulation and growth. *Nat. Biotechnol.* 1999, 17, 808–812.

74. Karhunen, P.; Rummakko, P.; Pajunen, A.; Brunow, G. Synthesis and crystal structure determination of model compounds for the dibenzodioxocine structure occurring in wood lignins. *J. Chem. Soc., Perkin Trans.* 1996, 1, 2303–2308.

75. Karhunen, P.; Rummakko, P.; Sipilä, J.; Brunow, G.; Kilpeläinen, I. Dibenzo- dioxocins; a novel type of linkage in softwood lignins. *Tetrahedron Lett.* 1995, 36, 169–170.

76. Karhunen, P.; Rummakko, P.; Sipilä, J.; Brunow, G.; Kilpeläinen, I. The formation of dibenzodioxocin structures by oxidative coupling. A model reaction for lignin biosynthesis. *Tetrahedron Lett.* 1995, 36, 4501–4504.

77. Quideau, S. Incorporation of *p*-hydroxycinnamic acids into lignins *via* oxidative coupling. Ph.D. Thesis, University of Wisconsin—Madison, USA, University Microfilms International #9428350, 1994. 1994.

78. Ralph, S. A.; Landucci, L. L.; Ralph, J. Available at http://ars.usda.gov/Services/docs.htm?docid=10429 (previously http://www.dfrc.ars.usda.gov/software.html), updated at least annually since 1993, 2005.

79. Nakatsubo, F.; Higuchi, T. Enzymic dehydrogenation of *p*-coumaryl alcohol. III. Analysis of dilignols by gas chromatography and NMR spectrometry. *Wood Res.* 1975, 58, 12–19.

80. Akiyama, T.; Kim, H.; Dixon, R. A.; Ralph, J. Dibenzodioxocin structures involving *p*-hydroxyphenyl units in C3H down-regulated lignins. In *10th International Congress on Biotechnology in the Pulp and Paper Industry, Madison, WI, USA, ICBPPI, University of Wisconsin Press*. 2007, pp 71; Poster PS PGMB 1.4.

81. Ede, R. M.; Main, L.; Ralph, J. Evidence for increased steric compression in *anti* compared to *syn* lignin model quinone methides. *J. Wood Chem. Technol.* 1990, 10, 101–110.

82. Ralph, J.; Elder, T. J.; Ede, R. M. The stereochemistry of guaiacyl lignin model quinonemethides. *Holzforschung* 1991, 45, 199–204.
83. Ralph, J.; Ede, R. M. NMR of lignin model quinone methides. Corrected carbon-13 NMR assignments via carbon-proton correlation experiments. *Holzforschung* 1988, 42, 337–338.
84. Miksche, G. E. Über das Verhalten des Lignins bei der Alkalikochung. VIII. Isoeugenolglykol-β-(2-methoxyphenyl)-äther über ein Chinonmethid. (Behavior of lignins during alkaline pulping. VIII. Isomerization of the phenolate anions of *erythro-* and *threo*-isoeugenolglycol-β-(2-methoxyphenyl) ether via a quinone methide.) *Acta Chem. Scand.* 1972, 26, 4137–4142.
85. Ralph, J. Lignin model quinone methides—facts and fallacies. In *Proceedings of the Third International Symposium of Wood and Pulping Chemistry, Vancouver, BC, Canada, Chemical Institute of Canada (CIC), and Canadian Pulp and Paper Association (CPPA), Canada.* 1985.
86. Cook, C. D.; Norcross, B. E. Oxidation of hindered phenols. V. The 2,6-di-*t*-butyl-4-isopropyl and -4-sec-butylphenoxy radicals. *J. Am. Chem. Soc.* 1956, 78, 3797–3799.
87. Leary, G.; Miller, I. J.; Thomas, W.; Woolbouse, A. D. The chemistry of reactive lignin intermediates. Part 5. Rates of reactions of quinone methides with water, alcohols, and carboxylic acids. *J. Chem. Soc., Perkin Trans.* 1977, 2, 1737–1739.
88. Hemmingson, J. A.; Leary, G. The chemistry of reactive lignin intermediates. Part II. Addition reactions of vinyl-substituted quinone methides in aqueous solution. *J. Chem. Soc., Perkin Trans.* 1975, 2, 1584–1587.
89. Ralph, J.; Helm, R. F. Rapid proton NMR method for determination of *threo:erythro* ratios in lignin model compounds and examination of reduction stereochemistry. *J. Agric. Food Chem.* 1991, 39, 705–709.
90. Ralph, J.; Marita, J. M.; Ralph, S. A.; Hatfield, R. D.; Lu, F.; Ede, R. M.; Peng, J.; Quideau, S.; Helm, R. F.; Grabber, J. H.; Kim, H.; Jimenez-Monteon, G.; Zhang, Y.; Jung, H.-J. G.; Landucci, L. L.; MacKay, J. J.; Sederoff, R. R.; Chapple, C.; Boudet, A. M. Solution-state NMR of lignins. In *Advances in Lignocellulosics Characterization;* Argyropoulos, D. S. Ed.; TAPPI Press: Atlanta, GA, 1999; pp. 55–108.
91. Nakatsubo, F.; Higuchi, T. Synthesis of 1,2-diarylpropane-1,3-diols and determination of their configurations. *Holzforschung* 1975, 29, 193–198.
92. Akiyama, T.; Sugimoto, T.; Matsumoto, Y.; Meshitsuka, G. *Erythro/threo* ratio of β-*O*-4 structures as an important structural characteristic of lignin. I: improvement of ozonation method for the quantitative analysis of lignin side-chain structure. *J. Wood Sci.* 2002, 48, 210–215.
93. Matsumoto, Y.; Ishizu, A.; Nakano, J. Studies on chemical structure of lignin by ozonation. *Holzforschung* 1986, 40, 81–85.
94. Landucci, L. L. Quinones in alkaline pulping. Characterization of an anthrahydroquinone-quinone methide intermediate. *Tappi* 1980, 63, 95–99.
95. Ralph, J.; Landucci, L. L.; Nicholson, B. K.; Wilkins, A. L. Adducts of anthrahydroquinone and anthranol with lignin model quinone methides. 4. Proton NMR hindered rotation studies. Correlation between solution conformations and X-ray crystal structure. *J. Org. Chem.* 1984, 49, 3337–3340.
96. Ralph, J.; Landucci, L. L. Adducts of anthrahydroquinone and anthranol with lignin model quinone methides. 2. Dehydration derivatives. Proof of *threo* configuration. *J. Org. Chem.* 1983, 48, 372–376.

97. Ralph, J.; Landucci, L. L. Adducts of anthrahydroquinone and anthranol with lignin model quinone methides. 3. Independent synthesis of *threo* and *erythro* isomers. *J. Org. Chem.* 1983, 48, 3884–3889.
98. Gierer, J.; Ljunggren, S. The reactions of lignin during sulfate pulping. Part 17. Kinetic treatment of the formation and competing reactions of quinone methide intermediates. *Sven. Papperstidn.* 1979, 82, 503–512.
99. Ralph, J.; Grabber, J. H. Dimeric β-ether thioacidolysis products resulting from incomplete ether cleavage. *Holzforschung* 1996, 50, 425–428.
100. Ralph, J.; MacKay, J. J.; Hatfield, R. D.; O'Malley, D. M.; Whetten, R. W.; Sederoff, R. R. Abnormal lignin in a loblolly pine mutant. *Science* 1997, 277, 235–239.
101. Savidge, R. A.; Forster, H. Coniferyl alcohol metabolism in conifers—II. Coniferyl alcohol and dihydroconiferyl alcohol biosynthesis. *Phytochemistry* 2001, 57, 1095–1103.
102. Lapierre, C.; Pollet, B.; MacKay, J. J.; Sederoff, R. R. Lignin structure in a mutant pine deficient in cinnamyl alcohol dehydrogenase. *J. Agric. Food Chem.* 2000, 48, 2326–2331.
103. Sederoff, R. R.; MacKay, J. J.; Ralph, J.; Hatfield, R. D. Unexpected variation in lignin. *Curr. Opin. Plant Biol.* 1999, 2, 145–152.
104. Ralph, J.; Kim, H.; Peng, J.; Lu, F. Arylpropane-1,3-diols in lignins from normal and CAD-deficient pines. *Org. Lett.* 1999, 1, 323–326.
105. Savidge, R. A. Dihydroconiferyl alcohol in developing xylem of *Pinus contorta*. *Phytochemistry* 1987, 26, 93–94.
106. Gang, D. R.; Fujita, M.; Davin, L. D.; Lewis, N. G. The 'abnormal lignins': mapping heartwood formation through the lignan biosynthetic pathway. In *Lignin and Lignan Biosynthesis;* Lewis, N. G.; Sarkanen, S., Eds.; American Chemical Society: Washington, DC, Vol. 697, *American Chemical Society Symposium Series;* 1998, pp 389–421.
107. MacKay, J. J.; Dimmel, D. R.; Boon, J. J. Pyrolysis MS characterization of wood from CAD-deficient pine. *J. Wood Chem. Technol.* 2001, 21, 19–29.
108. Holmgren, A.; Brunow, G.; Henriksson, G.; Zhang, L.; Ralph, J. Non-enzymatic reduction of quinone methides during oxidative coupling of monolignols: implications for the origin of benzyl structures in lignins. *Org. Biomol. Chem.* 2006, 4, 3456–3461.
109. Brunow, G.; Ronnberg, M. The peroxidatic oxidation of some phenolic lignin model compounds. *Acta Chem. Scand., Ser. B* 1979, 33, 22–26.
110. Aoyama, W.; Sasaki, S.; Matsumura, S.; Mitsunaga, T.; Hirai, H.; Tsutsumi, Y.; Nishida, T. Sinapyl alcohol-specific peroxidase isoenzyme catalyzes the formation of the dehydrogenative polymer from sinapyl alcohol. *J. Wood Sci.* 2002, 48, 497–504.
111. Ros Barceló, A.; Pomar, F. Oxidation of cinnamyl alcohols and aldehydes by a basic peroxidase from lignifying *Zinnia elegans* hypocotyls. *Phytochemistry* 2001, 57, 1105–1113.
112. Kobayashi, T.; Taguchi, H.; Shigematsu, M.; Tanahashi, M. Substituent effects of 3,5-disubstituted *p*-coumaryl alcohols on their oxidation using horseradish peroxidase–H_2O_2 as the oxidant. *J. Wood Sci.* 2005, 51, 607–614.
113. Önnerud, H.; Zhang, L.; Gellerstedt, G.; Henriksson, G. Polymerization of monolignols by redox shuttle-mediated enzymatic oxidation: a new model in lignin biosynthesis I. *Plant Cell* 2002, 14, 1953–1962.
114. Hatfield, R. D.; Vermerris, W. Lignin formation in plants. The dilemma of linkage specificity. *Plant Physiol.* 2001, 126, 1351–1357.

115. Takahama, U.; Oniki T.; Shimokawa H. A possible mechanism for the oxidation of sinapyl alcohol by peroxidase-dependent reactions in the apoplast: enhancement of the oxidation by hydroxycinnamic acids and components of the apoplast. *Plant Cell Physiol.* 1996, 37, 499–504.

116. Takahama, U.; Oniki, T. Enhancement of peroxidase-dependent oxidation of sinapyl alcohol by esters of 4-coumaric and ferulic acid. In *Plant Peroxidases, Biochemistry and Physiology;* Obinger, C.; Burner, U.; Ebermann, R.; Penel, C.; Greppin, H., Eds.; Université de Genève, Genève: Switzerland, 1996; pp. 118–123.

117. Takahama, U.; Oniki, T. A peroxidase/phenolics/ascorbate system can scavenge hydrogen peroxide in plant cells. *Physiol. Plantarum* 1997, 101, 845–852.

118. Christensen, J. H.; Baucher, M.; O'Connell, A. P.; Van Montagu, M.; Boerjan, W. Control of lignin biosynthesis. In *Molecular Biology of Woody Plants;* Jain, S. M.; Minocha; S. C., Eds.; Kluwer Academic Publishers: Dordrecht, 2000; Vol. 1 (Forestry Sciences, Vol. 64); pp. 227–237.

119. Ralph, J.; Bunzel, M.; Marita, J. M.; Hatfield, R. D.; Lu, F.; Kim, H.; Schatz, P. F.; Grabber, J. H.; Steinhart, H. Peroxidase-dependent cross-linking reactions of *p*-hydroxycinnamates in plant cell walls. *Phytochem. Rev.* 2004, 3, 79–96.

120. Fournand, D.; Cathala, B.; Lapierre, C. Initial steps of the peroxidase-catalyzed polymerization of coniferyl alcohol and/or sinapyl aldehyde: capillary zone electrophoresis study of pH effect. *Phytochemistry* 2003, 62, 139–146.

121. Sasaki, S.; Nishida, T.; Tsutsumi, Y.; Kondo, R. Lignin dehydrogenative polymerization mechanism: a poplar cell wall peroxidase directly oxidizes polymer lignin and produces *in vitro* dehydrogenative polymer rich in β-*O*-4 linkage. *FEBS Lett.* 2004, 562, 197–201.

122. Takahama, U. Oxidation of hydroxycinnamic acid and hydroxycinnamyl alcohol derivatives by laccase and peroxidase—interactions among *p*-hydroxyphenyl, guaiacyl and syringyl groups during the oxidation reactions. *Physiol. Plantarum* 1995, 93, 61–68.

123. Takahama, U.; Oniki, T. Effects of ascorbate on the oxidation of derivatives of hydroxycinnamic acid and the mechanism of oxidation of sinapic acid by cell wall-bound peroxidases. *Plant Cell Physiol.* 1994, 35, 593–600.

124. Hatfield, R. D.; Ralph, J.; Grabber, J. H. A potential role of sinapyl *p*-coumarate as a radical transfer mechanism in grass lignin formation. *Planta* 2008, 228, 919–928.

125. Dunford, H. B. Heme enzymes. In *Comprehensive Biological Catalysis;* Sinnott, M. Ed.; Academic press: San Diego, CA, 1998, Vol. 3, pp. 196–237.

126. Dunford, H. B. Peroxidases. *Adv. Inorg. Biochem.* 1982, 4, 41–68.

127. Kim, H.; Ralph, J.; Lu, F.; Ralph, S. A.; Boudet, A.-M.; MacKay, J. J.; Sederoff, R. R.; Ito, T.; Kawai, S.; Ohashi, H.; Higuchi, T. NMR analysis of lignins in CAD-deficient plants. Part 1. Incorporation of hydroxycinnamaldehydes and hydroxybenzaldehydes into lignins. *Org. Biomol. Chem.* 2003, 1, 268–281.

128. Ralph, J. What makes a good monolignol substitute? In *The Science and Lore of the Plant Cell Wall Biosynthesis, Structure and Function;* Hayashi, T., Ed.; Universal Publishers (BrownWalker Press): Boca Raton, FL, 2006; pp. 285–293.

129. Schatz, P. F.; Ralph, J.; Lu, F.; Guzei, I. A.; Bunzel, M. Synthesis and identification of 2,5-bis-(4-hydroxy-3-methoxyphenyl)-tetrahydrofuran-3,4-dicarboxylic acid, an unanticipated ferulate 8-8-coupling product acylating cereal plant cell walls. *Org. Biomol. Chem.* 2006, 4, 2801–2806.

130. Ralph, J.; Kim, H.; Lu, F.; Grabber, J. H.; Boerjan, W.; Leplé, J.-C.; Berrio Sierra, J.; Mir Derikvand, M.; Jouanin, L.; Lapierre, C. Identification of the structure and origin of a thioacidolysis marker compound for ferulic acid incorporation into angiosperm lignins (and an indicator for cinnamoyl-CoA reductase deficiency). *Plant J.* 2008, 53, 368–379.

131. Leplé, J.-C.; Dauwe, R.;]Morreel, K.; Storme, V.; Lapierre, C.; Pollet, B.; Naumann, A.; Kang K.-Y.; Kim, H.; Ruel, K.; Lefèbvre, A.; Josseleau, J.-P.; Grima-Pettenati, J.; De Rycke, R.; Andersson-Gunnerås, S.; Erban, A.; Fehrle, I.; Petit-Conil, M.; Kopka, J.; Polle, A.; Messens, E.; Sundberg, B.; Mansfield, S. D.; Ralph, J.; Pilate, G.; Boerjan, W. Downregulation of cinnamoyl coenzyme A reductase in poplar; multiple-level phenotyping reveals effects on cell wall polymer metabolism and structure. *Plant Cell* 2007, 19, 3669–3691.

132. Laskar, D. D.; Jourdes, M.; Patten, A. M.; Helms, G. L.; Davin, L. B.; Lewis, N. G. The *Arabidopsis* cinnamoyl CoA reductase *irx4* mutant has a delayed but coherent (normal) program of lignification. *Plant J.* 2006, 48, 674–686.

133. Jung, H. G.; Buxton, D. R.; Hatfield, R. D.; Ralph, J.,Eds.; *Forage Cell Wall Structure and Digestibility;* American Society of Agronomy, Crop Science Society of America, Soil Society of America: Madison, 1993; pp 794.

134. Jung, H.; Buxton, D.; Hatfield, R.; Mertens, D.; Ralph, J.; Weimer, P. Improving forage fiber digestibility. *Feed Mix* 1996, 4, 30–31, 33–34.

135. Pan, X. J.; Arato, C.; Gilkes, N.; Gregg, D.; Mabee, W.; Pye, K.; Xiao, Z. Z.; Zhang, X.; Saddler, J. Biorefining of softwoods using ethanol organosolv pulping: preliminary evaluation of process streams for manufacture of fuel-grade ethanol and co-products. *Biotechnol. Bioeng.* 2005, 90, 473–481.

136. Chapple, C.; Ladisch, M.; Meilan, R. Loosening lignin's grip on biofuel production. *Nat. Biotechnol.* 2007, 25, 746–748.

137. Toikka, M.; Sipilä, J.; Teleman, A.; Brunow, G. Lignin-carbohydrate model compounds. Formation of lignin-methyl arabinoside and lignin-methyl galactoside benzyl ethers via quinone methide intermediates. *J. Chem. Soc., Perkin Trans.* 1998, 1, 3813–3818.

INDEX